Applied Mathematical Sciences | Volume 60

Applied Mathematical Sciences

1. John: **Partial Differential Equations,** 4th ed.
2. Sirovich: **Techniques of Asymptotic Analysis.**
3. Hale: **Theory of Functional Differential Equations,** 2nd ed.
4. Percus: **Combinatorial Methods.**
5. von Mises/Friedrichs: **Fluid Dynamics.**
6. Freiberger/Grenander: **A Short Course in Computational Probability and Statistics.**
7. Pipkin: **Lectures on Viscoelasticity Theory.**
9. Friedrichs: **Spectral Theory of Operators in Hilbert Space.**
11. Wolovich: **Linear Multivariable Systems.**
12. Berkovitz: **Optimal Control Theory.**
13. Bluman/Cole: **Similarity Methods for Differential Equations.**
14. Yoshizawa: **Stability Theory and the Existence of Periodic Solutions and Almost Periodic Solutions.**
15. Braun: **Differential Equations and Their Applications,** 3rd ed.
16. Lefschetz: **Applications of Algebraic Topology.**
17. Collatz/Wetterling: **Optimization Problems.**
18. Grenander: **Pattern Synthesis: Lectures in Pattern Theory, Vol I.**
20. Driver: **Ordinary and Delay Differential Equations.**
21. Courant/Friedrichs: **Supersonic Flow and Shock Waves.**
22. Rouche/Habets/Laloy: **Stability Theory by Liapunov's Direct Method.**
23. Lamperti: **Stochastic Processes: A Survey of the Mathematical Theory.**
24. Grenander: **Pattern Analysis: Lectures in Pattern Theory, Vol. II.**
25. Davies: **Integral Transforms and Their Applications,** 2nd ed.
26. Kushner/Clark: **Stochastic Approximation Methods for Constrained and Unconstrained Systems.**
27. de Boor: **A Practical Guide to Splines.**
28. Keilson: **Markov Chain Models—Rarity and Exponentiality.**
29. de Veubeke: **A Course in Elasticity.**
30. Sniatycki: **Geometric Quantization and Quantum Mechanics.**
31. Reid: **Sturmian Theory for Ordinary Differential Equations.**
32. Meis/Markowitz: **Numerical Solution of Partial Differential Equations.**
33. Grenander: **Regular Structures: Lectures in Pattern Theory, Vol. III.**
34. Kevorkian/Cole: **Perturbation Methods in Applied Mathematics.**
35. Carr: **Applications of Centre Manifold Theory.**
36. Bengtsson/Ghil/Källén: **Dynamic Meterology: Data Assimilation Methods.**
37. Saperstone: **Semidynamical Systems in Infinite Dimensional Spaces.**
38. Lichtenberg/Lieberman: **Regular and Stochastic Motion.**

(continued on inside back cover)

M. Ghil S. Childress

Topics in Geophysical Fluid Dynamics: Atmospheric Dynamics, Dynamo Theory, and Climate Dynamics

With 143 Illustrations

Springer-Verlag
New York Berlin Heidelberg
London Paris Tokyo

M. Ghil
Courant Institute of
Mathematical Sciences
New York University
New York, NY 10012

S. Childress
Courant Institute of
Mathematical Sciences
New York University
New York, NY 10012

Institute of Geophysics
and Planetary Physics
University of California at Los Angeles
Los Angeles, CA 90024

Library of Congress Cataloging in Publication Data
Ghil, Michael.
Topics in geophysical fluid dynamics.
(Applied mathematical sciences ;)
1. Fluid dynamics. 2. Dynamic meteorology.
3. Dynamic climatology. 4. Dynamo theory
(Cosmic physics) I. Childress, Stephen.
II. Title. III. Series: Applied mathematical
sciences (Springer-Verlag New York Inc.)
QA1.A647 [QC809.F5] 510 s [551.5] 86-31387

Printed and bound by R.R. Donnelley & Sons, Harrisonburg, Virginia
Printed in the United States of America.

9 8 7 6 5 4 3 2 1

ISBN 0-387-96475-4 Springer-Verlag New York Berlin Heidelberg
ISBN 3-540-96475-4 Springer-Verlag Berlin Heidelberg New York

At first it was believed that Tlön was a mere chaos, an irresponsible licence of the imagination; now it is known that it is a cosmos and that the intimate laws which govern it have been formulated, at least provisionally.

Jorge Luis BORGES,
Tlön, Uqbar, Orbis Tertius, 1956
(transl. by J. E. Irby, 1961)

PREFACE

The vigorous stirring of a cup of tea gives rise, as we all know, to interesting fluid dynamical phenomena, some of which are very hard to explain. In this book our "cup of tea" contains the currents of the Earth's atmosphere, oceans, mantle, and fluid core. Our goal is to understand the basic *physical processes* which are most important in describing what we observe, directly or indirectly, in these complex systems. While in many respects our understanding is measured by the ability to predict, the focus here will be on relatively simple models which can aid our physical intuition by suggesting useful *mathematical methods* of investigation. These elementary models can be viewed as part of a hierarchy of models of increasing complexity, moving toward those which might be usefully predictive.

The discussion in this book will deal primarily with the Earth. Interplanetary probes of Venus, Mars, Jupiter and Saturn have revealed many exciting phenomena which bear on geophysical fluid dynamics. They have also enabled us to see the effect of changing the values of certain parameters, such as gravity and rotation rate, on geophysical flows. On the other hand, satellite observations of our own planet on a daily and hourly basis have turned it into a unique *laboratory* for the study of fluid motions on a scale never dreamt of before: the motion of cyclones can be observed via satellite just as wing tip vortices are studied in a wind tunnel.

The book includes, it is hoped, background material sufficient for a reader familiar with the elementary dynamics and mathematics of incompressible fluid motion. The topics selected reflect the research interests of the two authors: the dynamics of the atmosphere, ocean and ice sheets on a large span of time scales, and the dynamics of the Earth's interior with application to the dynamo theory of the magnetic field. These specific applications involve *approximations*, sometimes through formal asymptotic expansions in small parameters, sometimes through the judicious selection of dominant physical mechanisms, such as shallow-layer or geostrophic flow, sometimes through the restriction to the dominant degrees of freedom, such as mean flow and a few wave components.

An important theme of the book are the methods of model derivation and construction, as well as those of model analysis. In particular, systematic use is made of the *successive bifurcations* method, by which the complexity of a system's behavior can be made to increase or decrease as certain parameters, not necessarily small, are allowed to vary. The typical change from stationary to periodic to aperiodic behavior is pursued in a number of applications. Special attention is given to aperiodic, chaotic or weakly turbulent behavior, and to its theoretical predictability.

Part I introduces the basic concepts of geophysical fluid dynamics (GFD): the effects of *rotation* on fluid motion, the Rossby number which measures these effects, vorticity and its approximate conservation, the effects of *shallowness* on large-scale atmospheric and oceanic flow, the shallow-water equations, Rossby waves and inertia-gravity waves, geostrophic and *quasi-geostrophic* flow. We have not attempted more than a brief introduction to these topics. Those readers seeking a more complete exposition will probably want to read the relevant portions of a text such as Pedlosky (1979).

Part II deals with the dynamics of large-scale atmospheric flows. Chapter 4 is a quick review of the atmosphere's general circulation, of its vertical stratification and of the classical theory of *baroclinic instability*, as applied to cyclone waves. In Chapter 5 we introduce the method of successive bifurcations, as illustrated by an application to the phenomena in a *rotating*, differentially heated *annulus*. As the rates of rotation and of heating change, flow in the annulus exhibits various patterns, reminiscent of atmospheric and oceanic flows.

Chapter 6 considers a topic of current interest: the dynamics of large-scale, long-lived *anomalies* in atmospheric circulation, such as those connected with extremely cold or snowy winters, or with hot and dry summers. Resonant wave response to orographic and thermal forcing and weakly nonlinear wave-wave interactions are introduced. The method of successive bifurcations, via multiple equilibria, stable periodic solutions and transition to chaos, is applied.

Chaotic solutions have the tendency to dwell near some of the system's unstable equilibria. One of these shows the flow pattern associated with monthly means of normal, "zonal" flow. Another equilibrium is naturally associated with anomalous, "blocked" flow. The higher persistence of model solutions near the blocked equilibrium or near the zonal equilibrium depends on external parameters, such as sea-surface temperature, and could explain the empirically known dependence of flow *predictability* on the prevailing flow pattern.

Part III treats the dynamo theory of geomagnetism. The emphasis here is on those aspects of the *magnetohydrodynamics of rotating fluids* which are thought to provide the proper basis for understanding the existence and structure of the geomagnetic field. Chapter 7 provides a *survey* of the dynamo problem for the flow in the Earth's core, its history and various approaches to its solution. Paleomagnetic data are discussed and the analogy between "climate" models of the core and of the atmosphere is pointed out. The need for non-axisymmetric fields of velocity and magnetic flux to produce a dynamo effect is emphasized.

Chapter 8 deals with the *kinematic approach* to dynamo theory: the velocity field is given, and the growth of the magnetic field is studied. Two basic methods for simplifying the three-dimensional nature of the problem are presented. The *smoothing method* is based on scale separation between small-scale, periodic or turbulent asymmetries and the large-scale, symmetric mean fields. The second method assumes the asymmetries to be small in amplitude, rather than scale; it relies on Lagrangian transformations of coordinates and has recently been extended to atmospheric flows. Both methods give an *α-effect* of the asymmetric part of the field on the mean-field electrodynamics.

In Chapter 9, the full, interactive *hydromagnetic dynamo problem* is addressed. Weak-field approximations are shown to be unsatisfactory and coupled, macrodynamic model equations, with prescribed α-effect, are formulated. The approach of truncated, *simplified systems*, introduced in Chapter 5, is used to study fundamental effects which lead to sustained, finite-amplitude dynamo action with stationary, periodic and aperiodic character. Some aspects of model solutions are very suggestive of the *aperiodic* nature of *polarity reversals* in the paleomagnetic record, with shorter events and longer epochs of one polarity or another.

Part IV is a brief introduction to a rather novel field within the geophysical family. Climate dynamics as a quantitative discipline was essentially born in the sixties and developed rapidly in the seventies. Chapter 10 describes the basic phenomena governing the climatic system's *energy balance*, as well as a simple equilibrium model for this balance. Multiple equilibria, their stability and sensitivity to changes in insolation are studied. Stochastic perturbations give finite residence times near each deterministically stable equilibrium. Delays between model variables lead from equilibria to periodic and aperiodic solutions.

Chapter 11 introduces first the paleoclimatological evidence for the *glaciations* of the Earth's last 2 million years, the Quaternary. The plastic flow of ice sheets and the viscous flow of the upper mantle, important on this time scale, are described and modeled. Chapter 12 takes up the theory of climatic oscillations. Cryodynamics and geodynamics are combined with the radiation balance, the hydrological cycle and oceanic thermal inertia to produce a nonlinear climatic oscillator. This oscillator is then subjected to the small *astronomical forcing* due to changes of the Earth's orbital elements in its motion around the Sun. The result is *quasi-periodic* and *aperiodic oscillations* exhibiting certain known features of the paleoclimatic record. Some features which have received less attention until recently are emphasized, and some new features are predicted.

We believe the contents of the book to be suitable for an advanced graduate course in GFD. Depending on the particular interests and background of the teacher and the student body, various selections of material and combinations with more basic texts or with research papers are possible. Highlights of the entire material were actually covered in a one-semester course at the Courant Institute, although a more leisurely pace is probably to be recommended. For instance, Part I could be supplemented by selections from classical texts in fluid dynamics, GFD, dynamic meteorology, physical oceanography or geomagnetism. From there, a one-semester course could proceed through either Part II or Part III, with a healthy complement of current literature. Alternatively, a class thoroughly familiar with Part I could cover Parts II and III or Parts II and IV in one semester.

Part IV is self-contained to a larger extent than Parts II or III. For physical content, the preliminary reading of Part I, Chapter 4 and Chapter 6 is desirable, but not necessary. For the mathematical method, Chapter 5 provides a useful introduction, as it does for all subsequent chapters, but again is not essential. In fact, Part IV, supplemented by the necessary numerical methods and by current literature, was also used as lecture notes in a one-semester graduate course at the Courant Institute and in a two-quarter course at UCLA.

The reference list at the end of the book is as complete as we found it possible with respect to textbooks, research monographs and review articles. It is far form exhaustive as far as research papers are concerned. Specific readings on the material covered in the text are suggested at the end of each chapter. In Parts II, III and IV these suggestions are accompanied by a brief discussion which provides a few pointers to additional topics of interest.

Parts of the manuscript were read at various stages or used in class by A. Barcilon, R. Benzi, J. Childress, D. Dee, M. Halem, R. Hide, R. N.

Hoffman, G. Holloway, G.-H. Hsu, E. Isaacson, E. Källén, E. Kalnay, R. Kosecki, C. E. Leith, H. LeTreut, F. J. Lowes, D. Müller, C. Nicolis, G. R. North, J. Pedlosky, J. Rubinstein, J. Tavantzis and C. Tresser. All provided excellent comments which improved the book to the extent possible. The bibliographic notes benefited from conversations with A. Bayliss, R. M. Dole, J. E. Hart, I. M. Held, D. D. Joseph, D. Landau, P. D. Lax, R. S. Lindzen, K. C. Mo, A. P. Mullhaupt, I. Orlanski, J. Pedlosky, R. L. Pfeffer, R. T. Pierrehumbert, J. Shukla, H. R. Strauss, J. J. Tribbia, N. O. Weiss and W. Wiscombe.

Our debt to many colleagues and collaborators, direct and indirect, is expressed implicitly in every line of the book and explicitly in many references. Bernard Legras not only contributed to the blocking research summarized in Section 6.4, but also to the writing of all of Chapter 6. To him and all the others, our deep gratitude.

The copyright to the English translation of the Borges quote used as a motto belongs to New Directions Publ. Co., 1964. It is a pleasure to thank Academic Press, New York, for permission to reproduce Figures 11.4, 11.5 and 11.7, the American Geophysical Union, Washington, D.C., for Figures 11.9, 11.13 and 12.11 through 12.16, the American Meteorological Society, Boston, Mass., for Figures 6.2, 6.3, 6.9, 6.10, 6.11, 6.13 through 6.17, 6.19, 6.20, 10.8, 10.9. 10.10, 10.14, 10.15, 10.16, 11.8, 11.20, 12.1 and 12.3, and the Society for Industrial and Applied Mathematics, Philadelphia, Penna., for Figures 12.2, 12.4, 12.5 and 12.10.

The pieces of draft of the two authors would never have become a final manuscript without the incredibly fast typing and many good-humored corrections of Connie Engle and Julie Gonzalez at the Courant Institute; impeccable camera-ready copy was produced from this typescript by Kate MacDougall of the Springer-Verlag. The figures were drafted by Laura Rumburg and Brian Sherbs at the Laboratory for Atmospheric Sciences of the NASA Goddard Space Flight Center, and by the Production Department of Springer-Verlag.

Walter Kaufmann-Bühler, Mathematics Editor at Springer-Verlag, New York, was very helpful in turning the manuscript into a real book. During the research which led to this book and the writing itself, the authors enjoyed the support of the National Science Foundation, Division of Atmospheric Sciences and Division of Mathematics and Computer Science, and of the National Aeronautics and Space Administration, Atmospheric Dynamics and Radiation Branch.

Last, but far from least, nothing would have been possible without the understanding and support of our wives, Michèle and Diana.

Michael Ghil and Stephen Childress
Los Angeles and New York, May 1986

TABLE OF CONTENTS

PART I

FUNDAMENTALS

CHAPTER 1

EFFECTS OF ROTATION

The rotation of the Earth has an important dynamical effect on the fluid environment and we begin our discussion by studying the role of large-scale rotation in simple cases. In this chapter, the Rossby number is defined as a nondimensional measure of the importance of rotation. The usefulness of a rotating frame of reference is outlined.

Equations of fluid motion are first stated and discussed in an inertial frame of reference. The necessary modifications for passing to a rotating frame are derived. The Coriolis acceleration is shown to be the main term which distinguishes between the equations in a rotating and an inertial frame.

Vorticity is defined and approximate conservation laws for it, namely Kelvin's theorem and Ertel's theorem, are derived. The important distinction between barotropic and baroclinic flows is made. Finally, motion at small Rossby number is discussed, the geostrophic approximation is stated and the Taylor-Proudman theorem for barotropic flow is derived from it. Geostrophic flow is shown to be a good qualitative approximation for atmospheric cyclones.

1.1. The Rossby Number

Newtonian mechanics recognizes a special class of coordinate systems, called inertial frames, and by rotation we always mean rotation relative to an inertial frame. As we shall see, Newtonian fluid dynamics relative to a rotating frame has a number of unusual features.

For example, an observer moving with constant velocity in an inertial frame feels no force. If the observer, however, has a constant velocity with respect to a frame attached to a rotating body, a Coriolis force will generally be felt, acting in a direction perpendicular to the velocity. Such "apparent" forces lead to dynamical results which are unexpected and peculiar when compared with conventional dynamics in an inertial frame. We should therefore begin by verifying that this shift of viewpoint to a noninertial frame is really called for by the physical phenomena to be studied, as well as by our individual, direct perception of these flows, as rotating observers.

To examine this question for the dynamics of fluids on the surface of the Earth, we define Ω to be the Earth's angular speed, so $2\pi/\Omega \cong 24$ hr. Therefore, at a radius of 6380 km a point fixed to the equator and rotating with the surface moves with speed 1670 km/hr as seen by an (inertial) observer at the center of the Earth. On the other hand, the large-scale motion of the fluid relative to the solid surface never exceeds a speed of about 300 km/hr, which occurs in the atmospheric jet streams. Thus the speeds relative to the surface are but a fraction of inertial speeds, and it will be useful to remove the fixed component of motion due to the Earth's rotation, by introducing equations of motion relative to rotating axes.

In this rotating frame, we define U to be a characteristic speed relative to the surface, and define L as a length over which variations of relative speed of order U occur. From these parameters we then obtain a characteristic time $T = L/U$, which is a time scale for the evolution of the fluid structure in question. Since Ω is an inverse time, the product $T\Omega$, or more conventionally the *Rossby number*

$$\varepsilon = \frac{U}{2\Omega L} \tag{1.1}$$

is a dimensionless measure of the importance of rotation, when assessed over the lifetime or period of the structure.

The nature of this rotational effect will be given a dynamical basis presently. For the moment it is important only to realize that small ε means that the effects of rotation are important. With this in mind we show in Table 1.1 values of the Rossby number for a few large-scale flows on the Earth and Jupiter.

We conclude that rotation is important in all of these systems and that a rotating coordinate system is probably useful for their study. Minor changes in these estimates are needed to account for the fact that the proper measure of rotation is the local angular velocity of the

Table 1.1

	Feature	L	U	ε
Earth	Gulf Stream	100 km	1 m/sec	0.07
Ω - 7.3×10^{-5} sec^{-1}	Weather system	1000 km	20 m/sec	0.14
	Core	3000 km	0.1 cm/sec	2×10^{-7}
Jupiter				
$\Omega = 1.7\times10^{-4}$ sec^{-1}	Bands, Red Spot	10^4 km	50 m/sec	0.015

tangent plane to the planet, as we shall see in Section 3.2 below. Also, these estimates only mean that Coriolis forces, proportional to $2\Omega U$, are typically larger than the forces needed to produce the acceleration measured relative to the rotating frame, these being of magnitude U^2/L (or U/T) times the fluid density. There may be other forces, such as viscous or magnetic forces, which enter into the dynamical balance. In the Earth's fluid core, for example, magnetic and Coriolis forces are probably comparable, while in the free atmosphere and ocean it is predominantly the "pressure" forces which compete with the Coriolis force in the dominant dynamical balance.

1.2. Equations of Motion in an Inertial Frame

We consider a fluid of density ρ, moving under the influence of a pressure field $p(r,t)$ and a body force $F(r,t)$, where $r = (x,y,z) = (x_1,x_2,x_3)$ is the position vector and t is time. The equations describing *local conservation of momentum and mass are*

$$\rho\left(\frac{\partial u}{\partial t} + u\cdot\nabla u\right) + \nabla p = F, \tag{1.2a}$$

$$\frac{\partial \rho}{\partial t} + u\cdot\nabla\rho + \rho\nabla\cdot u = 0, \tag{1.2b}$$

where $u(r,t)$ is the velocity field and $\nabla = (\frac{\partial}{\partial x}, \frac{\partial}{\partial y}, \frac{\partial}{\partial z})$ is the gradient operator. We shall not use any special notation for vectors, unless confusion between a vector and one of its components or other symbols might arise, e.g., $\underset{\sim}{u} = (u,v,w)$. Eq. (1.2b) is often called the "continuity equation".

Given F, (1.2) comprises four scalar equations for the three velocity components, p, and ρ. Another relation is therefore needed to close the system, e.g., an equation of state connecting p and ρ. In many problems of geophysical interest, such as ocean dynamics, the approximate incompressibility of the fluid plays a prominent role. In these cases u is prescribed to be a solenoidal, i.e., divergence-free, vector field, thus closing the system. For the moment we leave the closure unspecified and deal with the incomplete system (1.2). We also take F at first to be a conservative field (such as gravitation) defined in terms of a potential ϕ by

$$F = -\rho\nabla\phi. \tag{1.3}$$

We are thus excluding here the possibly important contribution of the nonconservative viscous stresses.

The operation

$$\frac{\partial}{\partial t} + u\cdot\nabla \equiv \frac{d}{dt} \tag{1.4}$$

occurring in (1.2) is called the *material derivative*, since it is the time derivative calculated by an observer moving with the local or material velocity. To verify this last claim let $G(r,t)$ be a field and let $r = R(t)$ be the position of a fluid particle at time t. Then $\dot{R} = u$ is the velocity of the particle, and we have

$$\begin{aligned}\frac{d}{dt} G(R(t),t) &= \frac{\partial G}{\partial t} + \dot{R}\cdot\nabla G \\ &= \frac{\partial G}{\partial t} + u_i \frac{\partial G}{\partial x_i},\end{aligned} \tag{1.5}$$

where the summation convention of repeated indices is used. In particular

$$\frac{d^2R}{dt^2} = \frac{du}{dt}(R,t) \tag{1.6}$$

is the acceleration of the particle with position vector $R(t)$. A quantity Q (scalar, vector or tensor) which satisfies $dQ/dt = 0$ is said to be a *material invariant*.

1.3. Equations in a Rotating Frame

We would like to utilize the form taken by (1.2) relative to a rotating coordinate system. That is, we would like to work with the velocity as perceived by an observer fixed in the rotating frame, and with time differentiation as would be performed by such an observer. Spatial differentiation, involving a limit process at fixed time, is an operation which is invariant under transformation to a rotating frame. The essential problem is thus to deal properly with time derivatives.

Absolute and relative derivatives. Since differentiation involves a limit process, it is convenient to begin with an infinitesimal rotation of axes. Consider a constant vector P fixed rigidly relative to the rotating frame. Let the frame rotate with respect to the inertial frame through an angle $d\theta_3$ about the z or x_3-axis. If P has coordinates P_1, P_2, P_3 in the rotated frame, an *inertial* observer will measure in his frame, after rotation, coordinates $(P_1 - P_2 d\theta_3, P_2 + P_1 d\theta_3, P_3)$ for P. Thus, under this rotation, we have the invariant vector relation

$$dP = d\theta \times P. \tag{1.7}$$

Introducing time t, we write (1.7) as

$$\left(\frac{dP}{dt}\right)_0 = \frac{d\theta}{dt} \times P \equiv \Omega \times P, \tag{1.8}$$

where we have defined $\Omega(t) = \frac{d\theta}{dt}$ as the instantaneous angular velocity of the rotating frame.

Eq. (1.8) involves the subscript "o", which will stand for "as determined by an observer in the inertial frame", sometimes called *absolute*. Since P is here a constant vector in the rotating frame, (1.8) simply states that in this case the inertial observer sees a vector attached to the origin rotating (about the z-axis). Examples of such "rotating, fixed vectors" are a set of orthonormal basis vectors in the rotating frame, (i_1, i_2, i_3), i_k being the unit vector along the x_k axis. Thus we can recompute (1.8) as follows: Define

$$P = P_1 i_1 + P_2 i_2 + P_3 i_3 \tag{1.9}$$

as the representation of P in the rotating basis. Since P_1, P_2, and P_3 are independent of time we have, using (1.8) and the summation convention,

$$\left(\frac{dP}{dt}\right)_0 = P_k \frac{di_k}{dt} = P_k(\Omega \times i_k) = \Omega \times P_k i_k = \Omega \times P. \tag{1.10}$$

Thus we see that the basis vectors carry the effect of rotation in their time dependence.

Suppose now that the coordinates P_k in (1.9) depend on time, $P_k = P_k(t)$. Repeating (1.10) we obtain

$$\left(\frac{dP}{dt}\right)_0 = \left(\frac{dP}{dt}\right)_r + \Omega \times P, \tag{1.11a}$$

where

$$\left(\frac{dP}{dt}\right)_r = P_k'(t) i_k \tag{1.11b}$$

is the time derivative of P relative to the rotating observer, with $P'_k = dP_k(t)/dt$. The relation (1.11) is the basic result which must be used to transform equations of motion to rotating axes.

We remark that Ω is allowed to depend on time t in (1.11). This can be of interest in geophysical fluid dynamics when modeling the precession of the axis of rotation of the Earth. However this time dependence has negligible effects for most phenomena in the dynamics of the atmosphere, oceans, or fluid core of the Earth, and will be largely ignored below. We also note that, for the purpose of interpreting a formula such as (1.8) or (1.11), it is sometimes convenient to think of the relation as applying at an instant when both sets of axes coincide.

Actually, we do not need (1.11) for the continuity equation (1.2b), since it is obvious that the material derivative of a *scalar* field is independent of the coordinate frame. Indeed, its value is determined by the time dependence observed at a given particle of fluid. To verify the invariance explicitly, let $G(r,t)$ be a scalar field as observed in the inertial frame. Then, for *partial* time derivatives we have $P = \underset{\sim}{r}$ in (1.8) and thus

$$\left(\frac{\partial G}{\partial t}\right)_r = \left(\frac{\partial G}{\partial t}\right)_0 + (\Omega\times r)\cdot\nabla G. \tag{1.12a}$$

Now, applying (1.11) to the vector $r = R(t)$ which represents the particle position at time t, yields

$$\left(\frac{dR}{dt}\right)_0 = u_0 = \left(\frac{dR}{dt}\right)_r + \Omega\times r = u_r + \Omega\times r, \tag{1.12b}$$

where

$$\left(\frac{dR}{dt}\right)_r = u_r \tag{1.12c}$$

is the velocity vector relative to the rotating frame. Using (1.12b) in (1.12a) we then have

$$\left(\frac{\partial G}{\partial t}\right)_r + u_r\cdot\nabla G = \left(\frac{\partial G}{\partial t}\right)_0 + u_0\cdot\nabla G,$$

verifying the invariance of the material derivative.

Relative acceleration and Coriolis acceleration. The right-hand side of (1.12b) is another vector function of t, and we may repeat the process to obtain the acceleration:

$$\left(\frac{d^2R}{dt^2}\right)_0 = \frac{d}{dt}\,[(u_r + \Omega \times r)]_r + \Omega \times [u_r + \Omega \times r]$$

$$= \left(\frac{du_r}{dt}\right)_r + 2\Omega \times u_r + \dot{\Omega} \times r + \Omega \times (\Omega \times r) \tag{1.13}$$

$$= \left(\frac{du_0}{dt}\right)_0 .$$

where u_0 is the velocity perceived by the inertial observer. Using (1.13) in (1.2a) and recalling the invariance of spatial derivatives we thus obtain the desired system of equations relative to the rotating frame:

$$\left(\frac{\partial u_r}{\partial t}\right)_r + u_r \times \nabla u_r + 2\Omega \times u_r + \frac{1}{\rho}\nabla p$$

$$= -\nabla\phi - \dot{\Omega} \times r - \Omega \times (\Omega \times r), \tag{1.14a}$$

$$\left(\frac{\partial\rho}{\partial t}\right)_r + u_r \cdot \nabla\rho + \rho\nabla\cdot u_r = 0. \tag{1.14b}$$

If Ω is taken to be constant, and if we use the identity

$$\Omega \times (\Omega \times r) = -\nabla\left(\frac{|\Omega \times r|^2}{2}\right), \tag{1.15}$$

then Eq. (1.14a) takes the form

$$\left(\frac{\partial u_r}{\partial t}\right)_r + u_r\cdot\nabla u_r + \frac{1}{\rho}\nabla p + 2\Omega \times u_r = -\nabla\phi_c, \tag{1.16a}$$

$$\phi_c = \phi - \frac{1}{2}\,|\Omega \times r|^2, \tag{1.16b}$$

where ϕ_c now absorbs the *centripetal acceleration*. The latter is only about 1/300 of the gravitational acceleration at the surface of the Earth and so is negligible for most atmospheric and oceanographic purposes. The remaining term which distinguishes Eq. (1.16) for the relative acceleration from Eq. (1.2a) for the absolute acceleration is the *Coriolis acceleration* $2\Omega \times u_r$.

If U and L are the characteristic scales of Section 1.1 defined now in the rotating system, and if L/U is taken as the characteristic time, then in order of magnitude

$$\left(\frac{du_r}{dt}\right)_r \sim U^2/L, \quad 2\Omega \times u \sim 2\Omega U, \tag{1.17}$$

and so the Rossby number (1.1) has the dynamical meaning of a characteristic ratio of the relative acceleration of a fluid element to the Coriolis acceleration.

For future reference, we now drop the subscript "r" and rewrite the basic equations (1.14) as :

$$\frac{\partial u}{\partial t} + u\cdot\nabla u + 2\Omega \times u + \frac{1}{\rho}\nabla p = -\nabla\phi_c - \dot{\Omega} \times r, \qquad (1.18a)$$

$$\frac{\partial \rho}{\partial t} + \nabla\cdot(\rho u) = 0. \qquad (1.18b)$$

1.4. Vorticity

For any velocity field *u* we define the associated vorticity field by

$$\omega = \nabla \times u. \qquad (1.19)$$

For solid body rotation with angular velocity Ω, the velocity field is obtained by setting $u_r = 0$, so that (cf. (1.12b)) $u = u_\Omega = \Omega \times r$. The associated vorticity is

$$\nabla \times (\Omega \times r) = \Omega(\nabla \cdot r) - \Omega \cdot \nabla r = 2\Omega. \qquad (1.20)$$

Applied locally in an arbitrary velocity field, vorticity has the same meaning: it is equal to twice the angular velocity of a fluid element.

This interpretation of vorticity in terms of a local angular velocity is misleading in some respects, since the term "fluid element" is not really appropriate for describing what is basically a point property. More precisely, the rate-of-strain tensor given by the matrix of partial derivatives of the components of *u* at any point has a symmetric and an antisymmetric part. Vorticity is associated with the antisymmetric part, while the symmetric part gives rise to a *straining field*. The effect of the latter is to distort "fluid elements", by differentially stretching and compressing them. With this caveat, we return to the discussion of vorticity.

Since global solid body rotation can be characterized, according to (1.20), as a state of uniform vorticity, the mechanics of the fluid relative to a *rotating* coordinate system can alternatively be regarded as a mechanics of deviations from a state of uniform vorticity. As we have already noted, the practical usefulness of reference to a state of uniform rotation will depend on the magnitude of the deviations, and the Rossby number can also be regarded as a *measure of relative vorticity as a fraction of* 2Ω. The properties of vorticity relative to the rotating frame are thus clearly "inherited" from the mechanics in the inertial frame, so for the present discussion we return to (1.2) and omit subscripts.

Kelvin's theorem. Using a vector identity for $u\cdot\nabla u$, (1.2a, 1.3) may be written in the form

$$\frac{\partial u}{\partial t} + \nabla(\frac{1}{2}u^2) - u \times \omega + \frac{1}{\rho}\nabla p = -\nabla\phi. \tag{1.21}$$

The particularly simple case in which $u \times \omega \equiv 0$ is called a Beltrami flow and will play a role in Chapter 8.

Taking the curl of (1.21) we obtain an equation for the vorticity,

$$\frac{\partial \omega}{\partial t} + u\cdot\nabla\omega - \omega\cdot\nabla u + \omega\nabla\cdot u - \frac{1}{\rho^2}\nabla\rho \times \nabla p = 0. \tag{1.22a}$$

Using (1.2b) to eliminate $(\nabla\cdot u)$ from (1.22a) we have

$$\frac{\partial \omega}{\partial t} + u\cdot\nabla\omega - \frac{\omega}{\rho}\frac{d\rho}{dt} - \omega\cdot\nabla u - \frac{1}{\rho^2}\nabla\rho \times \nabla p = 0. \tag{1.22b}$$

Multiplying (1.22b) by $1/\rho$ and rearranging we obtain finally a simpler equation for ω/ρ,

$$\frac{d(\omega/\rho)}{dt} = \frac{\omega}{\rho}\cdot\nabla u + \frac{1}{\rho^3}\nabla\rho \times \nabla p. \tag{1.23}$$

It shows that the two terms on the right will tend to change ω/ρ as one moves with the fluid. In particular the vector ω/ρ is not in general a material invariant.

It would appear that the vorticity vector itself does not provide one with conservation laws having an intuitive appeal, basically because of the already mentioned non-local nature of physical reasoning based upon "fluid elements". It is advantageous, therefore, to recognize the divergence-free (solenoidal) property of the vorticity vector and to exploit the non-local concept of *flux* of a divergence-free vector field. This non-local approach is the basis for the fundamental theorem discovered by Lord Kelvin. *Kelvin's theorem* studies the evolution of the flux of ω through a surface S bounded by a simple closed contour C, when the latter consists of material points, i.e., of points which move with the fluid.

By Stokes' theorem, the flux we study is given by

$$\int_S \omega\cdot ds = \int_C u\cdot dr, \tag{1.24a}$$

where

$$\Gamma(t) = \int_C u\cdot dr \tag{1.24b}$$

is called the *circulation* on C. Let C be given by parametric equations $r = R(\alpha,t)$, $0 \leq \alpha < 1$; then a given material point on the curve corresponds to a given value of α. The parameter α therefore is a material or Lagrangian parameter. One obtains

$$\begin{aligned}\frac{d\Gamma}{dt} &= \frac{d}{dt}\int_C u\cdot\frac{\partial r}{\partial\alpha}\,d\alpha \\ &= \int_C \frac{du}{dt}\cdot dr + \int_C u\cdot\frac{d}{dt}\left(\frac{\partial r}{\partial\alpha}\right)d\alpha \\ &= \int_C \frac{du}{dt}\cdot dr + \int_C u\cdot\frac{\partial u}{\partial\alpha}\,d\alpha \\ &= \int_C \left[-\frac{\nabla p}{\rho} - \nabla\phi\right]\cdot dr,\end{aligned} \tag{1.25}$$

by using (1.2a, 1.3, 1.4), the Lagrangian definition of velocity, $u = dr/dt$, and $\int_C d(u^2/2) = 0$. From the last form of (1.25) we see that if p and ρ are functionally related, so that the integrand becomes a perfect differential, then the circulation is conserved. This is Kelvin's theorem and its proof, like Eq. (1.23), emphasizes the role of the pressure and density fields in the evolution of vorticity.

We shall refer to a fluid system for which surfaces of constant p can be defined and coincide with surfaces of constant ρ by the customary term *barotropic*, and define the *baroclinic vector*

$$B = \frac{\nabla\rho\times\nabla p}{\rho^2}. \tag{1.26}$$

Thus, Kelvin's theorem asserts the conservation of circulation in barotropic, inviscid flows. Flow situations in which $B \neq 0$ will be considered in Chapters 4 and 5.

Ertel's theorem. To probe deeper into the connection between Kelvin's theorem and local variations of vorticity it is useful to adopt a fully Lagrangian description of the motion, by defining $r(t;r_0)$ to be the position (x_1,x_2,x_3) at time t of a fluid particle initially at $r_0 = (x_{10},x_{20},x_{30})$. Since r_0 labels a particle, it is fixed following the motion and the Jacobian tensor $J = \{J_{ij}\} = \left\{\frac{\partial x_i}{\partial x_{j0}}\right\}$ satisfies

$$\begin{aligned} \frac{d}{dt} J_{ij} &= \frac{\partial}{\partial x_{j0}} \frac{dx_i}{dt} = \frac{\partial u_i}{\partial x_{j0}} \\ &= \frac{\partial u_i}{\partial x_k} \frac{\partial x_k}{\partial x_{j0}} = \frac{\partial u_i}{\partial x_k} J_{kj}. \end{aligned} \tag{1.27a}$$

For any differential $d\ell$ separating two nearby points r_0 and $r_0+d\ell$ on a material curve at time $t = 0$, the quantity $Jd\ell$ (with components $J_{ij}\, d\ell_j$) can be regarded as the vector connecting the two nearby points on the material curve at all subsequent times. By repeating the calculation (1.27a) for this latter quantity, we obtain

$$\frac{d}{dt} J_{ij} d\ell_j = \frac{\partial u_i}{\partial x_k} J_{kj}\, d\ell_j$$

or

$$\frac{d}{dt}(Jd\ell) = (Jd\ell)\cdot\nabla u. \tag{1.27b}$$

Comparing (1.27b) with (1.23) shows that ω/ρ and $dr \equiv Jd\ell$ satisfy the same equation, provided $B = 0$. Here ∇u is regarded as a given coefficient matrix for the equation. Now dr is a differential along a material curve. The instantaneous *trajectories* of the vector field $\omega/\rho = (\omega/\rho)(r,t)$ are defined analogously as the integral curves of the system

$$\frac{dr}{d\sigma} = \frac{\omega}{\rho}(r(\sigma),t).$$

We can, at any time, select such a trajectory and thereafter follow the evolution of this curve as a material curve. However, since ω/ρ and dr satisfy the same equation it follows, assuming uniqueness of the solution, that this material curve remains a trajectory of ω/ρ. Therefore, *the trajectories of* ω/ρ *are material curves* in a barotropic flow.

One usually refers to these trajectories as *vortex lines*, and they can be visualized as being carried about by the fluid. Thus the first term on the right of (1.23), which might have the appearance of a "source" of vorticity, is in reality part of the statement of the material nature of vortex *lines*, although ω/ρ itself is not a material invariant. The only true source of vorticity in (1.23) is the baroclinic vector B.

The material changes of ω/ρ which occur when $B = 0$ are due to changes in the geometry of the vortex lines, i.e., to the twisting and stretching of small elements threaded by material lines. As a small pencil of vortex lines, commonly called a *vortex tube*, is stretched out by the flow, the cross section diminishes and the local vorticity grows in order to maintain, in accordance with Kelvin's theorem, a fixed circulation about the tube.

The most concise local statement concerning vorticity is an immediate consequence of the material nature of vortex lines. If $B = 0$, the vector $(\omega/\rho)J^{-1} = (\omega/\rho)\cdot\nabla r_0$ *is a material invariant*. To see this note first, by differentiating $JJ^{-1} = I$, that

$$\frac{d}{dt} J^{-1} = -J^{-1} \frac{dJ}{dt} J^{-1}.$$

Thus, using (1.23) and (1.27a),

$$\begin{aligned}\frac{d}{dt}\left(\frac{\omega}{\rho} J^{-1}\right) &= \left[\frac{d}{dt}\left(\frac{\omega}{\rho}\right)\right]J^{-1} + \frac{\omega}{\rho}\frac{d}{dt}(J^{-1}) \\ &= \left(\frac{\omega}{\rho}\cdot\nabla u\right)J^{-1} - \frac{\omega}{\rho} J^{-1}(J\cdot\nabla u)J^{-1} = 0,\end{aligned} \tag{1.28}$$

which establishes the invariance. Since J is initially the identity tensor, we see that the invariant in (1.28) is just the initial value of ω/ρ, or

$$\frac{\omega}{\rho}(r,t) = \frac{\omega(r_0,0)}{\rho(r_0,0)} J(r,t). \tag{1.29}$$

Now suppose that λ is any materially invariant scalar. Then we may regard λ as a function of the Lagrangian labels r_0. Since (1.28) is a vector statement, each component vanishes and we have

$$\frac{d}{dt}\frac{\omega}{\rho}\cdot\nabla\lambda = \nabla_0\lambda\cdot\frac{d}{dt}\left(\frac{\omega}{\rho} J^{-1}\right) = 0. \tag{1.30}$$

This last result is *Ertel's theorem,* and $\frac{\omega}{\rho}\cdot\nabla\lambda$ is called *potential vorticity*.

Let us now apply Kelvin's theorem to a small contour C lying on a surface of constant λ, λ being again a scalar invariant (see Figure 1.1). Kelvin's theorem asserts the constancy of $\omega\cdot n\,dA$ where dA is the area enclosed by C, while Ertel's theorem asserts the constancy of $(\omega/\rho)\cdot n(d\lambda/d\ell)$. Since the mass $\rho\,dA\,d\ell$ is conserved, the two statements become identical in this local setting.

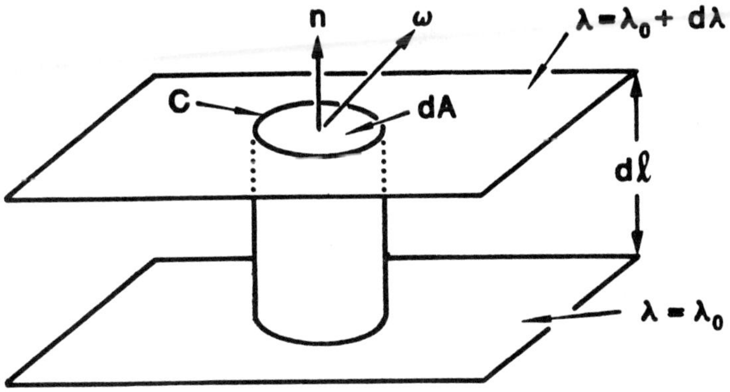

Fig. 1.1. Kelvin's theorem and Ertel's theorem for a material fluid element of cylindrical shape.

Intuitively, we see that as a tube of vortex lines is laterally compressed, the intensity of the field will increase, by Kelvin's theorem applied to the cross-section of the tube. The presence of ρ in Ertel's local invariant can be traced to the need to simultaneously satisfy Kelvin's theorem and conservation of mass in a small element undergoing deformation. In this sense Kelvin's theorem is the more fundamental characterization of vorticity. Nevertheless, Ertel's theorem will be useful to us, since in certain cases it is possible to *derive* an invariant λ and to cast the dynamical problem into a form equivalent to Ertel's equation. This is the case, for instance, in Section 2.1 below.

Our vorticity computations so far have been in the inertial frame, but in view of the correspondence

$$\omega_0 = 2\Omega + \omega_r$$

following from (1.12) and (1.20), one can write Ertel's equation relative to the rotating frame as

$$\frac{d}{dt}\left(\frac{2\Omega + \omega_r}{\rho} \cdot \nabla\lambda\right) = 0. \tag{1.31}$$

In contradistinction to Kelvin's theorem, Ertel's theorem can also be generalized to baroclinic flows. If $B \neq 0$, it suffices that the material invariant λ be a function of p and ρ only, $\lambda = \lambda(p,\rho)$, yielding $\nabla\lambda \cdot B = 0$. In barotropic flows, $B = 0$, and λ is unrestricted by this condition.

1.5. Motion at Small Rossby Number: The Geostrophic Approximation

From now on (unless we specify otherwise) equations of motion will be assumed to be relative to a rotating frame, and given by (1.18). To study motion at small Rossby number, we shall neglect the inertial terms $u_t + u\cdot\nabla u$ in (1.18) relative to the Coriolis force. Generally this requires that both $U/2\Omega L$ and $1/2\Omega T$ be small, where U, L, and T are scales of the velocity field.

The Taylor-Proudman theorem. The so-called *geostrophic approximation* to (1.18) is then

$$2\Omega \times u + \frac{1}{\rho}\nabla p = -\nabla\phi_c. \qquad (1.32)$$

The curl of (1.32) yields

$$-2\Omega\cdot\nabla u + 2\Omega(\nabla\cdot u) = B. \qquad (1.33)$$

If the baroclinic vector (1.26) vanishes, then with $\Omega = (0,0,\Omega)$ and $\underset{\sim}{u} = (u,v,w)$ the components of (1.33) reduce to

$$2\Omega\frac{\partial u}{\partial z} = 0, \quad 2\Omega\frac{\partial v}{\partial z} = 0, \quad 2\Omega\left(\frac{\partial u}{\partial x} + \frac{\partial v}{\partial y}\right) = 0. \qquad (1.34)$$

If, in addition, the fluid can be regarded as incompressible, the constraint $\nabla\cdot\underset{\sim}{u} = 0$ allows us to conclude from (1.34) that

$$2\Omega\frac{\partial}{\partial z}(u,v,w) = 0. \qquad (1.35)$$

This is the *Taylor-Proudman theorem*, stating that the velocity field in the geostropic approximation is invariant along the axis of rotation for a barotropic, inviscid flow. It highlights a striking feature of rotating flows, which was demonstrated experimentally by G. I. Taylor (Figure 1.2).

A cylindrical tank of water was rotated with constant speed, and on the bottom of the tank a small cylindrical obstacle was moved horizontally with very small relative speed. In ordinary inviscid fluid dynamics we would expect the fluid to move over and around the obstacle, the velocity vector being roughly tangent to the boundary. But on the top of the tank w must vanish, and the Taylor-Proudman theorem implies that in quasi-steady flow w must vanish everywhere. If fluid particles can only move horizontally, with a velocity independent of z, then the flow field is determined by its structure in a horizontal plane through the cylinder. Indeed, in the experiment, the obstacle was observed to

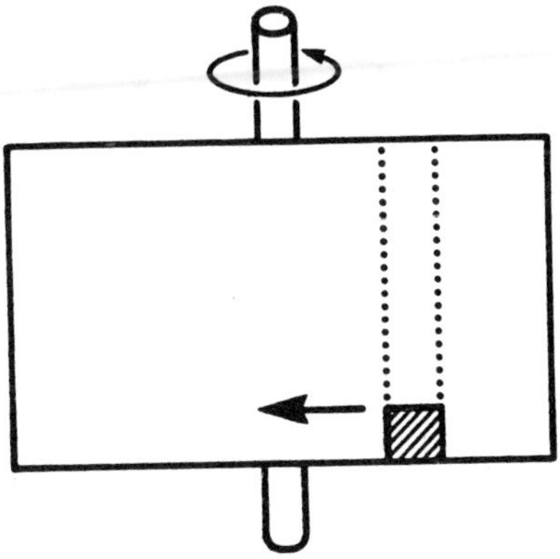

Fig. 1.2. The Taylor column, illustrating the Taylor-Proudman theorem in a rapidly rotating fluid.

carry an otherwise stagnant column of fluid with it. This "Taylor column" is an example of the nonlocal effect which boundary conditions can produce in a rapidly rotating fluid.

Geostrophic motion. Suppose now that we deal with a rotating layer of fluid of constant density with $\phi = gz$, and with $w = 0$, rather than $B = 0$. Then the components of (1.32) give

$$-2\Omega v = -\frac{1}{\rho}\frac{\partial p}{\partial x}, \quad 2\Omega u = -\frac{1}{\rho}\frac{\partial p}{\partial y}, \quad \frac{\partial p}{\partial z} = -\rho g. \tag{1.36}$$

This is an example of a *hydrostatic* as well as *geostrophic flow,* wherein the Coriolis force is in equilibrium with the horizontal pressure gradient.

Looking down on the layer, the flow in the vicinity of a minimum of pressure rotates in the same direction as the large scale rotation. In meteorology this would roughly approximate the wind field associated with a low in the pressure field, and is called a *cyclonic* circulation. Anticyclonic circulations correspond to local highs. Figure 1.3 is drawn for the northern hemisphere. In the southern hemisphere the cyclonic wind around a low moves in the opposite direction relative to the local normal.

The geostrophic approximation highlights the unusual features of the dynamics of a fluid viewed in a rotating coordinate system. To an observer in the inertial frame, geostrophic flows are predominantly solid body rotation. In order to understand the physical meaning of the dynamical balance represented in Figure 1.3, this background solid body

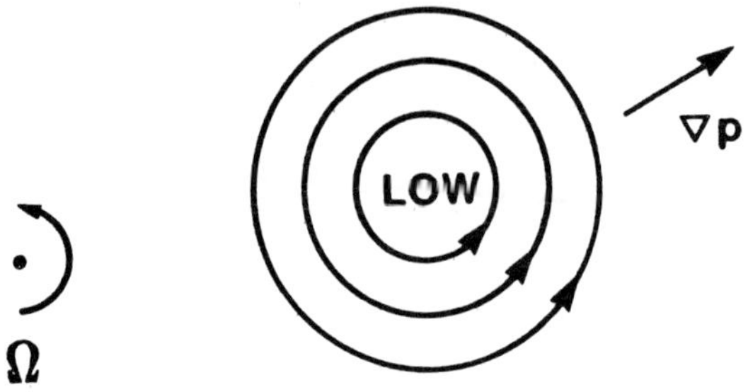

Fig. 1.3. Geostrophic flow in a northern hemisphere cyclone. Circles are streamlines, and arrows indicate the sense of rotation.

rotation must be kept in mind. However, it must also be remembered that a spatial pattern evolving on a moderate time scale relative to a rapidly rotating frame will be seen by the inertial observer as a slow modulation of the basic time dependence due to rotation of the pattern.

It is therefore helpful to deal directly with the relative motion in developing one's intuition concerning effects of rotation. For example, in interpreting Figure 1.3, the Coriolis acceleration associated with the eddy is seen to be directed toward the center of the low, and this acceleration is in equilibrium with minus the pressure gradient.

References

Batchelor (1967), Sections 3.1, 3.2, 3.5, 5.1-5.3, 7.6.

Greenspan (1968), Sections 1.1-1.5.

Landau and Lifschitz (1959), Sections 1-8.

Pedlosky (1979), Chapters 1 and 2.

CHAPTER 2

EFFECTS OF SHALLOWNESS

The most important *geometrical* approximation in geophysical fluid dynamics stems from the effective shallowness of the fluid layers relative to global horizontal scales of atmospheric and oceanic flow. In the present chapter we shall study this approximation in the simplest setting of an incompressible inviscid fluid of constant density, with uniform body force (gravitation) acting vertically downwards. The resulting approximate description of motion in a frame which rotates about a vertical axis is usually referred to as rotating shallow-water theory.

The equations of motion for constant rotation rate Ω are subjected to a systematic scale analysis, where H is the vertical scale, L is the horizontal scale and $\delta \equiv H/L$ is small. The shallow-water equations with rotation are thus derived, and shown to satisfy a particular form of Ertel's theorem. To wit, potential vorticity $(f_0+\zeta)/(h-h_0)$ is conserved, where ζ is the fluid's relative vorticity, $f_0 = 2\Omega$ the background vorticity, and $h-h_0$ the height of a fluid column.

Small-amplitude solutions of these equations are studied in basins with prescribed depth, using linear theory. Gravity waves are obtained as natural modes when $f_0 = 0$, and their modification when $f_0 \neq 0$ is investigated, yielding Poincaré's inertia-gravity waves. Related "edge waves", the so-called Kelvin waves, are also derived.

Finally, we study those modes with a low frequency which satisfy an approximate geostrophic balance between Coriolis force and pressure-gradient force. Stationary geostrophic modes, geostrophic contours of a basin and the infinite multiplicity of solutions they determine are analyzed. Rossby waves over sloping bottom topography are shown to remove this geostrophic degeneracy. Their relationship to the high-frequency Poincaré and Kelvin waves, and their physical structure are examined.

2.1. Derivation of the Equations for Shallow Water

For constant $\underset{\sim}{\Omega} = (0,0,\Omega)$ we write the governing equations (1.18) in the form

$$\partial_t u + u\cdot\nabla u + f_0(\hat{k} \times u) + \frac{1}{\rho}\nabla p = -g\hat{k}, \tag{2.1a}$$

$$\nabla\cdot u = 0, \tag{2.1b}$$

where $\partial_t = \partial/\partial t$, $f_0 = 2\Omega$ is the *Coriolis parameter*, $\hat{k} = i_3$ and the centripetal acceleration is neglected. We shall think of the fluid layer, which might represent an ocean basin, as occupying the region $h_0(x,y) \le z \le h(x,y,t)$. The upper surface $z = h(x,y,t)$ will be assumed to be at constant pressure p_0 (see Figure 2.1). If H is a typical value of h and h_0, and L is a typical horizontal scale of changes in h in p and in the velocity components (u,v,w), the layer is considered shallow if $H/L \equiv \delta \ll 1$. Since there is no preferred horizontal direction, we take u and v to be of possibly comparable size and of order U.

It is convenient to introduce the horizontal velocity $u_H = (u,v,0)$, as well as the horizontal gradient $\nabla_H = (\partial_x,\partial_y,0)$, and write (2.1) in the form

$$\nabla_H \cdot u_H + \partial_z w = 0, \tag{2.2}$$

where $\partial_z = \partial/\partial z$. The first term on the left is then of order U/L;

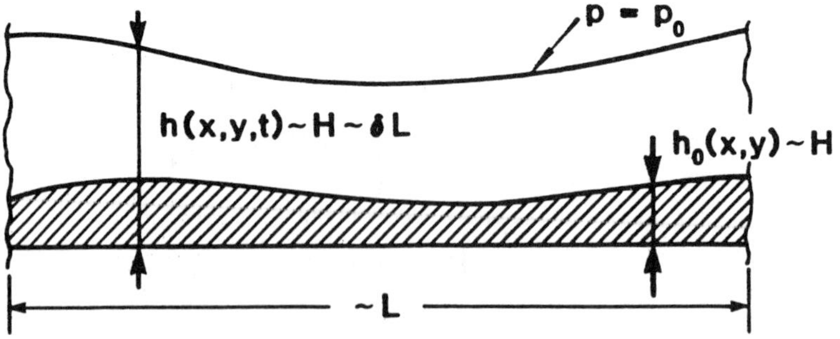

Fig. 2.1. Shallow layer of fluid.

if w varies by an amount of order W over the layer, the second term is of order W/H. These two terms must be comparable in order to account for conservation of matter in the presense of a divergence of horizontal velocity. Thus we take W to be of order δU. This small vertical component of velocity will be consistent with small inclination of both surfaces in Figure 2.1.

With this scaling, the material derivative is $O(U^2/L)$, and contains no negligible terms, provided that the time scale is taken as L/U. The horizontal pressure forces, which drive the horizontal motions, must be of order comparable to the acceleration. Taking the Coriolis and inertial accelerations to be also comparable to each other, so that the Rossby number $U/2\Omega L$ is $O(1)$, we have $p \sim \rho U^2$. Then the horizontal momentum balance is given by

$$\partial_t u_H + u_H \cdot \nabla_H u_H + w \partial_z u_H + f \hat{k} \times u_H = -\frac{1}{\rho} \nabla_H p. \tag{2.3}$$

On the other hand, the vertical momentum balance is

$$\frac{1}{\rho} \partial_z (p + \rho g z) = O(\delta U^2/L), \tag{2.4a}$$

yielding the *hydrostatic approximation*

$$p = -\rho g z + P(x,y,t) + O(\delta^2 \rho U^2). \tag{2.4b}$$

For a shallow layer this simplifies therefore to

$$p = -\rho g z + P(x,y,t), \tag{2.5}$$

where $P(x,y,t)$ is an arbitrary function. The latter is fixed by the dynamic condition $p(x,y,h,t) = p_0$ on the free surface. Thus

$$p = \rho g(h - z) + p_0. \tag{2.6}$$

We find that the pressure is also of order $\rho g H \sim \rho U^2$, yielding the characteristic horizontal velocity $U \sim \sqrt{gH}$.

According to (2.6), $\nabla_H p$ is independent of z. Hence we expect from (2.3) that u_H would remain independent of z were it initially so. Assuming that this is the case, Eq. (2.3) simplifies to

$$d_H u_H + f_0 \hat{k} \times u_H + g \nabla_H h = 0, \tag{2.7a}$$

where

$$d_H \equiv \partial_t + u_H \cdot \nabla. \tag{2.7b}$$

Equation (2.2) may be integrated to give

$$w = -z\nabla_H \cdot u_H + w_0(x,y,t) = \frac{dz}{dt}, \tag{2.8}$$

involving a second arbitrary function. To determine it, one imposes the kinematic conditions that the upper and lower fluid boundaries be material surfaces. If a boundary is defined implicitly by $S(x,y,z,t) = 0$, the latter condition is that $dS/dt = 0$ on $S = 0$. Applied to $z = h$ and $z = h_0$, this gives

$$0 = \frac{d}{dt}(z-h)_{z=h} = (w-d_H h)_{z=h} = -h\nabla_H \cdot u_H + w_0 - d_H h, \tag{2.9a}$$

$$\begin{aligned} 0 = \frac{d}{dt}(z-h_0)_{z=h_0} &= (-w-u_H \cdot \nabla_H h_0)_{z=h_0} \\ &= -h_0\nabla_H \cdot u_H + w_0 - u_H \cdot \nabla h_0. \end{aligned} \tag{2.9b}$$

Thus (2.9) determines w_0 and provides one equation to supplement (2.7). If we subtract one equation from the other, the supplemental equation is obtained in the form

$$d_H(h - h_0) = -(h - h_0)\nabla_H \cdot u_H. \tag{2.10}$$

Note that (2.10) has, in horizontal coordinates, the form of a continuity equation for a "density" $h-h_0$ (cf. Eq. (1.2b)). Physically, such an equation describes how a vertical fluid column changes in height in response to changes in horizontal divergence of the velocity field. Note, in this connection, that the absence of any vertical variation of u_H implies that vertical columns remain vertical.

The linear variation of w with z in fact implies that the fluid column is stretched uniformly and gives rise to an interesting material invariant of the shallow-water equations. Taking the exact material derivative of $z-h_0$ and using (2.8) and (2.9b), we obtain

$$\begin{aligned} \frac{d}{dt}(z-h_0) &= w - u_H \cdot \nabla_H h_0 \\ &= -z(\nabla_H \cdot u_H) + w_0 - u_H \cdot \nabla_H h_0 \\ &= -(z - h_0)\nabla_H \cdot u_H. \end{aligned} \tag{2.11}$$

But the left-hand side of (2.10) can also be written as $d(h-h_0)/dt$ (since $h-h_0$ is independent of z), so we may combine (2.10) and (2.11) to obtain

$$\frac{d\lambda}{dt} = 0, \quad \lambda = \frac{z-h_0}{h-h_0}. \tag{2.12}$$

Thus, a particle of fluid retains its position in a column as a given fraction of the height of the column.

We are now in a position to apply Ertel's theorem (Section 1.4) to the material invariant λ. To do this, we need the shallow-water approximation to the inertial or absolute vorticity ω_0. Taking into account that u_H is independent of z we have

$$\omega_0 = (w_y - v_z, u_z - w_x, f_0 + v_x - u_y) = (0, 0, f_0 + \zeta) + O(\delta\zeta), \tag{2.13}$$

where $\zeta = v_x - u_y$ is the relative vorticity of the flow, and subscripts are used to indicate partial derivatives with respect to x, y and z. Actually, to justify (2.13) it must be shown that there is no term of order δ in the expansion for u_H which could contribute $O(1)$ terms through the z-derivatives. But this follows from the estimate (2.4), which indicates that the shallow-water equations determine the leading terms in an expansion *in* δ^2. Thus, the contributions of u_z and v_z to (2.13) are in fact of order δ as stated.

Consequently, in the shallow-water approximation, Ertel's theorem, cf. Eq. (1.31), yields a potential vorticity equation in the form

$$d_H\left(\frac{f_0 + \zeta}{h - h_0}\right) = 0. \tag{2.14a}$$

We can interpret this equation quite easily in physical terms: since $h - h_0$ is inversely proportional to the horizontal cross-sectional area of a thin column, (2.14a) is in fact the shallow-water version of a local form of Kelvin's theorem in a rotating frame.

Some simple consequences of (2.14a) are obtained by writing it in the form

$$\frac{d_H\zeta}{f_0 + \zeta} = \frac{d_H(h - h_0)}{h - h_0}, \tag{2.14b}$$

where $d_H f_0 = 0$. Since planetary vorticity f_0 is positive, $f_0 > 0$, in the northern hemisphere, it follows from (2.14b) that relative vorticity ζ and layer depth $h - h_0$ increase and decrease together, at least when ζ is cyclonic, $\zeta > 0$. This monotonic dependence of ζ on $h - h_0$ will also hold for anticylonic ζ, provided $f_0 \gg |\zeta|$. The latter is typically the case for large-scale geophysical flows. The same conclusions apply in the southern hemisphere, where $f_0 < 0$, for cyclonic flows of arbitrary strength and for weak anticyclonic flows; notice that the southern hemisphere cyclonic flows are characterized by $\zeta < 0$ (cf. Fig. 1.3 and accompanying remarks).

Of course we do not know what the changes in ζ are until we find $h(x,y,t)$. In Chapter 3 we shall see that, for small Rossby number, relations connecting u_H (and therefore ζ) to h become quite simple. In this case (2.14) becomes a nonlinear partial differential equation in h, which completely determines the shallow-water dynamics.

The shallow-water theory described by Eqs. (2.7, 2.10) has a number of interesting features, both with and without the Coriolis force term. We study here some aspects of the relevant linear theory, and take up in the next chapter some aspects of the nonlinear equations. In other words, the next two sections will address the *small-amplitude* limit of shallow-water theory, while Chapter 3 contains results about the theory's *small Rossby number* limit.

2.2. Small Amplitude Motions in a Basin

Suppose that the fluid lies in a basin occupying a closed set D in the x-y plane (Figure 2.2), and that when at rest (with h constant) the depth of the fluid is $H(x,y) > 0$ over D. We then set

$$h-h_0 = H(x,y) + \eta(x,y,t), \tag{2.15}$$

where η and hence $u_H = (u,v)$ are small perturbations. Since nonlinear terms in the perturbations are neglected, (2.7) and (2.10) reduce to

$$u_t - f_0 v + g\eta_x = 0, \tag{2.16a}$$

$$v_t + f_0 u + g\eta_y = 0, \tag{2.16b}$$

$$\eta_t + (Hu)_x + (Hv)_y = 0. \tag{2.16c}$$

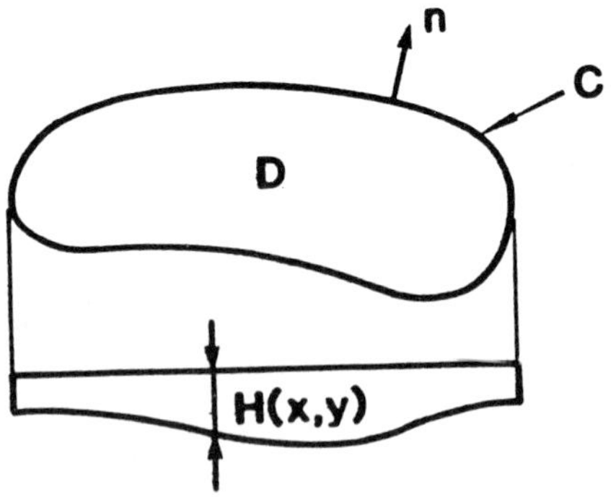

Fig. 2.2. Plan view and cross-section of a shallow basin.

Here and in the sequel subscripts t, x and y are used to indicate partial derivatives with respect to the given variable.

When $f_0 = 0$, we obtain the equations for small-amplitude oscillations in a nonrotating basin. In this case (2.16a-c) can be combined to give

$$\eta_{tt} - g(H\eta_x)_x - g(H\eta_y)_y = 0. \tag{2.17a}$$

If the contour C is an impenetrable barrier, the appropriate boundary condition for system (2.16) is $u_H \cdot n = 0$, where n is the normal to C. In this case, (2.16a,b) with $f_0 = 0$ yields

$$\nabla_H \eta \cdot n = 0 \tag{2.17b}$$

as the corresponding boundary condition for (2.17a).

If we look for eigenfunctions $\eta = \exp\{i\omega t\}N(x,y)$ of (2.17), where N has zero integral over D, then the resulting equation for N is equivalent to the Euler-Lagrange equation for the following variational problem:

$$\min_N \int_D H(\nabla_H N)^2 \, dxdy, \quad \int_D N^2 \, dxdy = 1.$$

Here the minimum is to be taken over all N which satisfy $\partial N/\partial n = 0$ on the boundary and have zero integral over D. The Lagrange multiplier in the problem may then be identified with ω^2/g. In this way, the minimum frequency ω_0 is determined variationally by

$$\omega_0^2 = \min_N g \frac{\int_D H(\nabla_H N)^2 \, dxdy}{\int_D N^2 \, dxdy} .$$

We note in passing that, for given $H(x,y)$, $\omega_0 > 0$, and the Rayleigh-Ritz procedure provides an effective means for constructing approximate eigenfunctions. These eigenfunctions represent standing gravity waves in a shallow basin.

The relative simplicity of this theory is a consequence of the self-adjoint property of the boundary-value problem (2.17a,b). Writing (2.17a) as

$$\eta_{tt} = L\eta,$$

it is easily seen that L is formally self-adjoint,

$$\int_D \phi \, L\psi \, dxdy = \int_D \psi \, L\phi \, dxdy,$$

for all functions ϕ,ψ satisfying the boundary condition (2.17b).

We turn now to the effect of rotation on small-amplitude motions. It can be shown from (2.16) that, if $f_0 \neq 0$, the associated linear operator is no longer self-adjoint. To exhibit this operator it is convenient to introduce the quantities

$$\sigma = (Hu)_x + (Hv)_y, \qquad s = (Hv)_x - (Hu)_y, \tag{2.19}$$

which may be thought of as representing the divergence and vorticity of the integrated mass flux across the layer. From (2.16a,b) we obtain by differentiation

$$\sigma_t - f_0 s + g[(H\eta_x)_x + (H\eta_y)_y] = 0, \tag{2.20a}$$

$$s_t + f_0\sigma + g(H_x\eta_y - H_y\eta_x) = 0. \tag{2.20b}$$

Eliminating s and using $\sigma = -\eta_t$, the equation for η becomes

$$(\eta_{tt} + f_0^2\eta)_t = g(H\eta_{xt})_x + g(H\eta_{yt})_y + f_0 gH_x\eta_y - f_0 gH_y\eta_x. \tag{2.21}$$

We first consider the simplest case, namely H = constant, and substitute $\eta = \exp\{i(\underset{\sim}{k}\cdot\underset{\sim}{x}_H + \omega t)\}$ to obtain a dispersion equation for traveling plane waves in an unbounded domain:

$$i\omega[f_0^2 - \omega^2] = -i\omega gHk^2. \tag{2.22}$$

Disregarding for the moment the wave corresponding to $\omega = 0$, these waves are known as Poincaré or *inertia-gravity waves*, and represent the generalization to the rotating case of familiar gravity waves. For every angular frequence $\omega \neq 0$, there are two waves with phase speeds $c_k = -\omega/k = \pm(f_0^2 + gHk^2)^{1/2}$. One propagates in the positive, the other in the negative direction relative to the wave vector $\underset{\sim}{k}$. They are observed in the atmosphere as well as in large bodies of water.

Seen from above, particle paths in the nonrotating case appear as line segments parallel to the direction of propagation of the wave (Figure 2.3). With rotation, the segments become ellipses whose eccentricity decreases with rotation. These facts can be most easily derived from a local Lagrangian description. If one considers analogously the Poincaré waves in a finite basin, one finds that for certain simple basins the boundary condition selects a discrete set of eigenfunctions, each exhibiting a dependence of frequency upon a (discrete) ordering parameter, which is similar to (2.22).

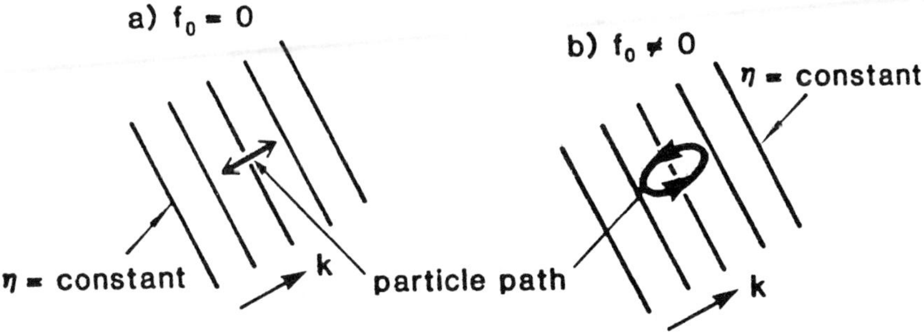

Fig. 2.3. Wave fronts and particle paths for gravity waves: a) without rotation, and b) with rotation.

From (2.22) we may write the dispersion relation for inertia-gravity waves in the form

$$(\omega/f_0)^2 = 1 + L_R^2 k^2, \tag{2.23}$$

where $L_R = \sqrt{gH}/f_0$ has the dimensions of a length, and is called the *external Rossby radius of deformation*. The physical importance of this length will be apparent in various problems studied below (cf. Sections 3.1 and 4.3). In the present context, $2\pi L_R$ is a horizontal scale below which the Poincaré waves become essentially gravity waves with small Coriolis effect. To take a typical case for the Earth's atmosphere, let $H = 10$ km, giving $L_R \cong 3 \times 10^3$ km. Thus, for horizontal scales of $2\pi/k < 3 \times 10^3$ km, we note that the gravity contribution to (2.23) dominates.

Notice that the frequency of these waves increases with wave number k. It is always larger than f_0, i.e., their period is smaller than the period of planetary rotation. A typical period of the inertia-gravity waves at length scales $L_R = O(10^3$ km) is about 2 hours. For most geophysical problems on such horizontal scales, this determines a short time scale. An important feature, therefore, of models for large-scale atmospheric and oceanic dynamics is an appropriate mechanism for filtering out the possible Poincaré-wave components of small period. Inertia-gravity waves are observed to have in fact small amplitudes over these length scales.

Kelvin waves. Another possible component of the oscillations of a basin is the family of Kelvin waves. These waves can be generated near a bounding contour and oscillate at a frequency which can be close to that associated with gravity waves. To take the simplest case, suppose that the basin is the upper half-plane $y > 0$.

Eliminating u from (2.16a,b) we obtain

$$v_{tt} + f_0^2 v + g\eta_{yt} - f_0 g\eta_x = 0. \tag{2.24}$$

If v vanishes on the boundary we must also have

$$\eta_{yt} - f_0\eta_x = 0, \quad y = 0. \tag{2.25}$$

With $\eta = \exp\{i(\omega t+kx)\}N(y)$, $k > 0$, Eq. (2.21) with $H = \text{constant}$ yields

$$i\omega(f_0^2-\omega^2 + gHk^2)N = i\omega gHN'', \quad y > 0, \tag{2.26a}$$

$$\omega N' = kf_0 N, \quad y = 0, \tag{2.26b}$$

where $(\ )' = d(\ )/dy$.

A new result can be obtained if we assume $f_0^2 - \omega^2 + gHk^2 > 0$, since then the y-dependence involves real exponentials rather than real trigonometric functions. To obtain a solution bounded in the upper half plane we must select the decaying exponential. Then (2.26b) requires $N = N_0 \exp\{f_0 k/\omega)y\}$, $\omega < 0$, in which case (2.26a) gives

$$(\omega^2 - gHk^2)(f_0^2 - 1) = 0. \tag{2.27}$$

Taking $\omega = -k\sqrt{gH}$, one obtains a wave with positive phase velocity identical to that of a pure gravity wave.

It can be checked from (2.16) that v vanishes identically for this wave, and that wave crests are parallel to the y-axis. On each line $y = \text{constant}$ the wave propagates as an ordinary gravity wave, but the x-motion is in geostrophic balance with the wave height η; the latter decreases in concert with u as y increases, at an exponential rate determined by the Rossby radius L_R.

Because of their exponential decrease away from the boundary, Kelvin waves can be regarded as "edge waves". They have vanishing frequency as $k \to 0$, a property which separates them from inertial-gravity waves (see Figure 2.6 below). But for moderate length scales in the x-direction they fall into the category of "fast" waves. Kelvin waves are particularly important in tropical meteorology and oceanography, where the equator plays the role of the "wall".

2.3. Geostrophic Degeneracy and Rossby Waves

The time scales associated with large-scale atmospheric phenomena are of the order of days, while those for the ocean are of the order of weeks or months. If these time scales are to be reflected in the small-amplitude solutions of the shallow-water equations, the relevant physics must be associated with the modes corresponding to the root $\omega = 0$ of the dispersion relation (2.22) in the constant-depth case.

Stationary geostrophic modes. Let us consider then the stationary modes in the small-amplitude theory. Eqs. (2.16) reduce for steady solutions to

$$(u,v) = (g/f_0)(-\eta_y,\eta_x), \tag{2.28a}$$

$$(Hu)_x + (Hv)_y = 0. \tag{2.28b}$$

This is a special kind of geostrophic flow (Section 1.5), and we refer to all solutions of (2.28) compatible with the condition on the boundary of the basin as stationary *geostrophic modes*. Eq. (2.28a) implies that $(g/f_0)\eta$ is a stream function for the flow's velocity field and that lines $\eta = \text{constant}$ are streamlines of the flow. In particular, the boundary has to be an isoline of η.

For the constant-depth problem we see that arbitrary divergence-free horizontal motions compatible with the condition that the boundary be a streamline are allowed. Eqs. (2.28) simply tells us what the fluid level must be to bring the system into geostrophic balance. From (2.21) and (2.28) we conclude that in the constant-depth case, with η proportional to $e^{i\omega t}$ in separated variables, the eigenvalue $\omega = 0$ is infinitely degenerate. Given any boundary contour, there are an infinite number of functions η which satisfy the boundary conditions.

If H depends upon x and y, the geostrophic modes may still be infinitely degenerate, since (2.28b) then implies that the stream function η may be an arbitrary function of H alone. Thus lines of constant depth and streamlines coincide. Although contours of constant H which intersect the boundary C must carry zero velocity, closed contours which do not (or nested islands of such contours) allow the construction of geostrophic modes (see Figure 2.4). Such contours, along with stationary geostrophic flow may occur, are often called geostrophic contours.

Sloping bottom topography. The question now arises as to what happens to the "disappearing" geostrophic modes when the depth function

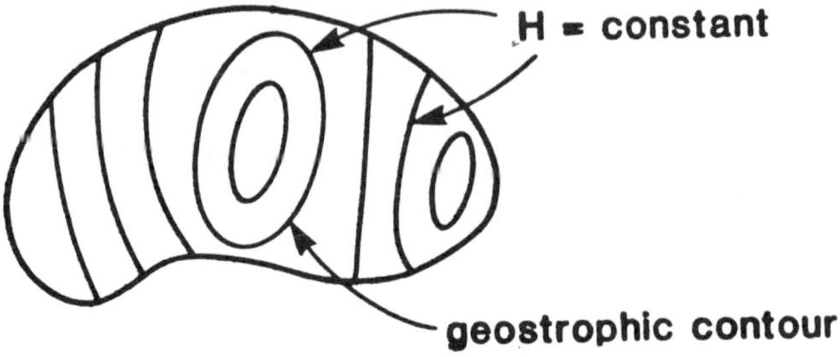

Fig. 2.4. Geostrophic contours for small-amplitude motion in a basin.

is perturbed, in such a way that the geostrophic degeneracy is partially or perhaps completely removed. To take a particularly simple case, consider a large basin whose depth function is locally

$$H = H_0(1 - \gamma y), \quad \gamma > 0. \tag{2.29}$$

It is helpful to consider the parameter γ as small, $\gamma = O(\varepsilon)$, to retain (2.29) over a sizeable region, and to disregard boundary effects.

The custom in dealing with large-scale flow of the atmosphere and oceans is to choose the x-coordinate pointing East, and the y-coordinate pointing North. The reason for considering a basin with depth decreasing northward will be discussed at the end of Section 3.2. We will see that such a geometry is able to simulate the important effect of latitudinal variation in the effective local rate of rotation.

With (2.29), the geostrophic contours are the lines y = constant, so the degeneracy of the constant-depth case has been significantly altered. To force recovery of the missing modes in (2.16) as a perturbation, we suppress completely the remaining geostrophic modes by requiring that η be of the form

$$\eta = \tilde{\eta} \exp\{i(\omega t + kx)\}, \tag{2.30a}$$

with similar expressions for u and v, and with small ω/f_0.

The smallness of ω guarantees slow changes in all the variables; in particular we shall have small u_t and η_t. Eqs. (2.16a,c) then imply an approximate *geostrophic balance*. Since $\eta_y = 0$, (2.16b) also states that u itself is small. The solutions we consider have therefore predominantly y-directed fluid velocity. That is, particle motion of the slowly-varying flow is essentially perpendicular to the geostrophic contours of the stationary flow discussed before, while the slow wave propagation is parallel to these contours. Moreover, the free surface η and velocity (u,v) of the flow (2.30a) are still in a nearly geostrophic, slowly shifting balance.

Substituting η from (2.30a) and H from (2.29) into (2.21) yields

$$i\omega(f_0^2-\omega^2) = -gH_0 i\omega(1-\gamma y)k^2 + H_0\gamma f_0 gik, \tag{2.30b}$$

which involves y explicitly in the first term on the right. This appears to be inconsistent with the assumed form of η. However, we are dealing only with small ω, and it is permissible without introducing any inconsistency to neglect the term γy in order to obtain a dispersion relation correct to $O(\epsilon)$. Rearranging terms, the latter takes the form

$$f_0^2 + gH_0k^2 - \omega^2 = \frac{H_0\gamma f_0 gk}{\omega} . \tag{2.31}$$

The solutions of (2.31) are indicated in Figure 2.5, which shows that there are (for small γ) two "perturbed" inertia-gravity modes, together with a third root of order γ. Since this root is small, an approximate dispersion relation for it is

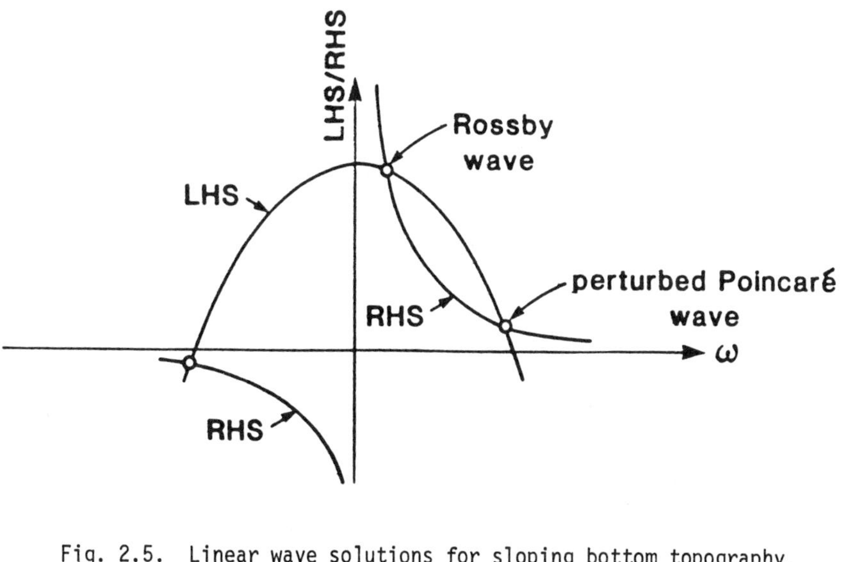

Fig. 2.5. Linear wave solutions for sloping bottom topography. The quantities LHS and RHS refer to the left- and right-hand side of Eq. (2.31), respectively.

$$\omega \simeq \frac{H_0 f_0 g k}{f_0^2 + g H_0 k^2} \gamma. \tag{2.32}$$

The waves associated with (2.32) are called *Rossby waves*, and in the present case of positive γ their phase velocity $c = -\omega/k$ is westward. Their frequency is always smaller than f_0, provided γ is not too large. They are further distinguished from both Poincaré and Kelvin waves by the fact that, for short waves, ω decreases with k (see Figure 2.6).

We might think of these waves as replacing, among the infinity of geostrophic modes in a basin of constant depth, the family of modes with y = constant as streamlines. In fact, to $O(\varepsilon)$, the flow is still non-divergent and in geostrophic balance

$$u = -\psi_y, \quad v = \psi_x, \tag{2.33a,b}$$

where $\psi = (g/f_0)\eta$. This also implies that, to $O(\varepsilon)$, the relative vorticity in the wave, cf. (2.13), is given by

$$\zeta = \nabla^2\psi. \tag{2.33c}$$

But it is precisely the small deviation from exact geostrophy, apparent

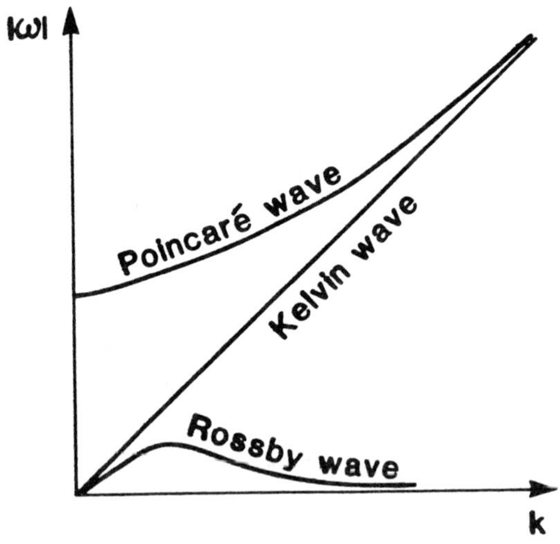

Fig. 2.6. Comparison of dispersion relations for small-amplitude shallow-water waves with rotation.

to O(1) only in the vorticity form of the equations, cf. (2.21, 2.30), which eliminates the geostrophic degeneracy and gives rise to the waves.

To understand the physical structure of Rossby waves, it is useful to consider their propagation in terms of a vorticity balance. According to (2.14), vorticity and depth along a material trajectory increase and decrease together. Examining the structure of the Rossby wave (2.30a) at any given time, we observe that ζ and v have the approximate variations shown in Figure 2.7. Since v advects ζ, wherever v is positive the relative vorticity at that point must be instantaneously decreasing with time, since advection would carry the fluid element from a deeper region into a shallower region ($\gamma > 0$). In order for this to happen the eddy pattern must drift to the West.

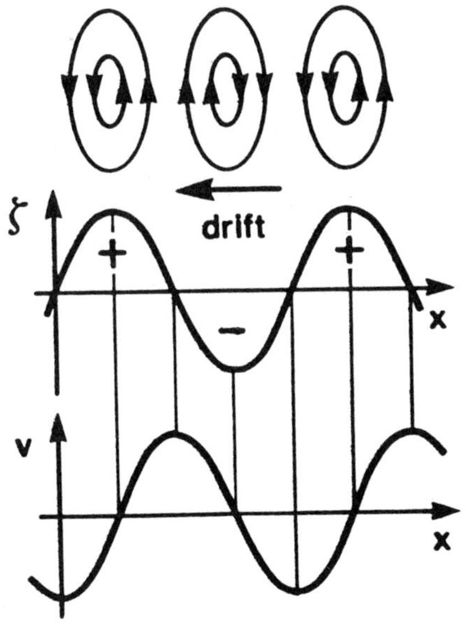

Fig. 2.7. The dynamics of linear Rossby waves. Streamlines in plan view are shown along with $\zeta(x)$ and $v(x)$.

References

Courant and Hilbert (1953), Chapters V and VI.

Gill (1982), Sections 7.1-7.6, 8.2 and 8.3, 10.4 and 10.5, 11.1-11.8.

Greenspan (1968), Sections 2.6 and 2.7

Landau and Lifschitz (1959), Section 100.

Pedlosky (1979), Sections 3.1-3.11.

Stoker (1957), Sections 2.2-2.4.

CHAPTER 3

THE QUASI-GEOSTROPHIC APPROXIMATION

Rossby waves are of geophysical interest because of their relatively long periods. These periods correspond to the slow time scale of the motions with large horizontal length scale in the atmosphere and oceans. The relevant dynamical balance is worthy of further study.

The point of departure in our discussion is the dominant geostrophic balance in the Rossby wave, which is evident from the simple, linear example discussed in Section 2.3. We are thus led to the question: what happens if we adopt a dominant geostrophic balance from the outset, as a guide to the selection of characteristic magnitudes? In what sense does the resulting version of shallow-water theory become governed by the dynamical balance exemplified by the Rossby wave? Is the resulting theory capable of exhibiting what must, to judge from the unpredictability of the weather, ultimately be a nonlinear dynamical balance? All of these questions can be answered in the simplest setting of constant-density shallow layers. The results point the way to a more general class of models of atmospheric dynamics, to which we shall turn in Chapter 4.

In this chapter we first introduce an expansion of quantities in the shallow-water equations with rotation into a power series in the small Rossby number $\varepsilon = UL/f_0$. Stopping at first order in this expansion yields the quasi-geostrophic potential vorticity equation. We show next the validity of these approximations for a small latitude band on the sphere, provided the constant local rotation rate $f_0/2$ used up to this point is replaced by a rotation rate varying linearly in the North-South direction with proportionality coefficient β.

Stationary solutions of this β-plane potential vorticity equation are studied in the presence of a North-South straight boundary. Onshore flow at infinity is seen to produce westward intensification of ocean currents, associated with the Gulf Stream, Kuroshio, and other narrow boundary currents. Finally, nonlinear Rossby waves are examined and shown to retrogress with respect to a mean zonal flow. Some comments are made about the interaction between Rossby waves, and the interaction between quasi-geostrophic and ageostrophic motions.

3.1. Scaling for Shallow Layers and Small Rossby Number

Let us suppose at the outset that the Rossby number $\varepsilon = U/Lf_0$ is small. In Chapter 2 we have not exploited this reasonable scaling in shallow-water theory, because of the relatively simple way in which inertial terms occur in the *linear* form of the theory studied there. In essence what follows will be implications of the smallness of ε, in combination with the low frequencies of Rossby-wave-like structures, for the *nonlinear* shallow-water theory. If U and L are again typical scales of horizontal velocity and horizontal structure, small ε will evidently make Coriolis forces strongly exceed inertial acceleration, so that the dominant dynamical balance in (2.7) must be between Coriolis and pressure forces. It is this balance which gives its name to the quasi-geostrophic approximation we introduce here.

To examine this balance, let us allow for an arbitrary amplitude of departures of layer height h (see Figure 2.1) from a constant value H_0, and write

$$h = H_0 + H_1\eta^*, \tag{3.1}$$

where H_1 is a second scaling parameter and η^* is a dimensionless perturbation. Defining other dimensionless variables and differential expressions by

$$u_H = U\,u^*, \quad \nabla_H = L^{-1}\nabla^*, \quad \partial_t = \frac{U}{L}\,\partial_t{}^*, \tag{3.2}$$

we see that, after substituting (3.1), (2.7) has the dimensionless form

$$\varepsilon(\partial_{t^*}u^* + u^*\cdot\nabla^*u^*) + \hat{k}\times u^* + \frac{gH_1}{f_0UL}\nabla^*\eta^* = 0. \tag{3.3}$$

To achieve the desired dynamical balance, we should therefore choose $H_1 = f_0UL/g$, a relation that can also be expressed in terms of the

Rossby radius of deformation $L_R = \sqrt{gH_0}/f_0$ as follows:

$$\frac{H_1}{H_0} = \frac{f_0UL}{gH_0} = \frac{U}{Lf_0} L^2\left(\frac{f_0^2}{gH_0}\right) = \varepsilon(L^2/L_R^2). \tag{3.4}$$

At this point we see the true dynamical significance of the Rossby radius of deformation as a fundamental horizontal length scale. It measures how extensive a structure must be for the Coriolis force to be comparable to pressure forces associated with the creation of hydrostatic equilibrium. Note that (3.4) means that a typical amplitude H_1 must increase with L/L_R to compensate for the relatively weaker effect of gravitation (expressed through a horizontal gradient) over an extensive structure. It is through the parameter L/L_R, that horizontal scale is explicitly introduced into quasi-geostrophic dynamics. Due to the presence of an intrinsic horizontal scale, L_R scale effects appear also in the dimensionless form of the governing equation (3.3).

We define the nondimensional radius of deformation λ by

$$\lambda^{-2} = (L/L_R)^2. \tag{3.5}$$

If λ is of order unity, as we shall now suppose, then we see from (3.4) that the layer height will actually depart only by order ε from the equilibrium height H_0. This might at first seem paradoxical, since the Coriolis force is made dominant, but it must be kept in mind that our scaling arguments have tended to focus on the relative importance of inertia, without reference to the absolute size of the (presumably essential) hydrostatic pressure gradient. Actually this pressure force is such that on structures of size L_R the Coriolis force is weaker by a factor ε, while inertia forces are weaker still by the same factor. Consequently only a small perturbation of height is needed to achieve geostrophic equilibrium. This is a key point since it allows some aspects of small-amplitude theory to be used to simplify a fundamentally nonlinear problem.

To allow for a bottom topography which is of the same order as the expected perturbation of H_0 we should rescale h_0 (the height of the bottom from the reference line in Figure 2.1) so that it is also of order ε in units H_0. The scaling of h and h_0 can then be summarized in the form

$$h = H_0(1 + \varepsilon\lambda^{-2}\eta^*), \tag{3.6a}$$

$$h_0 = H_0\, \varepsilon h_0^*. \tag{3.6b}$$

Eq. (3.3) has thus helped us establish the proper scaling (3.6) for the free surface and bottom topography. But letting $\varepsilon \to 0$ in (3.3) itself would merely lead to the geostrophic degeneracy discussed in Section 2.3. This degeneracy could only be avoided somewhat artificially in the linear theory by prescribing the form of the admissible perturbations. The analysis of these linear perturbations suggested the clue to a proper treatment of small deviations from geostrophy, namely to study the vorticity form of the equations.

We substitute therefore (3.6) into the Ertel-type equation (2.14a), which was derived in Section 2.1 using the scalar invariant $(z-h_0)/(h-h_0)$. In the present dimensionless variables this equation becomes

$$(\partial_{t^*} + u^*\cdot\nabla^*)\,\frac{\varepsilon^{-1}+\zeta^*}{1+\varepsilon\lambda^{-2}\eta^*-\varepsilon h_0^*} = 0, \tag{3.7a}$$

where

$$\zeta^* = \partial_{x^*}v^* - \partial_{y^*}u^*. \tag{3.7b}$$

Expanding in ε the Ertel potential vorticity we have

$$\frac{\varepsilon^{-1}+\zeta^*}{1+\varepsilon\lambda^{-2}\eta^*-\varepsilon h_0^*} = \varepsilon^{-1} + \zeta^* - \lambda^{-2}\eta^* + h_0^* + O(\varepsilon). \tag{3.8}$$

Neglecting terms of order ε or higher in (3.8), (3.7) may be written as

$$(\partial_t^* + u^*\cdot\nabla^*)(\zeta^*-\lambda^{-2}\eta^*+h_0^*) = 0. \tag{3.9}$$

Since we already neglect terms of order ε or higher in (3.9), the advection of the quasi-geostrophic potential vorticity may be evaluated in terms of the zeroth-order, geostrophic solution of (3.3), i.e.,

$$u^* = (-\eta^*_{y^*}, \eta^*_{x^*}), \tag{3.10a}$$

$$\zeta^* = \nabla^{*2}\eta^*. \tag{3.10b}$$

The expression for u^* shows that η^* is a stream function for the horizontal velocity. Note that, since $\nabla^*\cdot u^* = 0$ in this approximation, the vertical velocity component is $o(UH/L)$. Taking the three-dimensional curl of Eq. (2.3) and performing a scale analysis on its vertical component, we see that in fact $w = O(\varepsilon HL^{-1}U)$.

Substituting (3.10) into (3.9) leads to the nonlinear *potential vorticity equation* for shallow-water theory in the quasi-geostrophic approximation. Dropping superscript stars, this equation has the form

$$(\partial_t - \eta_y \partial_x + \eta_x \partial_y)(\nabla^2 \eta - \lambda^{-2} \eta + h_0) = 0. \tag{3.11}$$

This equation expresses the vorticity change which occurs in response to changes in bottom topography as well as in layer height, when the horizontal flow is in approximate geostrophic equilibrium with hydrostatic pressure.

3.2. The Beta-Plane Approximation

Before studying some solutions of the potential vorticity equation (3.11) it is appropriate to consider an observation made by Rossby, which considerably extends the physical scope of such model equations. Rossby observed that the appropriate surface for describing the dynamics of the atmosphere or ocean, in the vicinity of a point at latitude θ_0, is actually a patch of spherical surface. This patch would be seen to rotate about its normal, relative to an inertial observer, with angular velocity $\Omega \sin \theta_0$ (see Figure 3.1), Ω being the angular velocity of the planet. This "local" rotation rate is just the normal component of the vector angular velocity. Even more importantly, the local variation is significant over a portion of a *spherical* layer which is still accurately represented by shallow-water theory in Cartesian coordinates applied to a plane layer.

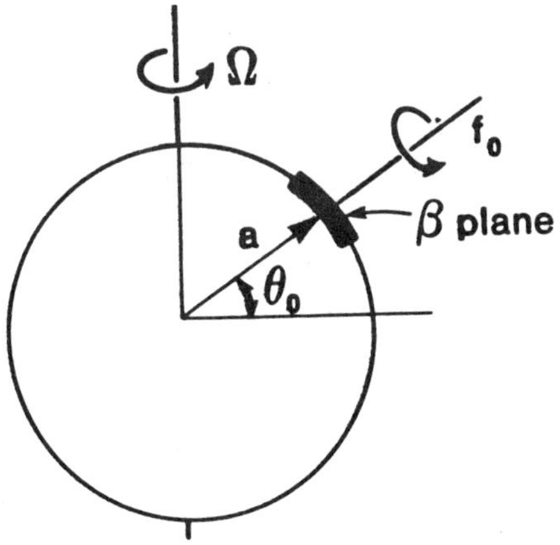

Fig. 3.1. Illustration of the beta-plane concept.

Rossby's point, which can be justified by formal asymptotics starting from shallow-water theory on a sphere, is that "sphericity" can be accommodated in the present formulation if the local planetary vorticity $f = 2\Omega \sin\theta$ is properly interpreted and allowed a simplified variation with latitude θ. If the latitude corresponding to $y = 0$ is $\theta = \theta_0$, then we have

$$f(\theta) = 2\Omega \sin\theta \cong 2\Omega \sin\theta_0 + 2\Omega(\theta - \theta_0)\cos\theta_0$$

$$\cong 2\Omega \sin\theta_0 + \frac{2\Omega \cos\theta_0}{a} y \equiv f_0 + \beta y$$

say, where a is the radius of the earth, $\beta = 2\Omega \cos\theta_0/a$ and $y = a(\theta - \theta_0)$. To incorporate this beta-plane approximation within the Cartesian shallow-water theory of the previous section, we need only replace f_0 by $f_0 + \beta y$. This is *not* a "tangent-plane" approximation to motion on a sphere. Rather, provided that the horizontal scale L is much smaller than a, the proper equations in *spherical* coordinates assume locally a Cartesian form.

More precisely, the β-plane effect can be justified in quasi-geostrophic theory because deviations from the Cartesian form of (3.7) make negligible contributions to the potential vorticity equation. For example the acceleration terms make contributions of order $\varepsilon L/a \sim O(\varepsilon^2)$ due to curvature; these are so small that our shallow-water version of Ertel's theorem is still correct through terms of order ε. On the other hand, by identifying dx and dy with $a\cos\theta d\phi$ and $a d\theta$ in spherical polar coordinates, with ϕ being longitude, the terms expressing geostrophic balance assume the two-dimensional Cartesian form.

To make its contribution to the potential vorticity equation of the same order as the other terms, β should be of order $f_0\varepsilon/L$, that is, $(L/a)\cot\theta_0$ should be comparable to ε. This is roughly the case, away from the equator, since a value of L/a of order 1/10 is appropriate in most problems of geophysical interest.

To effect the beta-plane approximation in the dimensionless form of the equations, we use $f_0 = f(\theta_0)$ in the definition of dimensionless variables, in which case (3.11) now becomes, after expansion in ε,

$$(\partial_t - \eta_y \partial_x + \eta_x \partial_y)(\nabla^2\eta - \lambda^{-2}\eta + \beta y + h_0) = 0, \tag{3.12}$$

with the nondimensional β equal to $\beta L^2/U$. The beta-effect is therefore mathematically equivalent to a linear term in y in the function

h_0 of the original formulation. In particular the Rossby wave studied in Section 2.3, if due entirely to the beta-effect, moves to the West in both hemispheres, since β is positive in both hemispheres. The equivalence between the β-effect and sloping bottom topography is useful, by the way, in the design of experiments involving Rossby waves, since variable depth is far easier to achieve in the laboratory than a spatially dependent rotation rate (compare Section 5.1).

3.3. The Inertial Boundary Layer

We consider an important example of steady solutions of the potential-vorticity equation, motivated by the occurence in ocean basins of intense currents such as the Gulf Stream or the Kuroshio. We suppose that h_0 is proportional to y so that its effect may be absorbed into an effective β.

For steady flow, Eqs. (3.10, 3.12) become equivalent to

$$\nabla^2\eta + \beta y = F(\eta), \tag{3.13}$$

where F is an arbitrary function of η. We suppose now that the y-axis is a barrier where $v = \eta_x$ must vanish, and consider a weak flow in the right half-plane which for large x is predominantly parallel to the x-axis, and directed westward. How does the flow "turn" as it approaches the barrier? When β is rather large, we shall see that the turning involves a strong current along the barrier.

For simplicity, let us assume that as $x \to \infty$, the flow tends to the uniform stream $(u,v) = (-1,0)$, and that $\eta(\infty,y) = y$. Assuming also that $\nabla^2\eta \to 0$ as $x \to \infty$, we see from (3.13) that $F(\eta(\infty,y)) = F(y) = \beta y$, which determines F completely. Thus

$$\nabla^2\eta + \beta y = \beta\eta. \tag{3.14a}$$

A solution of (3.14a) which has the appropriate behavior as $x \to \infty$, and also satisfies $\eta_x(0,\eta) = 0$, is

$$\eta = y\left(1 - e^{-\beta^{1/2}x}\right), \quad x \geq 0. \tag{3.14b}$$

We sketch the resulting flow in Figure 3.2. For large x the motion is near geostrophic, and it is only when a fluid particle is within a distance of order $\beta^{-1/2}$ from the barrier that this obstacle is sensed and turning of the flow begins. Near the origin we have

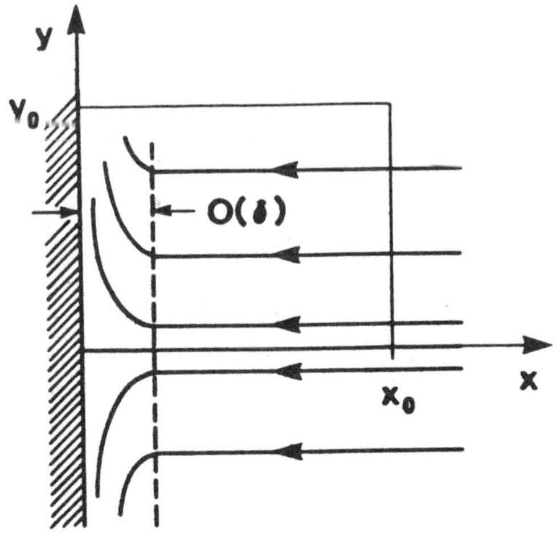

Fig. 3.2. Western boundary current intensification.

$\eta \cong \beta^{1/2}$ xy and, recalling that η is a streamfunction, the particle paths are hyperbolae.

As y increases, the flow entering the domain $0 \le x \le x_0$, $0 \le y \le y_0$, is turned into an intense current parallel to the y-axis, of width $\sim \beta^{-1/2}$. In physical variables, the width of the current, δ, is given by

$$\delta = \left(\frac{U}{\beta}\right)^{1/2} \quad \text{or} \quad \delta = \left(\frac{UH_0}{f_0\gamma}\right)^{1/2}, \tag{3.15a,b}$$

where γ is the bottom slope, depending on whether the beta we used was due to the Coriolis effect or to bottom topography. For the Gulf Stream a typical width is 100 km. Note that the parameter U is not the velocity of the narrow current itself, but rather some measure of the broad flow onto the shore, which is significantly smaller. Taking 10 cm/sec for the latter, we see that beta should be about $10^{-11}\ m^{-1}\cdot sec^{-1}$, if we neglect bottom topography. In fact this β is close to the actual value, although bottom topography does play a dynamical role in the structure of the Gulf Stream and other western boundary currents in the oceans.

A curious and important property of (3.13) is the change in the structure of solutions occurring when the direction of the East-West flow is reversed. One can think of this as being the situation for the

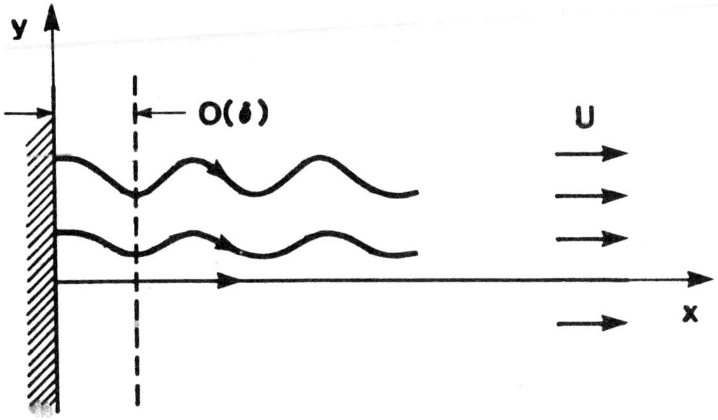

Fig. 3.3. Boundary-generated short waves.

Gulf Stream north of Cape Hatteras, while (3.14) represents the situation south of the Cape, roughly speaking.

If the flow at infinity is away from the western coastline, (3.14a) is changed to

$$\nabla^2\eta + \beta = -\beta\eta \, , \tag{3.16}$$

with solutions

$$\eta = -y[1 + A \sin(\beta^{1/2}x) + B \cos(\beta^{1/2}x)]. \tag{3.17}$$

The streamlines of this stationary flow are roughly sinusoidal in x and unevenly spaced in the y-direction (Figure 3.3). In fact, the u-velocity is independent of the long-shore variable y, while v is proportional to y. The amplification of the waves northward corresponds in a crude way to the presence of strong eddies in the northern branch of the Gulf Stream and Kuroshio.

We can view this solution as short waves riding on a flow to the East, seen as a steady structure in the present frame, fixed with respect to the boundary. This pattern is reminiscent of a westward-propagating Rossby wave as observed in a frame moving with the main current. The analogy is useful only up to a point, since the equations of motion are not actually equivalent under a Galilean transformation: the presence of the Coriolis term, according to Section 1.3, is due to the frame of reference being non-inertial. We shall turn therewith to an analysis of Rossby waves in quasi-geostrophic shallow-water theory.

3.4. Quasi-Geostrophic Rossby Waves

We have considered in Section 2.3 Rossby waves in linear shallow-water theory, and wish to see how such solutions look in the β-plane theory of Section 3.2. Surprisingly, the nonlinear equation (3.12) admits a family of plane waves as exact solutions, provided that h_0 is a linear function.

We set $\eta = N_0 \cos(k_1 x + k_2 y - \omega t + \alpha)$, where α is an arbitrary constant phase, and take ∇h_0 to be constant. Since $\nabla^2 \eta = -k^2 \eta$, with $k^2 = k_1^2 + k_2^2$, the nonlinear terms in (3.12) vanish for such a plane wave and the equation is satisfied provided that

$$\omega(k^2+\lambda^{-2}) + \beta k_1 - k_2 h_{0x} + k_1 h_{0y} = 0. \tag{3.18}$$

This is a slight generalization of the Rossby-wave dispersion relation obtained in Section 2.3. Of course, a superposition of two such Rossby waves is no longer a solution, due to their nonlinear interaction. Such interactions will be discussed in Sections 5.2 and 6.2.

The main point here is that inertia-gravity waves are no longer present at all, this being a consequence of the quasi-geostrophic dynamics. In effect the faster waves have been filtered out. Thus the present viewpoint is an extreme one, which does not indicate how the faster waves might interact with or otherwise influence the slow-time dynamics.

Finally, we note that (3.18), with $h_{0x} = 0$, is easily generalized to allow for a uniform mean flow with velocity $(U,0,0)$. The phase velocity component $c_1 = \omega/k_1$ then satisfies

$$c_1 = (Uk^2 - \beta - h_{0y})(k^2 + \lambda^{-2})^{-1}, \tag{3.19}$$

which shows that the wave moves westward, or retrogresses, with respect to the mean flow. This agrees with observations of planetary waves in the atmosphere's middle latitudes.

References

The general presentation has benefited from and can be complemented by Batchelor (1967), Section 7.7; Gill (1982), Section 12.6; Greenspan (1968), Section 5.4; Pedlosky (1979), Sections 3.12-3.19, 5.6-5.9, 6.1-6.3, and Robinson (1983), Chapters 1-6, 12-13 and 17-18.

Nonlinear interaction between slow, quasi-geostrophic, and fast, ageostrophic motions is a problem of current interest, both theoretical and practical. See for instance Baer and Tribbia (1977), Daley (1981), Errico (1982), Leith (1980), Lorenz (1980), Pedlosky (1979, Chapter 8) and Vautard and Legras (1986). With the present observational network (Gustafsson, 1981) and using the tools of sequential estimation theory (Ghil *et al.*, 1981), it is possible to determine the actual energy of fast, inertia-gravity waves and slow, low-frequency Rossby waves in the atmosphere (Dee *et al.*, 1985).

PART II

LARGE-SCALE ATMOSPHERIC DYNAMICS

CHAPTER 4

EFFECTS OF STRATIFICATION: BAROCLINIC INSTABILITY

4.1. A Perspective of the General Circulation

In Chapters 1 and 2, we have considered the influence of rotation and of shallowness on geophysical flows. The combined effects of the two led to the quasi-geostrophic shallow-water theory of Chapter 3. In this theory, the fluid has constant density and variations of the flow in the vertical are negligibly small. In spite of these idealizations, the theory was able to explain some features of atmospheric and oceanic flows with large horizontal scales, such as planetary waves in the atmosphere and western boundary currents in the ocean.

Large-scale atmospheric and oceanic flows show variations in the vertical which are responsible for some of their most interesting and intriguing features. To guide our study of these features, we shall take a closer look at the atmosphere of our planet.

The most simple-minded picture of the general circulation of the atmosphere would be like that of a room with an old-fashioned stove in winter, as shown in Figure 4.1. Cold air near the window sinks to the floor, while hot air near the stove rises to the ceiling. Conservation of mass closes the circuit by transporting the hot air along the ceiling towards the window, while cold air creeps along the floor towards the stove.

In this picture of the global atmosphere, the stove is the equator and the window is either pole. The picture even has a name, it is called a *Hadley circulation,* with air rising at the equator, traveling aloft towards the pole, sinking there and returning along the surface. Geostrophic balance would require the winds aloft to have a westerly (i.e.,

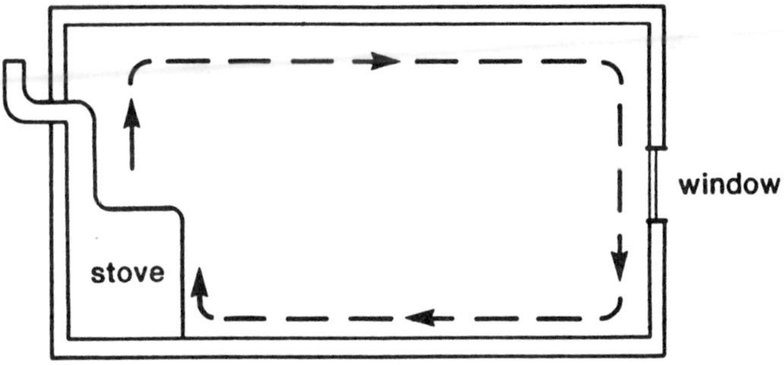

Fig. 4.1. Idealized picture of global atmospheric circulation.

eastward directed) component, while the surface winds would have an easterly (i.e., westward) component.

A nearly axisymmetric, single-cell circulation is actually observed on Venus, which has a much slower rotation rate than the Earth. More precisely, the solid surface of the planet has a clockwise (retrograde) rotation with a very long period of 243 terrestrial days. Its atmosphere, on the other hand, has a nearly solid-body easterly rotation which is much faster than that of the surface. This superrotation of the Venusian atmosphere is still quite slow compared to the Earth's atmosphere; it has a period of 4-5 terrestrial days.

On the Earth, a Hadley cell is observed in the tropics, its sinking branch being close to 30° of latitude, where the subtropical, arid zones are. Its surface easterlies are the *trade winds*. We have seen in Section 3.2 that, from the viewpoint of geophysical flows, it is the component of the planet's angular velocity vector along the local vertical which gives the local rotation rate and enters the local Rossby number, ε. Near the Equator, the local rotation rate $f/2$ is small, while the horizontal velocities in the Hadley cell are quite significant. Thus the local Rossby number $\varepsilon = U/fL$ is $O(1)$, and the effects of rotation are relatively small.

On Venus, the Rossby number, defined for deviations of its atmospheric flow from solid-body, retrograde rotation, is $O(1)$ or larger all over the surface. On Earth, for large-scale flow, i.e., $L \geq O(10^3\text{km})$, ε becomes small in middle and high latitudes. Apparently, the change in local Rossby number is related to a change in the character of the flow. Nonaxisymmetric, unsteady, aperiodic flow features are prevalent in mid-

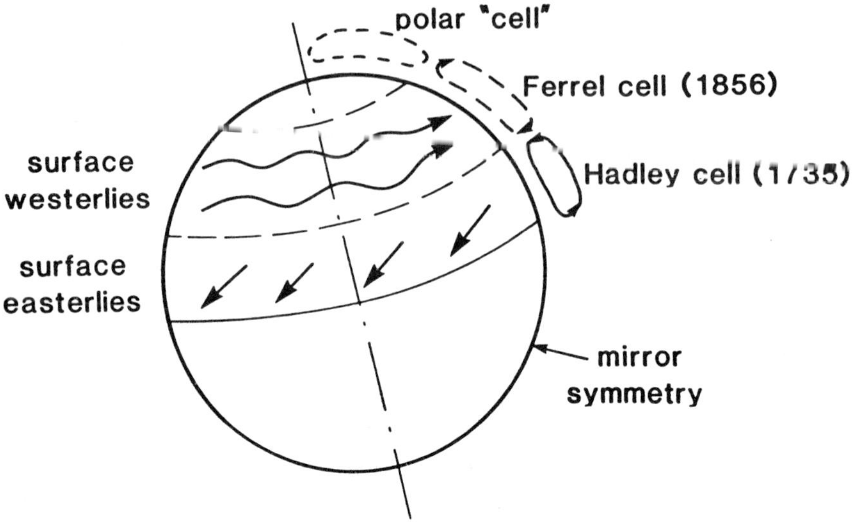

Fig. 4.2. A schematic diagram of atmospheric general circulation.

latitudes, as seen in Figure 4.2. It would seem that the decrease in f leads to an *instability* of the Hadley circulation.

Thermodynamic, as well as hydrodynamic effects are important in the destabilization of the axisymmetric, thermally driven Hadley circulation. We have seen in Section 1.4 the importance of the thermal structure of the mass field, as reflected by the baroclinic vector (1.26), in the evolution of the flow field's vorticity. The interaction between the thermal, baroclinic structure of the atmosphere and the large-scale midlatitude geostrophic effects creates a truly three-dimensional, rather complex instability, *baroclinic instability*.

In order to study this instability, we shall turn our attention in the next section to the atmosphere's vertical stratification. Simplified vertical structures, so called *model atmospheres*, will be discussed. and used to make plausible the prevailing westerlies of mid-latitudes. The observed structure of the lower atmosphere is outlined and pressure coordinates are introduced. Combining the equation of state for dry air with the equations of motion yields the *thermal wind equation*, which explains the intensification of the westerlies with height. The physical mechanism of baroclinic instability is discussed using a modification of Eady's wedge-of-instability argument.

In Section 4.3, a baroclinic, linear form of the quasi-geostrophic *potential vorticity equation* (PVE) is derived. The Boussinesq approximation for a stratified fluid in a rotating channel is introduced. The derivation of the PVE proceeds by scale analysis, giving special attention to the vertical static stability. The *internal Rossby radius of deformation* is defined and the role of the wedge of instability is highlighted. The derivation of the linearized PVE for baroclinic perturbations becomes straightforward after these considerations.

Section 4.4 outlines the classical normal-mode theory of baroclinic instability for the Eady problem, with constant planetary vorticity, $f = \text{const}$. Separation of variables leads to an eigenvalue problem for the vertical structure of waves. The stability condition, in terms of vertical stratification and of zonal and meridional wave numbers, is obtained, and growth rates of the unstable waves are derived. A rapid, preliminary comparison of theoretical results with observations is made. The need for a more refined analysis is pointed out, and the Charney problem, which includes the β-effect of nonconstant planetary vorticity, as well as a different vertical wind shear, is outlined. Finally, the energy conversion by the unstable waves of potential energy available in the mean flow into kinetic energy of the waves is discussed.

Section 4.5 reviews briefly various directions in which baroclinic instability theory has been extended. Two major problems not addressed by the theory in its present form are discussed: 1) the limitations on predictability of large-scale atmospheric flow imposed by its perpetual instability; and 2) the existence of anomalous, recurrent flow patterns which are relatively fixed in space and whose duration exceeds considerably the reciprocal growth rates of baroclinic, traveling waves. Implications for the success of numerical weather prediction are mentioned, and the importance of nonlinear phenomena in attacking these two problems is argued.

Section 4.6 provides some pointers for the interested reader to references on the material covered in the text, as well as to additional, related topics. References are given for the description and understanding of the general circulation, and the role played in this understanding by large-scale numerical models. Some references for the atmospheric circulation on Jupiter, Mars, and Venus are given. The standard textbooks concerning dynamics of the lower and upper atmosphere are mentioned. A few sources for the rich literature on baroclinic instability are provided. Some current references on the localized nature of synoptic disturbances and the question of dominant wave number in the atmosphere's energy spectrum are given. We close with bibliographic remarks on fronts and frontogenesis.

4.2. Vertical Stratification

We have already seen in Chapter 2, Eq. (2.4), how a small aspect ratio, $\delta \equiv H/L \ll 1$, implies that the vertical momentum equation reduces to the *hydrostatic equation*,

$$p_z = -g\rho; \tag{4.1}$$

here H is the typical vertical extent of the atmosphere, L the horizontal dimension of the motions under study, p the pressure, z the (locally) vertical coordinate, g the acceleration of gravity and ρ the density. We shall see that $H = O(10\text{ km})$, while we are interested in $L = O(10^3\text{km})$. Hence $\delta \ll 1$, and (4.1) is a *very* good approximation for large horizontal scales of motion.

Clearly, $g\rho > 0$, and thus p decreases *monotonically* with height z. Moreover, $g = g(z) \cong \text{constant.} = g_0 = g(z_0)$ for $|z-z_0| \leq H$. If ρ is given, $\rho = \rho(z,p)$, then (4.1) is an ordinary differential equation for $p = p(z)$. The solution of this equation, depending on the particular form of $\rho = \rho(z,p)$ chosen, is called a model atmosphere.

Model atmospheres. The true vertical structure of the Earth's atmoshere changes, of course, in time and with horizontal position. To get a better feel for it, we shall take a quick look at two simple models.

The simplest model atmosphere is the *homogeneous* one, $\rho = \text{const.} = \rho_0$. This would correspond to the true atmosphere being compressed by a lid to have a constant density equal to that at mean sea level (MSL), say, $z = z_0$. With $p_0 = p(z_0)$, the total pressure of the atmosphere above MSL, (4.1) immediately yields

$$p = p_0 - g\rho_0(z - z_0). \tag{4.2a}$$

Since $p \geq 0$, it follows that the homogeneous atmosphere has finite height,

$$H_0 = p_0/g\rho_0 = O(10\text{ km}). \tag{4.2b}$$

This height is called the *scale height* and, for given $p_0 \cong 1\text{ Atm} \cong 1000\text{ mb}$, it depends inversely on ρ_0.

At the equator, where the temperature at MSL is high and hence ρ_0 is low, it follows that H_0 is high; at the poles it is lower. The scale height thus decreases, monotonically on the average, from the tropics to either polar region. The barotropic, shallow-water theory of Chapter 2 applies strictly speaking only if ρ_0 is independent of the horizontal coordinate. Let us imagine this to be so, ρ_0 = const., but translate variations of temperature and density merely into variations of scale height, H_0, and identify this scale height with the free surface height h of Figure 2.1. In this idealized picture, the geostrophic approximation would yield at each latitude flow perpendicular to the average gradient of the free surface, i.e., westerly winds in both hemispheres. *Prevailing westerlies* are indeed observed throughout the mid-latitudes of either hemisphere (see Figure 4.2).

The commonly observed structure of the lower atmosphere is that temperature first decreases with height, then increases again. The point where T_z changes sign is called the *tropopause*. Below it lies the troposphere, above it, for as long as $T_z > 0$, the stratosphere. The pressure at the tropopause is about 200 mb, so that roughly 80% of the atmosphere's mass lies in the troposphere. This schematic situation is sketched in Figure 4.3.

The height of the tropopause decreases from equator to pole, in agreement with our previous discussion of the variation with latitude of the scale height. It is actually close in value to H_0, O(10 km).

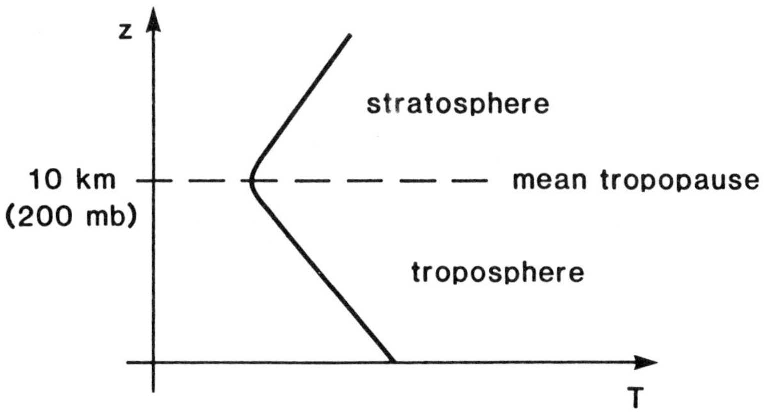

Fig. 4.3. Idealized vertical structure of the lower atmosphere.

To discuss a more realistic, nonhomogeneous model atmosphere, we have to introduce the *ideal gas law*,

$$p/\rho = RT, \tag{4.3}$$

where T is temperature and R the gas constant for air. To a very good approximation, (4.3) is the equation of state of dry air up to very low pressures.

The value of T_z in the troposphere is nearly constant with height. The U.S. Standard Atmosphere takes the so-called *lapse rate* $\Gamma \equiv -T_z =$ 6.5 K/km. We shall consider now the constant lapse-rate atmosphere.

Substituting (4.3) into (4.1), we obtain

$$p_z = -\frac{g}{R}\frac{p}{T(z)} . \tag{4.4}$$

For a constant lapse rate this yields

$$\frac{dp}{p} = -\frac{g}{R}\frac{dz}{T_0-\Gamma z} . \tag{4.5}$$

In particular, for $\Gamma = 0$, the *isothermal atmosphere*, $T = T_0$ has

$$p = p_0 e^{-z/H_0}, \tag{4.6}$$

where $H_0 = RT_0$ is again the scale height. Throughout the troposphere, p does decrease nearly exponentially with height. For $\Gamma \neq 0$, $\Gamma =$ 6.5 K/km say, (4.5) integrates to

$$p = p_0\{T(z)/T_0\}^{g/R\Gamma} .$$

Pressure coordinates. We have seen already that $p_z < 0$ and hence p and z are in one-to-one correspondence. This suggests the possibility of using p as a vertical coordinate. A practical reason for doing so is provided by the nature of upper-air measurements.

Until recently, most information about the atmosphere above the surface was collected by radiosondes. These are balloons whose instruments measure T, p and possibly winds as they ascend. To convert the vertical profile $T = T(p)$ into the three profiles $T = T(z)$, $p = p(z)$ and $\rho = \rho(z)$, it is convenient to transform the hydrostatic equation into the form:

$$z_p = -\frac{R}{g}\frac{T(p)}{p} .$$

This equation one can integrate numerically, given $T = T(p)$ as reported by the radiosonde, to yield $z = z(p)$. Thus one has $p = p(z)$ and hence $T = T(z)$, with $\rho = \rho(z)$ from (4.3).

In fact, it is better to change the point of view and introduce p as a vertical coordinate. Defining the *geopotential* $\phi = gz$, $\phi = \phi(p)$, the hydrostatic equation becomes

$$\phi_p = -\frac{RT}{p}, \tag{4.7}$$

or

$$\phi = \phi_0 - R\int_{p_0}^{p} T(p)\,d\log p. \tag{4.7'}$$

The last form of the equation justifies the use in meteorological parlance of the word *thickness* for temperature: T, or more precisely $-\int_{p_1}^{p_2} T(p)\,d\log p$, determines the thickness $\phi_2 - \phi_1 = g(z_2-z_1)$ of a layer of air between the isobaric surfaces $p = p_1$ and $p = p_2$, $p_1 > p_2$. As we shall see immediately, there are also theoretical reasons for changing from (x,y,z) to (x,y,p) coordinates.

In pressure coordinates, the equations of motion become

$$\partial_t u_p + u_p \cdot \nabla_p u_p + f\hat{k} \times u_p = -\nabla_p \phi - \hat{k}g, \tag{4.8a}$$

$$\partial_p \phi = -RT/p, \tag{4.8b}$$

$$\nabla_p \cdot u_p + \partial_p \omega = 0. \tag{4.8c}$$

As in Section 1.2, we leave for the moment the temperature equation needed for closure of the system unspecified. Here $\omega \equiv dp/dt$ replaces $w \equiv dz/dt$ as the "vertical" velocity component, and $(\ )_p$ replaces $(\ )_H$ as the "horizontal" part of the velocity vector or ∇ operator. In fact, pressure surfaces $\phi = \phi(x,y,\ p = \text{const.})$ are very nearly horizontal, so that u_p very nearly equals u_H. The vertical unit vector is denoted by $\hat{k} = (0,0,1)$, as before, $f = 2\Omega \sin\theta$ is the local Coriolis parameter, Ω is the angular velocity of the Earth and θ is latitude.

Notice that the flow is *incompressible* in pressure coordinates, cf. (4.8c). This at first appears as a miracle of the differentiation chain rule by which (4.8) follows from (1.14b) and (1.16). The physical explanation is that p-coordinates are mass coordinates, which is one of their theoretical advantages.

<u>Thermal wind shear</u>. Having considered the static stratification of the atmosphere and introduced pressure coordinates, we turn to a consideration of dynamic changes with height. The starting point is the geostrophic relation, which in p-coordinates becomes

$$f\hat{k} \times u_p = -\nabla_p \phi. \tag{4.9}$$

Equation (4.9) is derived from (4.8a) in the same way as (1.32) was from (1.18).

Dropping the subscript P and differentiating (4.9) with respect to the vertical coordinate *p* we obtain

$$f\hat{k} \times \partial_p u = -\nabla \phi_p .$$

Cross-multiplication on the left by $\frac{1}{f}\hat{k}$ yields

$$\partial_p u = \frac{1}{f}\hat{k} \times \nabla \phi_p .$$

Substituting ϕ_p from (4.7) leads to

$$\partial_p u = -\frac{R}{f}\hat{k} \times \frac{\nabla T}{p},$$

where we used the fact that ∇ operates at constant *p*. Multiplication by *p* finally yields the *thermal wind equation:*

$$\frac{\partial u}{\partial \log p} = -\frac{R}{f}\hat{k} \times \nabla T. \tag{4.10}$$

Equation (4.10) in fact gives the vertical *shear* of the p-horizontal wind, due to temperature gradients in the pressure surfaces. This is clearly a baroclinic effect, since $\nabla \rho \times \nabla p = 0$ and (4.3) together would imply $\nabla_p T = 0$.

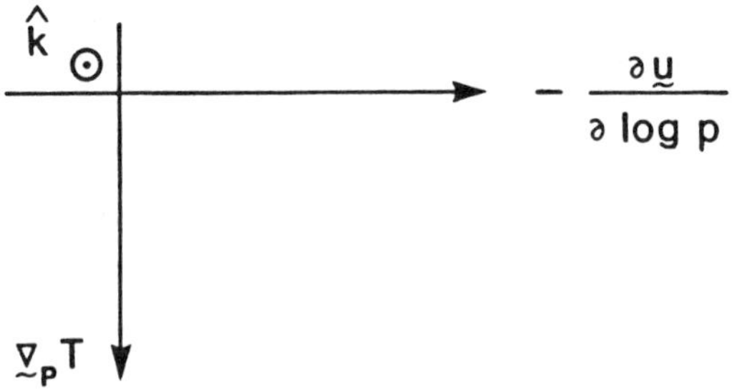

Fig. 4.4. The thermal wind shear.

Notice that for ∇T pointing equatorward (Figure 4.4), (4.10) states that the eastward velocity along isobaric surfaces *increases* with height ($p_z < 0$!). It is observed, in fact, not only that the height-averaged, barotropic wind in midlatitudes is westerly, but also that its magnitude increases with height.

The simple form (4.10) of the thermal wind equation in p-coordinates is another reason which makes them appealing in large-scale dynamic meteorology. In z-coordinates, this equation takes the form

$$f\hat{k} \times \partial_z u_H = -\frac{1}{\rho} \nabla_H p_z + \frac{\rho_z}{\rho^2} \nabla_H p.$$

Only the first term on the right-hand side can easily be interpreted using (4.1), and the connection to baroclinicity becomes much less obvious.

Baroclinic instability. It appears from our discussion of (4.10) that baroclinic effects do play an important role in explaining major features of the midlatitude atmospheric circulation. This role reinforces our earlier suspicion of baroclinicity contributing to destabilize the Hadley circulation, as discussed in Section 4.1.

A heuristic mechanism for this instability was suggested by Eady (1949). Let z point upward and y meridionally northward. Let isopycnic surfaces, ρ = const., tilt slightly upward with y, at an angle γ, while the stratification is statically stable, i.e., lighter fluid lies over the heavier layers. The tilt is balanced dynamically by a geostrophic wind or, in the ocean, a geostrophic current.

The following argument is a simplified mechanical version of Eady's original argument, which was based on the conversion of potential into kinetic energy by elementary particle displacements. Assume that a fluctuation occurs in the fluid, so that the particle P, originally at point P_0 on the surface $\rho = \rho_0$, is brought to the position P_H on the surface $\rho = \rho_1 > \rho_0$ (Figure 4.5). As a result of the density difference with respect to the surrounding fluid, P will then move up to a position P'. Due to the fact that the angle $\widehat{P_H P_0 y}$ is smaller than the angle of tilt, γ, the distance between P_H and P_0 is *increased* by this displacement, so $\overline{P'P_0} > \overline{P_H P_0}$. The situation is *unstable*, since a spontaneous displacement of a particle P tends to increase with time the distance from its original position P_0.

When P is pushed to the position P_L, on $\rho = \rho_{-1} < \rho_0$, with an angle $\widehat{P_L P_0 y} > \gamma$, it will move down to P''. Now $\overline{P''P_0} < \overline{P_L P_0}$ and the situation is *stable*, since P tends to be restored to its original position.

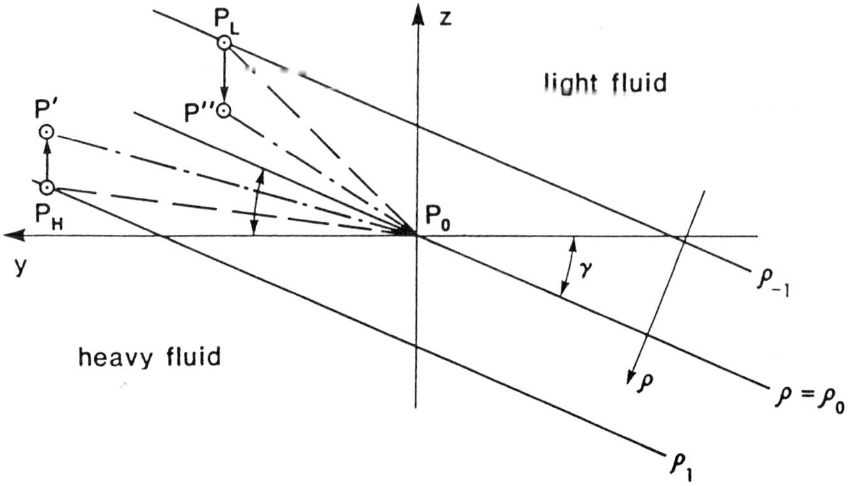

Fig. 4.5 Schematic diagram of the physical mechanism producing baroclinic instability.

This illustration does not explain the nature of the always present spontaneous fluctuations, nor that of the resulting motions. It suggests only that in a stratified fluid under rotation, motions with small vertical velocities superimposed on a strong current in the direction of rotation, i.e., westerly, might be unstable. To study the evolution in time of rotation-dominated, geostrophic motion, we have seen in Chapter 3 that the quasi-geostrophic potential vorticity equation (PVE) is a useful tool. Here, however, baroclinic effects compete with the geostrophic ones. It would seem therefore, that a baroclinic version of the quasi-geostrophic PVE is needed.

We have mentioned already in Part I some of the similarities and differences between oceans and atmospheres. Baroclinic instability theory was first derived with meteorological applications in mind. Its relevance to the oceans became quickly apparent. In the sequel, we shall actually derive it, for simplicity's sake, in the case of the ocean.

The role of ρ below and in Figure 4.5 would be taken in the atmosphere by the *potential temperature* Θ, defined by

$$s = c_p \log \Theta + \text{const.}, \tag{4.11}$$

where s is entropy and c_p the specific heat at constant pressure.

Potential temperature is a convenient and often used atmospheric variable. One reason is the useful analogy to ρ in the theory to follow. Another reason is that Θ = const., by definition, in isentropic flow, while Θ is at the same time quite close in value to the usual temperature T throughout the troposphere.

4.3. The Quasi-Geostrophic Baroclinic Potential Vorticity Equation

We shall consider an inviscid, nonconducting, Boussinesq fluid in an infinitely long rectangular channel D, with an $L \times H$ cross-section:

$$D = \{-\infty < x < \infty, \quad 0 \leq y \leq L, \quad 0 \leq z \leq H\}.$$

The channel rotates with constant angular velocity $(f/2)\hat{k}$ and is subject to the gravitational acceleration $-g\hat{k}$, as illustrated in Figure 4.6. Again, we think of x as being the *zonal*, East-West coordinate, and y the *meridional*, North-South coordinate, but ignore at first the β-effect of Sections 3.2-3.4.

The equations of motion are

$$\partial_t u + u\cdot\nabla u + f\hat{k} \times u = -\frac{1}{\rho_0}\nabla p + \alpha g(T-T_0)\hat{k}, \tag{4.12a}$$

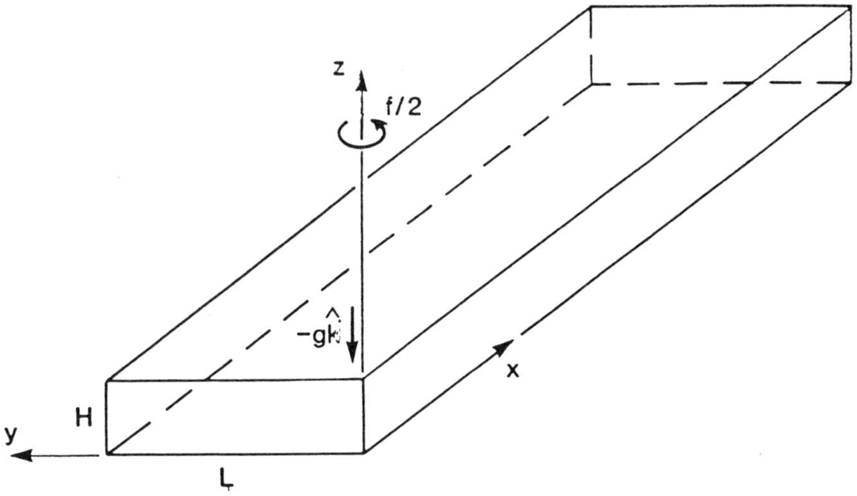

Fig. 4.6. Rotating channel in a gravitational field.

$$\nabla \cdot u = 0, \tag{4.12b}$$

$$\partial_t T + u \cdot \nabla T = 0. \tag{4.12c}$$

The fluid velocity relative to the rotating frame is $u = (u,v,w)$ and $\underset{\sim}{\nabla} = (\partial_x,\partial_y,\partial_z)$. Viscous forces are absent from the momentum equation (4.12a).

The *Boussinesq approximation* is that $\rho = \rho_0$ in the pressure-gradient term in (4.12a) and that flow is incompressible, cf. (4.12b); in the buoyancy term in (4.12a), however, the equation of state

$$\rho = \rho_0\{1 - \alpha(T - T_0)\} \tag{4.12d}$$

has to be used. Here ρ_0 is a reference density, T_0 a reference temperature and α the coefficient of thermal expansion. The energy equation (4.12c) states that all heat exchanges are by convection, no conduction or heat sources being present.

This approximation, often used also in the study of Rayleigh-Bénard thermal convection, is very reasonable in water. The corresponding approximation for the atmosphere is the *anelastic* one, and is most easily made in terms of the potential temperature Θ, as mentioned at the end of Section 4.2. The equations for the atmospheric case can be derived from system (4.8), augmented by a form of (4.12c) in which Θ replaces T.

The boundary conditions appropriate for (4.12) and for the geometry of Figure 4.6 are that the normal velocity components be zero at the walls

$$v = 0 \quad \text{at} \quad y = 0,\ L; \tag{4.13a}$$

$$w = 0 \quad \text{at} \quad z = 0,\ H. \tag{4.13b}$$

We shall study small perturbations of a basic flow state, which is chosen to be a simple idealization of the midlatitude westerlies discussed in Sections 4.1 and 4.2. This basic solution of (4.12) to be perturbed is $\underset{\sim}{u} = \underset{\sim}{U}$, $T = \overline{T}$, $p = P$, where

$$\underset{\sim}{U} = (\frac{U}{H} z,0,0), \tag{4.14a}$$

$$\overline{T} = T_0 + \frac{\Delta T}{H} z - \frac{fU}{\alpha g H} y, \tag{4.14b}$$

$$P = \rho_0\{\frac{\alpha g \Delta T}{2H} z^2 - \frac{fU}{H} zy\}. \tag{4.14c}$$

Thus the basic velocity is westerly and increases linearly with height, while basic isopycnic surfaces (cf. (4.12d) and (4.14b)), are planes tilting up with increasing y. The stratification is statically stable, since we take $\Delta T > 0$ and thus ρ decreases with height. Eqs. (4.14) show that the basic state is hydrostatically and geostrophically balanced. It is, however, a baroclinic state since T-surfaces, and hence ρ-surfaces, do not coincide with p-surfaces, $\nabla\rho \times \nabla p \neq 0$.

To obtain the proper scaling of the equations, we introduce dimensionless variables, denoted by $(\ )^*$:

$$(x,y) = L(x^*,y^*), \quad z = Hz^*, \quad t = \frac{L}{U} t^*,$$

$$(u,v) = U(u^*,v^*), \quad w = \varepsilon\delta U w^*,$$

$$T = T_0 + z^*\Delta T + \frac{fUL}{\alpha gH} T^*,$$

$$p = \frac{1}{2} \alpha g\rho_0 H\Delta T z_*^2 + fUL\rho_0 p^*.$$

In the scaling of p and T, the zeroth-order dependence on z was eliminated in defining T^* and p^*. The only scaling which is not the *a priori* obvious one is that of w, where $\delta = H/L$ is the aspect ratio and $\varepsilon = U/fL$ is the Rossby number.

From the continuity equation (4.12b) one would obtain, as for (2.2), the following order of magnitude relationship for a scale W of w:

$$\frac{U}{L} + \frac{U}{L} + \frac{W}{H} = 0.$$

This would yield merely the requirement that $W \leq O(\delta U)$. Why then the more stringent scaling $W = O(\varepsilon\delta U)$?

We have observed at the end of Section 3.1 that quasi-geostrophic theory for a constant-density fluid layer involves precisely this scaling of w. The absence of thermal effects distinguishes Eqs. (2.3, 3.3) from (4.12a). From (2.2) it immediately follows that the horizontal divergence is small, $\nabla_H \cdot u_H = O(U/L)$. This, combined with the vertical component of the curl of Eq. (2.3) yielded $W = O(\varepsilon\delta U)$. The presence of thermal effects in (4.12) renders matters more complex, so we shall consider first in some detail the scaling of vertical stratification.

Static stability. An often used nondimensional measure of vertical stratification is the Burger number

$$S = \frac{\alpha gH\Delta T}{f^2L^2} = \varepsilon^2 Ri, \tag{4.15}$$

where Ri is the Richardson number,

$$Ri = \frac{\alpha g H \Delta T}{U^2}, \tag{4.16a}$$

itself a nondimensional measure of static stability. Indeed, using $\rho = \bar{\rho}$ given by (4.12d) and (4.14b),

$$Ri = -\frac{gH^2}{U^2}\frac{1}{\rho_0}\frac{\partial\bar{\rho}}{\partial z}. \tag{4.16b}$$

so that Ri appears as a ratio of stratification to the kinetic energy of the basic flow.

The Burger number S can also be interpreted in terms of the *internal Rossby radius of deformation*, L_D,

$$L_D = (N/f)H, \tag{4.17a}$$

where N is the Brunt-Väisälä frequency of buoyancy oscillations,

$$N^2 = -\frac{g}{\rho_0}\frac{\partial\bar{\rho}}{\partial z}; \tag{4.17b}$$

this frequency increases with stratification. The internal radius L_D is usually much smaller than the *external* Rossby radius of deformation $L_R = \sqrt{gH}/f$, $L_D/L_R << 1$. L_R was introduced in Sections 2.2 and 3.1 (in connection with Eqs. (2.23) and (3.4), respectively); it is the scale of barotropic quasi-geostrophic motions. We shall see in this section and the next that L_D is the horizontal length scale of baroclinic disturbances in the atmosphere and ocean.

With (4.17a,b), Eq. (4.15) becomes

$$S = L_D^2/L^2. \tag{4.17c}$$

For the phenomena of interest, namely baroclinic free waves, it turns out that relatively *weak stratification* prevails, $S \leq O(1)$. When $S > O(1)$, then no such waves can arise and propagate in the fluid, and external forcing is necessary to change the static situation.

We shall consider, therefore, the parameter range of quasi-geostrophic, large-scale, weakly-stratified flows

$$\varepsilon << 1, \quad \delta << 1, \quad S = O(1); \tag{4.18a,b,c}$$

in particular, $L_D \cong L$. This means that the resulting motion, which will have a scale L_D in the zonal x-direction, is truly three-dimensional, since L_D is comparable to the scale L in the meridional y-direction.

Given (4.18), we return to the question of scaling w. Consider the slope of the isopycnic, viz., isothermal surfaces,

$$F(y,z) \equiv \frac{\Delta T}{H} z - \frac{fU}{\alpha g H} y = \text{const.}$$

They are given, according to the implicit function theorem, by

$$\frac{dz}{dy} = -\frac{F_y}{F_z} = \frac{fU}{\alpha g H} \cdot \frac{H}{\Delta T} = \frac{U^2}{\alpha g H \Delta T} \cdot \frac{fL}{U} \cdot \frac{H}{L}$$

$$= Ri^{-1}\varepsilon^{-1}\delta = \varepsilon\delta S^{-1}.$$

Eady's heuristic argument (Section 4.2) requires that the trajectory of a particle lie in the wedge $\{0 < z/y < \tan^{-1}\gamma\}$ for instability to occur. This suggests the scaling

$$W/U \le \tan^{-1}\gamma = (dz/dy)_{\overline{T}=\text{const.}} = \varepsilon\delta S^{-1}, \qquad (4.19)$$

or, using (4.18), $W = O(\varepsilon\delta U)$.

Perturbation equations. Having justified our choice of scaling, we introduce the nondimensional variables into (4.12-4.14) dropping the star superscripts $(\;)^*$. The basic state becomes

$$\underset{\sim}{U} = (z,0,0), \quad \overline{T} = -y, \quad P = -yz, \qquad (4.20\text{a,b,c})$$

and we wish to study nondimensionally small perturbations about it:

$$\underset{\sim}{u} = \underset{\sim}{U} + \underset{\sim}{u}', \quad T = \overline{T} + T', \quad p = P + p'. \qquad (4.21)$$

Substituting (4.21) into the nondimensional form of (4.12, 4.13), we shall have zeroth-order terms not containing the perturbation, terms linear in the perturbation, and terms quadratic in it. The zeroth-order terms will cancel identically, since (4.20) is a solution of the equations. The smallness assumption

$$|\underset{\sim}{u}'| \ll 1, \quad |T'| \ll 1, \quad |p'| \ll 1,$$

means that we can neglect the terms quadratic in perturbation quantities, i.e., linearize the resulting equations. Such a linear analysis is the first step in investigating the stability of the basic flow (4.20) to small perturbations and the structure of those perturbations, if any, which will grow.

The linearized equations, dropping primes, are

$$\varepsilon(u_t + zu_x + \varepsilon w) - v = -p_x, \tag{4.22a}$$

$$\varepsilon(v_t + zv_x) \qquad + u = -p_y, \tag{4.22b}$$

$$\varepsilon\delta^2(w_t + zw_x) \qquad = -p_z + T, \tag{4.22c}$$

$$T_t + zT_x - v + Sw \qquad = 0, \tag{4.22d}$$

$$u_x + v_y + \varepsilon w_z \qquad = 0. \tag{4.22e}$$

The term $\varepsilon^2 w$ in (4.22a) is due to the fact that $U_z = 1$, cf. (4.20a), and the terms $-v + Sw$ in (4.22d) are due to the linear dependence of $\overline{T}$ on y and z, cf. (4.14b).

Letting $\varepsilon \downarrow 0$ in (4.22) would lead to geostrophic degeneracy, cf. Section 2.3. The correct way of avoiding it and obtaining the *quasi-geostrophic approximation* for the baroclinic case is to work with the vorticity form of the flow equations; this was done in Section 3.1 for the barotropic case.

Taking the curl of the horizontal momentum equations (4.22a,b) eliminates p and yields the vorticity equation

$$\varepsilon(\partial_t + z\partial_x)(v_x - u_y) - \varepsilon^2 w_y + u_x + v_y = 0. \tag{4.23}$$

We recall now that the horizontal divergence $u_x + v_y$ is $O(\varepsilon)$ and substitute its value $-\varepsilon w_z$ from (4.22e) into (4.23), to yield

$$(\partial_t + z\partial_x)(v_x - u_y) - \varepsilon w_y - w_z = 0. \tag{4.23'}$$

At this point we are justified in letting $\varepsilon \downarrow 0$ in (4.22a-c) and in (4.23') to yield

$$u = -p_y, \quad v = p_x, \quad T = p_z, \tag{4.24a,b,c}$$

and hence

$$(\partial_t + z\partial_x)(p_{xx} + p_{yy}) = w_z. \tag{4.25a}$$

Furthermore, (4.22d) becomes

$$(\partial_t + z\partial_x)p_z - p_x + Sw = 0. \tag{4.25b}$$

Differentiating (4.25b) with respect to z and substituting w_z from (4.25a), we finally obtain

$$(\partial_t + z\partial_x)\{S(\partial_x^2 + \partial_y^2)p + p_{zz}\} = 0. \tag{4.26}$$

Although (4.26) was derived from (4.22) and not from Ertel's theorem, there is a clear analogy between it and the nonlinear, barotropic potential vorticity equation (PVE) of Section 3.1, Eq. (3.11). Eq. (4.26) is a linear form of the quasi-geostrophic PVE appropriate for baroclinic flows. The presence of the term p_{zz} takes into account vertical stratification, while $z\partial_x$ introduces the effect of vertical shear in the horizontal flow.

The boundary conditions (4.13) become in the quasi-geostrophic approximation

$$p_x = 0, \quad \text{at} \quad y = 0,1, \tag{4.27a,b}$$

and

$$(\partial_t + z\partial_x)p_z - p_x = 0 \quad \text{at} \quad z = 0,1. \tag{4.27c,d}$$

To solve (4.26) with (4.27) is the *Eady problem.* It treats the simplest model which exhibits baroclinic instability.

4.4. Baroclinic Instability

The theory. Problem (4.26, 4.27) is linear and homogeneous, so that $p \equiv 0$ is trivially a solution. We are interested in the existence of nontrivial eigenmodes, which we seek in the form of plane waves

$$p = \hat{p}(z)e^{ik(x-ct)}\sin \pi\ell y. \tag{4.28}$$

The connection between our results and the atmosphere will be found in recalling that x points eastward, i.e., k is a *zonal* wave number, while y points northward, so that ℓ is a *meridional* wave number. In this interpretation, there will be no distinction between an x-infinite and an x-periodic channel. Thus k will be arbitrary.

Boundary conditions (4.27a,b) state that, for a nontrivial plane wave (4.28) to exist, $\ell = 1,2,3,\ldots$. With integer ℓ and arbitrary k, substitution of (4.28) into (4.26) yields an equation for the vertical structure $\hat{p}(z)$ of the wave

$$\hat{p}_{zz} - S(k^2 + \pi^2\ell^2)\hat{p} = 0. \tag{4.29}$$

Its solution is

$$\hat{p} = A \sinh 2\kappa z + B \cosh 2\kappa z$$

where

$$4\kappa^2 = S(k^2 + \pi^2\ell^2) \tag{4.30}$$

for $\kappa^2 > 0$, $\ell \geq 1$ and k real.

Boundary condition (4.27c) yields $B = -2c\kappa A$, so that

$$\hat{p} = A(\sinh 2\kappa z - 2c\kappa \cosh 2\kappa z), \tag{4.31}$$

where A is the arbitrary amplitude of the eigenfunction $\hat{p}(z;\kappa)$. Finally, (4.27d) gives the characteristic equation for the eigenvalue c,

$$4\kappa^2 c^2 - 4\kappa^2 c + 2\kappa \coth 2\kappa - 1 = 0.$$

Using the half-angle formulae for hyperbolic functions, the solutions of this quadratic are

$$c = \frac{1}{2} \pm \frac{1}{2\kappa}\{(\kappa - \tanh \kappa)(\kappa - \coth \kappa)\}^{1/2}. \tag{4.32}$$

When the radicand of (4.32) is positive, both values of c are real. This corresponds to the existence of two *neutral* waves, with phase speeds c_1 and c_2, respectively. When the radicand is negative, c will be complex,

$$c = \frac{1}{2} \pm ic_i,$$

and this will lead to an exponentially growing solution (4.28); here $c_i = \operatorname{Im} c$ is real.

Since

$$\kappa - \tanh \kappa > 0$$

for $\kappa > 0$ (Figure 4.7), the radicand of (4.32) is nonnegative if and only if

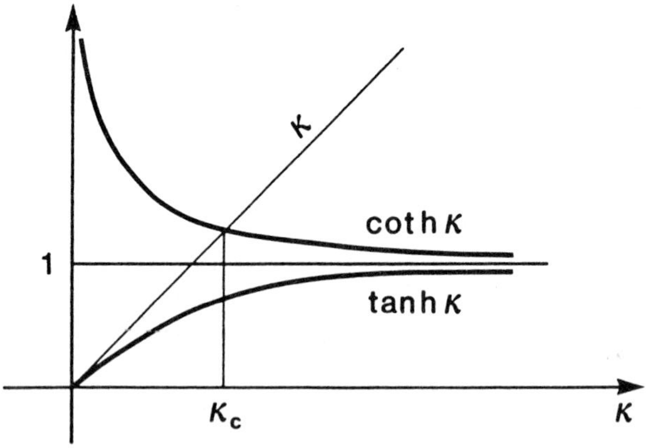

Fig. 4.7. Critical parameter value $\kappa = \kappa_c$ for Eady problem (see Eqs. (4.30) and (4.33)).

$\kappa \geq \coth \kappa.$

The function $\kappa - \coth \kappa$ is monotone increasing and crosses zero at a value $\kappa_c \cong 1.2$. Thus the *stability condition* for the Eady problem is that $\kappa \geq \kappa_c$ or

$$S(k^2 + \pi^2 \ell^2) \geq 4\kappa_c^2. \tag{4.33}$$

For each positive integer ℓ_0, one obtains a marginal stability curve $S = S_m(k;\ell_0)$

$$S_m = \frac{4\kappa_c^2}{\pi^2 \ell_0^2 + k^2}, \tag{4.34}$$

with stable states lying above it and unstable ones below. For $\ell_0 = 1$, the curve (4.34) becomes

$$S_m = S_c/(1 + k^2/\pi^2), \quad S_c = 4\kappa_c^2/\pi^2 \cong 0.58. \tag{4.35a,b}$$

Curve (4.35a) is shown in Figure 4.8. The curves for $\ell_0 = 2,3,...$ lie successively lower under curve (4.35), i.e., the region of instability they bound becomes successively smaller. We conclude that, for the same given static stability conditions S, and all zonal wave numbers k, the most unstable wave is the one with the largest meridional wave length. Notice also that, for any k or ℓ, $S < S_c \cong 0.58$ is a necessary condition for instability. This is consistent with our earlier scaling requirement of weak stratification (4.18c).

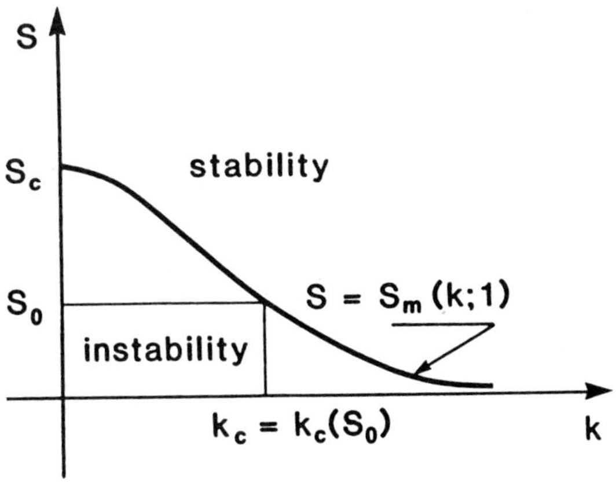

Fig. 4.8. Stability diagram for Eady problem.

It is essential for the physical interpretation of our results to remember that small perturbations of every wave number, zonal as well as meridional, will be present at any given time. Hence, if any of these perturbations is shown to be unstable by our linear analysis, it will grow, at least initially. It seems reasonable, furthermore, to assume at first that the one which grows fastest will reach finite amplitude, dominate the perturbation part of the flow and be most easily observable in nature.

The rate of growth of an unstable wave is given by

$$\sigma \equiv kc_i = \frac{k}{2\kappa}\{(\kappa - \tanh\kappa)(\coth\kappa - \kappa)\}^{1/2}. \tag{4.36}$$

Notice first, cf. (4.28), that the meridional wave number ℓ enters σ only via $\kappa = \kappa(k;S,\ell)$. We shall consider the most unstable value, $\ell = 1$. It would appear from Figure 4.8 that $k = 0$ is the most unstable zonal wave member, but in fact $\sigma(k;S,\ell) \to 0$ as $k \downarrow 0$. We show in Figures 4.9 and 4.10 the plots of the phase speed $c = c_r + ic_i$ as a function of κ, and of σ as a function of k for $\ell = 1$ and for a given $S < S_m$.

The growth rate σ is zero for $k = 0$ and again for $k = k_c$, where $k_c^2 = 4\kappa_c^2/S - \pi^2$, and $k_c = k_c(S) > 0$ for all $S < S_m \le S_c$. In between, $0 < k < k_c$, $\sigma(k)$ is convex upwards and has a unique maximum, at $k = k^*(S)$. The dimensional wave length of this wave is

$$\lambda_* = \frac{2\pi}{k^*(S)} L = \frac{2\pi}{k^*(S)\sqrt{S}} L_D. \tag{4.37}$$

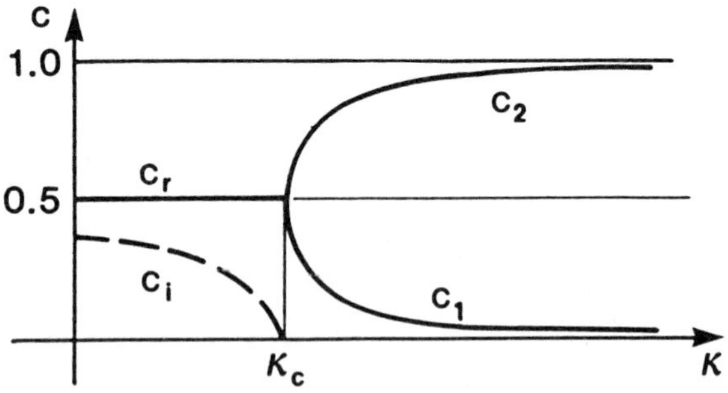

Fig. 4.9. Real phase speeds c_1 and c_2 of neutral waves and complex phase speed $c_r + ic_i$ of growing Eady wave.

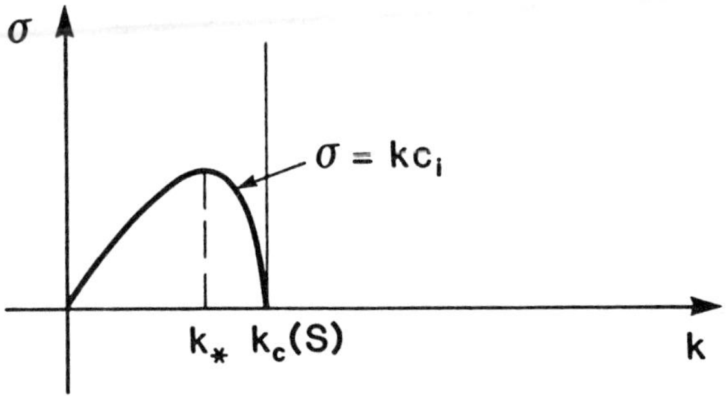

Fig. 4.10. Rate of growth of baroclinically unstable Eady waves.

Observational verification. It has been suggested in Section 4.1 that the baroclinic instability we have studied so far is largely responsible for determining the character of large-scale midlatitude flow. A critical test of this assertion is whether reasonable values of L_D and S will produce a value of λ_* close to the characteristic length of observed cyclone waves.

Substitution into (4.17) of a typical tropopause height $H = 10$ km, of $f = 10^{-4}\sec^{-1}$ corresponding to a latitude of 45°, and of $N = 10^{-2}\sec^{-1}$, corresponding to a lapse rate $\Gamma = 6.5$ K/km, yields $L_D = 10^3$km. The meridional extent L of the half-wave has to be larger than the lower bound $L \geq S_m^{-1/2} L_D$ given by (4.35), and less than the upper bound roughly given by 30° lat. between the Hadley cell and the arctic regions. This gives an approximate range of $1500 \text{ km} \leq L \leq 3000$ km and a corresponding range of S, cf. (4.17c).

The values of $\lambda_* = \lambda_*(S)$ which result from these choices are $O(10^3\text{km})$ and in good agreement with observed zonal ridge-to-ridge and trough-to-trough distances for midlatitude perturbations. The development times for cyclones, e.g., the doubling times of pressure perturbations at the center of a cyclonic low, which are $O(1$ day), are also in approximate agreement with the growth rate $\sigma^* = \sigma(k^*)$ given by (4.36). Furthermore, it is observed that shallower perturbations, for which H is smaller, have shorter wave lengths and faster growth rates, in accordance with the theory derived above.

It follows from this theory that, for given S, there is an upper bound on the zonal wave member k for which perturbations will be unstable. Short waves, therefore, are not baroclinically unstable; to the extent that they are present in the atmosphere's power spectrum, they must draw their energy from other sources. An important source is the nonlinear, turbulent cascade from interactions with the longer, baroclinically unstable waves.

The longest waves in the Eady problem are baroclinically unstable, but their growth rates are very small. An important energy source for these waves in the atmosphere is planetary-scale surface forcing, by orographic and thermal contrast between continents and oceans, rather than baroclinic instability.

The maximum amplitude of the baroclinically unstable Eady wave is at midheight, i.e., returning to dimensional variables, $z = H/2$. This corresponds to the so-called *steering level* of synoptic-scale disturbances. The speed of propagation is $U/2$, i.e., the average zonal speed of the basic flow. In fact, this is slightly incorrect, observed speeds of cyclone waves being slightly less: synoptic-scale weather systems are not so much stationary with respect to the mean current, as slightly *retrograde*. This retrogression was caused for the barotropic Rossby waves of Section 3.4 by the β-effect, cf. (3.19).

A model of baroclinic instability which includes fewer simplifying assumptions than Eady's model was analyzed by Charney (1947). Predictably, its mathematical structure is also slightly more complicated. For instance, the place of the vertical structure equation (4.29) is taken by a confluent hypergeometric equation

$$\zeta\chi_{\zeta\zeta} - \zeta\chi_{\zeta} + r\chi = 0, \tag{4.38}$$

where χ is a stream function, ζ is a scaled height and r a nondimensional number.

The β-effect, present in the Charney model, does indeed produce the desired retrogression. It also introduces a dependence of the instability on the vertical shear. In the Eady problem, mean velocity $U/2$ and vertical shear U were scaled by the same U. In the Charney problem, β introduces an independent scaling for shear, as well as removing the short-wave cut-off, so that all waves are unstable. The actual behavior of long, planetary waves and of short, sub-synoptic waves is in fact further affected by viscosity, variations in zonal flow, inhomogeneities of the lower boundary and the presence of other waves. This dependence on secondary effects, not taken into account in this highly simplified, linear analysis is clearly stronger than for the fastest growing, synoptic waves.

The observed structure of synoptic wave shows a backward tilt of the wave with height. For instance, the center of a cyclonic low on the 500 mb geopotential height surface will lie a few hundred kilometers behind the surface low. The corresponding tilt of troughs and ridges, with height as well as latitude, is exhibited by both Eady waves and Charney waves.

The basic mechanism of baroclinic instability is that, within the wedge of instability, light fluid is carried poleward and up, while heavy fluid is carried equatorward and down. Thus the perturbation extracts energy from the basic state, whose potential energy is reduced. We shall consider the energy conversion in detail in the Eady model.

The energy equation. Multiplying the linearized momentum equations (4.22a,b,c) by the nondimensional velocity components of the perturbation, u, v and w respectively, one obtains:

$$\varepsilon(\partial_t + z\partial_x)e'_K = -(up_x + vp_y + \varepsilon wp_z) + \varepsilon wT,$$

where we have introduced the kinetic energy density of the perturbation e'_K by

$$2e'_K = u^2 + v^2 + \varepsilon^2\delta^2 w^2. \qquad (4.39a)$$

Substituting w from (4.22d) yields

$$\varepsilon(\partial_t + z\partial_x)e'_K = -\nabla_\varepsilon \cdot p\underset{\sim}{u} + p\nabla_\varepsilon \cdot \underset{\sim}{u} + \varepsilon S^{-1}T(v - T_t - zT_x),$$

where

$$\nabla_\varepsilon = \partial_x + \partial_y + \varepsilon\partial_z$$

and (4.22e) states that $\nabla_\varepsilon \cdot \underset{\sim}{u} = 0$.

Remembering Eqs. (4.6, 4.7) and the role of temperature in determining the height of isobaric surfaces, it is natural to define the potential energy density e'_P of the perturbation by

$$2e'_P = S^{-1}T^2. \qquad (4.39b)$$

Using (4.39b), the total energy density $e'_K + e'_P$ will satisfy

$$\varepsilon(\partial_t + z\partial_x)(e'_K + e'_P) = -\varepsilon^{-1}\nabla_\varepsilon \cdot pu - S^{-1}vT.$$

Integrating this equation over a full period of the wave, $V = \{0 < x < 2\pi/k; 0 < y,z < 1\}$ yields

$$d_t(K' + P') = S^{-1}\int_V vT, \qquad (4.40)$$

where $K' = \int_V e'_K$ and $P' = \int_V e'_P$ are the kinetic and potential energies of the perturbation.

Eq. (4.40) states that the total energy of the perturbation feeds on the thermal Reynolds stress, vT; v and T in the wave are indeed positively correlated on the average. The dimensional form of (4.20b) and (4.22d) shows that in fact

$$\int_V vT = -\int_V vT \frac{d\overline{T}}{dy} .$$

Thus the perturbation will eventually lead to a decrease of the *available* potential energy of the basic state, as expressed in the tilt of the isopycnic surfaces.

4.5. Extensions of the Theory and Discussion

The observational verification of the main results from Charney's and Eady's stability analyses led to considerable work on more complex models. The stability of basic flows with a more complicated structure was investigated, allowing for meridional shear, as well as nonzonal flow. Cylindrical and spherical geometry, and more realistic boundary conditions were considered. Viscosity effects, as well as weak nonlinearity were studied, pushing the basically linear analysis to follow the growth of the waves to finite amplitude. Comparisons with atmospheric observations, laboratory experiments and detailed numerical model results were made. In short, baroclinic instability theory provided the core of dynamic meteorology for the last three decades, leading to a considerably improved understanding of midlatitude synoptic-scale motions.

There are two problems that the theory has not been able to deal with. The first problem, already recognized by Eady, is that the perpetual instability of midlatitude flow leads to a theoretical bound on its predictability. The theoretical predictability limit can be defined, for an atmospheric model, as the time at which two solutions, very close to each other at initial time, become as far apart as the mean separation of all model states. Estimates of this predictability limit have been made for various models and for various probability distributions of model states. The estimates range between one to two weeks (Lorenz, 1985).

The practical limit of usefulness of numerical weather forecasts is between three to seven days. It falls short of the theoretical limit by at least a factor of two. The question thus remains whether a different point of view than that taken heretofore could give some insight into this discrepancy between theory and practice. Such an insight could help produce also numerical forecasts beyond the current predictability limits, both practical and theoretical.

The second problem has to do with the observed existence of large-scale persistent anomalies of the midlatitude circulation. These anomalies have time scales of weeks to months, well beyond the life cycle of cyclones. Their climatic manifestations include excessively hot and dry summers, such as the European summer of 1976 or the North American summer of 1980, as well as very cold and/or snowy winters, such as the North American winters of 1976/77 and 1977/78.

Dynamically, these anomalies exhibit remarkably persistent, nearly constant patterns which are very different from the "normal", mostly zonal midlatitude flow. Meridional velocity components can be quite large, and the most striking type of anomaly is also called *blocking*: zonal flow is "blocked" by a stationary high, and has to find its way around it. The blocking high is frequently positioned close to the West Coast of the continent, so that flow over the continent will be from the north, rather than from the west, which explains some of the climatic effects noted above.

Blocked flow, in fact, seems to exhibit less synoptic variability than zonal flow as the blocking high deflects storm tracks away from its lee. The large-amplitude transitions between blocked and zonal flow regimes appear to have a well-defined distribution of occurrences. Episodes with enhanced duration of either blocked or zonal flow offer some hope of extended prediction, provided such episodes could be recognized in advance. A reasonable working hypothesis in explaining the distinctive features of the two types of atmospheric flow would seem to be that strong nonlinearities are at play, and that blocked and normal situations, each with its own stability properties, occupy well separated regions of the flow's phase space. We shall pursue this point of view in the next two chapters.

4.6. Bibliographic Notes

Section 4.1. The general circulation is of course one of the fundamental problems of atmospheric science. A very complete account of the physical principles useful in its understanding, such as those relating to energy and angular momentum transport, and zonally averaged statistics of the fields of motion, can be found in Lorenz (1967). The role of planetary waves in the circulation, and of idealized axisymmetric models, is discussed in Blackmon (1978). The history of the problem is updated by Lorenz (1983). An excellent description of many typical circulation features, both local and global, is given by Palmén and Newton (1969). The most up-to-date quantitative observations are carefully analyzed and summarized in Oort (1983).

The role of laboratory models in understanding features of the general circulation will be the topic of the next chapter. Detailed simulations of the general circulation using large-scale numerical models based on a discretization of the equations of motion, so-called general circulation models (GCMs), play an increasing role in understanding the present circulation and how it would change if certain conditions, such as solar radiation, the chemical composition of the atmosphere, or sea-surface heat fluxes were to change. Chang (1977), Corby (1969), and Gates (1979) give an account of the development of GCMs, the way hydrologic, radiative and other physical processes are represented in them, and the information they provide on the general circulation and its sensitivity to external changes. Numerical prediction models related to GCMs are covered by Haltiner and Williams (1980), Chapters 5-10, and by Phillips (1973).

For the circulation of Venus, Goody and Robinson (1966), Kalnay-Rivas (1973), and Stone (1975) provide two pioneering studies, and a review, respectively. Recent data on the middle atmosphere of Venus from the Pioneer orbiter mission are summarized in Taylor et al. (1980). A GCM for the atmosphere of Mars is developed and discussed by Pollack et al. (1981). The Jovian atmosphere is studied by Mitchell and Maxworthy (1985).

Section 4.2. The vertical structure of the terrestrial atmosphere and model atmospheres are discussed in detail by Dutton (1976), Chapter 3 and 4, and by Chamberlain (1978), Chapter 1. The upper atmosphere, above the tropopause and the stratopause, is studied by Craig (1965) and Holton (1975). The thermal wind equation and most of the material in our Chapters 1-4 is covered in detail by Holton (1979), Chapters 1-10.

Palmén and Newton (1969, Chapter 4) discuss the decrease of tropopause height with latitude. They show that where this height drops discontinuously, at approximately 30° lat. and 60° lat., the subtropical and polar jet streams occupy the corresponding "tropopause gap". The position and intensity of the jets changes with the seasons and, more strikingly, even within a given season. The latter change, from a strong to a weaker jet and then back to full strength, is called an *index cycle* (see Sections 5.3 and 6.1).

Sections 4.3 and 4.4. The original, fundamental articles on baroclinic instability are by Charney (1947), Eady (1949), and Fjørtoft (1950). The presentation here, aimed at mathematical brevity and physical clarity, rather than synoptic detail, has benefited most from Charney (1973), Chapters 7-9, Drazin (1978) and Pedlosky (1971). A definitive treatment is given by Pedlosky (1979), Chapters 6 and 7.

Section 4.5. The agreement between synoptic observations of cyclonic disturbances and the linear theory of Charney and Eady waves is far from being perfect. One important question still to be settled is that of the dominant scale of motion. Synoptic disturbances are actually local phenomena, rather than periodic waves (Simmons and Hoskins, 1979). To the extent that they correspond in their initial stages to a growing linear wave, this wave would have zonal wave number 12-15, rather than 6-9, as given by classical baroclinic theory, and grow faster than the normal modes of the latter.

One possible resolution of this discrepancy is that after a finite time of growth, nonlinear wave-wave interactions might limit the growth of the linearly most unstable wave and give a saturation spectrum dominated by a wave number with initially slower-growth (Pedlosky, 1979, Section 7.16; Pedlosky, 1981a): Eady (1949,1950) already recognized that the continuous generation of disturbances, dissipated by friction in the mid-atmosphere and planetary boundary layer (Holton, 1979, Chapter 5), will lead to a statistical equilibrium of the wave energy distribution, similar to that of particle energy distribution in statistical mechanics.

The statistical theory of large-scale atmospheric turbulence, two-dimensional (Fjørtoft, 1953) or quasi-geostrophic (Charney, 1971), provides a reasonable agreement with wave-energy distribution between spherical wave numbers 10 and 30 (Savijärvi, 1984; see also Section 6.6 below). The problem is that the peak in the energy spectrum is not near wave number 12, where the synoptic disturbances are, but near wave number 6.

A reasonable answer to this quandary appears to have two parts. Linear baroclinic instability theory can provide rapidly growing modes with wave numbers larger than ten, provided more realistic basic states are used (Frederiksen, 1983; Gall *et al.*, 1979). Moreover, absolute instability theory, as opposed to the classical normal-mode theory described here (Farrell, 1983; Merkine and Shafranek, 1980; Miyamoto, 1980, Section 11.6; Pierrehumbert, 1984), shows promise in explaining the localized character of the disturbances and their preferential location along storm tracks, off the East coast of Asia and North America for instance.

The second part of the answer is perhaps that the quasi-equilibrium spectrum of atmospheric energy is fed in a band at these wave numbers slightly above ten. The spectrum towards higher wave numbers k corresponds to an enstrophy-cascading process in that direction, yielding the k^{-3} inertial-range spectrum of energy in quasi-geostrophic theory (e.g., Sadourny, 1985). Towards lower wave numbers, the reverse energy-cascading inertial range would have a $k^{-5/3}$ spectrum. Clearly the lowest wave numbers are too much affected by boundary forcing to have a quasi-equilibrium spectrum totally given by such arguments, but the peak near $k = 6$ is at least consistent with this view.

A second question still not completely settled pertains to the most striking aspect of weather disturbances for the casual observer, namely fronts. Fronts are currently believed to be associated with cyclone waves and have one horizontal dimension comparable to the internal radius of deformation, O(1000 km). The second, transverse horizontal dimension is much smaller, O(100 km), and involves a large contrast across the front in air temperature and humidity, and in the velocity parallel to the front. A frontal passage can bring rain and a considerable drop in temperature to a given locality within a matter of hours. Similar phenomena are known to occur in the oceans, where the role of humidity is played by salinity.

The steepening of gradients in a cyclone wave near the surface can be explained by quasi-geostrophic and semi-geostrophic theory (Hoskins, 1982; Pedlosky, 1979, Section 8.4). Many features of the fine structure of fronts, such as rain bands, and their stability properties, remain however without a generally accepted explanation (Orlanski *et al.*, 1985). The conceptual frontal models used in operational forecasting are still based to a large extent on the polar front ideas of the pioneering Bergen school, which date back to the 1920s (Godske *et al.*, 1957, Chapters 7, 10, 14, 15 and 18; Petterssen, 1956, Chapters 8-18 and 26). Their updating could greatly benefit from recent progress in theory, observation and numerical simulation (Hoskins, 1983; Orlanski *et al.*, 1985).

CHAPTER 5

CHANGING FLOW PATTERNS AND SUCCESSIVF RIFURCATIONS

Out of Sections 4.2 through 4.4 emerges a certain view of midlatitude atmospheric circulation. It is dominated by essentially axisymmetric, zonal flow -- the prevailing westerlies. They represent the climatological mean of daily flow patterns. The zonal mean flow is baroclinically unstable, and cyclone waves grow on it, reaching finite amplitude and then decaying, to be replaced by new growth of instabilities.

The presence of contrasts between continents and oceans in topographic height and surface thermal properties introduces certain asymmetries in the predominantly axisymmetric, zonal flow. The mean flow, as modified by the asymmetry of surface features, as well as by the mean effect of the cyclonic eddies themselves, is still baroclinically unstable. To describe this picture in ever growing detail, it sufficed to study the linear instability of basic flows with more complicated structures, and to follow the growth of the instabilities thus obtained by asymptotic, weakly nonlinear methods.

The observed persistence of large deviations from nearly axisymmetric, zonal flow suggests the potential usefulness of genuinely nonlinear methods in studying atmospheric flow patterns. We shall start by illustrating the application of such methods to a laboratory idealization of the atmosphere's general circulation.

Section 5.1 opens with a description of the apparatus used in experiments with *rotating, thermally driven fluids*. The main nondimensional parameters, R and T, are defined in Eq. (5.2), and the flow field is shown to change with R and T from an axisymmetric, steady circulation to patterns with less and less spatial symmetry and with more and more temporal complexity. Analogies are drawn between the flow in a rotating annulus and large-scale atmospheric circulation in the tropics and in mid-latitudes.

In Section 5.2 the device of *spectral expansion* of quantities in the partial differential equations (PDEs) of motion is described, using a two-layer quasi-geostrophic model due to Lorenz. Arbitrary, *low-order truncation* leads to a system of ordinary differential equations (ODEs) of manageable size (14 equations). Competing requirements in applying this device to obtain simple nonlinear models are that the truncation preserve the main physical characteristics of the phenomena under consideration, while allowing a relatively complete mathematical investigation of their properties.

In Section 5.3 we analyze the system of equations obtained, and show that a *sequence of transitions* in solution complexity occurs which mirrors the transitions in rotating annulus flow. An axisymmetric steady Hadley regime (H) transfers its stability to a traveling Rossby wave regime, with either one or two waves (R_1 and R_2 or R'_{12} and R''_{12}). These periodic solutions become in turn unstable to doubly-periodic solutions (UV' and UV"), which are analogous to the *vacillation regime* in the annulus.

The transition from unsymmetric, UV vacillation to symmetric, SV vacillation is shown to be nonlocal in character, and associated with the existence of a homoclinic orbit in the system's phase-space flow. Numerical methods used to study the last two transitions are also discussed, and the SV solution is related to tilted-trough vacillation in the laboratory and to the midlatitude "index cycle".

Finally, the successive changes of flow pattern and exchanges of stability between them are summarized in terms of a *bifurcation diagram*, in which the character of the solution is shown as a function of one or more nondimensional parameters. Hopf bifurcation, in which stationary solutions transfer their stability to periodic ones, and secondary Hopf bifurcations, leading from simply-periodic to doubly-periodic solutions, are discussed. The final transition, from quasi-periodic to aperiodic flow, is more complicated and left for the next section.

Section 5.4 begins with the study of forced, dissipative systems of ODEs, concentrating on those which are obtained by spectral truncation of typical PDEs governing fluid flow. The concepts of a *dynamical system* and of its *attractor sets*, including strange attractors, are introduced.

The *Lorenz system* of three ODEs is derived from the PDEs governing two-dimensional thermal convection between two parallel plates. This system is shown to be a forced, dissipative dynamical system, whose phase flow is volume reducing. Pitchfork bifurcation of two stable convective stationary states from the unique conductive state occurs as the Rayleigh number r is increased passed $r = 1$, followed by subcritical Hopf bifurcation of unstable limit cycles from the convective states at $r = r_c$. These limit cycles are shown to saturate into two homoclinic orbits passing through the conductive state at $r = r_0$, $1 < r_0 < r_c$, and the system's complicated attractor set for $r > r_0$ is related to this peculiar configuration. The system's *strange attractor* is described geometrically, based on the numerical investigations of Lorenz (1963a) and subsequent partial analyses.

Section 5.5 describes some additional details of flow in rotating annuli and gives references on their experimental investigation and mathematical analysis. References for spectral expansion as a representation of solutions in detailed numerical computations are given. Extreme truncation as a tool for the study of qualitative nonlinear features is documented.

Remarks are made and references given on the study of dynamical systems and bifurcation theory. Floquet stability theory of periodic solutions is mentioned, with references. Some discussion is allotted to numerical methods for the investigation of dynamical systems, and their relation to the approximation and ergodic theory of these systems. "Preturbulence" and homoclinic explosion is mentioned in connection with the transition to symmetric vacillation and to turbulence in Section 5.3.

The relation between rotating annulus experiments, their mathematical analysis, and the atmosphere is further discussed. Eliassen-Palm fluxes, and related results on non-interaction between waves and mean zonal flow and on non-acceleration of the mean flow by the waves are mentioned. The complexity of various types of vacillation in the annulus is stressed, along with the wide range of vacillation periods obtained. Various theories to explain vacillation phenomena are briefly reviewed.

A few references for the Lorenz system and for the strange attractor concept are given. Aperiodic phenomena occurring in other sciences are touched upon. Other methods of deriving systems of ODEs equivalent to the Lorenz system from the PDEs of motion are mentioned, and further readings on thermal convection are suggested.

5.1. Rotating Annulus Experiments

The first experiments with a rotating tank of fluid seem to have been carried out by Vettin at the middle of the 19th century. He placed a lump of ice at the center of a cylinder filled with air and noticed that increasing the rotation rate led to a transition from an axisymmetric to an asymmetric, irregular flow pattern. The analogies with large-scale atmospheric flow, apparent to him, were somehow lost on his contemporaries.

Systematic experiments with rotating, thermally driven fluids started around 1950. The simplest apparatus for controlled experiments is the rotating annulus, illustrated in Figure 5.1.

The fluid at rest occupies the annular space $0 < a \leq r \leq b$, $0 \leq z \leq D$. The inner and outer walls are maintained at constant temperatures T_a and T_b, respectively, with $T_a < T_b$. The annulus is thought of as a simplified, idealized laboratory model of the midlatitude atmosphere, with the polar regions towards the axis of rotation, and the tropics towards the exterior.

In the absence of rotation, and for the imposed temperature differences $\Delta T \equiv T_b - T_a$ being small, $\Delta T \ll T_a$, the fluid will be at rest and the temperature distribution in its interior will be given by the heat equation governing conduction (see Section 10.2, Eqs. (10.7, 10.8)). Solving this equation for axisymmetric boundary conditions as indicated in

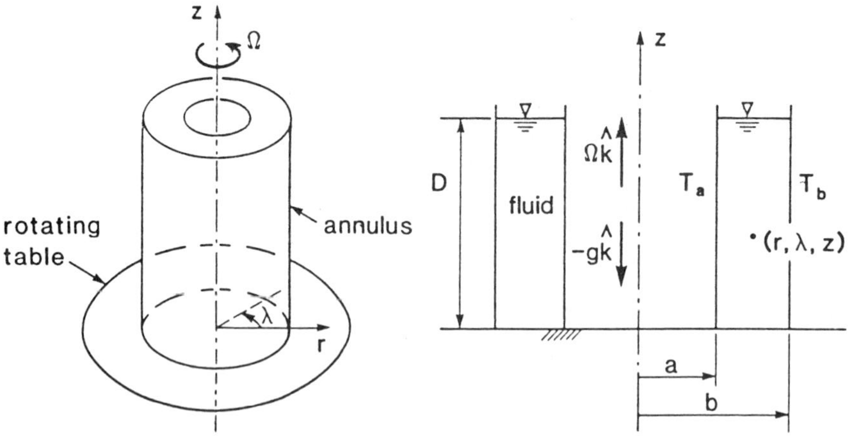

Fig. 5.1. Schematic diagram of a differentially heated, rotating annulus (after Hide, 1977).

Figure 5.1 yields an axisymmetric temperature distribution throughout the fluid,

$$T = T(r) = \frac{T_b \log(r/a) - T_a \log(r/b)}{\log b - \log a} . \tag{5.1}$$

As the angular velocity Ω and the temperature gradient ΔT are increased, fluid motions will appear and the temperature distribution will change.

It turns out to be convenient to describe the fluid's behavior patterns in terms of two nondimensional numbers: the thermal Rossby number $\mathcal{R}$,

$$\mathcal{R} = \frac{\alpha g D \Delta T}{\Omega^2 L^2} , \tag{5.2a}$$

and the Taylor number $\mathcal{T}$,

$$\mathcal{T} = \frac{4\Omega^2 L^5}{\nu^2 D} . \tag{5.2b}$$

Here α is the thermal expansion coefficient of the fluid, taken to be a liquid, whose density ρ is (approximately) linear in T,

$$\rho = \rho_0\{1 - \alpha(T - T_0)\} \tag{5.3}$$

(cf. also (4.12d)), $L = b-a$ is the horizontal length scale, and ν is the fluid's (approximately constant) kinematic viscosity. As many as 15 independent dimensionless parameters are required to specify an experiment of this type completely, but $\mathcal{R}$ and $\mathcal{T}$ are by far the most important ones.

$\mathcal{R}$ is a measure of the relative magnitude of thermal forcing, given by ΔT, and of the mechanical forcing, given by the rotation rate Ω. Notice that $\mathcal{R}$ is analogous to $4S$, S being the Burger number of Eq. (4.15). The analogy is not perfect, since in the model problem of Sections 4.3 and 4.4 ΔT was an imposed vertical temperature gradient, while here it is horizontal. Furthermore, D/L is not small, as in previous chapters. Still, due to Ω being much larger than the local rotation rates of the Earth, $f/2$ (see Figure 4.6), $\mathcal{R}$ here and S in Chapter 4 have comparable orders of magnitude, so that the flow patterns in the annulus are similar to those in the atmosphere. In particular, the "sloping convection" associated with baroclinic instability (Sections 4.2 and 4.4) does set in, translating the imposed horizontal ΔT into a vertical temperature difference between the free surface of the fluid and the bottom of the annulus.

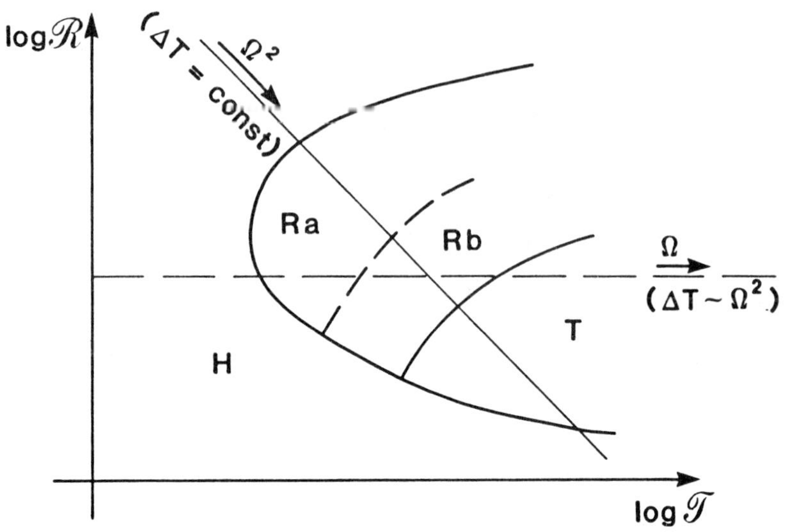

Fig. 5.2. Schematic regime diagram for a rotating, thermally-driven fluid (after Hide, 1977).

T is a measure of the mechanical forcing, given by Ω, as compared with the dissipation effects, provided by ν. The results of rotating annulus experiments can be best summarized in the schematic diagram of Figure 5.2.

For low R and T, motion is axisymmetric, with fluid rising at the outer wall and sinking at the inner wall; it is deflected in the sense of rotation at the top ("westerlies") and in the opposite sense at the bottom ("easterlies"). This flow regime has been labelled the *Hadley regime* (H in Figure 5.2).

Past a certain critical value of T, nonaxisymmetric flow patterns obtain. Notice that the actual external parameters which can be varied experimentally with relative ease are Ω and ΔT. For fixed ΔT and increasing Ω, the (R,T) log-log diagram is traversed along a straight line with slope (-1), parallel to the light solid line in Figure 5.2. For Ω^2 proportional to ΔT, one obtains horizontal straight lines in the figure (see light dashed line).

In either case, for increasing rotation rates, finite-amplitude Rossby waves with a low wave number (1-6) appear on the surface of the fluid, breaking the symmetry. The waves first travel with constant speed and shape around the annulus. This *steady Rossby regime* is designated by Ra in Figure 5.2.

As Ω is further increased (with ΔT = const. or $\Delta T \sim \Omega^2$), the waves start to change their shape, their amplitude, their phase relationship or even their wave number. These changes are relatively slow and smooth, however, preserving the general character of the wavy motion and the identity of the waves. In particular, they appear to be periodic in time or very nearly so. This unsteady, slowly varying Rossby regime is called *vacillation* and is labelled Rb in Figure 5.2.

Finally, as Ω increases even further, the motion becomes completely irregular and *turbulence*, denoted by T in the figure, sets in. All these types of motion have their counterpart in large-scale atmospheric flows. The study of transitions from one flow regime in the annulus to another seems to have, therefore, the potential of shedding some light onto the nature of these atmospheric flows and the conditions under which they arise.

As a matter of fact, the experiments themselves already give additional insight into the atmosphere's circulation patterns. In Section 4.1, we have seen that the *internal* Rossby number, $\varepsilon = U/fL$, of large-scale flow on the Earth is smaller in mid-latitudes than in the tropics, since U and L are comparable, while the local rotation rate $f/2$ is larger. The observed mean circulation in the tropics corresponds roughly to what we just called an H regime, while midlatitude flow patterns are, at different times and places, in an R or T regime (Figure 4.2).

From the point of view of Figure 5.2, $\mathcal{T}$ for midlatitude circulation is larger than for tropical circulation, due to the increase in $\Omega = f/2$. Furthermore, the temperature difference ΔT, at the same pressure level, say, between 30° lat. and 60° lat. is much larger than that between the Equator and 30° lat. (see Figure 10.5). Hence, the *external* thermal Rossby number $\mathcal{R}$ is of comparable magnitude in the two latitude belts, since $\Delta T/\Omega^2$ is of like magnitude.

The relative values of $\mathcal{R}$ and $\mathcal{T}$ for midlatitude and tropical circulations are consistent, therefore, with the relative positions of their analogs in Figure 5.2. On the other hand, the distinction between the externally prescribed parameters $\mathcal{R}$ and $\mathcal{T}$ in the laboratory, and the internally determined parameters ε and S in the atmosphere is an important one to keep in mind. The experiments, and their mathematical analysis below, might explain why, for given $\mathcal{R}$ and $\mathcal{T}$, the flow looks the way it does. They still do not explain why the real atmosphere internally chooses certain parameter values which are consistent with a certain flow pattern prevailing.

5.2. Simplified Dynamics, A Recipe

In order to describe theoretically the flow regimes observed in rotating annuli, one needs a model. The model has to be governed by a system of equations which satisfies two conflicting requirements: it should be (a) complex enough to have a variety of solutions resembling the regimes to be described, and (b) simple enough so that the solutions can be easily studied. A general recipe to obtain systems with some overlap between the two requirements became popular in fluid dynamics in the 1950s. This recipe was presented very clearly and exploited systematically by Lorenz in the early 1960s; it runs as follows.

Start with a system of *partial differential equations* (PDEs) which is relatively simple, but contains the essential physics and dynamics of the phenomena to be studied. In the instance at hand, such a system is the "two-layer" version of a baroclinic, quasi-geostrophic model of midlatitude atmospheric flow. This model itself is obtained by a judicious application of the quasi-geostrophic assumption of small Rossby number to the *nonlinear* form of Eqs. (4.12). The equations are then discretized in the vertical, as in Figure 5.3. The resulting system is

$$\partial_t \nabla^2 \psi = -J(\psi, \nabla^2 \psi) - J(\tau, \nabla^2 \tau) - k_0 \nabla^2 (\psi - \tau), \tag{5.4a}$$

$$\partial_t \nabla^2 \tau = -J(\psi, \nabla^2 \tau) - J(\tau, \nabla^2 \psi) + 2\Omega \nabla^2 \chi + k_0 \nabla^2 (\psi - \tau) - 2k_0 \nabla^2 \tau, \tag{5.4b}$$

$$\partial_t T = -J(\psi, T) + \sigma \nabla^2 \chi - h_1 (T - \sigma - T_*), \tag{5.4c}$$

$$\partial_t \sigma = \overline{T \nabla^2 \chi} + h_1 (\overline{T} - \sigma - \overline{T}_*) - 2h_0 \sigma, \tag{5.4d}$$

$$\nabla^2 \tau = \frac{\alpha g D}{8\Omega} \nabla^2 T. \tag{5.4e}$$

The flow domain $\mathcal{D}$ is a periodic channel, as in Figure 4.6, $\mathcal{D} = \{-\infty < x < \infty,\ 0 \le y \le \pi L\}$. It corresponds to a "straightening out" of the annulus in Figure 5.1, neglecting curvature effects. The horizontal Laplacian is $\nabla^2 = \partial_x^2 + \partial_y^2$. The dependent variables ψ, τ, χ and T are functions of horizontal position (x,y) and of time t, while $\sigma = \sigma(t)$. Hence only horizontal averages $\overline{(\ \ )} = \int_{\mathcal{D}_p} (\ \ )\,dxdy / \int_{\mathcal{D}_p} dxdy$ of the other variables appear in (5.4d), $\mathcal{D}_p$ being one x-period of $\mathcal{D}$.

The vertically integrated stream function is ψ, $(u,v,0) = \hat{k} \times \nabla\psi$, with the difference between the upper-layer and lower-layer stream functions being 2τ. The mean temperature is T, with 2σ the difference between layers; σ is a measure of static stability. The vertical

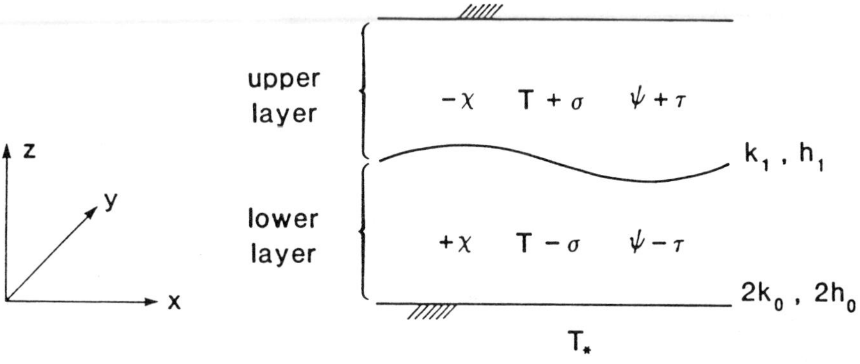

Fig. 5.3. Schematic diagram of the two-layer approximation for the baroclinic quasi-geostrophic model.

average of the horizontal flow (u,v) is nondivergent, $\nabla\cdot(u,v) = 0$ and $\pm\chi$ are the values of the velocity potential in the lower and upper layers respectively.

$J(f,g)$ is the Jacobian of the functions f and g, $J = f_x g_y - f_y g_x$. Thus, $J(\Psi,g)$ is the *advection* of a scalar g by the horizontal velocity field $(U,V) = \hat{k} \times \nabla\Psi$, $J = (U,V)\cdot\nabla g = (\nabla\Psi \times \nabla g)\cdot\hat{k}$. In meteorology and oceanography it is customary to use the word "advection" for large-scale quasi-horizontal transport, while the word "convection" is reserved for small-scale, three-dimensional transport phenomena.

The coefficients of friction between the two layers, and between the lower layer and the underlying surface, are k_1 and $2k_0$, respectively. The corresponding heating coefficients are h_1 and $2h_0$. The temperature T_* of the lower surface is constant in time, while the upper surface is insulated against heat losses.

Eqs. (5.4a,b) are the two-layer form of the quasi-geostrophic vorticity equation, i.e., of the nonlinear counterpart to (4.25a). Likewise, (5.4c,d) are the nonlinear counterpart of the thermodynamic equation (4.22d) or (4.25b). The appropriate form of the thermal wind equation (4.10) is (5.4e). Systems similar to (5.4), with various geometries, have actually been used at the beginnings of short-range numerical weather prediction, as well as in the long-term simulation of the atmosphere's general circulation. The model's two layers are the simplest possible way of introducing the stratification which plays such a key role in baroclinic instability, for both planetary atmospheres and rotating annulus experiments.

The next step in the recipe is to transform the equations into *spectral form*, i.e., expand them in terms of an orthogonal basis set $\{F_i\}$, preferably a denumerable one, $i = 1,2,\ldots$. The usual choice are the eigenfunctions of the Laplacian in the domain of interest, with appropriate boundary conditions:

$$\nabla^2 F_i = \lambda_i F_i \quad \text{in } \mathcal{D}, \tag{5.5a}$$

$$\partial_x F_i = 0 \quad \text{on} \quad y = 0,\ \pi L, \tag{5.5b}$$

and

$$F_i(x+p,y) = F_i(x,y), \tag{5.5c}$$

p being the period. More generally, (5.5b) is replaced by

$$\partial_s F_i = 0 \tag{5.5b'}$$

along the portion of the boundary of $\mathcal{D}$ which is a solid wall, where the normal velocity is zero, ∂_s being the tangential derivative.

These basis functions are orthonormal

$$\overline{F_i F_j} = \delta_{ij} = 0 \quad \text{if} \quad i \neq j, \tag{5.6a}$$

$$= 1 \quad \text{if} \quad i = j. \tag{5.6b}$$

For boundary conditions of the Neumann type (5.5b'), or for purely periodic flow, or for a combination of both, $\lambda_0 = 0$ and $F_0 = \text{const}$.

For a spherical geometry, F_i would be surface harmonics, while in a polar cylindrical geometry they would be Fourier-Bessel functions. In Cartesian coordinates they are simply sines and cosines, or complex exponentials.

The variables are now expanded in terms of F_i, and nondimensionalized,

$$\psi = 2\Omega L^2 \sum_1^\infty \psi_i F_i, \tag{5.7a}$$

$$\tau = 2\Omega L^2 \sum_1^\infty \tau_i F_i, \tag{5.7b}$$

$$T = 2A\Omega L^2 \sum_0^\infty \theta_i F_i, \tag{5.7c}$$

with a similar expansion for T_* in terms of θ_i^*,

$$\nabla^2 \chi = 2\Omega \sum_1^\infty \omega_i F_i, \tag{5.7d}$$

$$\sigma = 2A\Omega L^2 \sigma_0. \tag{5.7e}$$

In general, L is a characteristic length; in the present case, πL is the width of the channel, while $A = 8\Omega/\alpha g D$.

The only nonlinearity in (5.4) is provided by the advective terms, written as Jacobians. These can be expanded in terms of the basis functions using

$$L^2 J(F_j, F_k) = \sum_0^\infty c_{ijk} F_i, \tag{5.8a}$$

$$c_{ijk} = \overline{F_i J(F_j, F_k)}. \tag{5.8b}$$

The coefficients c_{ijk} in (5.8a) are antisymmetric in j and k, $c_{ikj} = -c_{ijk}$ and cyclically symmetric in all three indices, $c_{ijk} = c_{jki} = c_{kij}$.

Substituting (5.7) into (5.4) and using (5.6, 5.8) yields

$$\dot{\psi}_i = \frac{1}{2} \sum_{j,k=1}^\infty a_i^{-2}(a_j^2 - a_k^2) c_{ijk}(\psi_j \psi_k + \tau_j \tau_k) - k(\psi_i - \tau_i), \tag{5.9a}$$

$$\dot{\tau}_i = \frac{1}{2} \sum_{j,k=1}^\infty a_i^{-2}(a_j^2 - a_k^2) c_{ijk}(\tau_j \psi_k + \psi_j \tau_k) - a_i^{-2}\omega_i + k\psi_i - (k+2k')\tau_i, \tag{5.9b}$$

$$\dot{\theta}_i = \frac{1}{2} \sum_{j,k=1}^\infty c_{ijk}(\theta_j \psi_k - \psi_j \theta_k) + \sigma_0 \omega_i - h(\theta_i - \sigma_i) + h\theta_i^*, \tag{5.9c}$$

$$\dot{\sigma}_0 = -\sum_1^\infty \theta_i \omega_i + h\theta_0 - (h+2h')\sigma_0 - h\theta_0^*, \tag{5.9d}$$

and, for $i \neq 0$,

$$\theta_i = \tau_i. \tag{5.9e}$$

Here $a_i^2 = -\lambda_i/L^2$, with $a_i^2 > 0$ for $i \neq 0$, and $(\ )^{\cdot} = d(\ )/dt^*$, with $t^* = 2\Omega t$ the nondimensional time, in "rotation units" or "days". The coefficients k, k', h and h' are the nondimensional counterparts of the friction coefficients k_1 and k_0, and of the thermal coefficients h_1 and h_0, respectively; they all contain $(1/2\Omega)$ as a factor, e.g., $k = k_1/2\Omega$.

Notice that the domain-dependent basis functions F_i have disappeared from (5.9). The shape of the domain enters only through the scaled eigenvalues of the Laplacian, a_i, and through the interaction coefficients c_{ijk}. Thus (5.9) is the general spectral form of (5.4), independently of domain. In the cartesian geometry chosen for $\mathcal{D}$, the F_i are just products of sines and cosines, with appropriate values for the coefficients a_i and c_{ijk}.

The third and last step of the recipe is to *truncate* the infinite series in (5.9), yielding a finite system of *ordinary differential equations* (ODEs), (5.9') say. The level of truncation $N < \infty$ has to reflect a compromise between the two conflicting requirements: (i) wealth of solutions and fidelity to the physical behavior one is interested in, on the one hand, and (ii) amenability to study and understanding of solution behavior, on the other.

A serious quandary arises about the connection between the solutions of a truncated, low-order system of ODEs such as (5.9'), and those of the original system of PDEs, such as (5.4). One would hope that when a sufficient number of basis functions N has been retained, at least the qualitative behavior of solutions will be the same. Whether such a finite, sufficient number exists, and what it might be, are open questions for most problems to which this recipe has been applied. Still, the approach outlined here has furnished valuable information about what to look for in the solutions of PDE systems and in the phenomena they describe.

Orthogonal basis functions in the problem at hand can be chosen as

$$F_{00} = 1, \tag{5.10a}$$

$$F_{0m} = \sqrt{2}\cos m\eta, \tag{5.10b}$$

$$F_{nm} = 2\sin m\eta \cos n\xi, \tag{5.10c}$$

$$F_{-n,m} = 2\sin m\eta \sin n\xi, \tag{5.10d}$$

where $\xi = x/L$, $\eta = y/L$ and a double index (n,m) was used instead of i. The functions F_{0m} do not vary in the x-direction and will be called *zonal* components, x being throught of as pointing east. The functions F_{nm}, $F_{-n,m}$, $n \neq 0$, will be called *wave* components. The meridional index, m will be called the *mode*. For the functions (5.10) to be orthonormal, i.e., satisfy (5.6b) as well as (5.6a), they have to be divided by $(2\pi)^{1/2}$.

It turns out that a felicitous choice of truncation for the study of rotating annulus regimes is to retain a single, arbitrary value of n, i.e., a single wave, and two modes, $m = 1$ and $m = 2$. For convenience, indices are replaced by letters,

$$F'_0 = F_{00}, \quad F'_A = F_{01}, \quad F'_C = F_{02}, \tag{5.11a,b,c}$$

$$F'_K = F_{n1}, \quad F'_L = F_{-n,1}, \quad F'_M = F_{n2}, \quad F'_N = F_{-n,2}, \tag{5.11d,e,f,g}$$

and we let $F_0 = F'_0/(2\pi)^{1/2}$, $F_A = F'_A/(2\pi)^{1/2}$, and so on. Thus F_0, F_A and F_C represent the zonal portion of the flow, F_K and F_L are waves of mode 1, while F_M and F_N are waves of mode 2. The imposed surface temperature is chosen to be purely zonal and of mode 1, i.e., decreasing monotonically from $\eta = 0$ to $\eta = \pi$,

$$T_* = 2A\Omega L^2(\theta_0^* + \theta_A^* F_A). \tag{5.12}$$

The values of the eigenconstants $a_i = a_{\ell m}$, with $i = 0,A,C,K,L,M,N$, are given by

$$a_{\ell m}^2 = \ell^2 + m^2, \quad \ell = 0,n, \quad m = 1,2. \tag{5.13a}$$

The only interaction coefficients c_{ijk} which do not vanish are given by

$$\frac{c_{AKL}}{5} = \frac{c_{AMN}}{4} = \frac{c_{CKN}}{8} = \frac{c_{CML}}{8} = -\frac{4\sqrt{2}}{15\pi}n. \tag{5.13b}$$

It is also convenient to introduce some further notation

$$\beta = a_K^{-2}(a_L^2 - a_A^2) = n^2/(n^2+1), \tag{5.13c}$$

$$\beta' = a_M^{-2}(a_N^2 - a_A^2) = (n^2+3)/(n^2+4), \tag{5.13d}$$

$$\delta = a_K^{-2}(a_N^2 - a_C^2) = n^2/(n^2+1), \tag{5.13e}$$

$$\delta' = a_N^{-2}(a_L^2 - a_C^2) = (n^2-3)/(n^2+4), \tag{5.13f}$$

and

$$\varepsilon' = a_C^{-2}(a_M^2 - a_L^2) = \frac{3}{4}. \tag{5.13g}$$

Letting

$$\alpha = -C_{AKL}, \quad \alpha' = -c_{AMN}, \quad \alpha'' = -c_{CKN} = -c_{CML}, \tag{5.13h}$$

and eliminating τ_i in favor of θ_i by using (5.9e), one obtains the system

$$
\begin{aligned}
\dot{\psi}_A &= - k(\psi_A-\theta_A),\\
\dot{\psi}_K &= -\beta\alpha(\psi_L\psi_A+\theta_L\theta_A) - \delta\alpha''(\psi_N\psi_C+\theta_N\theta_C) - k(\psi_K-\theta_K),\\
\dot{\psi}_L &= \beta\alpha(\psi_A\psi_K+\theta_A\theta_K) + \delta\alpha''(\psi_C\psi_M+\theta_C\theta_M) - k(\psi_L-\theta_L),\\
\dot{\psi}_C &= \varepsilon'\alpha''(\psi_K\psi_N+\theta_K\theta_N) - \varepsilon'\alpha''(\psi_M\psi_L+\theta_M\theta_L) - k(\psi_C-\theta_C),\\
\dot{\psi}_M &= -\beta'\alpha'(\psi_N\psi_A+\theta_N\theta_A) - \delta'\alpha''(\psi_L\psi_C+\theta_L\theta_C) - k(\psi_M-\theta_M),\\
\dot{\psi}_N &= \beta'\alpha'(\psi_A\psi_M+\theta_A\theta_M) + \delta'\alpha''(\psi_C\psi_K+\theta_C\psi_K) - k(\psi_N-\theta_N),\\
\dot{\theta}_A &= - \omega_A + k\psi_A - (k+2k')\theta_A,\\
\dot{\theta}_K &= -\beta\alpha(\theta_L\psi_A+\psi_L\theta_A) - \delta\alpha''(\theta_N\psi_C+\psi_N\theta_C) - (1-\beta)\omega_K + k\psi_K - (k+2k')\theta_K,\\
\dot{\theta}_L &= \beta\alpha(\theta_A\psi_K+\psi_A\theta_K) + \delta\alpha''(\theta_C\psi_M+\psi_C\theta_M) - (1-\beta)\omega_L + k\psi_L - (k+2k')\theta_L,\\
\dot{\theta}_C &= \varepsilon'\alpha''(\theta_K\psi_N+\psi_K\theta_N) - \varepsilon'\alpha''(\theta_M\psi_L+\psi_M\theta_L) - (1-\varepsilon')\omega_C + k\psi_C - (k+2k')\theta_C,\\
\dot{\theta}_M &= -\beta'\alpha'(\theta_N\psi_A+\psi_N\theta_A) - \delta'\alpha''(\theta_L\psi_C+\psi_L\theta_C) - (1-\beta')\omega_M + k\psi_M - (k+2k')\theta_M,\\
\dot{\theta}_N &= \beta'\alpha'(\theta_A\psi_M+\psi_A\theta_M) + \delta'\alpha''(\theta_C\psi_K+\psi_C\theta_K) - (1-\beta')\omega_N + k\psi_N - (k+2k')\theta_N,\\
\dot{\theta}_0 &= -h\theta_0 + h\sigma_0 + h\theta_0^*, \qquad (5.14)\\
\dot{\theta}_A &= -\alpha(\theta_K\psi_L-\psi_K\theta_L) - \alpha'(\theta_M\psi_N-\psi_M\theta_N) + \sigma_0\omega_A - h\theta_A + h\theta_A^*,\\
\dot{\theta}_K &= -\alpha(\theta_L\psi_A-\psi_L\theta_A) - \alpha''(\theta_N\psi_C-\psi_N\theta_C) + \sigma_0\omega_K - h\theta_K,\\
\dot{\theta}_L &= -\alpha(\theta_A\psi_K-\psi_A\theta_K) - \alpha''(\theta_C\psi_M-\psi_C\theta_M) + \sigma_0\omega_L - h\theta_L,\\
\dot{\theta}_C &= -\alpha''(\theta_K\psi_N-\psi_K\theta_N) - \alpha''(\theta_M\psi_L-\psi_M\theta_L) + \sigma_0\omega_C - h\theta_C,\\
\dot{\theta}_M &= -\alpha'(\theta_N\psi_A-\psi_N\theta_A) - \alpha''(\theta_L\psi_C-\psi_L\theta_C) + \sigma_0\omega_M - h\theta_M,\\
\dot{\theta}_N &= -\alpha'(\theta_A\psi_M-\psi_A\theta_M) - \alpha''(\theta_C\psi_K-\psi_C\theta_K) + \sigma_0\omega_N - h\theta_N,\\
\dot{\sigma}_0 &= -(\theta_A\omega_A+\theta_K\omega_K+\theta_L\omega_L+\theta_C\omega_C+\theta_M\omega_M+\theta_N\omega_N) + h\theta_0 - (h+2h')\sigma_0 - h\theta_0^*.
\end{aligned}
$$

This completes the reduction of the system of PDEs (5.4) to a finite system of ODEs. Some such reduction is involved in many studies of physical problems, qualitative as well as quantitative. The choice of basis functions and truncation level is often important in maintaining the balance between physical realism, on the one hand, and amenability to understanding, on the other. Possible tests for the level of truncation being adequate are: to have captured the main features of the physical phenomena described, or to obtain similar results with sucessively augmented systems of ODEs. At least one of the two criteria above should be satisfied for this approach to claim credibility in the attack of any given problem. In other words, this is potent medicine for our lack of knowledge about real, nonlinear flow problems, and should be used with appropriate caution.

In the system (5.14) there are altogether 20 ODEs. The six variables $\omega_A, \omega_K, \ldots, \omega_N$ are easily eliminated, leaving a set of 14 ODEs in the 14 variables $\Psi = (\psi_A, \psi_K, \ldots, \psi_N; \theta_0, \theta_A, \ldots, \theta_N; \sigma_0)$. This equivalent system will be written symbolically as

$$\dot{\Psi} = Q(\Psi), \tag{5.14'}$$

where Q denotes the appropriate quadratic form obtained from the right-hand side of (5.14). In the next section, we shall investigate the solutions of this equation.

5.3. Analysis of Flow Regime Transitions

The purpose of formulating system (5.14) and studying its solutions is to provide an explanation for the various flow regimes described in Section 5.1, and in particular for the way transitions between the regimes might occur. We wish therefore (i) to find solutions of the system for various parameter ranges, which will resemble in character the flow regimes, and (ii) to study the way that the solutions change as the parameters are varied, as they were in the laboratory experiments.

The external parameters here will be taken to be the imposed temperature difference across the channel, θ_A^*, and the internal friction coefficient k. Both of them are nondimensional; θ_A^* is proportional to the thermal Rossby number R in (5.2a) and $k^{-2} = 4\Omega^2/k_1^2$ is proportional to the Taylor number T in (5.2b). The other friction coefficients are assumed to satisfy $h' = k' = h/2 = k/2$.

<u>Analytic model solutions</u>. A solution corresponding to the steady *Hadley* circulation (H in Figure 5.2) is easily found. In it, all wave components, i.e., those with subscripts K, L, M, N, as well as the zonal second-mode components ψ_C and θ_C, vanish. The remaining zonal components of mode 0 and 1 satisfy the steady-state relations

$$\psi_A = \theta_A, \tag{5.15a}$$

$$\theta_0 = \theta_0^* + \sigma_0, \tag{5.15b}$$

$$\sigma_0 = \theta_A^2, \tag{5.15c}$$

where θ_A is the single real root of the cubic equation

$$\theta_A + \theta_A^3 = \theta_A^*. \tag{5.15d}$$

Thus ψ_A, $\theta_0 - \theta_0^*$, θ_A and σ_0 are all given in terms of θ_A^*, independently of k.

The ratio

$$R = \sigma_0/\theta_A^2 \tag{5.16}$$

is proportional to the Richardson number, cf. (4.16a). For the Hadley regime in this model, $R \equiv 1$.

Steady *Rossby* circulations are also easy to find analytically. The key is to notice that if all second-mode variables, i.e., those with subscripts C, M and N vanish, the equations for those variables are satisfied identically. The remaining variables are given by (5.15a,b) and by

$$\theta_A = \alpha^{-1} k\, G(\sigma_0), \tag{5.17a}$$

$$\psi_K = B \cos \omega(t - t_0), \tag{5.17b}$$

$$\psi_L = B \sin \omega(t - t_0), \tag{5.17c}$$

$$\theta_K = p\psi_K + q\psi_L, \tag{5.17d}$$

$$\theta_L = -q\psi_K + p\psi_L, \tag{5.17e}$$

with σ_0 being implicitly defined in terms of θ_A^* and k in

$$\theta_A\{1 + \sigma_0(qG - p^2 - q^2)^{-1}(RqG - p^2 - q^2)\} = \theta_A^*. \tag{5.17f}$$

The quantities G, s, p and q introduced in (5.17) are in turn given by

$$G^2(\sigma_0) = p^{-1}(2+3s)^3\{(6-\beta) + (11-6\beta)s + 3(1-2\beta)s^2 - 5\beta s^3\}^{-1}, \tag{5.18a}$$

$$s = (1-\beta)^{-1}\sigma_0, \tag{5.18b}$$

$$p = (1+\beta^2G^2)^{-1}\{1 - (1-2s)(2+3s)^{-1}\beta^2G^2\}, \tag{5.18c}$$

$$q = (1+\beta^2G^2)^{-1}(3+s)(2+3s)^{-1}\beta G, \tag{5.18d}$$

which are also functions of σ_0 alone.

Solution (5.17, 5.18) represents Rossby waves of the first mode. They are characterized by an amplitude B, with

$$B^2 = (qG - p^2 - q^2)^{-1}\sigma_0(\sigma_0 - \theta_A^2), \tag{5.19a}$$

by a phase velocity ω/n, with

$$\omega = k(1+5s)(2+3s)^{-1}\beta G, \tag{5.19b}$$

and by an arbitrary phase, t_0 being undetermined. We call the flow regime associated with this solution "steady", in accordance with the terminology introduced in Section 5.1, and customary in fluid dynamics: the waves travel with constant shape and speed, cf. (5.19).

One notices next that the variables with subscripts C, K and L, if simultaneously equal to zero at some time, will stay zero at all other times. This gives another steady Rossby circulation, of the second mode. It is obtained from Eqs. (5.17-19) for the first-mode circulation by substituting the subscripts M and N for K and L, and the primed quantities α' and β' instead of α and β in (5.18,19), with G, s, p, q, B and ω being also modified accordingly.

We shall denote the model's two steady Rossby regimes by R_1 and R_2. They are both mathematical illustrations of the experimental regime Ra, differing by their meridional wave number, m.

The Hadley regime H of (5.14) exists for all values of k and of θ_A^*, while R_1 and R_2 also exist for large, overlapping ranges of these external parameters. The transition in the laboratory from H to Ra is likely, therefore, to be replicated by a change of the regimes' stability in the model.

Linear stability analysis. The stability of the purely zonal solution (5.15) to small waves of the first mode can be studied by linearizing the first-mode equations in (5.14') around the H solution. The resulting equations which govern small first-mode perturbations can be written as

$$\begin{pmatrix} \dot{K} \\ \dot{L} \end{pmatrix} = \frac{1}{s+1}\begin{pmatrix} M_1 & -M_2 \\ M_2 & M_1 \end{pmatrix}\begin{pmatrix} K \\ L \end{pmatrix}, \tag{5.20a}$$

where

$$K = \begin{pmatrix} \psi_K \\ \theta_K \end{pmatrix}, \quad L = \begin{pmatrix} \psi_L \\ \theta_L \end{pmatrix}, \tag{5.20b,c}$$

$$M_1 = \begin{pmatrix} -(s+1)k & (s+1)k \\ sk & -(2s+1)k \end{pmatrix}, \quad M_2 = \alpha \begin{pmatrix} (s+1)\beta\psi_A & (s+1)\beta\theta_A \\ (\beta s-1)\theta_A & (\beta s+1)\psi_A \end{pmatrix}, \tag{5.20d,e}$$

θ_A is given by (5.15d), s by (5.18b) and σ_0 by (5.15c).

The Hadley solution will be stable if all eigenvalues of the coefficient matrix in (5.20a), M say, i.e., all solutions λ of $\det(M-\lambda) = 0$, have negative real parts. If one or more λ's have positive real part, the H solution will be unstable.

The matrix M has four eigenvalues. Two of them equal the eigenvalues of the matrix $M_1 + iM_2$, the other two are their complex conjugates, which coincide with the eigenvalues of $M_1 - iM_2$. Since s and k are positive, the trace of M_1, and hence of M is negative; therefore, one pair of complex conjugate eigenvalues, at least, must have negative real part. The stability criterion, it follows, is that the other pair of λ's cross the imaginary axis.

The characteristic equation of $M_1 + iM_2$ is

$$\begin{aligned}(s+1)\lambda^2 &+ \{(3s+2)k - i(2\beta s+\beta+1)\alpha\theta_A\}\lambda \\ &+ \{(s+1)k^2 - 2\alpha^2\beta\theta_A - i(5s+1)\alpha\beta k\theta_A)\} = 0.\end{aligned} \tag{5.21}$$

The condition for this equation to have a purely imaginary root is just Eq. (5.17a), which we rewrite as

$$k^{-1}\theta_A = \alpha^{-1}G(\sigma_0); \tag{5.22a}$$

here, however, σ_0 is not given by (5.17f), but rather by

$$\sigma_0 = \theta_A^2, \tag{5.22b}$$

as in (5.15c).

Condition (5.22) is sharp, in the following sense. If $k^{-1}\theta_A$ is smaller than the value above, both roots of (5.21) will have negative real parts, while for a larger value exactly one root will have a positive real part. Hence (5.22) is precisely the condition for the zonal solution (5.15) to become unstable to wave perturbations of mode 1.

Loss of stability of a stationary solution like (5.15) to a periodic solution like (5.17) occurs in many other fluid dynamical problems. It will be discussed further at the end of this section and in Section 12.2. In the present case, it is striking that the linear instability condition (5.22a) for the axisymmetric regime H is formally the same as condition (5.17a) for the existence of a wave-like *finite-amplitude* periodic solution R_1.

It can be shown that, for fixed values of the external parameters θ_A^* and k, the growing part of a first-mode perturbation will eventually reach the values of ψ_K, ψ_L, θ_K and θ_L given by (5.17b-e), while at the same time modifying the zonal flow on which it started to grow so that eventually (5.17a,f) are satisfied. The formal similarity between (5.17a) and (5.22a) is due in part to the following peculiarity: the zonal-flow variables θ_0, σ_0, θ_A and ψ_A in the fully-developed R_1 regime are still constant in time, while the variables ψ_K, ψ_L, θ_K and θ_L, which define the actual mode-one wave, vary periodically in time.

Thus, solution R_1 develops, for a given parameter pair (θ_A^*, k) for which $k^{-1}\theta_A$ is large enough, from H. The graph of Eq. (5.22) is the leftmost solid curve, marked H/R_1, in the regime diagram shown in Figure 5.4.

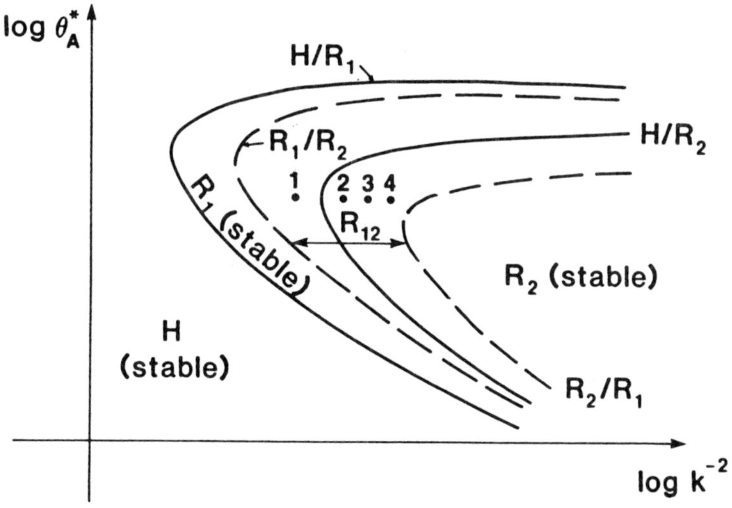

Fig. 5.4. Preliminary regime diagram for system (5.14) (after Lorenz, 1963b).

To the left of curve H/R_1, solution H is stable. It loses its stability across H/R_1 to R_1, although it continues to exist throughout the (θ_A^*, k^{-2})-plane. Solution R_1 changes gradually with θ_A^* and k as one moves into the concave part of H/R_1, from being nearly zonal to having higher and higher wave amplitude B, cf. (5.19a) and (5.17a,f). In other words, B equals zero for $R = 1$ (see Eq. (5.16)) and increases as R increases, away from the H/R_1 curve.

The instability of the zonal circulation we just described is the counterpart within the present model of the baroclinic instability studied in Section 4.4. The Rossby wave defined by (5.17-19) has the temperature field, given by (5.17d,e), lagging behind the stream function field (5.17b,c), since $q > 0$ (cf. (5.18) and (5.13c)). This is in accordance with the structure of baroclinic Rossby waves observed in both the laboratory and the midlatitude atmosphere.

The modification of the mean flow by finite-amplitude baroclinic waves in (5.14) and in the laboratory suggests a similar modification in the midlatitude atmosphere. Therefore, the proper basic state for a linear stability analysis like that of Section 4.4 is not simply the zonal average of an instantaneous atmospheric flow, such as that represented by a daily synoptic map. It is rather a hypothetical flow from which the propagating finite-amplitude waves have been subtracted, together with their mean zonal effect.

Unfortunately, it is hard to imagine what the global circulation, discussed in Sections 4.1 and 4.6, would be in the absence of cyclonic eddies. The references and descriptions of Sections 5.5 and 6.1 will add some information on this point.

Returning to the investigation of system (5.14'), it is possible to carry out the stability analysis of the zonal circulation (5.15) to small second-mode perturbations in a fashion entirely analogous to the preceding one. The only change is that the mode-dependent parameters α, β and s in Eqs. (5.20-22) have to be assigned the values α', β' and $s' = s(\beta')$ corresponding to $m = 2$. The resulting stability curve, which would be defined by (5.22'), is the second solid curve in Figure 5.4, designated H/R_2. It lies entirely to the right of H/R_1, on its concave side.

Consider then a change in external parameters θ_A^* and k in the direction of instability, e.g., increasing the rotation rate k^{-1} while θ_A^* is fixed. It follows from Figure 5.4 that the Hadley regime will lose its stability first to a growing mode-one wave. This might suggest that a second-mode wave could never become established in the rotating annulus, and be a mere mathematical figment of the model. But is the R_1 circulation itself stable to a small mode-two perturbation?

Again, one is tempted to proceed by the usual linearization approach, already applied to the stability analysis of the Hadley regime. Such an approach will almost succeed but not quite. We mentioned already that Eqs. (5.14') are quadratic in the fourteen variables $\psi_A, \psi_K, \ldots, \sigma_0$. Furthermore, the right-hand sides of the six equations for the second-mode variables $\psi_C, \theta_C, \ldots, \theta_N$ only contain these variables in a linear fashion. In other words, only interactions between the second-mode wave and the first-mode wave appear in these equations, with no interactions among the second-mode variables themselves being present.

Thus, the three equations in (5.14) for $\dot{\psi}_C$, $\dot{\psi}_M$ and $\dot{\psi}_N$, and the three equations for $\dot{\theta}_C$, $\dot{\theta}_M$ and $\dot{\theta}_N$ obtained by eliminating ω_C, ω_M and ω_N in (5.14) are already linear and homogeneous in the six second-mode variables, with coefficients depending on the two mode-zero variables θ_0 and σ_0, and on the six first-mode variables $\psi_A, \psi_K, \ldots, \theta_N$. The four wave components ψ_K, ψ_L, θ_K and θ_L, however, depend periodically on time. Hence the solution for second-mode perturbations cannot be expressed as the sum of purely exponential solutions, which are either growing or decaying, as in the constant-coefficient case.

There are infinitely many ways in which the periodic-coefficient case can be reduced to the simpler one, in order to determine stability by studying the sign of the real part of the eigenvalues of the constant-coefficient matrix. The physically most appealing way is to "freeze" the first-mode wave which propagates at constant speed, by choosing a system of coordinates moving with it. This corresponds to the following changes of variables:

$$\psi_K^0 = B^{-1}(\psi_K\psi_K + \psi_L\psi_L) = B, \tag{5.23a}$$

$$\theta_K^0 = B^{-1}(\psi_K\theta_K + \psi_L\theta_L) = pB, \tag{5.23b}$$

$$\psi_L^0 = B^{-1}(\psi_K\psi_L - \psi_L\psi_K) = 0, \tag{5.23c}$$

$$\theta_L^0 = B^{-1}(\psi_K\theta_L - \psi_L\theta_K) = -qB, \tag{5.23d}$$

$$\psi_M^0 = B^{-1}(\psi_K\psi_M + \psi_L\psi_N), \tag{5.23e}$$

$$\theta_M^0 = B^{-1}(\psi_K\theta_M + \psi_L\theta_N), \tag{5.23f}$$

$$\psi_N^0 = B^{-1}(\psi_K\psi_N - \psi_L\psi_M), \tag{5.23g}$$

$$\theta_N^0 = B^{-1}(\psi_K\theta_N - \psi_L\theta_M). \tag{5.23h}$$

The new second-mode variables ψ_C, θ_C, ψ_M^0, ..., θ_N^0 are governed by a linear, constant-coefficient system. The coefficients depend, as in (5.20), on θ_A^* and k.

The stability condition for the R_1 regime with respect to second-mode perturbations appears as the leftmost dashed curve in Figure 5.4, and is marked R_1/R_2. It lies between the H/R_1 and the H/R_2 curves. Hence the R_1 circulation within this model is stable between the H/R_1 and the R_1/R_2 curves.

One would be tempted to assume that R_2 is now stable all the way inside the concave part of R_1/R_2. But is R_2 in turn stable with respect to mode-one perturbations? The requisite analysis is entirely analogous to that of R_1 with respect to second-mode perturbations. A change of variables (5.23') will freeze the coefficients for the linear system describing the six variables ψ_C, θ_C, $\psi_{K'}^0$, $\theta_{K'}^0$, $\psi_{L'}^0$, $\theta_{L'}^0$, corresponding to a coordinate system moving with the velocity ω'/n of the R_2 circulation.

The stability condition obtained is plotted as the second dashed curve in Figure 5.4, marked R_2/R_1. It lies on the concave inside of the H/R_2 curve. The R_2 regime is stable to its right, for increasing k^{-2}, unstable to its left.

It follows that between the two dashed curves, R_1/R_2 and R_2/R_1, neither the R_1 nor the R_2 flows are stable. It is natural to assume that in this part of the regime diagram, both a first-mode and a second-mode wave coexist. The two waves could be combined with fixed amplitude ratio and equal phase speeds in a mixed-mode Rossby regime, which we shall designate by R_{12}. Or they could exchange energy with each other and with the zonal flow, in a periodic, quasi-periodic, or irregular fashion.

Conceivably, one could still find a mixed-mode R_{12} regime analytically, and study its stability by the previous coefficient-freezing methods, as long as it is not more complicated than time-periodic. In practice, it turns out to be more convenient from this point on to proceed by numerical methods. They are the only ones so far by which irregular, aperiodic solutions can be studied in detail.

Such studies were carried out in the early 1960s by Lorenz at MIT on a Royal McBee LGP-30 computer. Their interesting and instructive results are the object of the rest of this section and of the next section.

Numerical study of vacillation. An essential tool in studying time-dependent model solutions, as well as their stability, is a *numerical* integration *method* of the governing equations. A method which has been widely used in the work discussed in this section and the next is of the *predictor-corrector* type.

Let the system of ODEs under study be written symbolically as

$$\dot{X} = F(X), \tag{5.24}$$

where X and F are vectors. The time t^* is discretized in increments Δt, with $t_n^* = n\Delta t$. Let $\tilde{X}_n$ approximate $X(t_n)$ and $\tilde{F}_n = F(\tilde{X}_n)$.

Two predictor steps are taken:

$$\tilde{X}^p_{n+1} = \tilde{X}_n + \Delta t \tilde{F}_n, \tag{5.25a}$$

$$\tilde{X}^p_{n+2} = \tilde{X}^p_{n+1} + \Delta t\ \tilde{F}^p_{n+1}. \tag{5.25b}$$

The final, corrected value of X at t^*_{n+1} is

$$\begin{aligned}\tilde{X}_{n+1} &= \tilde{X}_n + (\Delta t/2)(\tilde{F}_n + \tilde{F}^p_{n+1}) \\ &= (1/2)(\tilde{X}_n + \tilde{X}^p_{n+2}).\end{aligned} \tag{5.25c}$$

The integration scheme (5.25) is numerically stable for Δt not too large and it is second-order accurate: Eq. (5.25c) approximates to $O(\Delta t^2)$ the time derivative of X at $t^*_{n+1/2}$.

The scheme's *stability* guarantees, roughly speaking, that numerical solutions of the discretized problem (5.25) will have the same stability properties as the true solutions of the ODE system (5.14). Its *accuracy* means that, for the same initial data, $X_0 = X(t_0)$ the time-discrete numerical solution X_n will lie close to the time-continuous one, $\tilde{X}_n = X(t_n) + O(\Delta t)$, for a fixed time, $t_0 \leq t \leq t_0+T$.

In this type of study one is interested mostly in the asymptotic behavior of solutions, $T \to +\infty$, independently of the detailed initial data X_0. It is this behavior which, in the laboratory experiments themselves, is stationary, periodic or irregular. One hopes, therefore, that the numerical solutions will also provide the correct information on asymptotic behavior of the continuous solutions.

The numerical exploration of system behavior proceeds by a judicious choice of parameter values for which the various possible types of asymptotic behavior, or regimes, occur. First, values of θ_A^* and k are chosen which lie between the dashed stability curves of the single-mode regimes R_1 and R_2, closer to the R_1/R_2 curve (point 1 in Figure 5.4).

Initial transients, which take the solution from the arbitrarily chosen initial data to regime behavior, die out rather quickly as they do in the laboratory. For the parameter values of point 1, the asymptotic solution which is obtained numerically is a Rossby circulation of mixed mode, R_{12}: the second-mode wave moves at the same phase speed as the first-mode wave, and the amplitude of the two modes are constant in time.

Actually, Eqs. (5.14) are invariant under a change of sign of the second-mode variables ψ_C, θ_C, ψ_M, ..., θ_N. The same is true for the "frozen" variables, $\psi_M^0, \ldots, \theta_N^0$, cf. (5.23). Hence there are two stable Rossby circulations of mixed mode, R'_{12} and R''_{12}, which differ by a change of sign of the second-mode variables.

The stability of R'_{12} and R''_{12} can be determined numerically in a very simple way, as in nature itself. An asymptotic solution regime, stationary or time-dependent, is stable if numerical solutions with initial data close to it tend to the same regime, unstable if initial perturbations from it grow in time. Thus an unstable regime is never observed either in nature or in a time-marching numerical integration. It follows from the way R'_{12} was obtained numerically that it is stable. Small perturbations applied to R''_{12} also die out in numerical integrations, so that it is stable too.

There are various ways of computing unstable regimes numerically, in spite of the difficulty just mentioned. The simplest one is integrating backwards in time. For many problems, such as (5.14), forward stability or instability is equivalent to backward instability or stability, respectively.

The mixed-mode flow regimes R'_{12} and R''_{12}, like the single-mode regimes R_1 and R_2, have one degree of freedom -- the phase of the propagating, finite-amplitude wave. The amplitudes are fixed. The Hadley circulation has zero degrees of freedom, since all its variables are constant in time, their values depending only on the external parameters.

The "frozen wave" device of (5.23) might suggest that the steady Rossby circulations R_1, R_2, R'_{12} and R''_{12} can also be thought of as stationary solutions with zero degrees of freedom, like H. Recall, however, that the equations of motion (5.4) are *not* invariant under a change of the angular velocity of the frame of reference, cf. Section 1.3 and end of Section 3.3. Hence we are justified in calling these Rossby circulations periodic solutions with one degree of freedom. For the remainder of this section, we shall be mostly concerned with flow regimes having two and three degrees of freedom.

Choosing the same nondimensional heating rate difference across the channel, θ_A^*, as before, and a higher rotation rate k^{-1} (point 2 in Figure 5.4) produces a solution which is doubly-periodic in the long run. Waves of the first mode are very strong, then weaken, their energy being taken up by those of the second mode. Those in turn become weaker, returning the energy to the first mode, while at the same time interacting with the zonal flow. In this circulation, the zonal flow is no longer constant in time, as it was in the previous regimes, including the *Ra* circulations R'_{12} and R''_{12}. The zonal flow here is itself periodic; its period equals that of the energy exchanges between first-mode and second-mode waves, and is much longer than the traveling period of the waves themselves.

The doubly-periodic solution corresponding to the parameter values at point 2 is clearly a mathematical analog of the *vacillations* occurring in region Rb of Figure 5.2. Both the shape of the wave, as given by the amplitude ratio of the two modes, and its phase vary periodically, with two different periods. This solution can therefore be described by two temporal degrees of freedom.

The existence of two mixed-mode steady flows, R'_{12} and R''_{12}, leads us to suspect that the same might be true of vacillatory mixed-mode flows. Indeed, two such flows exist, which differ by the sign of ψ_C. Both these flows are stable at point 2 of Figure 5.4.

The *phase space* of system (5.14') has 14 dimensions. The vacillatory solutions under discussion have only two degrees of freedom. Hence one can essentially follow the behavior of their phase-space *trajectory* by projecting it onto any two-dimensional coordinate plane. One convenient choice is the (ψ_C, ψ_K^0)-plane. The corresponding projections for the two vacillatory regimes are shown in Figure 5.5a.

The two closed trajectories at the left of the figure represent the two regimes of interest. They are mirror images of each other in the $\psi_C = 0$ hyperplane, including the direction of the orbit. Since each one lacks a symmetry plane by itself, one can call them *unsymmetric vacillations*, labeling them UV' and UV", respectively.

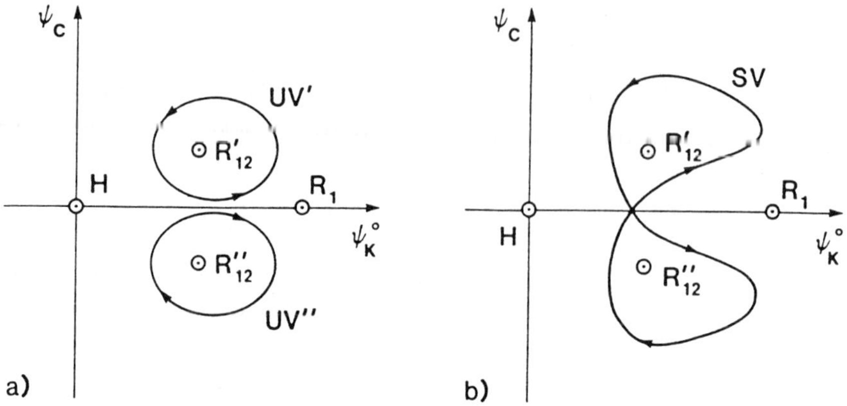

Fig. 5.5. Doubly-periodic solutions of system (5.14). a) Unsymmetric vacillation; b) symmetric vacillation (after Lorenz, 1963b).

The points in the figure are H, at the origin, R_1 on the ψ_K^0 axis, and R_{12}' and R_{12}'', in the first and fourth quadrant. They all exist for the parameter values under discussion, but are unstable. The point R_2 (not shown in the figure) lies in the hyperplane $\psi_K^0 = 0$, and is also unstable.

As one moves in the (θ_A^*, k^{-2})-plane from point 1 to point 2, both the steady mixed-mode circulations R_{12}' and R_{12}'' become unstable. The vacillations UV' and UV'' appear, their trajectories growing out of the points R_{12}' and R_{12}'', respectively, until they reach the size and shape in the figure.

To be more precise, the steady Rossby circulations R_{12}' and R_{12}'', R_1 and R_2 appear as points only in the "wave-freezing" coordinates (5.23). In general, they will appear as closed trajectories, or *limit cycles*. The unique degree of freedom they possess corresponds in physical space to the phase of the single-mode (cf. (5.17b,c)) or mixed-mode wave, and in phase space to the position along the limit cycle.

We have seen that for the same parameter values, both R_{12}' and R_{12}'' are stable. Which one obtains in a given numerical experiment depends on the initial data. Each one attracts trajectories, or *orbits*, starting in its neighborhood. This is why closed orbits are called limit cycles: they describe the limiting, or asymptotic behavior of all orbits nearby. Stable limit cycles describe the asymptotic behavior of forward orbits, as $t \to +\infty$, unstable limit cycles that of backward orbits, as $t \to -\infty$.

Moving further to the right, to point 3 in Figure 5.4, another type of vacillation obtains. The projection of its trajectory onto the (ψ_C, ψ_K^0)-plane is shown in Figure 5.5b. It is symmetric about the ψ_K^0-axis, deserving the name *symmetric vacillation*, SV. Its period is approximately twice as long as that of the UV vacillation.

The change in time of the contour plots for the stream function field, $\psi = \psi(x,y,t)$, shows similarities between SV and the "tilted-trough" vacillation observed in the laboratory; the trough and ridge lines change their tilt from SW-NE to SE-NW. This change is accompanied by a shift of the meridional position of the maximum velocity in the "westerlies", corresponding to a change in the sign of ψ_C. Such a shift in the latitude of the westerly jet, along with a change in its amplitude, occurs during atmospheric *index cycles*, as well as in the laboratory (see also Sections 4.6 and 6.1).

Comparison of Figures 5.5a and 5.5b, as well as the relative lengths of the periods for SV and UV, would suggest that SV results from the "soldering" together of the two separate limit cycles of UV' and UV". That this is so is further borne out by Figure 5.6.

In this figure two solutions are shown, both obtained for the same value of θ_A^*. The dashed curve represents the UV' solution which

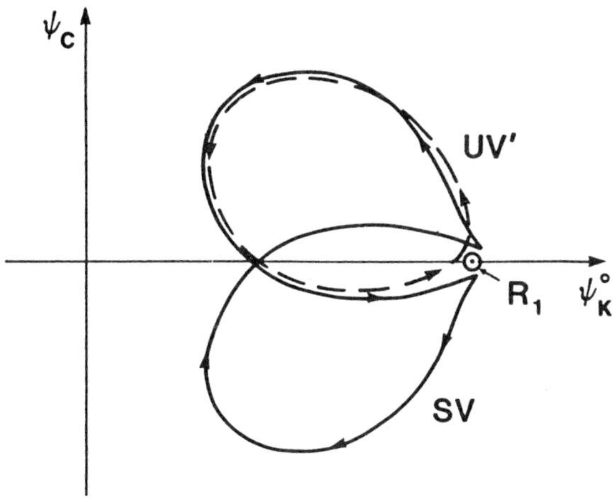

Fig. 5.6. The transition between two types of vacillation produced by a homoclinic orbit through point R_1 (after Lorenz, 1963b).

obtains for a value of k^{-1} between points 2 and 3 in Figure 5.4. The solid curve is the SV orbit for a very slightly higher rotation rate, i.e., higher k^{-1}, still to the left of point 3. Both curves pass very close to the point R_1, which is unstable.

Small perturbations from R_1 are governed by a system of 13 linear equations with "frozen", constant coefficients. For values of (θ_A^*, k) close to those used in Figure 5.6, the coefficient matrix of this system has a single positive eigenvalue, and 12 eigenvalues with negative real parts. The twelve eigenvectors corresponding to the latter span a hyperplane P through R_1, while the single unstable eigenvector spans a line L through R_1 which does not lie in P. Orbits can approach R_1 if they are nearly tangent to P, but eventually have to leave the neighborhood of R_1 becoming tangent to L. Thus R_1 is a generalized saddle point, or *hyperbolic point*: trajectories passing close to it resemble locally hyperbolae.

In general, a trajectory approaching R_1 along P leaves along L, never to return. For a special value of θ_A^* and k, very close to those used in Figure 5.6, the trajectory starting precisely at R_1, in the direction L, meets the nonlinear extension of P and is steered back into R_1. Such a closed orbit is called *homoclinic*.

For values of k^{-1} slightly below the critical, trajectories close to the homoclinic one always leave the neighborhood of R_1 along the same half-line, either L^+ or L^-. Let us say that the UV' trajectory (dashed in Figure 5.6) always leaves along L^+. Its mirror image, UV" (not shown) always leaves along L^-. For slightly supercritical values of k^{-1}, orbits approaching R_1 close to the homoclinic one leave this neighborhood in alternating directions, first L^+ and then L^-, say. Among these lies the SV trajectory (solid).

It follows that UV' and SV do not coexist for the same (θ_A^*, k)-values. Either vacillation type occupies a certain region of the parameter plane, to the exclusion of the other. The transition from the one to the other does not occur due to a local change of stability, as was the case for the H/R_1, R_1/R_{12} and R_{12}/UV transitions. The UV/SV transition is due to a global, genuinely nonlinear property of the system: the existence of a homoclinic orbit.

Both SV and UV are illustrations of the vacillatory Rb regime in Figure 5.2, as R_1, R_2 and R_{12} were descriptions of the steady Ra regime. It remains to find the mathematical counterpart of the irregular, chaotic T regime. Choosing θ_A^* as before, and k^{-1} still larger (point 4 in Figure 5.4) yields a numerical solution with the required properties. The variables sometimes take values close to certain values they had previously assumed. These near repetitions, however, are not exact, and they occur at irregular intervals. Such a solution has at least three degrees of freedom.

The classification of solutions to (5.14) can be summarized in the regime diagram shown in Figure 5.7, which is a further elaboration of Figure 5.4.

For various parameter values, ten types of solutions are possible and asymptotically stable: one axisymmetric, steady flow, H, four asymmetric steady flows, R_1, R_2, R'_{12} and R''_{12}, four vacillatory flows, UV', UV", SV and S'V', and one type of irregular flow, T.

The fourth type of vacillatory flow, S'V', is a vacillation which resembles somewhat SV but is not perfectly symmetric. It arises from SV by the latter's loss of stability.

The aperiodic solutions corresponding to the T regime are difficult to study in detail for the relatively complicated model governed by Eqs. (5.14). We shall discuss such asymptotically aperiodic behavior and the transition to it from periodic behavior in the next section, for an even simpler model.

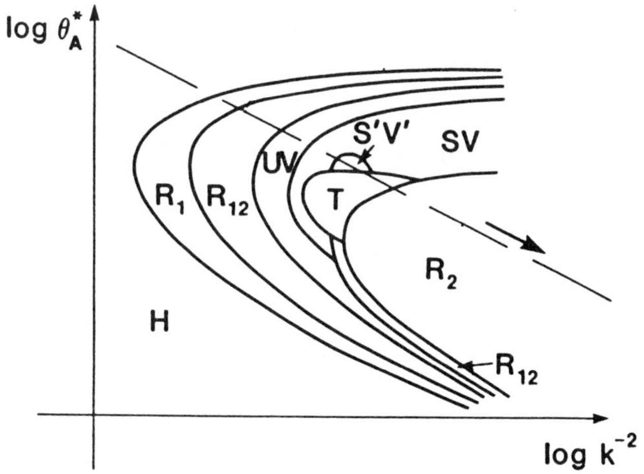

Fig. 5.7. Complete regime diagram for system (5.14) (after Lorenz, 1963b).

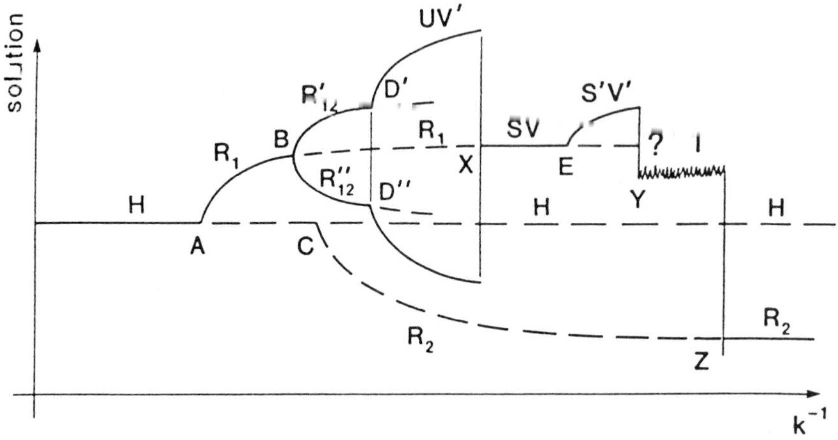

Fig. 5.8. Bifurcation diagram for system (5.14), with parameter varying along light dashed line in Figure 5.7.

The transitions between the various types of solutions are shown schematically in Figure 5.8. It represents the light dashed cross-section through Figure 5.7 for increasing rotation rate k^{-1}.

Each line in the figure indicates a solution *branch,* in other words a single solution type, whose characteristics vary smoothly as parameters change. Solid lines indicate stable branches, dashed lines unstable ones.

Points in the solution-parameter space, such as A, B and C, in the neighborhood of which more than a single solution branch exists, are called *bifurcation* points. The mathematical phenomenon by which branches split or fork at such points is called itself bifurcation. Point A lies on the transition line H/R_1, B on R_1/R_{12}, etc. The bifurcation of simply-periodic solutions, such as R_1, from stationary solutions, such as H, occurs frequently in fluid dynamics.

When the transition from stationary to periodic solutions is due to the change of a single parameter this phenomenon is called Poincaré-Andronov-Hopf bifurcation, or *Hopf bifurcation* for short. In it, two complex conjugate eigenvalues of the system's steady-state form, linearized around the stationary solution, move across the imaginary axis as the bifurcation parameter increases, so that their real part becomes positive. The stationary solution thus loses its stability at the bifurcation point, and a limit cycle starts growing as the parameter increases further. According to Eqs. (5.20 - 5.22), this is what happens at point A, with the periodic, wavy part of R_1 slowly growing as k^{-1} increases away from A.

At point B, the limit cycle R_1 becomes unstable, and transfers its stability to the limit cycles R'_{12} and R''_{12}. At point D' the limit cycle R'_{12}, which has one degree of freedom, transfers its stability to the doubly-periodic solution UV', which has two degrees of freedom: the phase and amplitude of the mode-one wave, or the intensity of the zonal circulation and the phase of the wave, say. There are thus two periodicities associated with UV', rather than the single periodicity of R_1 or R'_{12}. The second period is much longer than the first.

What is the geometric picture in phase space of UV'? In Figure 5.5, R'_{12} appears as a point, while in the original coordinates we saw it is a limit cycle. This limit cycle can be thought of as obtained by rotating the plane (ψ_K^0, ψ_C) around the ψ_C-axis, with angular velocity ω (cf. (5.17)).

UV' would appear, in the rotation above, as a spiral winding around the circle described by R'_{12}. One degree of freedom, like for R'_{12}, corresponds to the position along this spiral. The second degree of freedom corresponds to having started the spiral at a different point on the curve marked UV in Figure 5.5, and then rotating the (ψ_K^0, ψ_C)-plane. There are therefore an infinity of spirals associated with the doubly-periodic circulation UV', all lying on the doughnut-shaped surface, or generalized *torus*, which the curve UV' describes when rotated around the ψ_C-axis.

The regimes UV', as well as UV" appear in general in phase space as an invariant torus. Individual solutions belonging to regime UV' spiral on the surface of the torus; the spirals differ only by a phase shift along the azimuthal coordinate of the torus. It is the torus, however, rather than individual spirals, which attracts as a whole. In other words, any orbit near the torus will be asymptotic to a given spiral on the torus; which one depends on the initial point of the orbit. Any trajectory originating on the torus itself stays on the torus: the torus is *invariant* under the flow in phase space governed by Eqs. (5.14).

The sequence of bifurcations A-B-D leads to solutions of increasing complexity, as the simpler solutions lose their stability. This picture of successive bifurcations is often encountered in the study of nonlinear problems in biology and chemistry, as well as in continuum physics and fluid dynamics. It provides a general strategy for the description of the more complicated types of behavior in the phenomenology being modeled, while the transitions occurring *en route* are usually of interest in themselves.

The transition X from UV to SV is of a different type, which could be called spontaneous or global bifurcation. We saw that it is associated with the existence of a homoclinic orbit in the system, rather than with local loss of stability. The detailed structure of such a transition can be quite complicated, and is difficult to study numerically as well as analytically.

The transitions to T and from T, denoted by Y and Z in the figure, have not been investigated. The final stability recovery of the simple solution R_2 has no counterpart in the laboratory, and is probably due to the model's truncation. It is remarkable that, aside from this last transition, such a simple model could qualitatively reproduce the increasing complexity of flow patterns observed in the rotating annulus experiments.

5.4. Deterministic Aperiodic Flow

We have seen that in differentially heated rotating annuli the flow eventually becomes irregular, more or less chaotic, as it often is in the atmosphere and in the ocean. The successively more complicated stationary and time-periodic motions are but steps in an evolution of complexity which leads to the onset of turbulence.

Forced, dissipative systems. The onset of turbulence is an important phenomenon in many types of fluid flow. A reduction of the equations of fluid motion according to the scheme of Section 5.2 will in general produce a system of ODEs with the following structure:

$$\dot{X}_i = a_{ijk}X_jX_k - b_{ij}X_j + c_i, \quad i = 1,2,\ldots,N, \tag{5.26}$$

where we use the summation convention for repeated indices. Here $X = (X_1, X_2, \ldots, X_N)$ is an N-vector and $A = (a_{ijk})$, $B = (b_{ij})$ and $C = (c_i)$ are constant tensors of appropriate dimensions, with i, j and k ranging between 1 and N. The right-hand side of (5.26) gives the form of $F(X)$ in (5.24) which usually results from the spectral expansion and truncation of the governing equations in most fluid dynamical problems, as well as in many chemical and biological ones.

Any system of ODEs in which F depends on $X = X(t)$, but not explicitly on t, is called *autonomous*. System (5.26) is thus autonomous, since A, B and C are not functions of time. Furthermore, due to the nice continuity properties of $F(X)$, the solutions of (5.26) exist and are unique for any initial data $X(t_0) = X_0$, up to some time t_0+T. The same is true if t is replaced by $-t$, i.e., if we consider the system backward in time.

Due to the autonomous character of F, the graph of the solutions of (5.26) in phase space, i.e., its trajectories, are only defined up to a shift in time: the trajectory of the solution with X_0 as "initial" data for $t = t_1 > t_0$, say, $X(t_1) = X_0$, is the same as that of the solution with $X(t_0) = X_0$. We cannot visually distinguish between solutions which pass through the same point, $X = X_0$, at one time or another. As a result of the uniqueness of solutions and of their time-translational invariance, trajectories of (5.26) can never intersect or touch each other in finite time. All these facts are easiest to check and visualize for $N = 2$, but hold for arbitrary N.

In many cases, the coefficients of (5.26) have the following additional properties, which we shall call (C) and (D), respectively:

$$\text{(C)} \quad a_{ijk}X_iX_jX_k \equiv 0, \tag{5.27a}$$

$$\text{(D)} \quad b_{ij}X_iX_j > 0; \tag{5.27b}$$

they hold for arbitrary values of the X's, with $X \neq 0$ in (5.27b). Property (C) states that, in the absence of the other terms, $B = 0$ and $C = 0$, system (5.26) would be *conservative*. Indeed, let us define its generalized energy $E = E(X)$ by

$$2E(X) = X_iX_i. \tag{5.28}$$

Then multiplying each equation in (5.26) by X_i and summing, with $B = 0 = C$, yields $\dot{E} = 0$ or $E = \text{const}$. We also denote by x the length of the vector X, $x^2 = X_iX_i = 2E$.

Property (D) is equivalent to the statement that matrix B is positive definite. Physically it means that the system contains a *dissipative* mechanism. Indeed, for the full system,

$$\dot{E} = -b_{ij}X_iX_j + c_iX_i. \tag{5.29}$$

This shows that B causes the energy of the system to decrease, if (D) holds.

It is easy to verify, in particular, that Eqs. (5.14), after multiplication by $a_i^2 = -\lambda_i/L^2$, satisfy conditions (5.27a,b). The quadratic terms correspond to advection of heat and momentum, viz., the Jacobians in (5.4), the linear terms to friction and heating at the surface and between layers. The multiplication by $a_i^2 > 0$ in the momentum equations corresponds to consideration of kinetic energy $E_k = \overline{(\nabla\psi)^2}$ as their proper contribution to the total energy in (5.28).

The rate of change $\dot{E}$ of $E(X)$ along a system trajectory $X = X(t)$ is given by (5.29). On any energy surface in phase space, E = const., this rate attains its algebraically largest value for X pointing in the direction of the vector C, $\max c_i X_i = cx$, where $c^2 = c_i c_i$. Thus

$$\dot{E} \leq -\beta_1 x^2 + cx, \tag{5.30}$$

where $\beta_1 = \min_{x=1} b_{ij} X_i X_j > 0$ is the lowest eigenvalue of the matrix B. Clearly $\dot{E} = x\dot{x}$ will be negative for any $x > c/\beta_1$.

It follows that trajectories of a *forced dissipative* system (5.26), satisfying (5.27), will all eventually enter a ball $\mathcal{B} = \{X: x \leq c/\beta_1\}$, never to leave it again. In the absence of forcing, $C = 0$, this ball is reduced to a point, the origin $X = 0$. It is the competition between the forcing C and the dissipation B which renders the behavior of these systems interesting. Part of the interest, both physical and mathematical, derives from the fact that this competition has to be played out within the bounded ball $\mathcal{B} = \{E(X) \leq c^2/2\beta_1^2\}$.

One consequence of the boundedness of all forward trajectories, already stated implicitly in the preceding paragraph, is that they exist for all times, $T \to +\infty$. Backward trajectories, although unbounded, also exist for all times, $T \to -\infty$, due to the conservative character (C) of the system's nonlinearity.

A system of ODEs with the property that all its solutions exist for arbitrarily long time intervals is called a *dynamical system*. The contemporary theory of dynamical systems is mostly concerned with the asymptotic properties of their solutions.

In particular, bounded trajectories must have certain accumulation points, or limit points. The set of limit points of a dynamical system is called its *attractor set*. For a scalar system, $N = 1$, even if $F(X_1)$ is more complicated than (5.26), the attractor set has to be made up of stable steady states. For $N = 2$, the Poincaré-Bendixson theorem essentially states that the only components of the attractor set can be stable equilibria, limit cycles or trajectories joining equilibria.

For $N \geq 3$, much more complicated attractor sets are possible. They have been called *strange attractors* by Ruelle and Takens (1971). The word "strange" here has to be taken with a grain of salt. It reminds us of "irrational" numbers and of "imaginary" or "complex" ones. If strange attractors prove to be half as important as they appear at present in explaining chaotic behavior, we had better get used to their strangeness rather quickly.

In the sequel, we shall study the simplest known example of a forced dissipative dynamical system (5.26,27) in which a strange attractor rears its ugly head. It was first used by Lorenz to describe transition to turbulence in thermal convection. Later, versions of the "Lorenz system" have been shown to govern many other phenomena. In particular, one form of the system is analogous to a spectrally truncated version of the nonlinear, barotropic quasi-geostrophic potential vorticity equation which was introduced in Chapter 3 and will be studied in the next chapter.

The Lorenz system. In the Rayleigh-Bénard problem, a horizontal layer of fluid of depth D is heated from below, maintaining the difference between top and bottom temperature at a fixed value ΔT. For low ΔT, heat is transferred upwards by conduction only, the temperature decreases linearly with height, and the fluid is at rest. As ΔT is increased, convection sets in and the fluid overturns in parallel "rolls". The importance of convection in explaining geomagnetism will be discussed in Sections 7.2, 7.3 and 9.1.

Taking the y-direction along the rolls, it is reasonable to assume that at first the flow is y-independent and occurs in a vertical (x,z)-plane. The governing equations, in vorticity form, are

$$\partial_t \nabla^2 \psi = -J(\psi, \nabla^2 \psi) + \nu \nabla^4 \psi + g\alpha\theta_z, \tag{5.31a}$$

$$\partial_t \theta = -J(\psi, \theta) + \frac{\Delta T}{D} \psi_x + \kappa \nabla^2 \theta. \tag{5.31b}$$

Here J is the Jacobian with respect to the x and z variables, $J(f,g) = f_x g_z - f_z g_x$, ψ is the stream function for the planar vertical motion, θ is the departure of temperature from its linear, conductive profile, g, α and ν are as in Section 5.1, the Boussinesq approximation being made, and κ is the thermal conductivity. The boundary conditions which are easiest to treat, although not the most realistic ones, are that $\psi = \nabla^2 \psi = 0$ at $z = 0, H$, and that $\theta = 0$ at the upper and lower boundary.

Fluid properties are characterized by the nondimensional *Prandtl number*,

$$\sigma = \nu/\kappa, \tag{5.32a}$$

whose physical significance is obvious. Its value is O(1) for air, O(10) for water, much lower for liquid metals and much higher for organic fluids. Fluid motions depend most strongly on the nondimensional *Rayleigh number*,

$$R = \frac{g\alpha D^3 \Delta T}{\nu\kappa}. \tag{5.32b}$$

The purely conductive rest state becomes unstable to rolls with a vertical wave length 2H and a horizontal wave length 2H/a for a Rayleigh number $R > R_a$, where

$$R_a = \pi^4(1 + a^2)^3/a^2.$$

The minimum value of R_a is given by $R^* = \min_a R_a = 27\pi^4/4 \cong 657$; it occurs for the horizontal wave number $a = 1/\sqrt{2}$.

The spectral time-dependent variables, X, Y and Z are introduced now by

$$\psi = \frac{(1+a^2)\kappa\sqrt{2}}{a} X \sin(\pi a\xi) \sin \pi\zeta, \tag{5.33a}$$

$$\theta = \frac{R_a \Delta T}{\pi R} \{\sqrt{2}\, Y \cos(\pi a\xi) \sin \pi\zeta - Z \sin 2\pi\zeta\}, \tag{5.33b}$$

where $\xi = x/H$, $\zeta = z/H$, and all other terms in a trigonometric expansion are omitted. Thus X measures the rate of convective overturning, while Y and Z measure the fundamental temperature variations in the horizontal and vertical. Substituting (5.33) into (5.31) yields the system of ODEs

$$\dot{X} = -\sigma X + \sigma Y, \tag{5.34a}$$

$$\dot{Y} = -XZ + rX - Y, \tag{5.34b}$$

$$\dot{Z} = XY - bZ. \tag{5.34c}$$

Here differentiation is with respect to the nondimensional time $\tau = \pi^2(1+a^2)\kappa t/H^2$, while $r = R/R_a$ and $b = 4/(1+a^2)$, so that σ, r and b are all positive parameters. For this case of (5.26) for which N = 3, it is convenient to make the identifications $X_1 = X$, $X_2 = Y$, $X_3 = Z$.

It is immediately obvious from (5.34) that the quadratic terms are conservative. To see that the linear terms are dissipative, it is easiest to introduce a linear change of variables

$$X = X', \quad Y = Y', \quad Z = Z' + r + \sigma. \tag{5.35a}$$

This will change the rX term in (5.34b) into $-\sigma X'$ and thus produce the quadratic form $b'_{ij}X'_iX'_j = \sigma X'^2 + Y'^2 + bZ'^2$ which is clearly positive for nonzero X', Y' or Z'. In addition, the forcing term $c'_3 = -b(r+\sigma)$ will be put in evidence in (5.34c), with no other changes to the system.

It follows from (5.29,30) that trajectories of (5.34) are bounded inside an energy surface S given by

$$\dot{E}' \equiv -\sigma X'^2 - Y'^2 - bZ'^2 - b(r+\sigma)Z' = 0. \tag{5.35b}$$

This is a sphere centered at $X' = Y' = 0$, $Z' = -(r+\sigma)$, or $X = Y = Z = 0$, and of radius $r + \sigma$.

We proceed to study the attractor set within the ball B bounded by the sphere S, for various parameter values. The state of rest, $X = Y = Z = 0$, which is the center of B, is obviously an equilibrium solution of (5.34). Whether it actually lies in the system's attractor set depends on its stability.

The local stability of any solution $W_0 = (X_0,Y_0,Z_0) = W_0(\tau)$ to small perturbations $w(\tau) = W-W_0$ depends on the properties of the linearized system

$$\dot{w} = \left.\frac{\partial(F_1,F_2,F_3)}{\partial(X\ ,Y\ ,Z\ )}\right|_{W=W_0} \cdot w\ . \tag{5.36a}$$

We denote the Jacobian matrix $(\partial F/\partial W)(W=W_0)$ in (5.36a) by L_0,

$$L_0 = \begin{pmatrix} -\sigma & \sigma & 0 \\ r-Z_0 & -1 & -X_0 \\ Y_0 & X_0 & -b \end{pmatrix}. \tag{5.36b}$$

In general, L_0 will depend on time τ, via the time dependence of $W_0(\tau)$.

An interesting property of L_0 is that X_0, Y_0 and Z_0 appear only in off-diagonal position. In fact, the sum of the constant diagonal elements, the trace of L_0, is negative,

$$\text{tr}\ L_0 = -(\sigma + b + 1) < 0. \tag{5.37a}$$

This algebraic fact has an important geometric consequence.

Eqs. (5.34) can be thought of as the Lagrangian description of a *flow* in (X,Y,Z)-space, whose velocity components are $\dot{X}$, $\dot{Y}$ and $\dot{Z}$. The trajectories of the system are the pathlines of this flow. They not only take each point X_0 uniquely into a point $X(\tau;X_0)$, where $X(0;X_0) = X_0$, but also take any region V_0 of phase space into a region $V(\tau)$, with $\mathcal{V}(0) = \mathcal{V}_0$. In particular, under the flow, the volume V_0 of the region $\mathcal{V}_0$ becomes the volume $V(\tau)$ of $\mathcal{V}(\tau)$. The rate of change of the volume, $\dot{V}$, is given by the divergence of the flow,

$$\dot{V} = (\partial_X \dot{X} + \partial_Y \dot{Y} + \partial_Z \dot{Z})V = (\mathrm{tr}\ L_0)V. \tag{5.37b}$$

It follows from (5.37) that the flow induced by (5.34) in phase space is volume reducing. Before discussing further the consequences of this fact, let us compute, for comparison purposes, the divergence of the flow of the more general system (5.26),

$$\begin{aligned}\dot{V}/V &= \partial_{X_i}(a_{ijk}X_jX_k - b_{ij}X_j + c_i)\\ &= a_{iik}X_k + a_{iji}X_j - b_{ii}.\end{aligned} \tag{5.38}$$

Clearly the forcing C does not affect the divergence.

We have seen in Section 5.2 that for a system whose quadratic terms are derived from Jacobians, i.e., from nondivergent advection terms in the original flow equations, the "interaction coefficients" a_{ijk}, given by (5.8), have the antisymmetry property

$$\text{(A)}\quad a_{ijk} = -a_{ikj}, \tag{5.39a}$$

as well as the cyclical symmetry property

$$\text{(S)}\quad a_{ijk} = a_{jki} = a_{kij}. \tag{5.39b}$$

In particular, systems (5.14) and (5.34) share these properties. The same will be true of all the systems to be studied in the next chapter.

Property (A) immediately implies (C)

$$a_{ijk}X_iX_jX_k = -a_{ikj}X_iX_jX_k = -a_{ikj}X_iX_kX_j = 0.$$

The converse is not true in general. For a system (5.26) with property (A), it immediately follows from (5.38) that

$$\dot{V} = -(\mathrm{tr}\ B)V. \tag{5.40}$$

Thus (5.37) is but a special instance of this more general result.

If the system had no dissipative terms, $B = 0$, the phase-space flow it governs would be volume preserving, $\dot{V} = 0$, $V = \text{const}$. This is the form that *Liouville's theorem* for Hamiltonian systems takes in the context of our theory. For dissipative systems (5.26), however, (D) implies in particular $-\mathrm{tr}\ B < 0$, so that their flow, given (A), is volume reducing.

Returning to the Lorenz system (5.34) and to (5.37), in this case $\mathrm{tr}\ L_0 = -\mathrm{tr}\ B = \text{const.} < 0$. Hence the flow will shrink the volume V_0

of any initial region V_0 exponentially with time, $V(\tau) = V_0 \exp\{-(\sigma+b+1)\tau\}$. It follows that the attractor set of system (5.34) has to have zero volume with respect to its three-dimensional phase space. This will allow, as we shall see, not only zero-dimensional points and one-dimensional limit cycles to be present in the attractor set, but also two-dimensional surfaces.

The general result on volume reduction was extracted from (5.36) due to the constancy of $\operatorname{tr} L_0$, and is thus independent of the solution $W_0 = W_0(\tau)$ about which (5.34) was linearized. The sum of the real parts of the eigenvalues of L_0 having to be negative does not mean that each one individually has to be, as we saw in the analysis of (5.20). Furthermore, when $W_0 = W_0(\tau)$ and hence $L_0 = L_0(\tau)$, it is not these eigenvalues alone which determine the properties of growth or decay of the solutions $w(\tau)$ of (5.36).

For $W_0 = (0,0,0)$, however, it is easy to ascertain its stability. The characteristic equation $\det(L_0-\lambda) = 0$ in this case is

$$(\lambda+b)\{\lambda^2 + (\sigma+1)\lambda + \sigma(1-r)\} = 0. \tag{5.41}$$

It has always the real root $\lambda = -b < 0$, whose eigenvector lies along the Z-axis. The other two roots are also real for $r > 0$. For $0 < r < 1$ they are both negative, with one becoming positive as r crosses $r = 1$ and staying so for all $r > 1$.

The conductive state of rest W_0 becomes unstable at $r = 1$, and one expects convection to set in. Indeed, for $r > 1$, Eqs. (5.34) possess two additional steady-state solutions, $W_{1,2}$, with $X = Y = \pm\sqrt{b(r-1)}$, $Z = r-1$, which correspond to convective rolls rotating in opposite directions. They clearly bifurcate out of the rest state $W_0 = (0,0,0)$ as r crosses the critical value 1. This phenomenon is called *pitchfork* bifurcation; in it, two symmetric stable steady states arise from a single steady state, which loses its stability but continues to exist as the bifurcation parameter increases. The pitchfork bifurcation of branches W_1 and W_2 from W_0 is illustrated in Figure 5.9. From (5.35) it follows that the two steady convective states will always be inside the ball B which bounds solutions for large τ, provided $\sigma > b-1$.

Substituting $W_{1,2}$ into (5.36) instead of W_0 will produce constant-coefficient matrices $L_{1,2}$, respectively. Both of these have the same characteristic equation,

$$\lambda^3 + (\sigma+b+1)\lambda^2 + (r+\sigma)b\lambda + 2\sigma b(r-1) = 0. \tag{5.42a}$$

For $r > 1$, (5.42a) possesses one real negative root and two complex

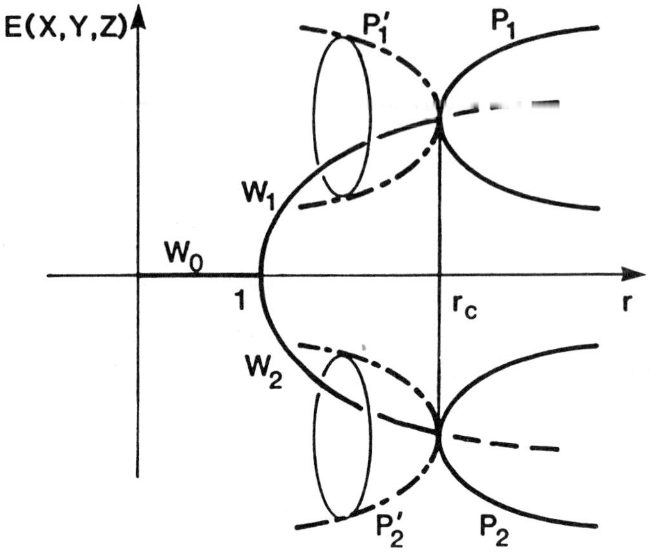

Fig. 5.9. Subcritical Hopf bifurcation for the Lorenz system.

conjugate roots. The real part of this pair changes sign as r passes through r_c,

$$r_c = \sigma(\sigma + b + 3)/(\sigma - b - 1). \tag{5.42b}$$

For $\sigma < b+1$, there is no positive $r_c(\sigma,b)$ which yields instability of the convective states $W_{1,2}$. When $\sigma > b+1$, there is always a positive r_c so that $W_{1,2}$ are both stable for $1 < r < r_c$, and both unstable thereafter. This is where the fun begins.

A strange attractor. What happens when $\sigma > b+1$ and $r > r_c$? Evidently the steady convective states could lose their stability to periodic solutions, as was the case in Section 5.3 for the transition from the steady Hadley solution H to the periodic Rossby solution R_1.

One would suspect that, as the pair of complex conjugate eigenvalues of $L_{1,2}$ crosses the imaginary axis at $r = r_c$, Hopf bifurcation of the stable periodic solutions P_1 and P_2 from W_1 and W_2, respectively, occurs. This is indeed the case for a certain domain of the (σ,b)-plane.

The branching of solutions with higher complexity in the direction of successive bifurcations is called *supercritical* bifurcation. This is the only type we have encountered thus far. It is called supercritical since the physical parameter which determines the change in complexity, such as r here, is usually chosen to increase in the direction of higher complexity. It makes for a pretty picture in which more branches grow as the "height" of the tree, i.e., the parameter r, increases.

Examples of supercritical bifurcation in Figure 5.9 are the pitchfork bifurcation of the steady states W_1 and W_2 from W_0, and the Hopf bifurcations we have just discussed. For most values of σ and b, however, the periodic solutions P_1' and P_2' bifurcate in the direction of decreasing r and furthermore they are unstable. This type of behavior is called *subcritical* or *reverse* bifurcation.

What happens then for $r > r_c(\sigma,b)$ in those cases in which the Hopf bifurcation at $r = r_c$ is subcritical? To find out, it is time again for some numerical exploration.

Choosing first the horizontal wave number a to give the lowest value of the critical Rayleigh number R_a, $a^2 = 1/2$, yields $b = 8/3$. A value of $\sigma = 10$ then gives $r_c = 470/19 \cong 24.74$. The supercritical value $r = 28$ places the unstable steady convective states at $W_1 = (6\sqrt{2}, 6\sqrt{2}, 27)$ and $W_2 = (-6\sqrt{2}, -6\sqrt{2}, 27)$. We recall that the rest state $W_0 = (0,0,0)$ is stable along the Z-axis and in one additional direction, unstable in the third. Each convective state is stable in one direction, with trajectories spiraling out in two other directions.

The following diagram, Figure 5.10, sketches the stability properties of the three steady states for $\sigma = 10$, $b = 8/3$ and $r = 28$. The directions of approach to an equilibrium W_k span its stable *tangent subspace*, denoted by $T^s(W_k)$. Only the tangent plane $T^s(0)$ for the origin is shown. The directions of flight at W_k span its unstable tangent plane, $T^u(W_k)$. $T^u(W_1)$ leans forward -- in the positive X-direction -- with increasing Z, $T^u(W_2)$ leans back. The line of intersection of $T^u(W_1)$ with $T^s(0)$ is shown solid, that of $T^u(W_2)$ with $T^s(0)$ is dashed. The unstable spiral around W_1 is counterclockwise, that around W_2 is clockwise. To prevent exaggerated cluttering of the sketch, the stable tangent lines at $W_{1,2}$, into the planes $T^u(W_{1,2})$, have not been shown.

In the direction of the unstable tangent $T^u_+(0)$, within the first (X,Y)-quadrant, starts a trajectory we denote by $M^u_+(0)$. In the opposite direction, tangent to $T^u_-(0)$, there is a trajectory $M^u_-(0)$. Together, these two trajectories form the *unstable manifold* $M^u(0)$ of W_0. $M^u_+(0)$ approaches $T^u(W_1)$, circles around W_1 counterclockwise for about half a turn, then leaves the neighborhood of $T^u(W_1)$; it intersects $T^s(0)$

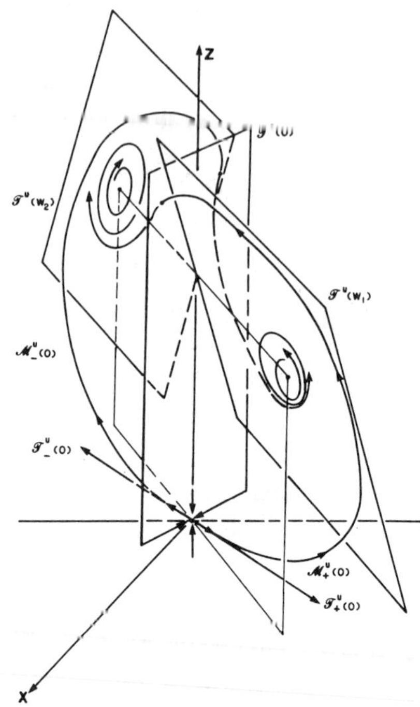

Fig. 5.10. Stable and unstable manifolds for the three stationary solutions of the Lorenz system at $r = 28$ (after Guckenheimer and Holmes, 1983, Figs. 2.3.2, 2.3.3 and 6.4.3).

and approaches $T^u(W_2)$, on the other side of $T^s(0)$, where it describes half a turn clockwise around W_2. $M_-^u(0)$ moves towards $T^u(W_2)$ and around W_2, crosses $T^s(0)$ behind the (Y,Z)-plane and approaches $T^u(W_1)$ from behind, in the direction of increasing X. In Figure 5.10, lines hidden by planes are dashed, and apparent intersections of lines show the one behind interrupted, to create some illusion of spatial perspective.

We have seen at the end of Section 5.3 that the unstable manifold of a hyperbolic point could "come back" as part of its stable manifold, creating a homoclinic orbit. Already this interaction between the stable and unstable manifolds of a single point created some complexity in the system, causing the appearance of a more complicated attractor set (see also Section 5.5). In fact, for $r = r_0 \cong 13.926 < r_c$, the two orbits starting at the origin, $M_\pm^u(0)$, both return into it tangent to the positive Z-axis. These two homoclinic orbits are the limit of the two unstable periodic orbits denoted by P_1' and P_2' in Figure 5.9, as $r \downarrow r_0$. Furthermore, for $r > r_0$, the unstable manifolds of all three critical

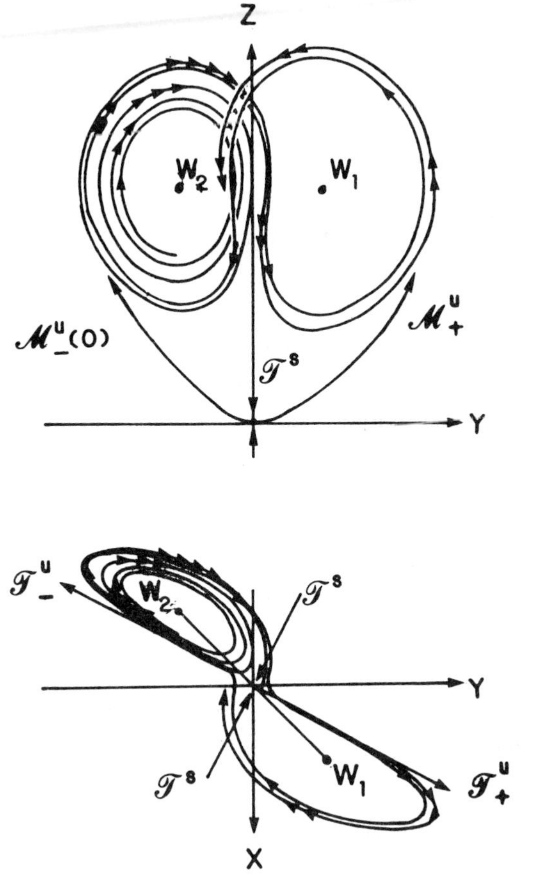

Fig. 5.11. Aperiodic orbits of the Lorenz system for $r = 28$ (after Lorenz, 1963a).

points interact in an entirely nontrivial manner, leading us to expect a quite complicated attractor set.

Figure 5.11 shows the projection onto the (X,Y) and (Y,Z)-plane of a numerical solution to (5.34) obtained by the predictor-corrector method (5.25). Also shown, for the state of rest only, are its unstable eigenvector, $T^u_\pm$, in (X,Y), and the trace of the plane T^s spanned by the two stable eigenvectors onto (X,Y) and (Y,Z); the projection onto (Y,Z) of the unstable manifold, $M^u_\pm$, is indicated too.

Only a segment of the solution is shown. In the (Y,Z)-plane, we can easily follow it as it loops clockwise around W_2 three times (single, double, triple and quadruple arrows on the left of Figure 5.11), and then crosses the (X,Z)-plane. After one counterclockwise loop around W_1 (single arrows on the right of the figure), it returns for one loop around W_2 (single arrows with tails), then still another one around W_1 (double arrows). In the (X,Y) plane, one can clearly see the backward tilt of the loops around W_2 going over near the X-axis into the forward tilt of the loops around W_1, and vice versa.

All numerical solutions are attracted by a certain portion of the two planes $T^u(W_1)$ and $T^u(W_2)$. Having gotten close to one of them, a solution starts to loop around the corresponding critical point, W_1 or W_2. No solution can ever penetrate a certain neighborhood of either point, since both are unstable within the corresponding tangent plane. More precisely, within the unstable manifold $M^u(W_1)$ locally approximated by $T^u(W_1)$, there is an excluded neighborhood of W_1. Within $M^u(W_2)$ there is also an excluded neighborhood of W_2.

In fact, due to the instability of W_1 and W_2, the loops around either point increase in amplitude, until they reach a critical size, apparent in the upper part of Figure 5.11. When this threshold is reached, the orbit will transit across the YZ-plane and start oscillating around the other unstable fixed point. The number of loops around W_1 or around W_2, until the threshold is reached, varies irregularly in time for each solution, and from one solution to another.

We have seen earlier that arbitrary domains of phase space are deformed by the flow (5.34) in time so that their volume tends to zero. Since stable equilibria and limit cycles are conspicuously absent from the numerically obtained attractor set, one might conjecture that the attractor is a smooth surface, or two-dimensional manifold. Figure 5.12 is a sketch of a surface which seems to be a good numerical approximation to the attractor set.

The surface is shown by the level curves of X as a function of Y and Z, X = X(Y,Z). The "holes" around W_1 and W_2 appear clearly in this representation. A trajectory looping around W_2 clockwise would be taken onto the sheet of the surface which "leans back", close to $T^u(W_2)$, and thus lies behind the sheet close to $T^u(W_1)$. There appear to be, therefore, two sheets. The isopleths, or level curves, of X = X(Y,Z) on the hidden part of the second sheet are dashed, the visible ones are solid thin lines.

The heavy solid line tangent to the Y-axis at the origin is $M^u_{\pm}(0)$. Its hidden continuations are dotted. The two sheets appear to merge along $M^u(0)$ in the lower portion of the figure, as well as near W_1 and

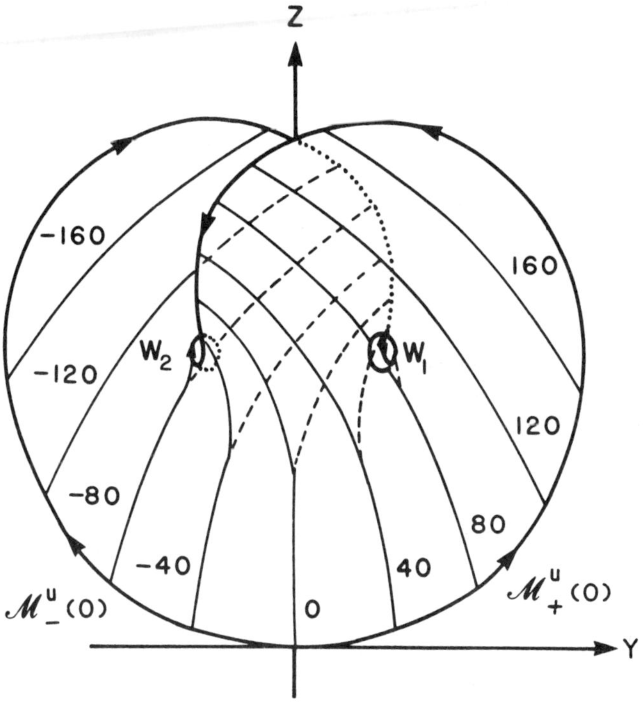

Fig. 5.12. A simplified, smooth sketch of the Lorenz attractor (after Lorenz, 1963a).

W_2. There is, however, a slight catch with this apparent two-sheeted structure of the attractor.

Since (5.34) is an autonomous system, its trajectories cannot intersect. If the attractor were really made up of only two smooth sheets, merging along portions of $M^u(0)$, trajectories circling around W_1 and around W_2 would have to either intersect, or leave the attractor. Neither one of these alternatives is possible.

The dilemma is resolved by the strangeness of the attractor. Consider a loop of a trajectory around W_1, say, lying in a given sheet of the attractor. As the trajectory approaches the Z-axis, it will flow parallel to it, cf. Figure 5.11, toward the origin. At that point, the given trajectory can either leave the neighborhood of the axis in the direction of $M^u_+(0)$, returning to loop with larger amplitude around W_1, or it can leave in the direction of $M_-(0)$, to start looping around W_2.

We know already that of two trajectories arbitrarily close to each other, one below the critical threshold will go to the right, back to W_1, the one above the threshold will go to the left, on to W_2. Thus the given sheet of the attractor we started with splits near $M^u(0)$ into two sheets, one to the right, the other to the left of the YZ plane. Each such sheet splits again, yielding a total of four sheets, and so on *ad infinitum*.

Each one of the two "sheets" in Figure 5.12 is thus made up of an infinite number of smooth sheets. These sheets are very closely packed together, and hence well approximated by the two original ones. The double infinity of sheets making up the attractor appears to merge along the portions of $M^u(0)$ in the neighborhood of the three critical points.

Each of the sheets is smooth; in the neighborhood of a point P lying on it, it can be represented locally by a nice function $X_P = X_P(Y,Z)$ and these local functions X_P can be continuously patched together. But what does the attractor look like in the direction perpendicular to the sheets, i.e., in the direction in which it is attracting?

Drilling a hole through the attractor at P, parallel to the X-axis, say, we know already that we encounter infinitely many sheets, within a rather small interval of X, $[X_1,X_2]$ say. In fact, it can be shown that the number of intersection points X_α between X_1 and X_2 is nondenumerable: the natural numbers $\mathbb{N} = \{1,2,3,\ldots\}$ are not sufficient to count the indices α, and all the real numbers $\mathbb{R}$ are needed.

On the other hand, the total volume of the attractor has to be zero. Hence the infinitely many points $\{X_\alpha\colon \alpha \in \mathbb{R}\}$ cannot occupy any subinterval $[X_\beta,X_\gamma]$ of $[X_1,X_2]$. Otherwise, using a neighborhood of P in its sheet, described by $X_P = X_P(Y,Z)$, as the base, we could construct a little cylinder of height $X_\gamma - X_\beta$. This cylinder would be entirely within the attractor and have nonzero volume.

The structure of the intersection set $\{X_\alpha\}$ within $[X_1,X_2]$ is similar to that of the classical *Cantor set* C on the unit interval $[0,1]$. C is the set of numbers x between 0 and 1 whose ternary expansion, $0.a_1a_2a_3\ldots$, $x = \Sigma_1^\infty a_n 3^{-n}$, does not contain any $a_n = 1$. This set is nondenumerable, contains all its limit points, and has measure zero.

The way in which the many sheets of the attractor are "sown" together at the edges is in fact quite complicated. Furthermore, the nature of the sewing changes with the value of r. Clearly, this attractor is very strange. System (5.34) has the minimally required number of degrees of freedom, $N = 3$, to have a nontrivial attractor, and appears quite innocuous. The fact that its attractor looks so strange is rather remarkable.

What about periodicity? The attractor contains a denumerable infinity of periodic orbits, all of which are unstable. All other orbits, nondenumerably many, are *aperiodic*. The orbit in Figure 5.11 is typical: it looks locally smooth, but alternates irregularly between loops around W_1 and loops around W_2.

To which extent the properties of this particular attractor are inherited by the original system of PDEs (5.31), and how well they explain actual irregular flows, is still an open question. More generally, extensive experimental work has been inspired by the new deterministic theories about the onset of turbulence. Some features of the theoretical models, concerning the dimensionality of attractor sets and the Fourier spectrum of flow quantities, seem to be borne out by measurements. The situation appears hopeful, but unsettled, which together mean exciting.

5.5. Bibliographic Notes

Section 5.1. This section is based largely on Hide (1977). The role of laboratory experiments in understanding the general circulation of the atmosphere is discussed by Lorenz (1967), Chapter 6. A closer analogy to the atmosphere is actually provided by flow in a rotating dishpan (Fultz *et al.*, 1959), but experimental parameters are more easily controlled in the annulus, and the results are more reproducible therefore in the latter device. Notice that in fact Hide's original interest was modeling the flow in the Earth's core (see Part III).

Connections between Rossby waves in an annulus and finite-amplitude baroclinic instability theory are discussed by Hart (1979). Many details of the axisymmetric and asymmetric flows in the annulus are influenced by the existence of boundary layers near the bottom and the sidewalls (McIntyre, 1968; Robinson, 1959). The spatial and spectral structure of flow fields in the nonsteady regimes of a fully instrumented annulus is described by Pfeffer *et al.* (1980).

Section 5.2. Representing solutions of *linear* differential equations, ordinary and partial, by a series of special functions or by an integral with a special kernel is a standard method for obtaining such solutions and analyzing their properties. Truncation in this case can sometimes be justified mathematically by a rigorous estimate of the remainder (Courant and Hilbert, 1953, Chapters V and VII).

Many numerical methods for the solution of linear and nonlinear PDEs are based on truncated series expansions. The functions used in the expansion either have compact support, i.e., they are nonzero only on a small portion of the entire domain for which the problem is defined (Strang and Fix, 1973) or they are classical special functions (Gottlieb and Orszag, 1977). Such methods are also gaining acceptance in the formulation of general circulation and numerical weather prediction models (Haltiner and Williams, 1980, Chapter 6; Leith, 1978).

The idea of using extreme truncation in *nonlinear* problems, for the purposes of a preliminary qualitative investigation of solution properties, appeared in the late 1940s and the 1950s. The term "low-order truncation" was coined by Platzman (1960), and Saltzman (1962) gives many early references. The particular truncated system discussed in this section was developed by Lorenz (1960a,b, 1963b), based on earlier work of Phillips (1956).

The extent to which the solutions of such low-order truncated systems do reflect the properties of the full equations is a matter of active research (Guckenheimer and Holmes, 1983, p. 92). Rigorous results on finite-dimensionality of attractor sets for the Navier-Stokes equations were obtained by Constantin and Foias (1985), and by Ruelle (1981a). Additional references, including careful numerical studies of the minimal truncation compatible with the qualitative behavior of interest, are given in Section 6.5. A related question is that of the similarity of qualitative behavior across a hierarchy of approximations in the PDEs themselves (Gent and McWilliams, 1982).

<u>Section 5.3</u>. This section is based on Lorenz (1962, 1963b). A condensed version of the material, and a review of more detailed numerical computations for rotating annulus flow appear in Ghil (1978).

The qualitative study of differential equations, and the role of bifurcation theory in understanding their geometric, phase-space behavior, is a major theme of 20th century mathematics, from Poincaré and Birkhoff through the Russian school of the 1930s to the present day. No book can be singled out to give both an adequate historical account and an overview of the theory as it stands today. For the prospective reader of this chapter, Guckenheimer and Holmes (1983) could be most useful, serving both as an introduction and a handbook. Applications of this qualitative geometric approach to various fluid flow problems, from experiment through theory, are covered by Swinney and Gollub (1981).

The reduction of a linear system of equations with periodic coefficients to one with constant coefficients, as in Eq. (5.23), is a particular way of looking at perturbation theory for periodic solutions of a nonlinear system of equations. The latter is often called Floquet theory (Arnold, 1983, Section 26; Hartman, 1964, Section 4.6; Iooss and Joseph, 1980, Section 7.6; Stoker, 1950, Section 6.2).

Numerical methods are an essential, indispensable tool in the study of nonlinear problems. Numerical methods for ODEs are analyzed in Isaacson and Keller (1966), Chapter 8. Classical numerical analysis deals with stability and accuracy of numerical solutions to the initial-value problem for finite time. Examples show, however, that a numerical scheme which is stable and accurate in the usual sense for most initial data can behave in a chaotic manner over long times, even for an ODE which has only asymptotically stationary solutions (Ushiki, 1982). A classical analysis of round-off error as random perturbations to the trajectories of an ODE system can be found in Henrici (1963, Sections 5.4 and 5.5).

For certain dynamical systems it can be proved that nice approximation properties exist, in the presence of arbitrary errors along an orbit (the Anosov-Bowen shadowing theorem: Katok, 1981; Ruelle, 1981b). Numerical methods with good stepwise accuracy can be applied safely to the study of such systems (Benettin *et al.*, 1980). Most systems of physical interest, however, do not satisfy conditions known to be sufficient for the requisite approximation results to hold. In using numerical methods for these systems, a little optimism and a lot of skill is required (Lichtenberg and Liebermann, 1983, Section 5.2b). See also Section 6.4 for an example of numerical verification of stability to random errors.

The role of the homoclinic orbit of Figure 5.6 is discussed by Lorenz (1963b), in connection with the transition from the UV' to the SV regime. In all likelihood the SV regime is "preturbulent" (Kaplan and Yorke, 1979), and the transition to the fully chaotic T regime is also associated with the existence of this orbit, which induces a Smale horseshoe in the model's phase flow (Shilnikov, 1965). Recent experimental investigations of this transition in the annulus include Buzyna *et al.* (1984), Farmer *et al.* (1982), Guckenheimer and Buzyna (1983), and Hart (1984).

Returning thus to the annulus and the atmosphere, we noticed in the transition from the Hadley regime H to the single-wave Rossby regime R_1 that the mean flow is modified by the unstable growth of the wave to finite amplitude. Of course, once the wave growth saturates, and a steady situation is reached in this extremely simple flow regime, wave and mean flow no longer interact.

The observation that in certain steady, non-dissipative flow situations small-amplitude waves do not interact with the mean zonal flow goes back to Charney and Drazin (1961). Eliassen and Palm (1960) already noticed that the divergence of a special mean flux is zero in these situations. Various generalizations of Eliassen-Palm fluxes and of related non-interaction or non-acceleration results have been derived recently (Andrews, 1983; Pedlosky, 1979, Section 6.14). The Lagrangian-mean methods used in some of these generalizations are related to Braginsky's method for weakly diffusive, nearly axisymmetric kinematic dynamos (Sections 8.1 and 8.5; see also Andrews and McIntyre, 1978).

It is not clear, as flux formulations become more and more complicated, and the non-acceleration results more and more circumspect, whether the waves and the mean flow are not better off after all being considered together, rather than separately (Rhines, 1981; see also Section 9.1). This appears to be the case in studying the simple vacillatory regimes UV and SV, where two waves exchange energy periodically with each other and with the zonal flow. In the presence of quasi-periodic interactions of two or more modes, the zonal flow is no longer constant in time, as it was in the R_1 regime.

What actually happens in the annulus during vacillation is considerably more complicated (Buzyna *et al.*, 1984; Hart, 1979; Hide and Mason, 1975; Pfeffer *et al.*, 1980) than described in this section; hysteresis effects in the transition between wave numbers occur (wave-number vacillation), the characteristics of amplitude vacillation and of shape, structural or tilted-trough vacillation differ, while various secondary effects make the distinction between steady, periodic, quasi-periodic and truly irregular motion rather difficult.

Typical periods of vacillation for a rotating annulus are of the order of 10-100 rotation periods (Hide and Mason, 1975; Pfeffer *et al.*, 1980), according to the type of vacillation and to the value of the many secondary nondimensional numbers in the experiment. Sometimes the vacillation periods are integer multiples of the basic wave drift period (Hide and Mason, 1975; see also Section 12.7 below), and sometimes they are not.

The vacillation period itself is nearly independent of the upper surface boundary condition. The basic wave drift period for the dominant wave, however, is not. In the presence of a rigid lid, the mean zonal flow rate is small, and the wave drift is correspondingly slow, its period exceeding the vacillation period. When the fluid surface is free, the mean zonal flow is rapid, and the wave drift period is accordingly much shorter than the vacillation period (Hignett, 1983).

The complexity of the situation has stimulated many theoretical analyses aside from the one presented here. Weakly nonlinear theories, based on amplitude equations for a single wave or wave packets (Barcilon and Pfeffer, 1979; Boville, 1980; Hart, 1979; Pedlosky, 1979, Section 7.16; Pedlosky, 1981a), explain various features of the flow, especially near the curve of neutral stability in the regime diagram (curve H/R_1, say in Figure 5.4).

Linear interference (Lindzen *et al.*, 1982) of non-interacting, baroclinically unstable waves appears to match some of the data for the period of amplitude vacillation, given certain assumptions about the *steady* mean flow. But direct measurements show that the mean flow is *not* steady when vacillation occurs. They also suggest that barotropic effects and wave-wave Reynolds stresses are important, at least in experimental tilted-trough vacillation, and hence probably in the atmospheric index cycle (Hart, 1979; Hide and Mason, 1975; Pfeffer *et al.*, 1980; compare also Lorenz, 1963b, Fig. 3, with Pfeffer *et al.*, 1980, Fig. 3b, for instance). Furthermore, Hignett (1983) shows strong evidence for combination tones with a certain structure between the wave frequency and the vacillation frequency in amplitude vacillation, concluding in particular that non-linear wave-mean flow interaction does occur.

Finally, Rand (1982) has formulated a theory of rotating and modulated waves in an abstract setting, connecting the spatial symmetry properties of solutions with their temporal behavior. Such a theory could be applied to the full PDEs governing the motion, subject to certain technical caveats, and hence, eventually, to the annulus experiments and other rotating flows. Unfortunately, Rand's formulation of this theory for doubly-periodic waves with circular symmetry predicts that no frequency locking can occur (see Sections 12.5 and 12.7 for definitions and further references), in contradiction to the experimental results of Buzyna *et al.* (1984), Hide and Mason (1975) and Hignett (1983).

Section 5.4. This section is based on Lorenz (1963a, 1964, 1980) and on part of the considerable literature dealing with the Lorenz system. Guckenheimer and Holmes (1983), Sections 2.3, 5.7 and 6.4, Marsden and McCracken (1976), Sections 4B and 12, and Sparrow (1982) treat the Lorenz system in mathematical and numerical detail, and give many more references.

The paradigm (Kuhn, 1970, Chapter V) of strange attractors as a useful concept in explaining fluid turbulence is due to a large extent to Ruelle and Takens (1971) and to McLaughlin and Martin (1975). Recent reviews include Lanford (1982) and Ruelle (1985).

Aperiodic behavior in biological (May, 1976), chemical (Kuramoto, 1984), electrical and mechanical (Guckenheimer and Holmes, 1983, Sections 2.1 and 2.2, respectively), and geomagnetic (see Section 7.4 below) phenomena is tentatively explained, at least in part, by the same paradigm (see also Sections 12.3 and 12.7 for other types of *conservative*, Hamiltonian chaos). Many details, if not most, still have to be worked out in each application, but the contribution of dynamical system theory to our understanding of aperiodicity in general and fluid dynamical turbulence in particular will certainly be important and lasting.

The Lorenz system can be obtained in many other ways than by model truncation of the convection equations (see also Saltzman, 1962). The theory of Poincaré normal forms (Arnold, 1983, Chapters 5 and 6) can be applied to the PDEs of fluid dynamics (Coullet and Spiegel, 1983) to yield certain canonical systems of ODEs, characteristic of one or more concomitant instabilities of a certain type (see also Iooss and Joseph, 1980, p. 156). Such an approach produces in particular the Lorenz system, and its attendant chaotic behavior, as a set of amplitude equations from weakly nonlinear baroclinic instability theory (Pedlosky and Frenzen, 1980). Another realization of the Lorenz system, as a shunted disc dynamo, will be studied in Section 7.4.

The Lorenz system is in fact a good approximation to what happens in thermal convection only for r close to 1. The full phenomenology of convection experiments, and other theoretical approaches, are reviewed by Busse (1981) and Libchaber (1985).

CHAPTER 6

PERSISTENT ANOMALIES, BLOCKING AND PREDICTABILITY

(with B. Legras)

6.1. Phenomenology of Blocking

The rotating annulus experiments presented in the last chapter reproduce some important qualitative features of the atmosphere's highly complex general circulation. Sequences of daily weather charts show large-scale wave-like patterns retrogressing with respect to a westerly mean zonal flow. These traveling planetary waves have many similarities with the corresponding Rossby wave regime in a rotating annulus. The atmospheric traveling waves, however, are much less regular than the laboratory waves. They are superimposed, moreover, on standing waves forced by zonal asymmetries of the Earth's surface, as discussed toward the end of Section 4.4, and in Section 4.6.

In the atmosphere, the perennially traveling waves sometimes give way to an entirely different regime. The latter is characterized by the presence of a large-scale high pressure pattern which will persist at or near the same location for a duration which may exceed ten days. The pattern usually consists of a strong warm ridge, which occurs initially just off the west coast of a continent. The phenomenon is called *blocking*, since the ridge impedes mean zonal flow and the traveling of cyclone waves through it, deflecting them poleward.

Figure 6.1 provides a relatively rare, but instructive case of a block occurring simultaneously over both the Atlantic and the Pacific oceans. The level lines are monthly mean height contours of the 500mb pressure surface for the month of January 1963. The ridge typical of Pacific blocking extends along the West Coast of North America, while the Atlantic pattern exhibits a characteristic Ω-shape, with a (cold) low (L) positioned just South of the (warm) blocking high (H).

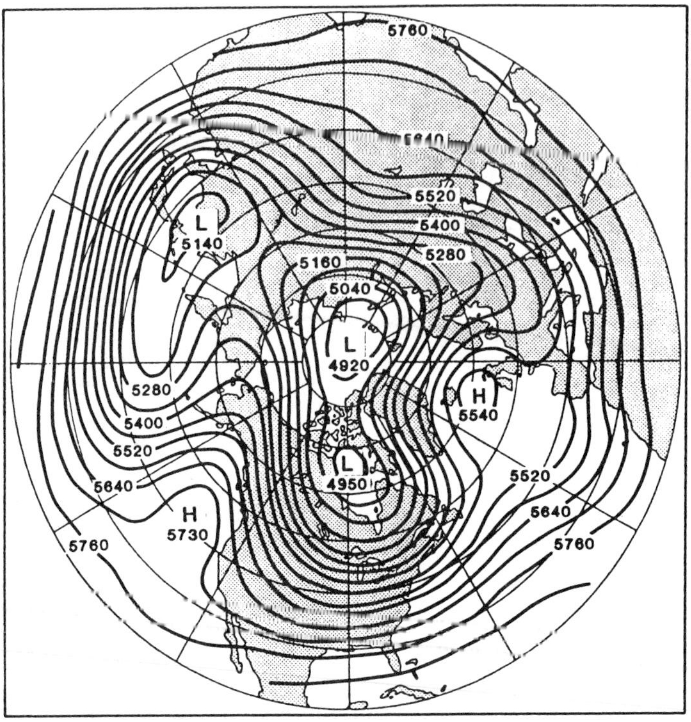

Fig. 6.1. Monthly mean map of 500 mb geopotential heights for January 1963 (courtesy of K. C. Mo). Contour intervals are 20m.

(L) positioned just South of the (warm) blocking high (H).

The contours between 40°E and the dateline show the shape of planetary waves typical of the average winter circulation. This average circulation is a superposition of the forced quasi-stationary waves and of the traveling waves typical of the Rossby regime, the latter appearing through their monthly mean effects on the quasi-stationary flow.

Blocking does occur also in the Southern Hemisphere, but is climatologically less important and hence less well documented there and will only be mentioned in the bibliographic notes (Section 6.6). All subsequent considerations in the main text apply thus to the Northern Hemisphere.

There appears to be a connection between blocking and the index cycles mentioned in Sections 4.6 and 5.3. This connection is somewhat obscured by the fact that blocking is usually confined to a restricted sector, Atlantic or Pacific, while the index reflects the zonally-averaged intensity of the westerly wind inaa given latitude band (Namias, 1950; Rossby *et al.*, 1939; see also Sections 6.4 and 6.6).

Blocking events are most evident during the winter months, with fewer pronounced cases in summer and in the transition seasons. The strongest, most persistent developments do occur usually in winter. Over a period of 15 years (1963-1977), one can estimate that there were 200-400 days, out of a total of 1350 days of winter, on which blocking occurred over either the North Atlantic or the North Pacific. The number of days, and their distribution with season, depends on the exact criteria used to define blocking. Such criteria are not generally agreed upon at present, not even for the Northern Hemisphere.

Although the atmosphere seems to spend on the whole relatively little time in a blocked state, the climatological effects of blocking are of primary importance. The large-scale flow departs considerably from the *normal*, i.e., from the climatic average of the same month over a number of years. Any such departure from the normal in large-scale atmospheric fields is called an *anomaly*.

During blocking events, cold Arctic air is advected over Eastern North America or Europe, while the western part of the continent is cut off from its normal supply of moist maritime air. The attendant negative anomalies of temperature in winter and of precipitation in summer have a significant impact on people's lives and on the economy.

Clearly, anomalies varying in spatial extent, flow pattern and dur - tion occur as part of the atmosphere's normal variability. The striking fact about certain blocking events is their persistence, well beyond the life cycle of a cyclone, and their nearly recurring spatial pattern. The life cycle of traveling cyclones lasts about 5-7 days in mid-latitudes. Any anomaly of duration longer than one week can therefore be termed *persistent*. Blocking is one, but not the only type of persistent anomaly observed in the atmosphere, as we shall see in Sections 6.5 and 6.6.

At this point, the connection with the discussion of *predictability* in Section 4.5 should become clear. For the typical zonal flow regime, with its cyclone waves growing, traveling, interacting and decaying, the predictability limit of synoptic events is probably one-to-two weeks, as many studies indicate. For the slowly varying planetary flow regime a associated with a persistent anomaly, this limit appears to be longer.

It is our hope that the differing stability properties of various atmospheric states can be used to determine the intrinsic predictability of flows starting from these states: synoptic events during a relatively quiescent blocking sequence, once established, might be more predictable than during a zonal flow sequence, with its high instability. Additional classes of states, intermediate in stability between blocked and zonal, will perhaps be diagnosed, and their predictability assessed.

Due to the *zonally asymmetric* nature of blocking, it is reasonable to assume that asymmetries of the atmosphere's lower boundary play a role in the phenomenon. We shall study, therefore, the effect of forcing by surface topography and by land-ocean thermal contrasts on planetary-scale flows. The line of inquiry will follow that in the previous chapter. The simplest models capable of reproducing the distinctive features of the phenomena of interest will be considered. The intricate behavior of these models will be examined by the method of *successive bifurcations*.

Fair warning is in order at this point: the subject matter of this chapter is a topic of active research. Questions are far from settled. Alternatives to our point of view are possible and lead to important results which complement the ones presented here. The purpose of writing this chapter is to stimulate that elusive creature, the interested reader, to roll up his or her sleeves and join the enterprise under way. We can only hope that they shed as much light and add as little heat as possible.

In Section 6.2, we introduce interactions between wave solutions of the linearized barotropic vorticity equation, and the effect of bottom topography on such waves. The analogies between resonances in wave-wave and wave-topography interactions and the usual concepts of resonance in mechanical oscillators are outlined. A numerical experiment is described in which a resonant triad of barotropic Rossby waves, forced by topography, was obtained. The flow pattern associated with this triad resembles blocking and persists for over ten days.

In Section 6.3, we pass on to low-order spectral truncations of the nonlinear barotropic vorticity equation in a β-channel, with simplified forcing and dissipation. Two levels of truncation are considered. For a three-mode version of the model, three stationary solutions coexist at the same value of the forcing parameter. Two of these are stable, one resembling zonal, the other blocked flow. The instability of the third solution, which separates the two stable ones, is due to the interaction of the mean flow with the orography.

In a six-mode version of the model, the blocked solution is no longer stationary, but transfers its stability to a periodic solution with small amplitude. Results for both spectral versions of the model are confirmed by calculations with a finite-difference discretization of the basic equation, with moderate resolution.

In Section 6.4, we consider the same vorticity equation on the sphere, discretizing it into 25 spherical harmonics. This permits the multiple equilibria of the previous section to appear for more realistic values of the forcing, and to bear a more marked resemblance to synoptically defined zonal and blocked flows.

Wave-wave interactions influence strongly the stability properties of the equilibria and the time evolution of nonequilibrium solutions. Time-dependent solutions show persistent sequences which occur in the phase-space vicinity of the zonal and blocked equilibria. Composite flow patterns of the persistent sequences are similar to the equilibria nearby, which permits the unambiguous definition of quasi-stationary flow regimes, zonal and blocked, respectively. The number of episodes of blocked or zonal flow decreases monotonically with their duration, in agreement with observations.

The statistics of transitions between the two types of planetary flow regimes are computed from the model's deterministic dynamics. These transitions, called breaks in statistical-synoptic long-range forecasting, are shown to be influenced by changes in model parameters.

In Section 6.5, we discuss the significance of these results in explaining the midlatitude atmosphere's low-frequency variability. The forcing and dissipation parameters of Sections 6.3 and 6.4 correspond to the effect of changing boundary data for the real atmosphere, such as sea-surface temperature anomalies, snow cover or soil moisture. A change in these data does not call forth a simple, unique atmospheric response, but rather a change in the probability distribution of possible flow regimes and of expected persistence times in each regime.

The bibliographic notes of Section 6.6 start with references on the problems of long-range forecasting, followed by papers on blocking and on the index cycle, both classical and recent, as well as papers on the appearance of blocking events in large-scale numerical models. Papers on resonant triad interactions for surface gravity waves and planetary waves are cited together with those on statistical studies of turbulence, with and without topography. The role of thermal forcing in the maintenance of quasi-stationary waves is mentioned, along with other studies of resonant wave response to lower boundary effects.

Baroclinic effects, regional aspects of blocking and comparisons with observations are reviewed next. Some of the literature on transition to chaotic flow is cited, and the study of phase-space structure in the chaotic regime is emphasized over the study of the possible transitions to such a regime. The complementary nature of the local and global points of view, in phase space and in physical space, is discussed, with additional references. The chapter concludes with a brief review of the 30-50 day oscillation in the global atmospheric circulation.

6.2. Resonant Wave Interactions and Persistence

We introduce here some of the heuristic ideas which underlie the fluid dynamical content of this chapter. The models we deal with are all based on the barotropic *potential vorticity equation* (PVE), Eq. (3.11), to which some form of forcing and dissipation is added. We start by considering a β-plane approximation of this equation, similar to (3.12),

$$\partial_t \nabla^2 \psi + J(\psi + Uy, \nabla^2 \psi) + \beta \psi_x = F(x,y). \tag{6.1}$$

Here $U = \text{const.}$ is a mean zonal flow, the departure (u,v) from which is nondivergent, but not necessarily small, $(u,v) = (-\psi_y, \psi_x)$, ψ being its stream function. J and ∇^2 are the Jacobian and Laplacian with respect to the Cartesian horizontal coordinates x and y, and $\beta = f_y$ is assumed constant, f being the Coriolis parameter. The flow domain is an x-periodic channel of width B and period L. The meridional velocity v is zero at the rigid walls $y = 0$ and $y = B$.

In the absence of forcing, $F = 0$, a flow governed by Eq. (6.1) with the given boundary conditions has constant *kinetic energy* E,

$$E \equiv \iint (\nabla\psi)^2 dx\, dy = \text{const}, \tag{6.2a}$$

and constant *enstrophy* or mean square vorticity, Z,

$$Z \equiv \iint (\nabla^2\psi)^2 dx\, dy = \text{const}. \tag{6.2b}$$

To see this, it suffices to multiply (6.1) by ψ and $\nabla^2\psi$, respectively, and integrate by parts over the periodic β-channel. The property of conserving both energy and enstrophy plays an important role in two-dimensional flows and one of its consequences will be encountered in the immediate sequel.

The forcing F considered at first corresponds to the direct effect of bottom topography on the mean flow in (3.12), $F = -fUh_x$. The interaction of the finite-amplitude perturbation flow (u,v) with the topography $h(x,y)$ is neglected in this section, and will be reintroduced in Section 6.3. Our considerations about waves forced by topography, and their interactions, will be fairly general and speculative in this section. They only serve as background and motivation for the rest of the chapter, in which they will be developed and rendered more precise.

The stream function ψ can be expanded, according to the recipe in Section 5.2, as a double Fourier series,

$$\psi = \sum_{\underset{\sim}{k}} \left(\psi_{\underset{\sim}{k}} e^{ik_x x} + \psi^*_{\underset{\sim}{k}} e^{-ik_x x}\right) \sin k_y y, \tag{6.3a}$$

where $(\ )^*$ denotes the complex conjugate of a quantity ( ). The wave vector $\underset{\sim}{k} = (k_x, k_y)$ is defined as

$$\underset{\sim}{k} - 2\pi(m/L,\ n/2B), \tag{6.3b}$$

with m and n integer wave numbers. The sign of m and n is immaterial, and wave numbers will be defined only up to sign throughout this section.

The spectral form of Eq. (6.1) is

$$\{d_t + ik_x(U-\beta/k^2)\}\psi_{\underset{\sim}{k}} = i \sum_{\underset{\sim}{p},\underset{\sim}{q}} c_{\underset{\sim}{kpq}} \psi_{\underset{\sim}{p}} \psi_{\underset{\sim}{q}} - k^{-2} F_{\underset{\sim}{k}}, \tag{6.4}$$

where $k^2 = k_x^2 + k_y^2$. The summation in (6.4) occurs in fact only over wave triplets $(\underset{\sim}{k},\underset{\sim}{p},\underset{\sim}{q})$ which satisfy the *selection rule*

$$\underset{\sim}{k} + \underset{\sim}{p} + \underset{\sim}{q} = 0. \tag{6.5a}$$

For these interacting triplets, the coefficients $c_{\underset{\sim}{kpq}}$, cf. (5.8b), are nonzero and given by

$$c_{\underset{\sim}{kpq}} = (1/2)(p^2-q^2)(p_x q_y - q_x p_y)/k^2, \tag{6.5b}$$

with $\underset{\sim}{p} = (p_x,p_y)$, $\underset{\sim}{q} = (q_x,q_y)$, $p^2 = p_x^2+p_y^2$ and $q^2 = q_x^2+q_y^2$; for all other triplets, the interaction coefficients are zero. Wave triplets which satisfy (6.5a) are called *triads*.

Forced standing waves and resonant triads. As our starting point, we consider the response to a weak forcing at wave number $\underset{\sim}{p}$ with amplitude $F_{\underset{\sim}{p}}$. The response to this forcing is assumed to be confined initially to the same wave number, and to be approximated by the linearized form of Eq. (6.4), namely

$$(d_t + ip_x c_p)\psi_{\underset{\sim}{p},\ell} = A_{\underset{\sim}{p}}, \tag{6.6a}$$

with

$$c_p = U - \beta/p^2 \tag{6.6b}$$

and $A_{\underset{\sim}{p}} = -p^{-2}F_{\underset{\sim}{p}}$. The linear response $\psi_{\underset{\sim}{p},\ell}$ is given by

$$\psi_{\underset{\sim}{p},\ell}(t) = \psi_0\, e^{-ip_x c_p t} + (A_{\underset{\sim}{p}}/ip_x c_p)(1-e^{-ip_x c_p t}), \tag{6.7}$$

where ψ_0 is the initial amplitude of the wave. Both p_x and c_p are real, and the free solution $\psi_0 e^{-ip_x c_p t}$ corresponds to a traveling wave, cf. (6.3), with phase speed c_p. By choosing ψ_0 to equal $A_{\mathbf{p}}/ip_x c_p$, the traveling part of (6.7) can be eliminated, retaining only the standing part. In the atmosphere, the dissipation not present in (6.6) will damp the traveling wave, justifying our retention of the standing part only.

The linearly forced *standing wave* has spectral coefficient $\psi_{\mathbf{p},s}$, whose amplitude is

$$|\psi_{\mathbf{p},s}| = \left| \frac{F_{\mathbf{p}}}{p_x(Up^2 - \beta)} \right| . \tag{6.8}$$

The phase of this wave is orthogonal to that of the forcing, so that in physical space the same amount of energy is absorbed by the wave over one quarter wavelength as is returned over the adjacent quarter wavelength.

The linear amplitude (6.8) is a fair approximation to the amplitude of the true standing $\mathbf{p}$-wave excited by the topography, provided $Up^2-\beta$ is not too small. In nearly-resonant cases, $|Up^2-\beta| \ll \beta$ say, the approximation (6.8) clearly breaks down, and saturation effects have to limit the amplitude of the wave. These effects include dissipation and nonlinear interaction with other waves, cf. (6.4).

To relate this concept of resonance with the familiar one from linear vibration theory, recall that the equation

$$\ddot{x} + \omega_0^2 x = A \cos \omega t, \tag{6.9a}$$

with $(\ )^{\cdot} = d_t(\ )$, exhibits *near resonance* when $|\omega-\omega_0| \ll \omega_0$. If $\psi_{\mathbf{p},\ell} = y$, $\omega_0 = p_x c_p$ and $i\omega_0 A_{\mathbf{p}} = A$, then (6.6) becomes

$$\dot{y} + i\omega_0 y = A/i\omega_0 . \tag{6.9b}$$

This can be transformed into the usual linear oscillator equation (6.9a) by letting $\omega = 0$ and

$$\dot{x} - i\omega_0 x = y . \tag{6.9c}$$

Thus constant forcing can be thought of as periodic forcing of infinite period. Physically, for near-resonant wave response to topography to occur, the phase velocity of the wave has to be such, cf. (6.6b) and (6.8), that the wave be close to the periodicity of the forcing, i.e., nearly stationary.

Notice moreover that, for t fixed and $c_p \to 0$, rather than c_p fixed and $t \to \infty$, the forced response in (6.7) is proportional to t, since $(1 - e^{-ip_x c_p t})/c_p \to ip_x t$. Thus the stationary response here exhibits the same *secular growth* in time as the amplitude of the oscillatory response in (6.9a) for $\omega \to \omega_0$. In the case of a simple resonance, like here, secular growth is linear in t; for multiple resonances, the growth will be polynomial in t.

Consider now two standing waves with wave vectors $\underset{\sim}{p}$ and $\underset{\sim}{q}$, their amplitudes given by (6.8). The two forced waves will interact trilaterally with another wave $\underset{\sim}{k}$, which satisfies the triad selection rule (6.5a). The wave $\underset{\sim}{k}$ is assumed to be free of direct forcing, $F_{\underset{\sim}{k}} = 0$, and no waves but $\underset{\sim}{k}$, $\underset{\sim}{p}$ and $\underset{\sim}{q}$ are present, at least initially.

In the absence of forcing, conservation of energy and of enstrophy for the spectral equations (6.4) becomes

$$E \equiv \sum_{\underset{\sim}{k}} k^2 \psi_{\underset{\sim}{k}} \psi^*_{\underset{\sim}{k}} = \text{const.} \tag{6.10a}$$

and

$$Z \equiv \sum_{\underset{\sim}{k}} k^4 \psi_{\underset{\sim}{k}} \psi^*_{\underset{\sim}{k}} = \text{const.} \tag{6.10b}$$

In fact, energy and enstrophy are conserved separately within each interacting triad, if all other waves have null amplitudes. Applying Eqs. (6.10) to the isolated triad $(\underset{\sim}{k},\underset{\sim}{p},\underset{\sim}{q})$, it can be shown that the energy transfer from $\underset{\sim}{p}$ and $\underset{\sim}{q}$ to $\underset{\sim}{k}$ will be efficient, exciting the initially quiescent free wave, provided k lies between p and q, say

$$p < k < q. \tag{6.11}$$

We shall assume this to be the case, assuming further that p, q and hence k are all of the same order of magnitude, $O(\kappa)$, and that the common order of magnitude of $\psi_{\underset{\sim}{p}}$ and $\psi_{\underset{\sim}{q}}$ is ψ_κ.

The triad $(\underset{\sim}{k},\underset{\sim}{p},\underset{\sim}{q})$ will interact resonantly, dominating other possible interactions, provided the *interaction number* $R(\underset{\sim}{k};\kappa) = \psi_\kappa \kappa / c_{\underset{\sim}{k}}$ is large. R is the ratio of particle velocity, or typical fluid velocity, $\psi_\kappa \kappa$, to the phase velocity $c_{\underset{\sim}{k}}$ of the free wave in the triad. The condition $R(\underset{\sim}{k};\kappa) >> 1$ is equivalent to the velocity of the fluid described by our triad being much larger than the phase velocity of the traveling wave,

$$\psi_\kappa \kappa >> c_{\underset{\sim}{k}}. \tag{6.12}$$

This condition has a simple heuristic interpretation, which is the nonlinear counterpart to the discussion of Eqs. (6.9a,b,c), and which follows presently.

The nonlinear *interaction time* T_N can be estimated from dimensional considerations in Eq. (6.4) to be

$$T_N \cong 1/(\psi_\kappa \kappa^2). \tag{6.13a}$$

In our case, this is the time it takes the free wave $\mathbf{k}$ to grow to the common size ψ_κ, before it modifies considerably the forced waves $\mathbf{p}$ and $\mathbf{q}$, and before other free waves are generated.

On the other hand, the linear effects of advection by the mean flow U and of meridional changes in planetary vorticity $\beta = f_y$ cause the free wave $\mathbf{k}$ to travel at phase speed $c_k = U - \beta/k^2$, shifting away from the sustained, phase-synchronized excitation by the standing waves, $\mathbf{p}$ and $\mathbf{q}$. The characteristic time T_L of this effect, which limits growth, is the linear period of the free wave, given by

$$T_L \cong 1/(\kappa c_k). \tag{6.13b}$$

For the nonlinear excitation to be effective, the phase of the free wave has to be nearly constant for a time of order T_N,

$$T_N/T_L \ll 1. \tag{6.13c}$$

Substituting (6.13a,b) into (6.13c) yields the requirement (6.12) that $R(\mathbf{k};\kappa) \gg 1$. Notice that the interaction number $R(\boldsymbol{\ell};\kappa)$ may be very different for a wave $\boldsymbol{\ell}$ close to $\mathbf{k}$ and satisfying (6.5), due to the rapid variation of c_ℓ with ℓ.

If the atmospheric mean flow U is such that $c_k = 0$, and the $\mathbf{k}$-wave is stationary, it is clear that this wave will initially grow by nonlinear interaction with topographic waves $\mathbf{p}$ and $\mathbf{q}$ satisfying (6.5). It is also to be suspected that other free waves with small phase velocities might be excited rather rapidly, thus complicating the picture. However, there is hope that the triad $(\mathbf{k},\mathbf{p},\mathbf{q})$ might be relatively stable for some time, *after* the growth of $\mathbf{k}$, and *before* further complication takes over, destabilizing the configuration.

A persistent blocking pattern. Numerical experiments of Egger (1978) have shown that this is so, for a combination of waves $\mathbf{k}$, $\mathbf{p}$, $\mathbf{q}$ with two important properties: (i) $\mathbf{p}$ and $\mathbf{q}$ correspond to large spectral components of Northern Hemisphere topography, and (ii) the fully developed triad exhibits some of the synoptic, phenomenological features of blocking. Egger integrated numerically Eq. (6.1), using a spectral expansion in the x-direction, truncated at $k_x \leq 3$, and a finite-difference discretization in the y-direction, using 12 grid points between the walls $y = 0, B$. Such a discretization also corresponds roughly to $k_y \leq 3$. The coefficients of the x-expansion, represented at the 12 points, were advanced in time by Heun's scheme, a single-step, second-order accurate numerical method for systems of ODEs. This scheme also includes, when applied to a discretized PDE such as (6.1), numerical dissipation, i.e., selective damping of waves with higher spatial wave numbers.

The topography was restricted to $\mathbf{p} = (1,1)$ and $\mathbf{q} = (3,1)$, where we use the integers (m,n) of (6.3b) to designate the corresponding components $(\pm k_x, \pm k_y)$ of any wave vector $\mathbf{k}$. To obtain the high-low pair of pressure cells characteristic of Atlantic blocking (Figure 6.1), the third member of the triad expected to build the "block" should be $\mathbf{k} = (2,2)$. This triad satisfies both (6.5a) and (6.11). The mean zonal flow U was chosen so as to render $\mathbf{k}$ linearly stationary, $U = \beta/k^2$, which for $\beta = 1.6 \times 10^{-11} m^{-1} s^{-1}$, its value at 45°N, and $L = 2B = 1.6 \times 10^7 m$, yields $U = 13$ m/s.

The height of the topography was taken equal to 1000 m in $\mathbf{p}$ and to 300 m in $\mathbf{q}$. Initial values of $\psi_{\mathbf{p}}$ and $\psi_{\mathbf{q}}$ were chosen equal to those given by (6.8), with all other waves initially zero. The evolution of the wave components which were not negligible is shown in Figure 6.2.

The Fourier components shown in the figure are $\psi_{mnc} = \text{Re}\, \psi_{\mathbf{k}}$ and $\psi_{mns} = -\text{Im}\, \psi_{\mathbf{k}}$, for $\mathbf{k} = (m,n)$. The forced waves are (1,1) and (3,1). The large-amplitude free waves are (2,2) and (1,3), the latter being

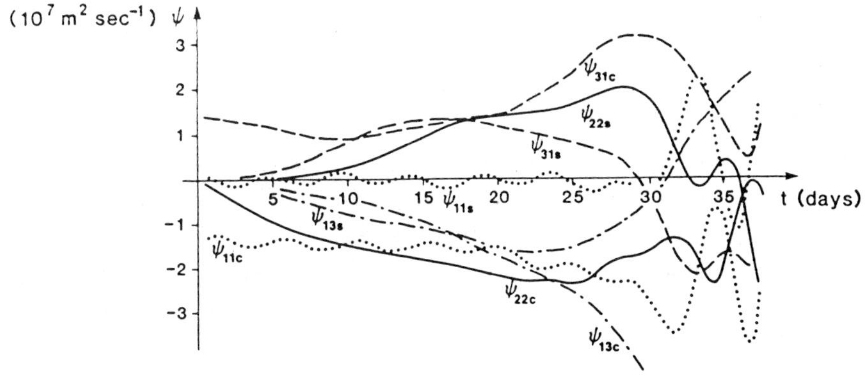

Fig. 6.2. Evolution of two traveling waves interacting with two forced standing waves (after Egger, 1978).

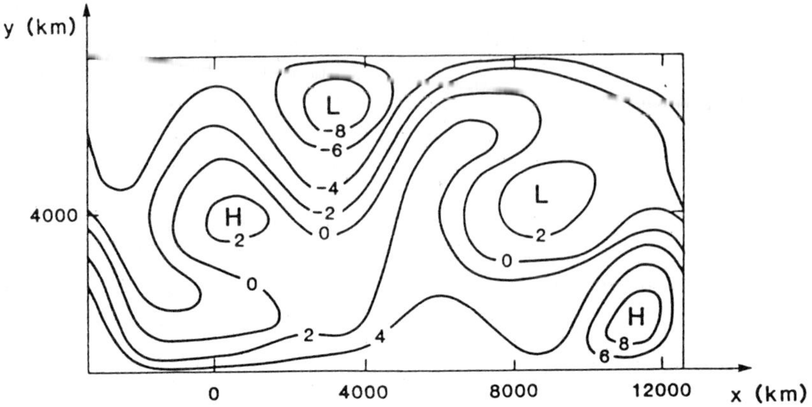

Fig. 6.3. Persistent blocking structure associated with four interacting planetary waves (after Egger, 1978).

almost stationary as well; in fact its linear phase velocity is c_{13} = 3 m/s << U.

An amplitude $O(10^7 m^2 s^{-1})$ is typical of forced standing waves in Northern Hemisphere winter. Wave (2,2) grows slowly to such an amplitude over the first ten simulated days or so, and then flattens out till about day 26. Between days 15 and 25, although ψ_{13c} grows steadily, all other waves are nearly constant in amplitude, $(\psi^2_{mnc} + \psi^2_{mns})^{1/2}$, and in phase, $\tan^{-1}(\psi_{mns}/\psi_{mnc})$.

Figure 6.3 shows the synoptic pattern at day 25, which is actually quite similar to that at day 20. The plot gives contour lines of the perturbation stream function ψ in units of $10^7 m^2 s^{-1}$.

The pattern at $x = 0$, with its well-developed high and the splitting of the westerly flow around it, is strikingly similar to the Atlantic Ω-block in Figure 6.1. A second ridge appears at $x \cong 6000$ km. This pattern did persist in the experiment, with little change, for over ten days. After day 25, it is clear from Figure 6.2 that ψ_{13c} grows rapidly and, after day 30, wave (1,1) is no longer standing still. At that time, all waves interact strongly with each other, while additional waves are generated and cannot be neglected any longer.

Egger's work has shown that nonlinear wave-wave interactions are potentially important in establishing a blocking pattern, and are consistent with its persistence for one-to-two weeks. It also allows us to conclude that certain aspects of blocking activity can be studied in a barotropic model.

Additional work, mentioned in Section 6.6, has contributed to understanding the two essential aspects of the blocking mechanism outlined in this section: the existence of standing planetary waves forced by topography and by land-sea thermal contrasts, on the one hand, and their resonant interaction with slowly traveling free waves, on the other. The amplitude of nonresonantly forced standing waves is determined essentially by the forcing, while that of the free waves is determined by the competition of nonlinear interactions with dissipation. Nonlinear interactions, as well as dissipation, also play a role in limiting the amplitude of standing waves forced close to resonance, as we shall see in the next section.

6.3. Multiple Stationary States and Blocking

In the previous section we have introduced the weakly nonlinear dynamics of growing and interacting free and stationary waves, and discussed their potential relevance to the establishment and duration of persistent anomalies. We are ready now to pick up the thread of successive bifurcations which we hope will lead us from the perturbation of mean zonal flows to the fully nonlinear, global picture of different classes of planetary flows with varying stability properties.

Zonal and blocked flow equilibria. The first step on the successive bifurcation route, as we know already from Section 5.4, is multiple steady states. Charney and DeVore (1979) have shown that it is possible to obtain two stable stationary solutions of the barotropic PVE, one of which has synoptic features of zonal flow, the other of blocking. Their starting point was the equation

$$\partial_t(\nabla^2 - L_R^{-2})\psi + J\{\psi, (\nabla^2 - L_R^{-2})\psi + f_0 h/H_0 + \beta y\} = \alpha\nabla^2(\psi^* - \psi). \qquad (6.14)$$

The free-surface term in the potential vorticity, $-L_R^{-2}\psi$, neglected in (6.1), has been reintroduced, with $L_R = \sqrt{gH_0}/f_0$ being the Rossby radius of deformation, cf. Eqs. (3.4,3.5,3.11). Interaction of the total flow $(u,v) = (-\psi_y, \psi_x)$, $\psi = \psi(x,y,t)$, with the topography $h = h(x,y)$, as given by $J(\psi, f_0 h/H_0)$ is allowed in this model.

An (x,z) cross-section through the flow is shown in Figure 6.4. This figure is closely related to Figure 2.1, with some additions and slight changes of notation, cf. also Eq. (3.1).

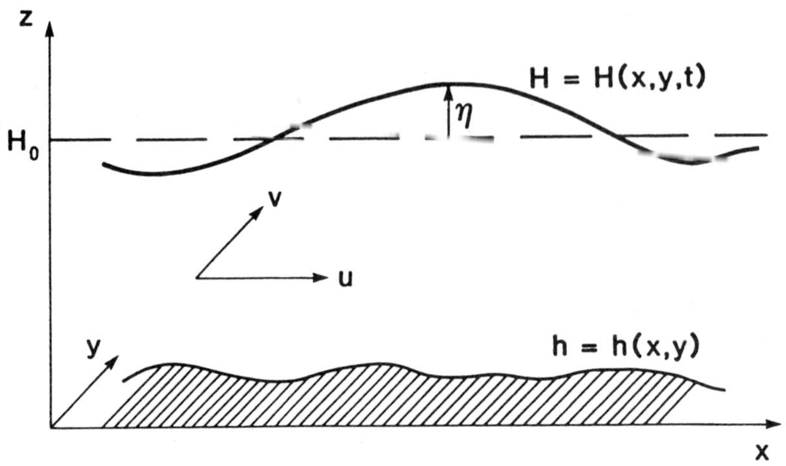

Fig. 6.4. Quasi-geostrophic, barotropic flow over topography.

The height of the free surface is H, with a typical mean value of H_0; $\eta = H-H_0$ is the deviation of H from H_0, which has a small, but finite amplitude. By the quasi-geostrophic approximation, $\psi = g\eta/f_0$, where f_0 is a typical value of the Coriolis parameter, taken here at mid-channel. The width of the β-channel is $B = \pi L$, with the x-period equal to πL as well.

The left-hand side of (6.14) conserves the potential vorticity $Q = (\nabla^2 - L_R^{-2})\psi + f_0 h/H_0 + \beta y$ for each fluid particle. It also conserves the sum of the kinetic energy K, with density $(H-h)(u^2+v^2)/2$, and of the potential energy P associated with the free surface, whose density per unit area is $g\eta^2/2$.

In this nonlinear formulation, no distinction is made between mean flow and perturbation. The topography enters here as a variable coefficient, which leads to exchanges of momentum and energy between the waves present, without creating or destroying either. The zonal flow participates fully in these exchanges.

The *forcing* $\alpha\nabla^2\psi^*$ represents, within this barotropic model, the thermal driving of midlatitude circulation by the pole-to-equator temperature contrast, as discussed in Sections 4.2 and 4.6 (Eqs. (4.2) and (4.10); notice that ψ^* here and below does *not* denote complex conjugation). The *dissipation* $-\alpha\nabla^2\psi$ represents frictional effects in the planetary boundary layer.

A spectral expansion similar to (5.11) is applied, retaining the modes

$$F_A = \sqrt{2}\cos(y/L), \tag{6.15a}$$

$$F_K = 2\cos(nx/L)\sin(y/L), \quad F_L = 2\sin(nx/L)\sin(y/L). \tag{6.15b,c}$$

F_A is a zonal mode, while F_K and F_L are two waves with the same, arbitrary zonal wave number *n*. The topography is restricted to wave mode K, with nondimensional coefficient h_0',

$$h/H_0 = 2h_0'F_K, \tag{6.15d}$$

and $H_0 = 10^4$m. The forcing is restricted to zonal mode A,

$$\psi^* = L^2 f_0 \psi_A^* F_A, \tag{6.15e}$$

with $\pi L = B = 5 \times 10^6$m, $f_0 = 10^{-4}s^{-1}$. The time *t* is nondimensionalized by f_0^{-1}, the coordinates *x* and *y* by L. The resulting nondimensional value for α is chosen equal to 10^{-2}, which corresponds to a linear relaxation time of ψ to ψ^* of about eleven days. The nondimensional value of L_R^{-2}, $\lambda^{-2} = f_0^2L^2/gH_0$, was actually taken equal to zero, thus replacing the free surface by a solid lid.

The nondimensional spectral system obtained from Eqs. (6.14,6.15) is

$$\dot{\psi}_A = h_0\psi_L - k(\psi_A - \psi_A^*), \tag{6.16a}$$

$$\dot{\psi}_K = -(\alpha_n\psi_A - \beta_n)\psi_L - k\psi_K, \tag{6.16b}$$

$$\dot{\psi}_L = (\alpha_n\psi_A - \beta_n)\psi_K - h_n\psi_A - k\psi_L. \tag{6.16c}$$

The coefficients are all positive with $k = \alpha = 10^{-2}$; h_0 and h_n are proportional to h_0', and β_n to β. The proportionality constants depend on the mode $m = 0,1$, and on the arbitrary zonal wave number *n*; *m* and *n* also determine α_n, with $m = 1$. System (6.16) is of type (5.26): the quadratic terms arising from advection are conservative. Moreover, the linear terms arising from the β-effect and from the topography are also conservative. The forcing is $k\psi_A^*$, and the dissipative terms are proportional to *k*.

The stationary solutions of (6.16) are given by

$$\psi_L = -\frac{h_nk}{k^2+b^2}\psi_A, \tag{6.17a}$$

$$\psi_K = \frac{h_nb}{k^2+b^2}\psi_A, \tag{6.17b}$$

and ψ_A is one of the roots of the cubic equation

$$F(\psi_A) = \psi_A^*, \tag{6.17c}$$

where

$$F(\psi_A) = \psi_A\left\{1 + \frac{h_0 h_n}{k^2 + b^2}\right\} \tag{6.17d}$$

and

$$b(n;\psi_A) = \alpha_n\psi_A - \beta_n. \tag{6.17e}$$

A graph of the solutions of (6.17c) is shown in Figure 6.5.

Notice also that for ψ_A, and hence b, fixed, Eqs. (6.16b,c) represent a linear, forced and damped oscillator with one degree of freedom. In the terminology of point-mass mechanics, ψ_k can be interpreted as the position and ψ_L as the velocity of a particle. The forcing is constant in time, as in Eqs. (6.6-6.8). The resonance condition for this oscillator, cf. (6.9), is $b(n;\psi_A) = 0$.

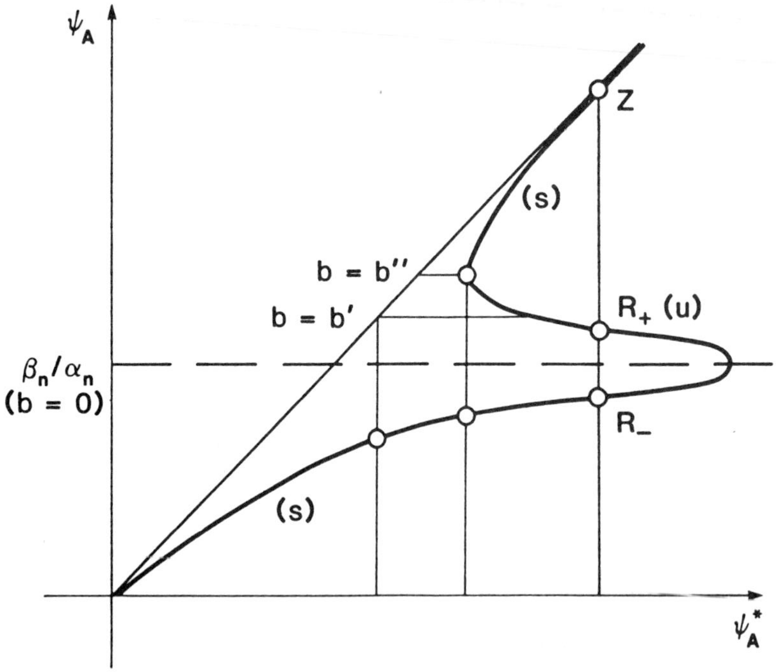

Fig. 6.5. Multiple equilibria of a three-mode quasi-geostropic system with forcing and orography (after Charney and DeVore, 1979).

For an appreciable range of the zonal forcing ψ_A^*, there are three equilibrium solutions (ψ_A, ψ_K, ψ_L), corresponding to the three values of ψ_A in Figure 6.5. Two of them, denoted by R_- and R_+ in the figure, are close to the quasi-linear resonance $b(n) = 0$ of the n-waves K and L. For these two equilibria, the wave component ψ_L is large compared to the zonal component ψ_A, and ψ_K is small, cf. (6.17a,b). The third equilibrium solution, Z in the figure, is an almost zonal flow close to ψ_A^*. The latter solution is very little influenced by the interaction of the zonal flow with the topography, while the former two are dominated by it.

The stream function fields for solutions Z and R_- are shown in Figures 6.6a and 6.6b, respectively. The contours of ψ are shown in

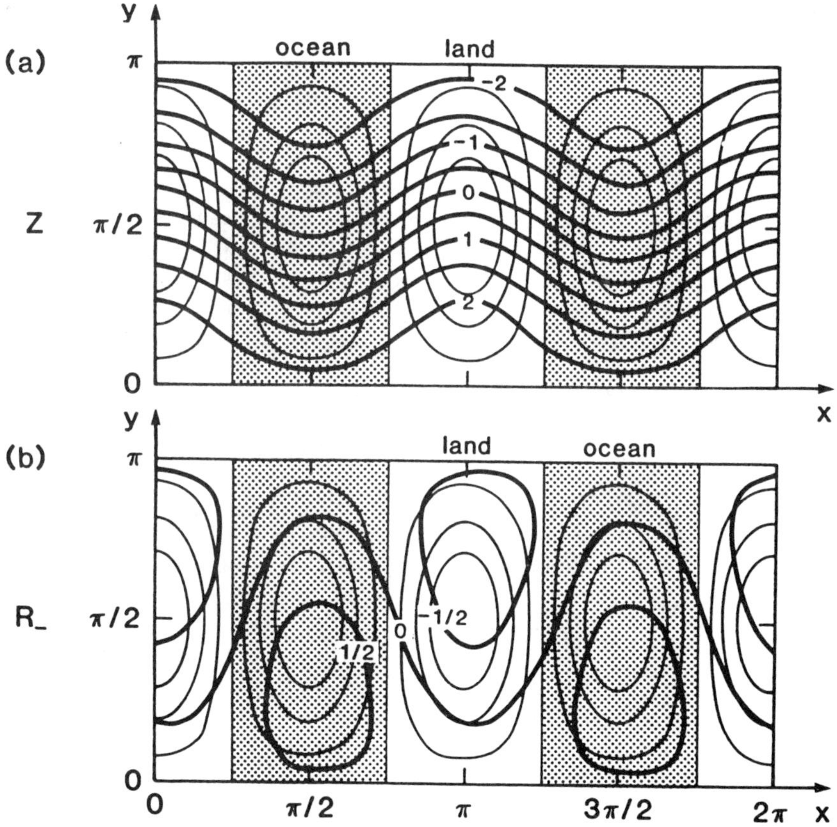

Fig. 6.6. Flow patterns of Z equilibrium (a) and R_- equilibrium (b) (after Charney and DeVore, 1979).

heavy lines, the topographic heights in light lines. The figure corresponds to $n = 2$, $h_0' = 0.2$ and $\psi_A^* = 0.2$. The regions of negative topography -- the "oceans" -- are shaded. Recall that the topography is restricted to the K-wave, and that the L-wave has a 90° phase lag with respect to it.

The flow field in Z is nearly zonal, with the weak wave almost entirely in phase with the topography. In R_-, by contrast, there is a strong meridional component of the flow. Moreover, the strong wave is shifted considerably westward, its K-component being almost zero near resonance, cf. (6.17b). This feature is in agreement with the position of blocking ridges in Figure 6.1.

Stability analysis of the resonance. To study the stability of the stationary flows R_+, R_- and Z, one linearizes (6.16) about the steady states given by (6.17). The resulting linear system of ODEs has coefficient matrix $M = M(\psi_A, \psi_K, \psi_L)$,

$$M = \begin{pmatrix} -k & 0 & h_0 \\ \alpha_n \psi_L & -k & -h \\ \alpha_n \psi_K - h_n & b & -k \end{pmatrix}. \tag{6.18a}$$

The eigenvalues σ of this matrix determine the linear stability of the equilibria $R_\pm$ and Z; they are given by the characteristic equation

$$\det\,(M - \sigma) = 0. \tag{6.18b}$$

It would appear that (6.18b), being only a cubic equation, can be easily solved in closed form, by Cardan's formulae. These formulae, however, are rather complicated and do not allow one to deduce easily the properties of the roots. Instead, it is simpler to carry out the complete stability analysis by a perturbation approach, since the dissipation coefficient k is small. This approach is similar to the perturbative study of the cubic dispersion relation in Sections 2.2 and 2.3 (Figures 2.5 and 2.6). The perturbation parameter here will be $\varepsilon = k/b$, and special attention has to be given to the neighborhood of the resonance, where b is also small. The subscript n will be dropped below from α, β and h, for convenience.

In the limit $k \to 0$, independently of $b \neq 0$, $\psi_L = 0$, $\psi_K = h\ {}_A/b$, and ψ_A is given by the real roots of $F^{(0)}(\psi_A) = \psi_A^*$, where

$$F^{(0)}(\psi_A) = (1 + hh_0 b^{-2})\psi_A. \tag{6.17'}$$

In this limit, the graph of ψ_A as a function of ψ_A^* in Figure 6.5 has a horizontal tangent at $\psi_A = \beta/\alpha$, where $b = 0$.

The stability of the solutions is obtained by solving (6.18) with $k = 0$, i.e.,

$$\sigma^3 + b^3\sigma - h_0(\alpha\psi_k - h)\sigma = 0. \tag{6.19a}$$

The roots are given by $\sigma = 0$ and by $\sigma^2 = -(b^2 - hh_0\beta b^{-1})$. The first root yields marginal stability, while the other two change from real to complex at $b = 0$ and at $b = b'$, where

$$b' = (hh_0\beta)^{1/3} \tag{6.19b}$$

lies on the R_+ branch, since $F'_{(0)}(\psi_A) < 0$ at this point.

For $0 < b < b'$, the pair of nonzero roots of (6.19a) is real and the two roots have opposite signs, so that this portion of the R_+ branch is unstable, independently of the value of k. For b outside this interval, the pair of nonzero roots is purely imaginary, yielding marginal stability to zeroth order in $\varepsilon = k/b$. In particular no change of stability to this order occurs at the turning point $b = b''_{(0)}$, given by $F'_{(0)}(\psi_A) = 0$, where the branches $Z^{(0)}$ and $R_+^{(0)}$ are smoothly joined.

It is necessary therefore to proceed to the study of stability to first order in ε. Writing $\sigma_j = \sigma_j^{(0)} + \varepsilon\sigma_j^{(1)} + O(\varepsilon^2)$ for $j = 1,2,3$, we have $\sigma_1^{(0)} = 0$ and $\{\sigma_{2,3}^{(0)}\}^2 = -(b^3 - b'^3)b^{-1}$. No correction is needed for $\sigma_{2,3}$ in the interval $0 < b < b'$, but $\sigma_{2,3}^{(1)}$ have to be determined outside this interval, and $\sigma_1^{(1)}$ everywhere.

Expanding (6.18) in ε near $\sigma = 0$ yields to $O(\varepsilon^2)$

$$\sigma_1 = -k\left\{1 + \frac{1+b/\beta}{1-(b/b')^3}\right\} .$$

The expression above changes sign for $b = b''$, where $F'(\psi_A) = 0$. For $b > b''$, along the Z branch, $\sigma_1 < 0$, while for $b' < b < b''$, along the R_+ branch, $\sigma_1 > 0$. Hence this portion of the R_+ branch is also unstable. Another change of sign occurs at $b = b'$, but $\sigma_1^{(1)}$ passes through infinity, rather than zero, at this point. Therefore the neighborhood of $b = b'$, as well as that of $b = 0$, require a separate investigation.

To determine the behavior of $\sigma_{2,3}$ for $b < 0$ and $b > b'$, we expand (6.18) in ε near $\sigma_2^{(0)}$ and $\sigma_3^{(0)}$, respectively. The result, to $O(\varepsilon^2)$, is

$$\sigma_{2,3} = \pm i\left\{b^{-1}(b^3 - b'^3)\right\}^{1/2} - k\left\{1 + \frac{1}{2}\frac{b+\beta}{\beta}\frac{b'^3}{b^3 - b'^3}\right\}.$$

The real part is $O(k)$, negative for $b < 0$ and positive for $b > b'$.

The stability results to $O(\varepsilon^2)$ are summarized in the middle panel of Figure 6.7. The upper panel shows schematically the solution branches R_-, R_+ and Z of ψ_A as a function of ψ_A^*, with the quasi-linear, damped resonance at $b = 0$ and the turning point at $b = b''$. The neighborhoods of $b = 0, b', b''$ are left open. It remains to fill in the behavior of $\sigma_{1,2,3}$ in these neighborhoods, and the way that the roots of the characteristic equation are joined smoothly through the respective points.

We shall only indicate briefly how this is done near the resonance, $b = 0$. The characteristic equation at $b = 0$ becomes

$$(\sigma+k)^3 + (\sigma+k)b'^3\beta^{-1} = 0,$$

with the three stable roots $\sigma_1 = -k$ and $\sigma_{2,3} = -k \pm i(hh_0)^{1/2}$.

The derivative with respect to b of (6.18b) at $b = 0$ is

$$\{3(\sigma+k)^2 + b'^3\beta^{-1}\}\frac{d\sigma}{db} - b'^3k^{-2}(\sigma+2k) = 0.$$

Hence $d\sigma/db = \beta/k$ at $\sigma = -k$, and $d\sigma/db = (\beta/4k^2)\sigma^*$ at $\sigma = -k \pm i(b'^3\beta^{-1})^{1/2}$. These values of $d\sigma/db$ allow one to follow the motion of the roots $\sigma_{1,2,3}$ in the complex plane for b near resonance. The result is that σ_1 becomes positive and large for b slightly positive, while σ_2 and σ_3 become real, negative and of unequal order.

A similar analysis, only more complicated, applies near $b = b'$. The complete behavior of the characteristic roots of (6.18) is shown in the lowermost panel of Figure 6.7. The order of magnitude of each root is also indicated on the ordinate. The continuous branches of roots are marked $\sigma_{a,b,c}$ in this panel, and the fact that $\sigma_b = \sigma_c^*$ for certain b-intervals is indicated by the initials cc, for "complex conjugate".

It is clear from the lowermost panel of Figure 6.7 that the stability of both Z and R_- is due to the dissipation mechanism, represented by k. The instability of R_+ is due to two different causes, one for $0 < b < b'$, the other for $b' < b < b''$, respectively; see the middle panel of the figure also for comparison. The more interesting one is the instability in the first interval, which is independent of dissipation.

A necessary condition for barotropic instability of the mean zonal flow $\bar{u} = \bar{u}(y)$ in the absence of topography is that its absolute vorticity, $\beta - \bar{u}_{yy}$, vanish. This is not the case. It is not a resonant instability

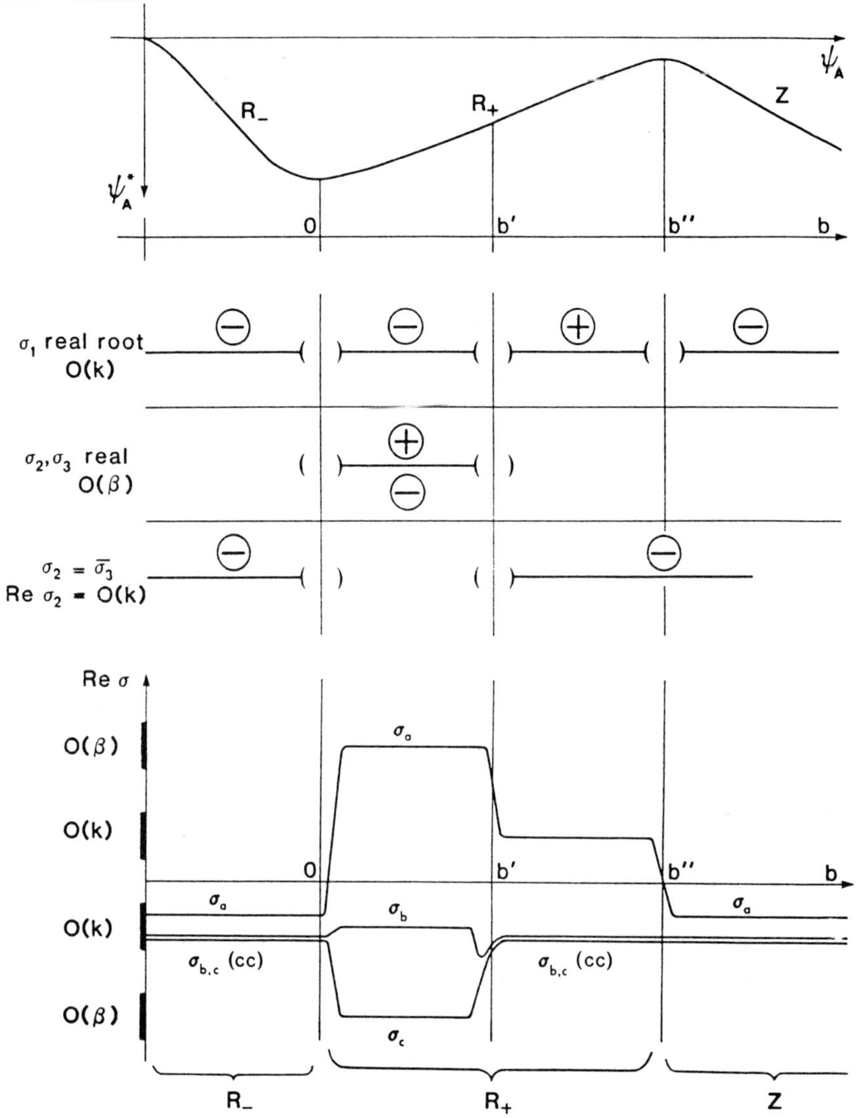

Fig. 6.7. Stability results for the Charney and DeVore model of Figures 6.5 and 6.6.

of a triad interaction either, for which a second y-mode, as in Section 5.3 or in (6.21) below, would be required. The present instability is due solely to the interaction of the mean flow with the topography, and was called therefore by Charney and DeVore *topographic instability*.

In the nomenclature of linear vibration theory, alluded to before, the type of resonance associated with this instability is called *parametric resonance*. It is due to periodic variations in the coefficients of the oscillator,

$$\ddot{x} + \omega^2(t)x = 0, \tag{6.20a}$$

where

$$\omega^2 = \omega_0^2(1 + \delta \cos \gamma t), \tag{6.20b}$$

rather than to a nonzero, periodically varying right-hand side. Resonance occurs in (6.20) when δ is small and $\gamma = 2\omega_0/n$, for n integer. In the presence of slight dissipation, the strength of the resonance decreases with n.

Parametric resonance makes the high-amplitude motion of a child's swing possible, starting with small, well-timed impulses. It plays an important role in studying the stability of periodic solutions to nonlinear problems, as discussed in Section 5.5. The present case is degenerate, as seen from Eqs. (6.9), and (6.17), since the period $2\pi/\gamma$ of the variation in the coefficients is infinite.

Upon introducing, in addition to (6.15a,b,c), the second-mode spectral components

$$F_C = \sqrt{2}\cos(2y/L), \tag{6.21a}$$

$$F_M = 2\cos(nx/L)\cos(2y/L), \quad F_N = 2\sin(nx/L)\cos(2y/L), \tag{6.21b,c}$$

into (6.14), the zonal solution remains stable. The "blocked" solution R_-, however, is destabilized for a certain range of parameter values, and replaced by a stable limit cycle.

Numerical solutions of (6.14), using a finite-difference discretization on a grid of 16×16 points, confirm the presence and stability of the equilibria R_- and Z, for the same geometry and parameter values as in the highly-truncated spectral model. The transition from the stationary flow R_- to a stable periodic solution with properties similar to R_- is also present in the grid-point model.

The simple model governed by Eqs. (6.14,6.15,6.21) exhibits the existence of several flow regimes, depending on the values of external parameters. The near-resonant solution R_- presents some of the features of blocking: strong meridional flow and location of the geopotential height ridges west of the main topographic barriers. The stable limit cycle around R_- might be a reasonable representation of the slight vacillation in the position and intensity of a blocking high.

What is missing is the strong instability of the zonal flow. The latter is not a statistical, say monthly, mean of the atmosphere's most prevalent regime, upon which at any time cyclone waves and other transients are superimposed. The transition from the weakly vacillating, persistent, blocking anomalies to the strongly aperiodic "normal" regime also needs to be explained. The next step in our successive bifurcation scheme will, it is hoped, take us there. An increase in the number of degrees of freedom allowed, and the introduction of genuine wave-wave coupling, as opposed to interactions via the mean flow only (Eqs. (6.16b,c)), should make the appearance of aperiodic behavior rather likely.

6.4. Multiple Flow Regimes and Variations in Predictability

Governing equations. In this section, we study the barotropic PVE (3.11) on the spherical Earth, following a series of articles by Legras and Ghil (1983, 1984, 1985; the last one will be referred to as LG3, for short). The β-plane approximation is dropped and the full variation of the Coriolis parameter f with latitude is taken into account,

$$f = 2\Omega\mu; \tag{6.22a}$$

here Ω is the angular velocity of the Earth's rotation and μ the sine of latitude, $\mu = \sin\theta$. As a result, the term $f_0 h/H_0 + \beta y$ in (6.14) is replaced by $f(1+h/H_0)$. Moreover, the free-surface term in the potential vorticity Q is kept, i.e., $L_R^{-2} \neq 0$. The $L_R^{-2}\psi$ term has the effect of shifting the eigenvalues of the Laplacian, and thus reducing the linear phase speed of free Rossby waves. From Section 6.2 we know that this favors their interaction with the stationary forcing.

The variations are scaled by a typical length a, the radius of the Earth, a typical height H_0, the scale height (cf. Sections 4.2 and 6.3), a typical time $(2\Omega)^{-1}$, and a typical velocity U:

$$L_R = a\lambda, \quad h = H_0 h', \quad t = (2\Omega)^{-1} t', \tag{6.22b,c,d}$$

$$(\psi,\psi^*) = aU(\psi',\psi_*'), \quad \alpha = 2\Omega\alpha'. \tag{6.22e,f,g}$$

The nondimensional form of the equation, dropping primes, is thus given by

$$\partial_t(\nabla^2-\lambda^{-2})\psi + \rho J\{\psi,(\nabla^2-\lambda^{-2})\psi\} + J(\{\psi,\mu(1+h)\} = \alpha\nabla^2(\psi^*-\psi). \qquad (6.23)$$

Here ∇^2 and J are the Laplacian and Jacobian with respect to the longitude ϕ and the meridional variable μ, on the surface of the unit sphere.

The nondimensional parameter ρ,

$$\rho = U/2\Omega a, \qquad (6.24)$$

is similar to a Rossby number, and is associated with the strength of the forcing ψ^*, cf. (6.22f). It multiplies, cf. (6.22e), the sole nonlinear term in (6.23), and plays, as we shall see, the role of a critical parameter in the behavior of its solutions.

The basis functions used for the discretization of (6.23), cf. (5.5), are the surface spherical harmonics $Y_n^m = Y_n^m(\phi,\mu)$, which satisfy

$$\nabla^2 Y_n^m = -n(n+1)Y_n^m; \qquad (6.25a)$$

they are given by

$$Y_n^m(\phi,\mu) = P_n^m(\mu)\, e^{im\phi}, \qquad (6.25b)$$

where $P_n^m(\mu)$ are the associated Legendre functions of the first kind, which are bounded at the poles. They are normalized according to

$$P_n^m(\mu) = \left\{\frac{(2n+1)}{2}\frac{(n-m)!}{(n+m)!}\right\}^{1/2} \frac{1}{2^n n!} \times (1-\mu^2)^{m/2}\, d_\mu^{n+m}(\mu^2-1)^n. \qquad (6.25c)$$

A surface harmonic Y_n^m has zonal wave number m, cf. (6.25b), and meridional wave number $n-m$, cf. (6.25c). For each eigenvalue of the Laplacian, $-n(n+1)$, there are $2n+1$ eigenfunctions Y_n^m.

The stream function is expanded in a finite series,

$$\psi(\phi,\mu,t) = \sum_{n=0}^{n_0}\sum_{m=-n}^{n} \psi_n^m(t)\, Y_n^m(\phi,\mu), \qquad (6.26a)$$

truncated at $n_0 = 9$. Figure 6.8 displays the triangular truncation in n and m which was used.

The modes present in (6.26) are circled in the figure. Symmetry of the flow with respect to the Equator is assumed, as well as a sectorial periodicity (mod π) in longitude. In other words, only components with odd n and even m are retained, resulting in $N = (n_0+1)^2/4 = 25$ real modes. Between them, 132 nonlinear, triadic interactions operate. It

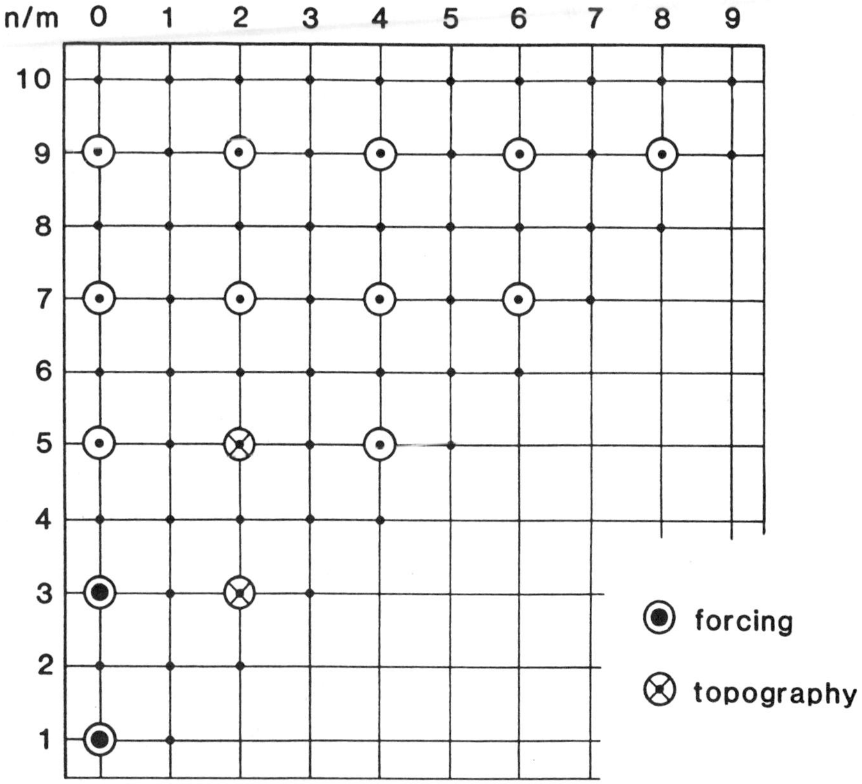

Fig. 6.8. Spectral truncation of Eq. (6.23) in spherical harmonics. The modes retained are circled, topographic modes present are crossed, and forcing modes are indicated by a full circle (after Legras and Ghil, 1983).

should be noticed that the present spatial resolution of the global observational network is only twice as high as the one afforded by (6.26a) with $n_0 = 9$, i.e., mean observational resolution corresponds roughly to $n_0 = 18$.

The topography is taken, as in the preceding section, to represent to lowest order the Northern Hemisphere, with two equal continental masses separated by two oceans, so that

$$h = 4h_0\mu^2(1-\mu^2)\cos 2\phi; \qquad (6.26b)$$

we take $h_0 = 0.1$, cf. Sections 6.2 and 6.3. Multiplication of (6.26b) by the nondimensional Coriolis parameter $f = \mu$ leads to the term fh retaining the modes marked by crosses in Figure 6.8.

The mean forcing flow ψ^* is a zonal jet

$$\psi^* = \overline{\psi}_1^0 Y_1^0 + \overline{\psi}_3^0 Y_3^0 = -\kappa\mu^3; \tag{6.26c}$$

κ is a nondimensional constant chosen in such a way that the maximum forcing speed, which occurs at 50° N, has a dimensional value of 60 m/s for $\rho = 0.20$. The forcing modes are marked by a full circle in the figure.

Although a barotropic model is used, we would like its solutions to share as many characteristics as possible with the real atmosphere, which has both barotropic and baroclinic features. In particular, the length scale and speed of propagation of model waves should be close to that of observed Rossby waves. The discussion in Section 4.3, Eqs. (4.17, 4.18), and in Section 4.4, Eq. (4.37) ff., indicates that the radius of deformation appropriate for baroclinic waves is the internal Rossby radius, while barotropic waves are characterized by the external Rossby radius. Hence we take $L_R = 1100$ km, as a heuristic interpolation between an internal, baroclinic radius of deformation and the external, barotropic one.

Substituting (6.26) into (6.23) yields a system of 25 ODEs for the vector $\Psi(t)$ with components $\psi_n^m(t)$, cf. (6.26a). This system is similar to (5.14); writing it out explicitly is tedious and uninformative so we give it merely in compact vector-matrix rotation:

$$\dot{\Psi} = \rho\Psi^T A\Psi - B\Psi + C. \tag{6.27}$$

System (6.27) is forced dissipative, with quadratic nonlinearity, having the general form (5.26).

Stationary solutions. The steady states $\Psi(t) \equiv \Psi_s$ of (6.27) are given by a system of $N = 25$ algebraic equations, depending on the parameters $r = (\rho,\alpha)$. This does not permit an analytic solution as in the previous section or in Chapter 5; it is furthermore out of the question to obtain periodic and aperiodic solutions of the full system in closed form.

An investigation of the behavior of system (6.27) starts, according to the usual program, by examining its stationary solutions. For α large, tending to infinity, and ρ fixed, the stationary solution of (6.23) is clearly unique and tends to ψ^*. For ρ large, tending to infinity, and α fixed, there still exists a stationary solution tending to ψ^*, which is not necessarily unique. In the limit $\rho \to 0$, an equilibrium is reached which stays near ψ^* if α is at least of the same order as h. These three asymptotic forms of the solution are stable with respect to time-dependent perturbations. The same results hold for the discrete system (6.27).

Between these limits lies the interesting domain of parameter space. As it happens, the presence of the parameters can be put to good use in order to follow solutions around numerically, by a *continuation method*. Knowing a solution Ψ_0 of

$$G(\Psi;r) \equiv \rho\Psi^T A\Psi - B\Psi + C = 0 \tag{6.28}$$

for a given parameter value r_0, $G(\Psi_0;r_0) = 0$, one searches for solutions $\Psi(r)$ near the point (Ψ_0,r_0) by using

$$\frac{\partial G}{\partial \Psi}(\Psi - \Psi_0) + \frac{\partial G}{\partial r}(r-r_0) = 0, \tag{6.29}$$

where $\partial G/\partial\Psi$ and $\partial G/\partial r$ are matrices of partial derivatives evaluated at $(\Psi,r) = (\Psi_0,r_0)$.

The particular method employed is *pseudo-arclength continuation*. It solves (6.29) as an ordinary differential equation in the arclength s given by $ds^2 = ||d\Psi||^2 + ||dr||^2$, using a predictor-corrector method. Here $||X||$ stands for the length of the vector X, in the appropriate dimension of Euclidean space, namely $N = 25$ for the *phase space* of Ψ and $p = 2$ for the *parameter space* of r. The correction step uses a Newton-type technique for Eq. (6.28).

This continuation method allows one to explore completely a one-parameter solution branch of stationary solutions. It elminates the difficulties encountered in other methods at regular turning points of a branch, where the rank of either matrix in (6.29) is less than maximal.

Still, no method provides automatically the entire picture of all possible stationary solutions for the full (N+p)-dimensional *phase-parameter space*. The picture may be filled in by carrying out a large number of explorations following different directions in parameter space. This approach, given some empirical ingenuity, provides redundant information and allows one to connect sheets of solutions which may appear as separate branches in certain one-parameter cross-sections of the parameter space.

Dependence on parameters. In order to describe the distribution of stationary solutions as a function of the parameters, we plot in Fig. 6.9 the square root of their potential energy E as a function of ρ for different values of α. For $\alpha^{-1} = 1.1$ days (Curve A in Fig. 6.9a) we are in the asymptotic domain of large α and the unique solution branch

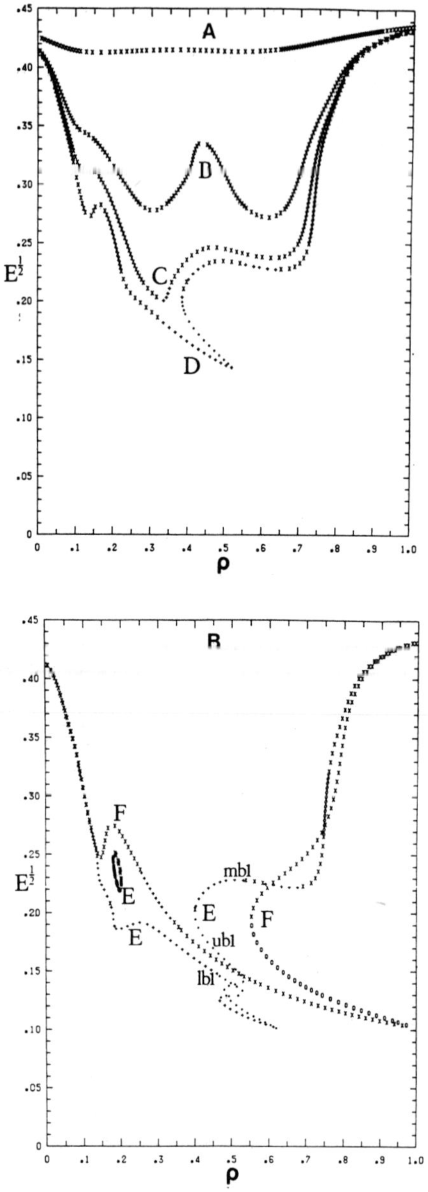

Fig. 6.9, A – B

Fig. 6.9. Potential energy E of stationary solutions as a function of the forcing parameter ρ for several fixed values of the dissipation parameter α. The stability of the solutions is denoted by the symbols: ×, 0, +, *, and ·; see text below for details. Values of α^{-1} in days:

(a) A, α^{-1} = 1.1, B, α^{-1} = 3.3; C, α^{-1} = 5.0; D, α^{-1} = 6.7. (b) E, α^{-1} = 10.0; F, α^{-1} = 10.0, quasi-linear model. (c) G, α^{-1} = 20.0, (d) H, α^{-1} = 20.0, quasi-linear model. The labeling of various branches is explained in the text (from LG3).

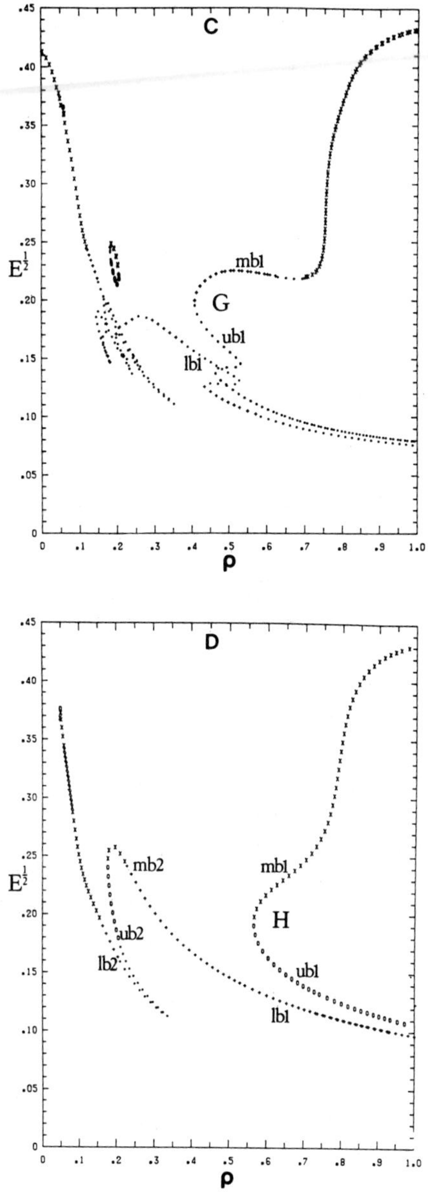

Fig. 6.9, **C – D**

differs little from ψ^* for all ρ. In the sequel, α will always be given in day^{-1}, while ρ is nondimensional.

As α decreases, the relaxation no longer compensates for the destabilization of the mean flow by the orography, and waves are produced which interact with each other and feed back energy to the mean flow. As a result, the solution branch above is continuously distorted into a family of stationary solutions which has notably smaller energy than ψ^* and is characterized by a strong flux of energy extracted by the waves from the mean flow (curves B and C).

For $\alpha^{-1} \simeq 5.25$ days, a fold develops in this solution branch near $\rho = 0.5$, analogous to the one in Fig. 6.5. This fold leads to the existence of three stationary solutions for the same value of the parameters (curve D of Fig. 6.9a).

As α decreases further, we see (curve E, Fig. 6.9b) an isolated closed branch, or *isola*, which detaches itself from the main branch for $\alpha^{-1} \geq 8.5$ days and is present for $0.18 \leq \rho \leq 0.20$, down to small values of α. Next, another fold appears for $\alpha^{-1} \simeq 11$ days at $\rho \simeq 0.20$ and develops for $\alpha^{-1} \geq 15$ days (curve G, Fig. 6.9c) into a complicated structure.

In order to compare these results with (6.16), we perform a similar analysis on a *quasi-linear* version of (6.27), in which all wave-wave interactions were suppressed. In this new system, the nonzonal components of a stationary solution satisfy a linear system with coefficients depending on the zonal state. This system shows nonlinear resonances for some values of the zonal flow due to the existence of small denominators.

For arbitrary parameter values, the quasi-linear stationary solutions obtained have all components with zonal wave number $m \geq 4$ equal to zero. Time-dependent integrations of the quasi-linear model show that perturbations with $m \geq 4$ in a nonstationary solution decay to zero in time. Since the only forced modes have $m = 0$ and $m = 2$, and since no interactions between $m = 2$ and $m \geq 4$ are present, it follows that in the quasi-linear model the mean flow is *barotropically stable*.

Figs. 6.9b (curve F) and 6.9d (curve H) show the cross-sections $E = E(\rho)$ of the quasi-linear model for $\alpha^{-1} = 10$ and 20 days. The large resonance observed in both figures for $\rho \geq 0.5$ is the image of the Charney-DeVore orographic instability mechanism in the present model.

The difference between the upper branch of the resonance (marked by UB1 in Fig. 6.9b) and the main branch (marked MB1) consists essentially of a decrease of the zonal component of the flow. Between the upper branch (UB1) and the lower branch (LB1), on the other hand, there is little change in the zonal component, but a global westward phase shift of the energy containing components $(n,m) = (3.2)$ and $(5,2)$ without modification of their amplitude, as in (6.16).

The *second resonance*, noticed already for the fully nonlinear model, is also visible, especially in Fig. 6.9d. This second resonance, appearing at more realistic values of ρ, $\rho \leq 0.30$, is due to the addition of more degrees of freedom in the meridional direction: the number of folds in the resonance pattern increases with the number of meridional modes, but most of them accumulate near $\rho = 0$.

This resonance is different in character from the first one. In the quasi-linear model, a uniform westward phase shift between its upper branch (UB2 on curve H) and its lower branch (LB2) affects the modes (5,2), (7,2) and (9,2), but the largest-scale mode (3,2) remains unchanged. This indicates that *the resonance develops on a basic flow which*, far from being zonal, *possesses a strong wave component.*

In the fully nonlinear model (Fig. 6.9c), the first resonance keeps the appearance of a unique distorted fold. The second one exhibits a more complicated structure where, as α decreases, new folds develop on the already existing ones in a seemingly endless cascade. The ρ-extent of the folds grows considerably as α goes to zero. Comparing Figure 6.9c with Figure 6.9d, we see that the isolated branch, associated with the second resonance in the nonlinear model, is due to the reconnection of branches MB2 and UB2 of the quasi-linear model.

For a detailed study, both analytic and numerical, of the way in which these two resonances, and additional ones, arise, we refer to LG3 (Figure 4 and Appendix C there).

Flow patterns. The spatial resolution provided by the model allows comparatively complex flow patterns to appear. The description of the ensemble of patterns appearing as stationary solutions is facilitated by the fact that these depend relatively little on α along a given sheet of solutions. This result was expected for large and for small values of ρ, but holds also true for intermediate values.

It turns out that the flow patterns appearing for large values of ρ, $\rho \geq 0.4$, are very similar in character to those of Fig. 6.6. This is true of the fully nonlinear, as well as the quasi-linear version of the model, near the first resonance. In the quasi-linear model, the features of both the zonal and the blocked flow patterns are simply more pronounced, due to the absence of interaction with higher wave numbers.

The second resonance, which occurs in (6.27) at lower, more realistic values of the forcing parameter ρ, has a more complicated structure than the previous one, involving modes of degree $n \geq 5$. The detailed structure of the enlarged $E^{1/2}$ vs. ρ curve for $\alpha^{-1} = 20$ days (curve G in Fig. 6.9c) is shown for the neighborhood of $\rho = 0.20$ in Fig. 6.10.

In this region of parameter space, the solutions' flow patterns can be classified into four families -- Blocking, Zonal 1, Zonal 2 and Double Block -- which correspond to the various branches as denoted in Fig. 6.10. Representative examples of the solution families identified in Fig. 6.10 are shown in Fig. 6.11. Inside each family there exist amplitude and phase variations, but the general pattern of the solution remains unchanged. Transitions along branches between Blocking, the Double Block and Zonal 2 occur quite sharply in the circled segments of Fig. 6.10.

The projection used in Figure 6.11 and in subsequent figures of flow fields (Figures 6.15 and 6.16) is a conformal conical projection of ratio 2/3. It maps a sector of the Northern Hemisphere located between 0° and 270° longitude onto the half disk shown in the figure; this is the same as showing only that part of Figure 6.6 with x between 0 and $3\pi/2$, mapped onto the rectangle $\{0 \leq x \leq \pi,\ 0 \leq y \leq \pi\}$.

The Zonal 1 flow (Fig. 6.11a), associated with the isolated branch, has a high energy level: the maximum intensity of its zonally-averaged jet is 50 ms^{-1}. The Zonal 2 flow (Fig. 6.11b) is less intense, with a 35 ms^{-1} jet in zonal average. It exhibits in fact a ridge on the west side of the orography and a trough on the east side, similar to the averaged winter circulation of the Northern Hemisphere. Both types of zonal flow may thus be associated with the regular planetary flow regime, and we shall do so in the sequel.

The west-coast ridge is strongly intensified in the blocking solution (Fig. 6.11c), which shows a well developed high center on the west side of the orography, splitting the zonal flow into two jets. The averaged zonal wind is reduced to 18 ms^{-1} and the geopotential height difference between the trough and the ridge is about 1000 m. The exaggerated amplitude of this flow feature would presumably be reduced by thermal damping and by interactions with higher wave numbers to the correct order of magnitude of 500 m.

We show also in Fig. 6.11d one solution of the quasi-linear model (curve H, Fig. 6.9d), located on the lower branch of the second resonance for the same value of the parameters as in Fig. 6.11c. Both patterns are quite similar, although the quasi-linear solution is shifted eastward and exhibits a weaker zonal wind.

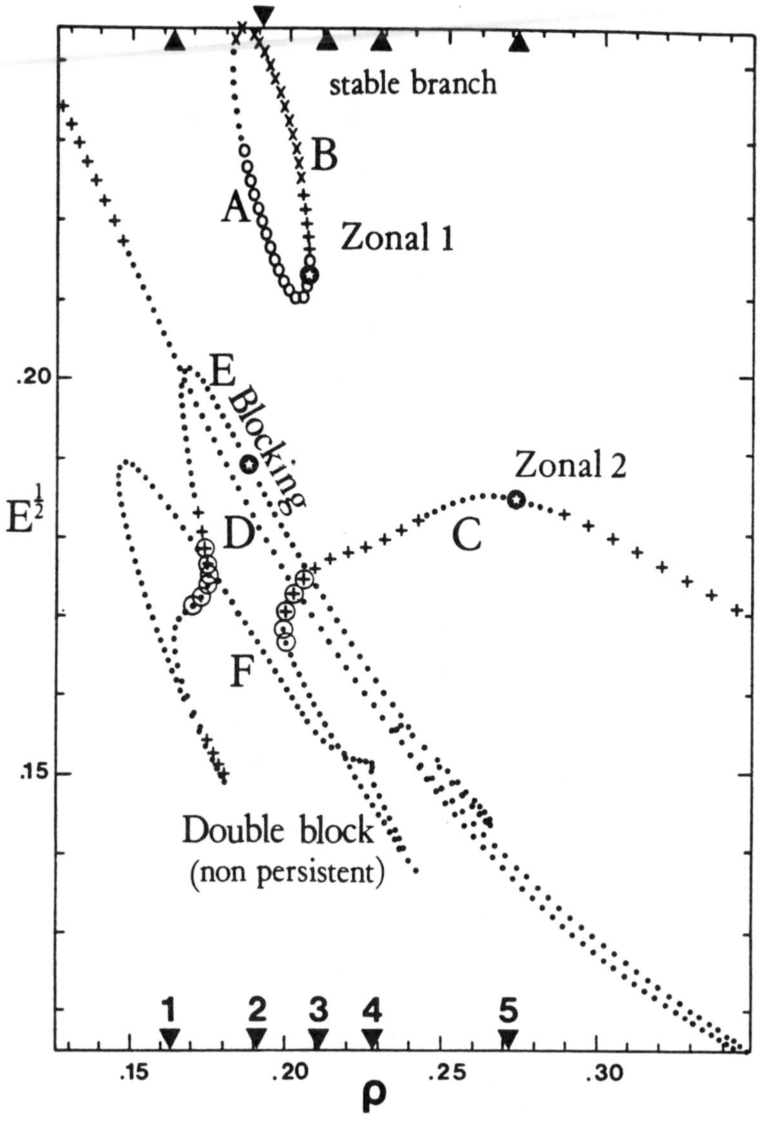

Fig. 6.10. Blow-up of a portion of Fig. 6.9c. Stationary solution branches associated with the second resonance for α^{-1} = 20 days. Numbered pointers on the abscissa indicate values of ρ for which time-dependent model solutions are investigated in detail (from LG3).

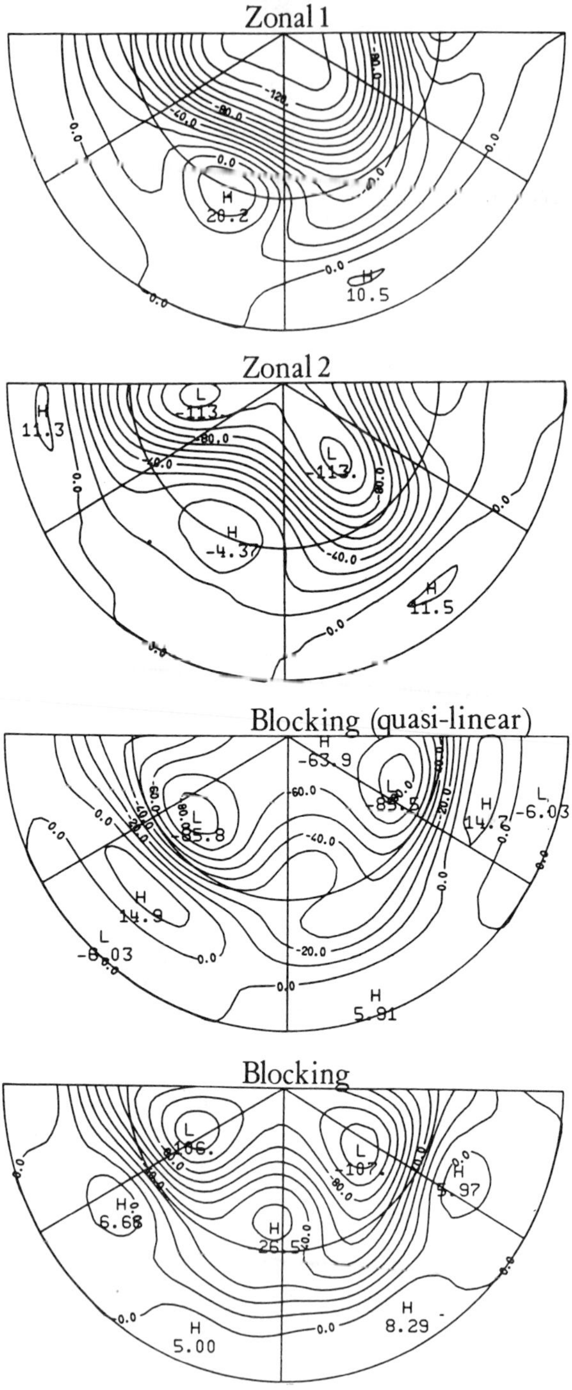

Fig. 6.11. Contours of the stream function Ψ for the stationary solutions indicated by stars in Figure 6.10: (a) Zonal 1; (b) Zonal 2; (c) Blocking; and in Figure 6.9d: (d) Blocking in quasi-linear model (curve H) (from LG3).

The flow patterns apparent for $\rho \simeq 0.2$ will have easterly winds in the tropics. The maximum zonally-averaged easterly speed occurs at the equator and lies between 2 ms^{-1} and 8 ms^{-1}. This maximum speed is higher for the zonal than for the blocked stationary solutions, in qualitative agreement with observations.

Stability of stationary solutions. Linear stability of a stationary solution Ψ_s is determined by the spectrum of the linear operator $L_s = L(\Psi_s)$, given by

$$L_s\chi = (\Delta_T - \lambda^{-2})^{-1}\{\rho\, J_T(\Psi_s,\Delta_T\chi) + J_T[\chi,\rho\Delta_T\Psi_s + \mu(1+h)] + \alpha\Delta_T\chi\}, \tag{6.30a}$$

where Δ_T and J_T are the truncated Laplacian and Jacobian, respectively. L_s is easily obtained as an $N \times N$ matrix from Eq. (6.27),

$$L_s\chi = \rho(\Psi_s^T A\, \chi + \chi^T A\, \Psi_s) - B\chi. \tag{6.30b}$$

Eigenvalues are then computed by a standard algorithm.

Let Ψ_s be a stationary solution for which L_s has no eigenvalue with zero real part, i.e., Ψ_s is a generalized saddle, or *hyperbolic point,* like R_1 in Fig. 5.6. Then the quadratic character of the non-linearity in Eq. (6.27) insures that there exists in phase space a finite, small neighborhood $U = \{||\Psi - \Psi_s|| < \varepsilon\}$ of Ψ_s, such that the following dichotomy holds: if Ψ_s is stable, then solutions $\Psi(t)$ of (6.27) with initial data in U, $||\Psi(0) - \Psi_s|| < \varepsilon$, stay in U and tend to Ψ_s as $t \to \infty$; if Ψ_s is unstable, then almost all solutions of (6.27) starting in U will leave it in finite time, so that there exist times $0 < t_1 \leq t \leq t_2$ for which $||\Psi(t) - \Psi_s|| \geq \varepsilon$. The excluded set of initial data in the latter case refers to the *stable manifold* $M^s(\Psi_s)$ of Ψ_s (see Section 5.4), which has, by the definition of Ψ_s as an unstable fixed point, dimension $s < N$ and hence volume zero; s being just the number of eigenvalues with negative real part. We shall see later that trajectories of (6.27) do return after some time, $t > t_2$, to the neighborhood of certain linearly unstable solutions by following closely the stable manifold M^s of the solution, and dwell there for long times. The persistence properties of such neighborhoods in phase space, and of the associated planetary flow regimes in physical space, will be explored numerically below.

As a first, analytical step in this exploration, we return now to the linear stability properties of stationary solutions. These are shown in Figs. 6.9a-d: stable solutions are marked by x; unstable solutions are labeled to indicate the number u of eigenvalues with positive real parts, as follows: 0, one real eigenvalue ; +, two complex conjugate eigenvalues; *, two real eigenvalues; ·, three eigenvalues or more.

The number u of unstable eigenvalues is important for the nonlinear dynamics. The eigenvectors associated with the unstable eigenvalues span the solution's unstable tangent space T^u, and $s+u = N$ for almost all points in phase-parameter space, i.e., almost all stationary solutions $\Psi_s = \Psi_s(r)$, are hyperbolic. The nonlinear extension of $T^u(\Psi_s)$ is the *unstable manifold* $M^u(\Psi_s)$, which has the same dimension u in a finite neighborhood of the solution. If $0 < u << s$, then one may hope that the given unstable solution plays a significant role in the global, nonlinear dynamics. This situation is illustrated in Figure 6.12 for $u = 1$, $s = 2$ and $N = 3$.

In the figure, trajectories $\Psi(t)$ approach T^s near Ψ_s and spiral in on Ψ_s. If the trajectory were exactly in M^s, it would take an infinite time to reach Ψ_s and never leave it again. Chances for that,

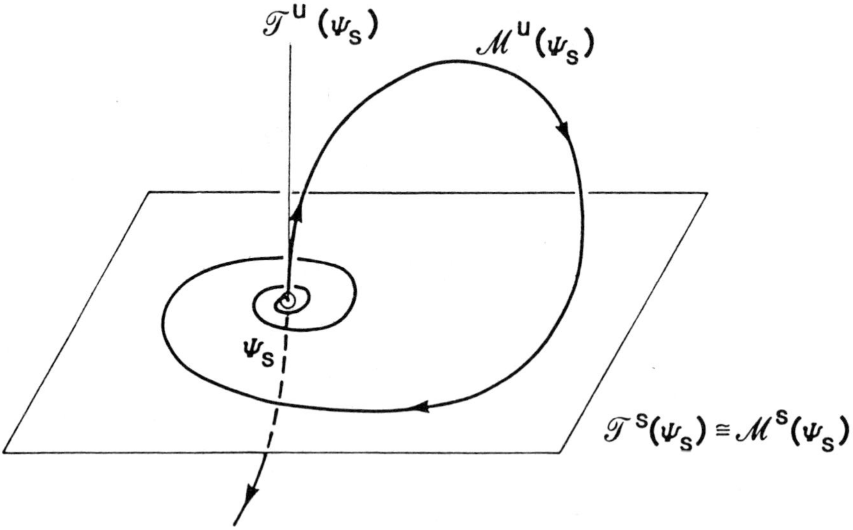

Fig. 6.12. Behavior of system trajectories in the neighborhood of a hyperbolic point of saddle-focus type.

however, are zero, so that the trajectory spends actually a long, but finite time near Ψ_s. Very close to Ψ_s, however, trajectories are pulled away along T^u, and ejected entirely from the neighborhood of Ψ_s.

The case in which $\Psi_s = \Psi_s(r)$ is such that some eigenvalues of L_s have zero real part, and hence $s+u < N$, corresponds to lower-dimensional manifolds, i.e., to points or curves in our two-dimensional parameter space, with $r = (\rho,\alpha)$. The set of r for which $s(r) + u(r) < N$ has in general zero volume, or measure, in parameter space and is known as the *bifurcation set*.

Turning points on solution branches, also called *saddle-node bifurcations*, are associated with a real eigenvalue passing through zero (see Section 10.2, Figure 10.6). Likewise, stationary solutions lose their stability to periodic solutions at points of *Hopf bifurcation*, where a pair of complex conjugate eigenvalues crosses the imaginary axis from the left into the right half-plane (Section 12.2). The presence of three or more unstable eigenvalues indicates the possibility of aperiodic solutions, cf. Section 5.4.

As a rule, we find that the most stable solution of (6.27) at a given value of the parameters is the one with the highest energy, i.e., the one closest to Ψ^*. Solutions associated with resonances exhibit a relatively large number of unstable eigenvalues. This number increases along the resonant branches and as α decreases. The dimension u of the unstable manifold may exceed 10 on certain branches for $\alpha < 0.05$ day^{-1}.

Figure 6.13 shows the linear stability of the most stable stationary solution in the (α,ρ)-plane of parameters with $h_0 = 0.1$. Due to the remark above, this figure can also be seen as a plan view from the top of the $E = E(\rho,\alpha)$ surface whose cross-sections appear in Figures 6.9 and 6.10. It is analogous to the regime diagrams for flow in the rotating annulus appearing in Figures 5.2, 5.4 and 5.7.

For $\alpha > 0.205$, a nearly zonal solution, close to the forcing flow Ψ^*, is unique and stable. Solutions at very small and very large ρ are also stable for all α, in agreement with the statements with Hopf bifurcation.

As α decreases, instabilities associated with Hopf bifurcation develop for $0.1 < \rho < 0.75$. Each one of these instabilities gives rise to one of the hatched lobes in Fig. 6.13. These lobes grow in the (α,ρ)-plane in the direction of decreasing α, and eventually merge into one large hatched area, where no stationary solution at all is stable. The fact that stability decreases with decreasing dissipation is in accordance with our general fluid-dynamical intuition, and should hold

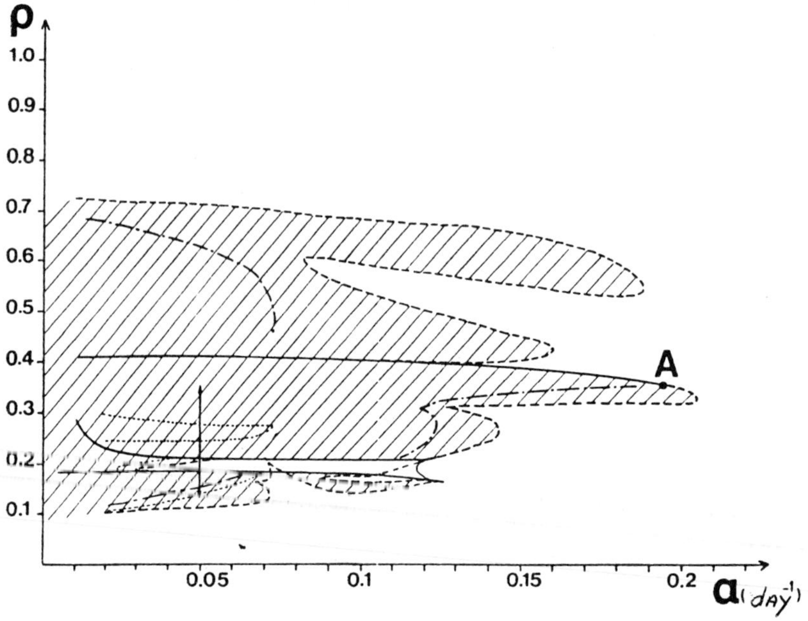

Fig. 6.13. Stability domain of the most stable solution as a function of ρ and α. Solid lines: regular turning line of the sheet of stationary solutions. Dashed lines: Hopf bifurcation of a stable stationary solution. Dotted lines: second Hopf bifurcation. Dash-dotted line: the onset of chaotic regimes. The instability domain where no stable stationary solution exists is hatched (from LG3).

for the continuous Eq. (6.23), as well as for the discrete, truncated model (6.27).

Within the hatched instability area, the turning line of the main branch MB1 into the upper branch UB1, and the two boundaries of the isolated, Zonal 1 branch are shown as solid lines, for $\rho \simeq 0.4$ and near $\rho \simeq 0.2$, respectively. The folding of the first resonance (solid) merges at point S with a Hopf bifurcation line (dashed in the figure). According to numerical evidence, S is a triple bifurcation point. The chaotic behavior in the neighborhood of this point corresponds to the analysis of *saddle-foci* referred to in Sections 5.5 and 6.6 (see also Figure 6.12).

The isolated branch with Zonal 1 flow patterns is always more stable than the other solutions which coexist with it at the same parameter values.

and at lower energies. We shall not discuss here the complicated bifurcation patterns to which the second resonance, lying under the isola, gives rise. Heuristically, the multiple folds of this nonlinear resonance seem to be related to the accumulation of linear resonances.

Finally, we notice that multiple stable equilibria occur in this model only in a very small parameter range, near $\rho = 0.18$ and $\alpha = 0.07$. On the contrary, such multi-stable solutions are characteristic of wide parameter ranges in simpler, more highly truncated or quasi-linear models (cf. Figures 6.5 and 6.9b).

For multiple flow regimes to exist in the present model over a wider range of parameter values, they must therefore be related to the existence of higher-dimensional attractors. The discussion of stable and unstable manifolds following Eq. (6.30) and Figure 6.12, and the study of bifurcation patterns in Fig. 6.13, leads us to expect that parts of such complicated attractors might lie close to unstable fixed points or even contain such points (as in the case of the Lorenz attractor, Section 5.4). We shall also see that very persistent flow pattern evolutions in physical space might be explained by the phase-parameter space proximity of a saddle-node bifurcation.

Periodic and aperiodic solutions. In the previous subsection, we have seen how stationary solutions lose their stability, and we expect periodic and aperiodic solutions to arise instead as stable flow regimes. To study the flow patterns associated with these flow regimes, a large number of numerical integrations of the evolution equation (6.27) were carried out. Numerical solutions were computed for hundreds of values of the parameters and various initial data, each solution being computed for thousands of simulated days. This numerical study, while not exhaustive in all parts of parameter space, provides a pretty good qualitative picture of model behavior in certain regions of this space; these regions are either physically most realistic or else are of interest in order to complete the global knowledge of possible types of behavior.

The transition from stable stationary solutions to *stable* periodic solutions occurs by *supercritical* Hopf bifurcation along most of the boundary of the hatched area in Fig. 6.13. Inside the hatched area, the stable *limit cycles* arising at the boundary grow in size and keep their stability for a finite distance in parameter values.

Above the turning line of the first resonance, at $\rho \simeq 0.4$ in Fig. 6.13, there exists a limit cycle arising from the, now unstable, main branch solution MB1. This limit cycle, which is stable in a large region of parameter space, has small amplitude throughout, so that the flow pattern associated with it always resembles closely the pattern of the stationary solution MB1; such behavior was encountered in system (6.14, 6.15,6.21) for the lower branch, R_-, of this resonance. Transition from this periodic solution to aperiodic solutions occurs for $\alpha \leq 0.07$ along the dash-dotted line indicated in Fig. 6.13. The aperiodic solutions to the left of this line have much larger amplitude and hence greater variations in flow pattern than the periodic solution manifold from which they arise.

Below the first turning line at $\rho \simeq 0.04$, transition from periodic to chaotic behavior occurs much closer to the boundary of stability of stationary solutions. A detailed study of solution behavior in the neighborhood of the triple bifurcation point S, along the line segment $\rho = 0.35$, $0.16 \leq \alpha \leq 0.18$, shows, for α decreasing, a cascade of *period-doubling bifurcations* from the originally stable limit cycle. Chaos begins at the accumulation point of the sequence of successive bifurcation values, which is $\alpha \simeq 0.1624$. Windows of regular behavior, and associated intermittency phenomena, are also observed inside the chaotic domain, in agreement with the full period-doubling scenario. On the other hand, direct transition from stationary to chaotic behavior is observed numerically at point S itself, in agreement with the saddle-focus scenario.

At lower values of α, transition to chaos appears to be more complicated. At $\alpha = 0.05$ day^{-1}, and starting with small values of ρ, the chaotic regime is entered along a branch of stationary solutions with blocking-type flow patterns (see Figures 6.10 and 6.11c). Hopf bifurcation occurs at $\rho \simeq 0.123$, leading to a stable limit cycle with initial period of approximately 40 days, and with rapidly increasing amplitude. As ρ increases further, period-doubling bifurcations occur, along with the growth of background noise and with intermittency. We refer to Section 6.6 for a possible interpretation of this noisy periodicity of 40 days.

For the parameter values of $\alpha^{-1} = 20$ days and $\rho = 0.149$, the solutions exhibit a nearly recurrent sequence of fixed duration. This sequence is about 140 days long and is not repeated identically from one occurrence to another, due to the presence of a significant amount of spectrally-continuous noise. Still, each occurrence is easily identifiable by the similar evolution of several diagnostic variables. The recurrent sequences can succeed each other one or more times -- three-to-five successive appearances are typical -- or be interrupted by other sequences (LG3, Figure 6.9).

As ρ increases above 0.149, the recurrent sequence undergoes period doubling, the noise level increases and more intermittency is observed, i.e., longer separating, nonrecurrent sequences occur. Above $\rho \simeq 0.16$, solution behavior seems completely chaotic. Unlike the transitions near point S, significant changes in flow pattern appear for the time evolutions occurring in this part of parameter space.

Inside the chaotic regime which has dash-dotted boundaries in Fig. 6.13 new instabilities obtain, leading system trajectories to visit new regions in phase space. At this point, the trajectories exhibit sequences of flow patterns associated in succession with different planetary flow regimes familiar to synoptic meteorologists, such as a zonal regime and more meridional regimes (see Section 6.1).

This type of behavior appears to be characteristic of large-scale atmospheric dynamics. Relatively little is known theoretically about transitions between chaotic regimes, or *crises*. To explore the multiplicity of flow regimes which suggest themselves to our attention, we shall study the main features of these regimes numerically, guided by existing theory of phase-space structure on the one hand, and by our synoptic-predictive concerns on the other.

Persistent regimes. We study here the existence of recurrent, persistent flow-pattern sequences in the evolution of chaotic solutions. The closest analogy between such persistent sequences and actual planetary flow regimes obtains in the present model for the region of the second resonance, at relatively small values of α. Therefore, we shall investigate the question of persistent flow patterns for the parameter domain shown in Figure 6.10: $0.13 \leq \rho \leq 0.35$, with $\alpha = 0.05 \text{ day}^{-1}$ and $h_0 = 0.1$. In this domain, the only stationary solutions which are stable lie on the isola, and most numerically observed solutions are aperiodic.

As a measure of persistence, we adopt the quantity $C(t)$,

$$C(t) = \frac{||\Psi(t + \tau) - \Psi(t)||}{\tau}, \tag{6.31}$$

which measures the speed of the solution point $\Psi(t)$ along its orbit in phase space. Here τ is a sampling time interval, taken equal to two days in dimensional units. When $C(t)$ is small for a given length of time, the evolution of the flow pattern in physical space is observed to be slow, and the pattern persists for that length of time.

Figures 6.14a-e show the evolution of $C(t)$ for several solutions computed at $\rho = \rho_1, \rho_2, \ldots, \rho_5$, with ρ increasing from Figure 6.14a to Figure 6.14e, as indicated in the caption (see also Fig. 6.10 for orientation). Notice first that the total range of variation in $C(t)$ increases from $\Delta C \simeq 0.10$ in Figure 6.14a to $\Delta C \simeq 0.50$ in Figure 6.14e. Hence the irregularity of the motion in phase space increases with ρ.

The minima of $C(t)$ are not as sharp as the maxima. The width of the minima is larger and their flatness more pronounced for larger values of ρ, both width and flatness being most striking for $\rho = \rho_4$. It turns out that all persistent sequences belong to the blocking family of flow patterns for $\rho = \rho_1$ and $\rho = \rho_2$. They have exclusively zonal patterns for $\rho = \rho_4$ and $\rho = \rho_5$.

We shall call therefore the *flow evolutions* associated with $\rho_1 \leq \rho \leq \rho_2$ blocking-dominated or blocked evolutions, and the ones for $\rho_4 \leq \rho \leq \rho_5$ zonal-flow dominated, or near-zonal evolutions. Persistent sequences with either pattern are present for the intermediate parameter value $\rho = \rho_3$. Thus, the near-zonal flow evolutions are generally more agitated, and the blocked flow evolutions more quiescent, in agreement with synoptic evidence.

A more precise definition of persistence can be given by choosing an offset value, C_0. Averaging the flow patterns during a given sequence where $C(t)$ remains below the offset, $C(t) \leq C_0$, produces a *composite pattern*. Such a composite is shown in Figure 6.15a for numerical experiment 2 at $\rho = \rho_2$ and the sequence centered at $t = 2640$ days with an offset value $C_0 = 0.0045$. It possesses strong similarities to the stationary blocking solution shown in Fig. 6.10c. All other composites obtained in this experiment are very close to the one shown. The same results hold for experiment 1 at $\rho = \rho_1$.

In both experiments 1 and 2, two consecutive blocked sequences are separated by rather complicated episodes. Most of these episodes possess zonal transients. One of them, marked by an arrow in Fig. 6.14b is shown in Figure 6.15b. These transients, however, do not bear any similarity with the Zonal 1 family, since they remain at low energy.

We must mention here two points which will be developed later. First, that all along the two experiments at $\rho = \rho_1$ and $\rho = \rho_2$, the time-evolving solution remains closer to the blocked branch of stationary solutions than to Zonal 1 or Zonal 2-type solutions. Second, the Zonal 1-type solutions are hardly observed numerically in the parameter range where they exist as stable stationary solutions. The basin of attraction

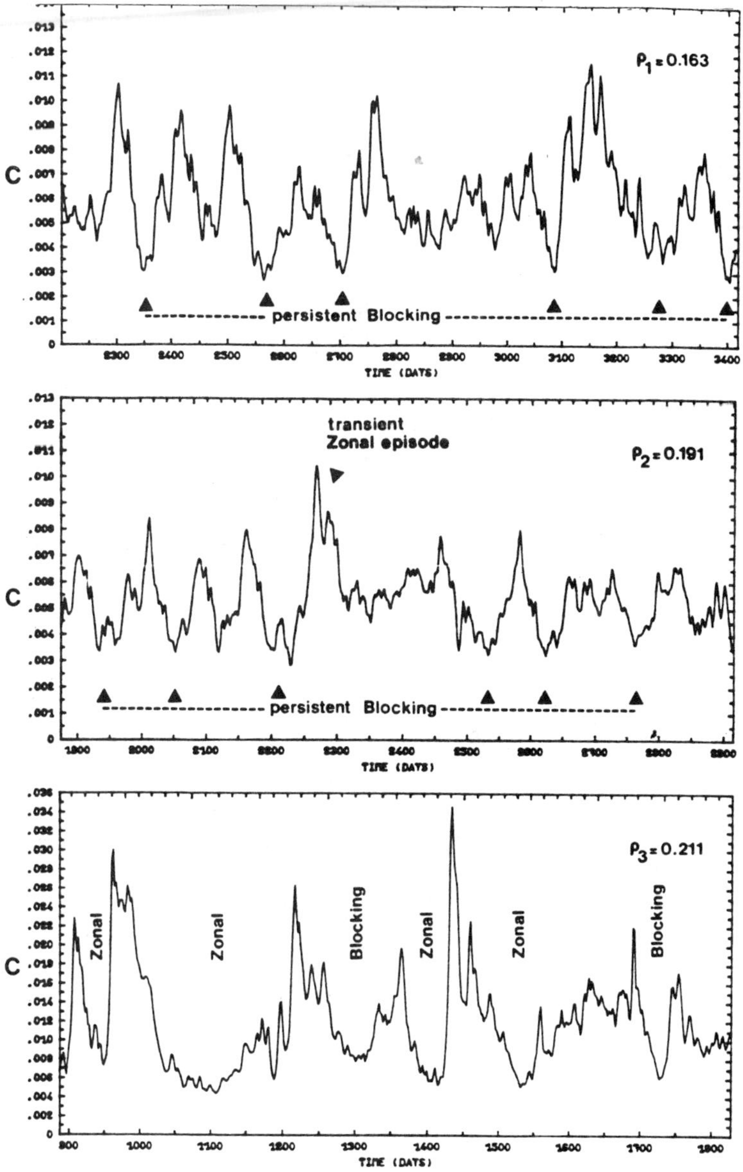

Figs. 6.14a-c

Fig. 6.14. Variations in time of the speed of evolution of solutions C(t), Eq. (6.31). (a) $\rho_1 = 0.163$, (b) $\rho_2 = 0.191$, (c) $\rho_3 = 0.211$, (d) $\rho_4 = 0.228$, (e) $\rho_5 = 0.272$. Persistent sequences correspond to low values of C (from LG3).

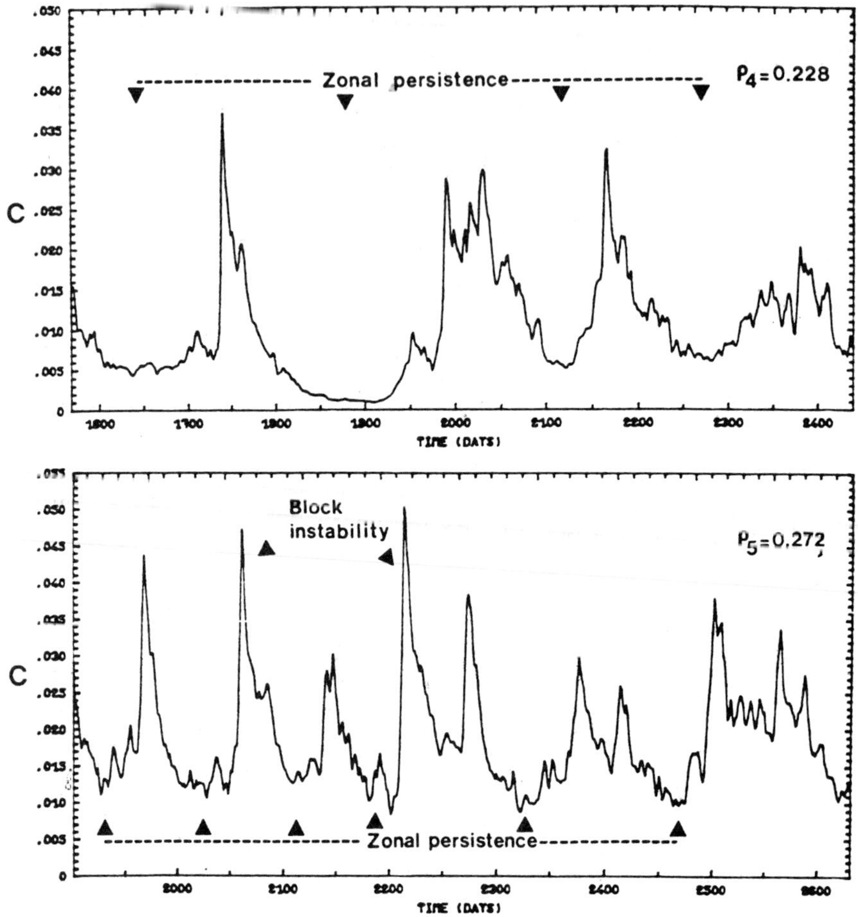
Zonal persistence
$P_4 = 0.228$
C
TIME (DAYS)
Block
instability
$P_5 = 0.272$
C
Zonal persistence
TIME (DAYS)

Figs. 6.14d,e

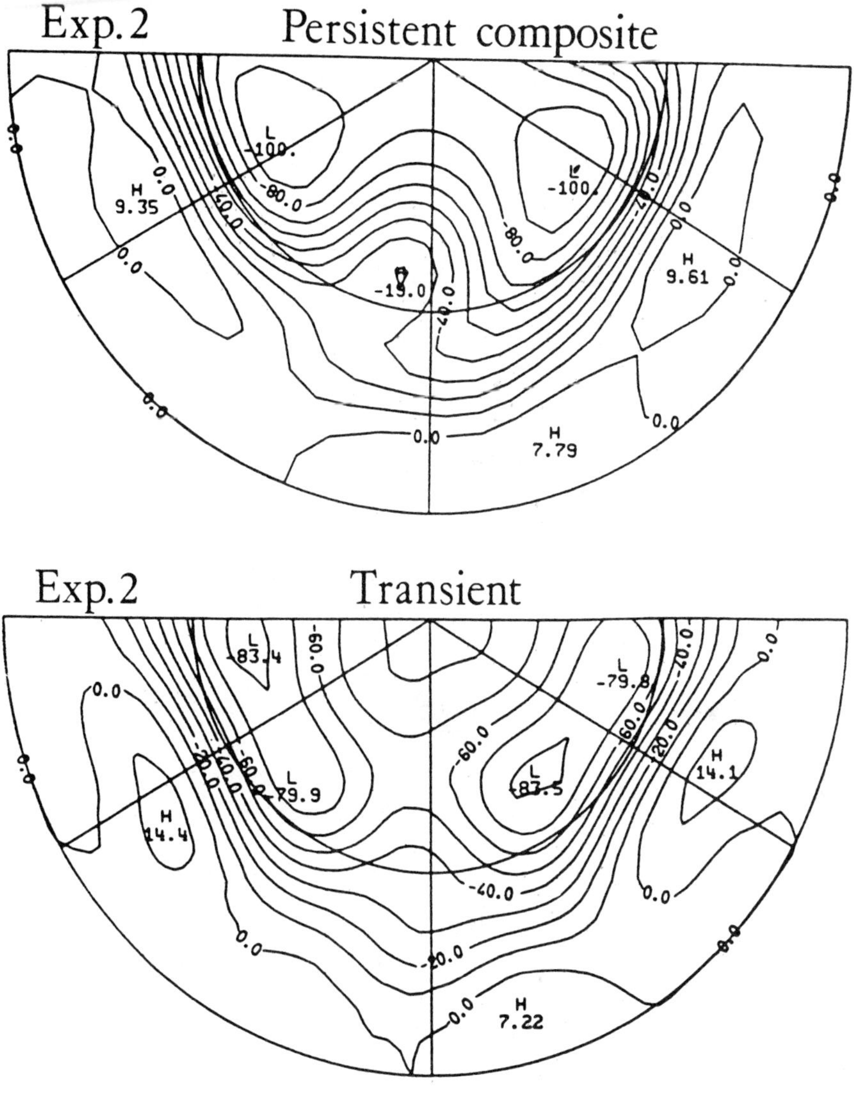

Fig. 6.15. Two stream function fields from experiment 2 at $\rho_2 = 0.191$: (a) composite of a persistent sequence; (b) instantaneous field taken during a transient zonal episode (from LG3).

of the branch appears to be very small in volume and limited to the immediate vicinity of the branch in distance.

The persistence behavior in Figures 6.14d and 6.14e is rather different from Figures 6.14a and 6.14b. The minima of $C(t)$ are quite flat compared to the maxima, thus defining sequences of very persistent flow patterns. But sharp instability peaks are in evidence, showing that the solution point in phase space is violently expelled from the region of persistence. It then relaxes slowly to a quieter evolution, producing a reversed, irregular sawtooth profile in the $C(t)$ time series. Furthermore, the observed flow characteristics are reversed with respect to the previous experiments, 1 and 2: persistent sequences show zonal patterns and transient blocks occur during episodes of rapid flow evolution, together with other complicated types of highly meridional circulations.

This succession of events is strongly reminiscent of the classical synoptic description of an individual *index cycle* (see also Section 6.1). During such a cycle, high-index flow with strong midlatitude westerlies and long superimposed waves is followed by a shortening of the upper-air wavelength pattern, then by a sudden and complete breakup of the westerly flow into north-south oriented pressure cells, with occlusion of stationary cyclones and strong anticyclogenesis, and finally by a gradual reestablishment of the zonal flow. The low-index portion of the cycle, where meridional flow is strong, includes blocking patterns, but their duration varies between a few days and a few weeks.

The whole sequence of events is called a cycle because a high-index pattern occurs at its beginning, as well as at its end, but no exact periodicity, i.e., no constant duration of the cycle, is implied by observations. Furthermore, the occurrence of marked cycles is favored at certain seasons, but is not the same from year to year, suggesting therewith a dependence on thermal forcing; the intensity of this forcing is represented in the model of this section by the parameter ρ.

To study in further detail the character of flow patterns in (6.27) for this index-cycle-like evolution, we composite persistent sequences as before. Figure 6.16a shows one composite pattern obtained from the sequence centered at $t = 2330$ dyas in Figure 6.14e with the offset value $C_0 = 0.009$. This value of C_0 is equivalent to the one used in experiment 2, at $\rho = \rho_2$, after normalization by the mean value of $C(t)$ in either experiment.

Surprisingly, this composite is more similar to the Zonal 1 type of solution, which does not exist as a stationary solution at this value of parameters, than to the stationary solution on the Zonal 2 branch at

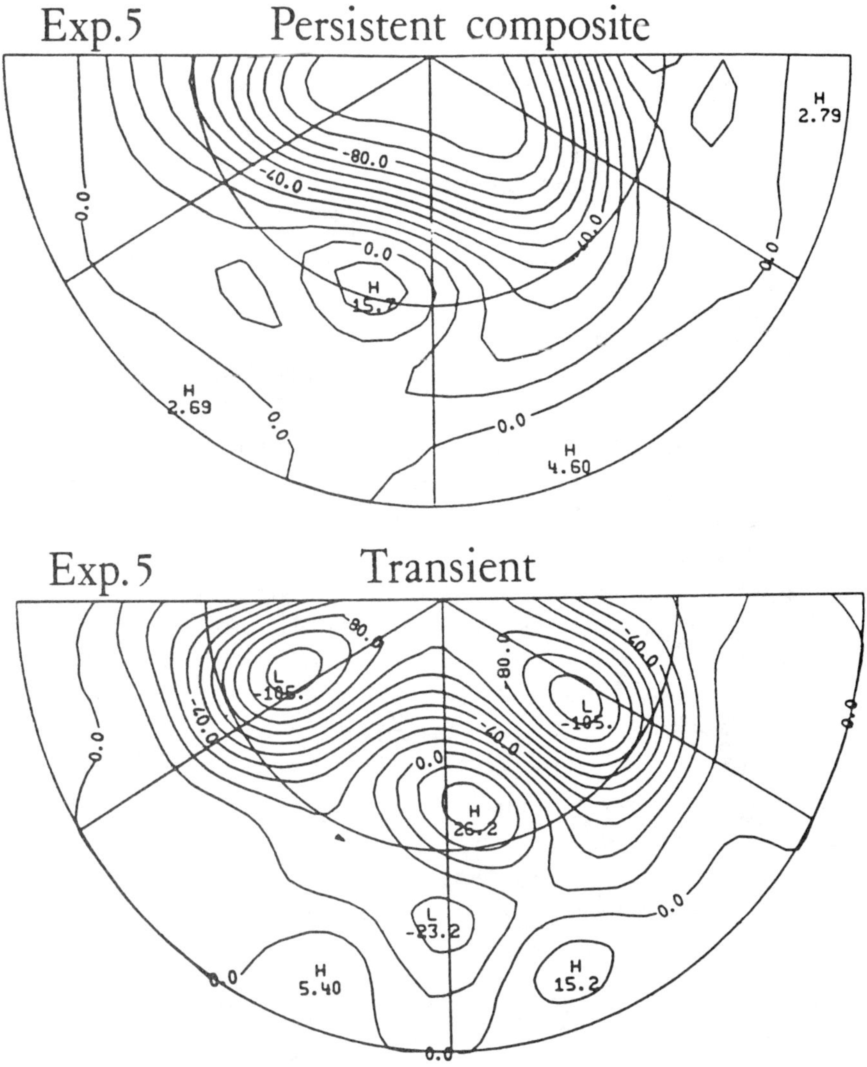

Fig. 6.16. Stream function fields from experiment 5 at $\rho_5 = 0.272$: (a) composite of a persistent sequence, (b) instantaneous snapshot taken during a transient Block instability (from LG3).

$\rho = \rho_5$. The same observation holds for all composites obtained from experiment 4. In experiment 5, some composites are found to be closer to Zonal 2 stationary solutions.

One transient solution, marked by the upper right arrow in Figure 6.14e is shown in Figure 6.16b. This dipole block resembles certain *modon* patterns which appear as spatially localized stationary solutions of the free, conservative barotropic vorticity equation. In the present, forced-dissipative model, this feature is present only as a very transient, unstable flow pattern. But the question of stability of isolated coherent structures in realistic large-scale flows is under active investigation, and the interplay between the two points of view, local and global in physical space, will certainly contribute greatly to a better understanding of the question of atmospheric persistent anomalies. Moreover, synoptic evidence indicates that blocking events with such a dipole structure are often quite short-lived.

The intermediate experiment 3, Fig. 6.14c, exhibits both types of persistent sequences, zonal and blocked. The flow pattern characteristic of each sequence, easily identifiable by visual inspection of microfilm output, is indicated in Figure 6.14e. Four zonal sequences and two blocking sequences appear in the numerical solution segment displayed in the figure, but the relative frequency and length of the sequences depends on the exact value of ρ, and on the segment of aperiodic solution chosen.

The persistent sequences discussed here, like the stationary solutions they resemble, exhibit easterlies in low latitudes. For persistent zonal flow, at $\rho = \rho_5$, the maximum zonally-averaged easterly speed at the Equator is 8 ms^{-1}, while the maximum westerly speed is 50 ms^{-1} at 60 N and mean zonal flow is nearly null between 7 N and 20 N. For persistent blocked flow, at $\rho = \rho_2$, the maximum easterlies are 3 ms^{-1}, the maximum zonally-averaged westerlies 25 ms^{-1}, and a more sharply defined critical belt of near-zero mean zonal wind speed occurs near 10 N.

In general, the critical line of null zonal wind speed is ill defined. Its latitudinal position changes with a characteristic time comparable to that of all other changes in low wave numbers, more slowly in a blocked regime and faster in a zonal regime. The distribution of zonal momentum with latitude in the model changes in qualitative agreement with observed changes between high-index and low-index situations. But the profiles of mean zonal wind are not good indicators, by themselves, of transitions between persistent episodes and more transient ones.

In order to check more directly the connection between persistent sequences and stationary solutions, we compute for all experiments the deviations $D_i(t) = ||\Psi(t) - \Psi_i||$ to a prescribed stationary solution Ψ_i. Each label i corresponds to one particular branch of stationary solutions as indicated in Figures 6.10 and 6.17. For each experiment at a given value of ρ, we consider stationary solutions present at this value of ρ. An exception is made for Zonal 1-type solutions: for values of ρ where this branch does not exist, we consider the solution located at the nearest turning point of the isolated branch, marked by a star in Figure 6.10. The correlation coefficient between the pair of time series $C(t)$ and $D_i(t)$ is also computed and indicated in Figure 6.17.

Figure 6.17a shows the variations of $D_i(t)$ for experiment 2 at $\rho = \rho_2$. The trajectory remains close to both blocked branches D and E, with some preference for branch E. Larger distances are maintained to the Double-Block solution (F), Zonal 2 solution (C) and to the Zonal 1 point closest to the solution (B). A general modulation of period close to 40 days contains a significant part of the autocorrelation power spectrum (not shown here) of all deviation time series, $D_i(t)$. This modulation is reminiscent of the original limit cycle which develops along the blocked branch D when it loses its stability at lower ρ (see also previous discussions of this limit cycle and subsequent, 140 day-long recurrent sequence, and Section 6.6).

The deviations from branches B, C and F are poorly correlated with $C(t)$. There is better correlation with blocked branches D and E, although not as good as expected from persistent composite patterns and examination of the detailed microfiche film of the experiment. The reason for this imperfect correlation lies in the fact that our measure of deviation $D_i(t)$ does not distinguish between phase displacements of a given planetary flow pattern and actual change in pattern. Essentially the same results with respect to $D_i(t)$ and correlation with $C(t)$ hold for experiment 1.

Figure 6.17c shows the variations of $D_i(t)$ for experiment 4. The trajectory is now closest to the Zonal 1 (B) and Zonal 2 stationary solutions (C), while the blocked solution (E) remains at larger distances. The deviation from the Zonal 1 solution is strongly correlated with $C(t)$, cf. Figure 6.14d. The weaker, but still apparent correlation with deviation from the Zonal 2 branch comes entirely from episodes of rapid variations, during which the trajectory is actually far from any stationary solution. The deviation from the blocked branch (E) is now clearly

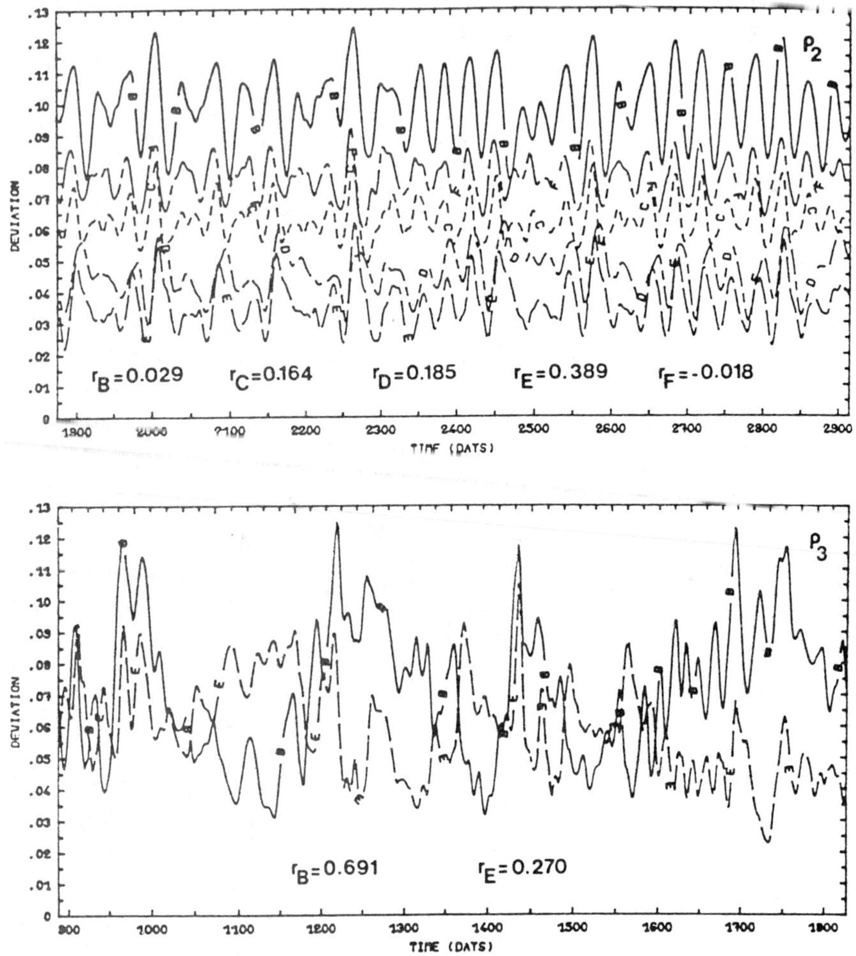

Figs. 6.17a,b

Fig. 6.17. Variations in time of the deviations $D_i(t)$ of the trajectory in phase space from various branches of stationary solutions. Correlations r_i between the time series C(t) and the deviations $D_i(t)$ are indicated on each panel (from LG3).

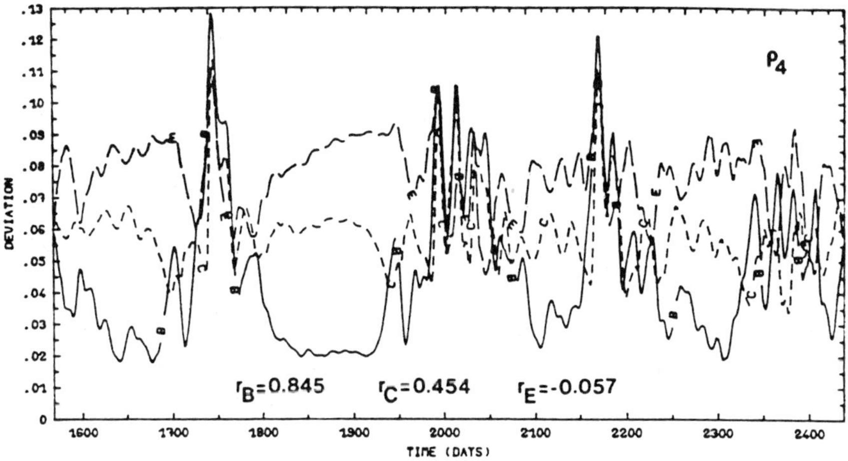
DEVIATION
TIME (DAYS)
ρ_4
r_B=0.845
r_C=0.454
r_E=-0.057

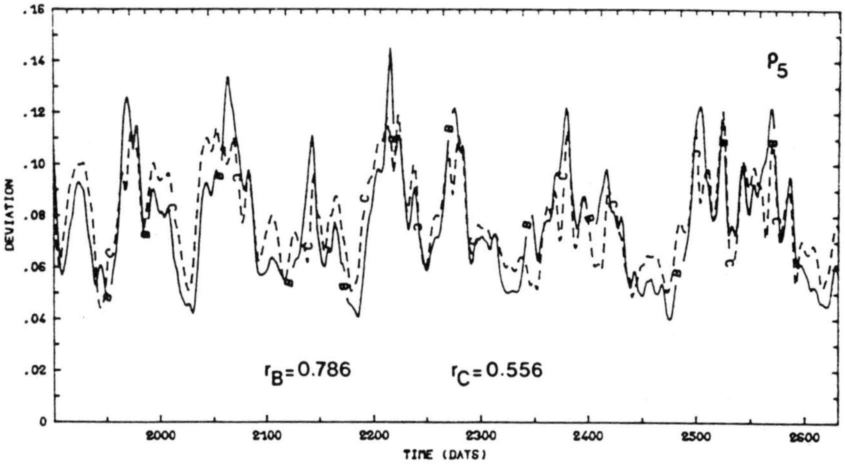
DEVIATION
TIME (DAYS)
ρ_5
r_B=0.786
r_C=0.556

Figs. 6.17c,d

anticorrelated with $C(t)$ during persistent sequences. However the overall correlation between $D_E(t)$ and $C(t)$ is close to zero due to the positive contribution of transients.

For experiment 6, Figure 6.17d, the average deviations from stationary solutions increase. The correlation between $C(t)$ and the deviation $D_B(t)$ from Zonal 1 solutions is very good (see also Figure 6.14e), and remains higher than between $C(t)$ and the deviation $D_C(t)$ from the Zonal 2 solution. However, $D_C(t)$ is also well correlated now with $C(t)$. There is still no correlation between $C(t)$ and deviations from the blocking branch (not shown in the figure). As ρ increases beyond ρ_5, the deviations from the Zonal 1 and Zonal 2 branches tend to correlate equally well with $C(t)$, until new planetary flow regimes appear due to the Charney-DeVore resonance.

Finally, for experiment 3, Figure 6.17b, the analysis of deviations $D_i(t)$ confirms our previous observations. The minima of deviation from Zonal 1 solutions are well correlated with minima in $C(t)$ during episodes designated zonal in Figure 6.14c, while persistent sequences of blocking, as indicated in Figure 6.14c, are associated with closer proximity to the blocked branch of stationary solutions (E). The overall correlation is better between $D_B(t)$ and $C(t)$ than between $D_E(t)$ and $C(t)$, being again a function of the exact value of ρ, and solution segment.

For all zonal regimes, we have seen that the Zonal 1 type of solution plays a very special role in attracting the trajectories into its neighborhood. On the other hand, there is an apparent paradox in the fact that Zonal 1 regimes are observed for parameter values for which Zonal 1 stationary solutions do not exist, while blocking regimes obtain when Zonal 1 solutions do exist. A few qualitative arguments may help us understand, at least roughly, this behavior.

The first part of the paradox is probably a consequence of the saddle-node bifurcation, which terminates the isolated branch of Zonal 1 stationary solutions at ρ just below $\rho_3 = 0.211$. In such a bifurcation, illustrated in Figure 6.18 (see also Figure 10.6), a pair of stationary solutions, one stable, the other one unstable, coalesce and leave behind an arc of one-dimensional manifold along which the motion in phase space is slow (null at the bifurcation point itself) compared to the rapid convergence onto this arc from other directions.

This topological structure traps the trajectory along the slow manifold, leading to persistent sequences of Zonal 1-like flow patterns.

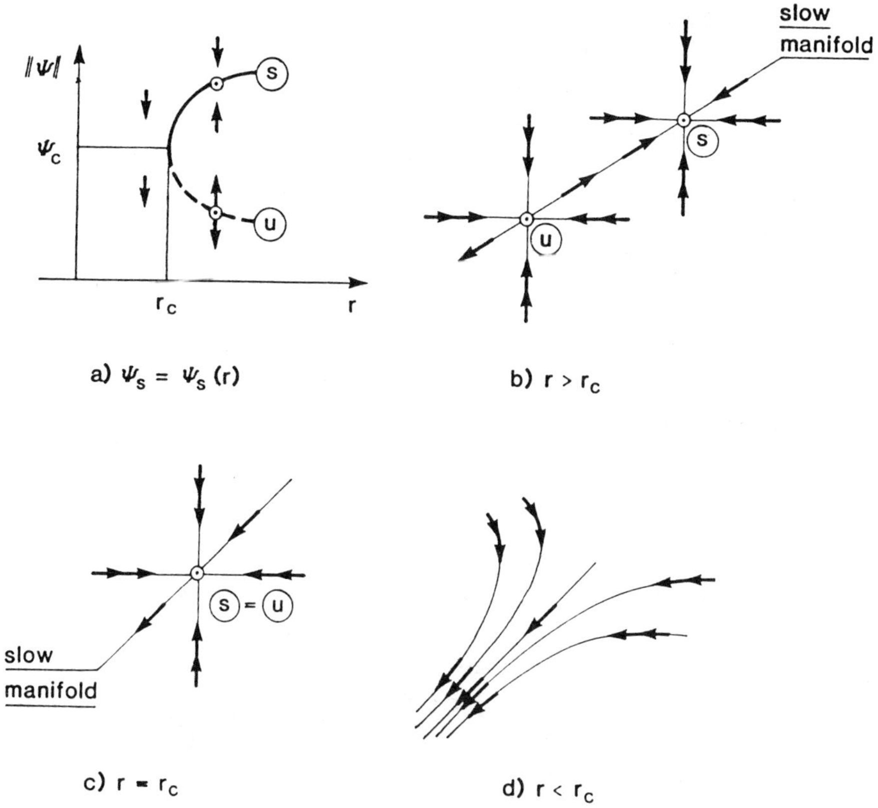

Fig. 6.18. Sketch of a saddle-node bifurcation in which two branches of stationary solutions, one stable and the other unstable, coalesce at the parameter value $r = r_c$ (after LG3).

However, the efficiency of this mechanism to generate such sequences depends on the ability of the entrance funnel into the neighborhood of the given arc to span a large portion of phase space, in order to ensure recurrence of the phenomenon. Further consequences of the action of this mechanism on the model's predictability properties are presented in the next subsection.

The explanation of the paradox's second part is somewhat more complex. In addition to the two saddle-node bifurcations points at its ρ-extremities, several Hopf bifurcations from the branch of Zonal 1 solutions occur along the most energetic, upper part of the isola; the number of unstable eigenvalues for L_S varies there from 0 to 4. At least one of the Hopf bifurcations is *subcritical*, leading to an *unstable* limit cycle which surrounds the stable solution when it exists and repels most of the incident trajectories (compare Section 12.2, Figures 12.8-12.10). This may explain the smallness of the attracting basin for stable Zonal 1 stationary solutions and the nonexistence of stable attractors like limit cycles in the vicinity of the entire isolated branch.

On the other hand, there exists numerically for $\rho \lesssim 0.205$ a stable attractor associated with the blocking-type branches of stationary solutions. Almost all trajectories converge rapidly to this attractor, so that no return is observed to the Zonal 1 branch. In the neighborhood of $\rho \simeq 0.205$, a complicated transition occurs, after which the preceding attractor loses its stability and trajectories are allowed to return recurrently close to the Zonal 1 branch. A link appears possible between such transitions and the existence of heteroclinic orbits connecting the different branches of stationary solutions (see again Figure 12.10, for comparison).

Several results corroborate this hypothesis; for instance the evolution at $\rho = 0.204$ of deviations $D_B(t)$ from the Zonal 1 stationary solution, starting from the neighborhood of the stable upper branch of the isola. Instead of going directly to the stable zonal solution nearby, the trajectory is first ejected to a large distance, displaying an episode of transient block, which is out of phase with the unstable stationary blocking solution; only after this large excursion does the trajectory return to the stable zonal solution near its starting point. Since heteroclinic orbits are structurally unstable, it is difficult to compute them precisely and a more detailed study, like in Section 12.2, would be required.

Previous results are sensitive to the value of the Rossby radius of deformation L_R. L_R has no effect on the distribution of stationary solutions, but influences their stability and the dynamics of time-dependent solutions. A small value of L_R favors the excitation of a large-scale stationary response to orographic forcing, like blocking, since it reduces the phase speed of planetary waves. For a discussion of results when $L_R = \infty$, and hence also for L_R very large, i.e., for a purely barotropic atmosphere, we refer to LG3 (Figure 16, in particular).

Regime-dependent predictability. Recalling Figure 6.12, we have brought so far strong numerical evidence for the connection between persistence of certain flow patterns in physical space and the proximity of the trajectory in phase space to stable or unstable stationary solutions with similar flow patterns. We investigate now the potential usefulness of this connection in determining the predictability of the associated planetary flow regimes.

At first, this leads us to examine the statistics of persistent sequences in the numerical experiments described above. Figure 6.19 shows the number of events per simulated year plotted against the duration of events, for solutions computed at $\rho = \rho_1, \rho_2, \ldots, \rho_5$. In order to get enough confidence in the statistics, long-time integrations were performed over time intervals 24 times as long as those displayed in Figures 6.14a-e. For $\rho = 0.20$, this corresponds to roughly 65 years of dimensional time, a record over four times as long as the 15-year data set at the basis of many current observational studies, and used here for each separate experiment. The sampling interval is 1.5 nondimensional time units, or 3 days at $\rho = 0.20$.

The offset value for each experiment is taken as the time average of $C(t)$ minus one half of its standard deviation. This choice of offset means that we compare here properties of persistence relative to the prevailing flow evolution imposed by the parameter values. It is clear from Figure 6.14, and was mentioned already before, that flow evolution in the blocking-dominated parameter range of $\rho_1 \leq \rho \leq \rho_2$ is on the average much more persistent than in the zonal-flow dominated range $\rho_4 \leq \rho \leq \rho_5$. Nevertheless, the minima and maxima of $C(t;\rho)$ for given ρ still correspond to persistent and rapidly changing flow patterns, respectively.

All curves in Figure 6.19 exhibit a smooth decrease of number of persistent sequences with the duration of persistence. The lack of a

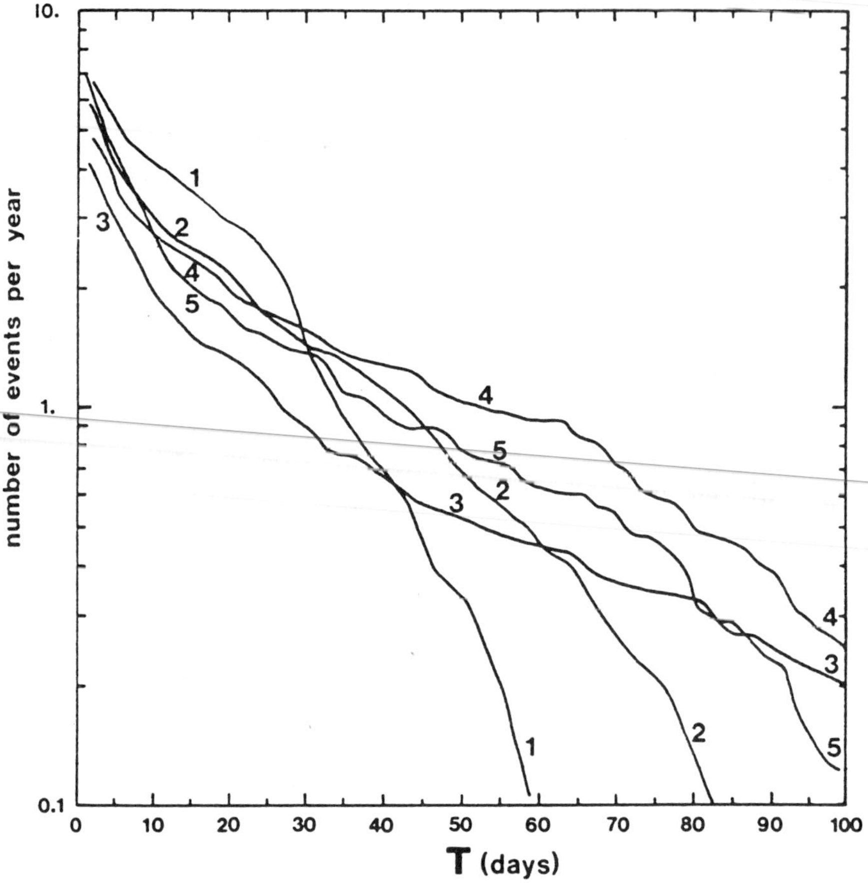

Fig. 6.19. Number of persistent sequences per year whose duration exceeds the number of days indicated on the abscissa, for experiments 1 to 5. The offset values for persistence C_0 are: $C_{01} = 5.0 \times 10^{-3}$, $C_{02} = 5.0 \times 10^{-3}$, $C_{03} = 7.0 \times 10^{-3}$, $C_{04} = 8.8 \times 10^{-3}$, $C_{05} = 12.6 \times 10^{-3}$ (from LG3).

maximum, or even of a pronounced plateau at a given duration, indicates that neither relatively long zonal sequences, appearing for $\rho_4 \leq \rho \leq \rho_5$, nor blocking sequences appearing for $\rho_1 \leq \rho \leq \rho_2$, have a preferred duration.

In the log-linear coordinates of the figure, the duration of *runs*, or persistence of a given sign for a linear first-order Markov process with mean zero (cf. Eqs. (10.28-10.32), is given by a straight line. The slope $-\nu$ of such a line indicates the exponential relaxation time, or e-folding time, τ, of the process, often called *red noise* for short, with $\tau = 1/\nu$. We have fitted a straight line by least squares to the curves in Figure 6.19, ignoring the first few points where uncertainty is due to the finite sampling interval, and the last few which are uncertain due to the finite length of each experiment. The results are shown in Table 6.1.

Table 6.1. The characteristic exponent λ_L, its inverse λ_L^{-1}, and the e-folding time τ for experiments 1 through 5*.

Exp.	1	2	3	4	5
ρ	0.149	0.191	0.211	0.228	0.272
$\lambda_L(\mathrm{day}^{-1})$	0.028	0.033	0.058	0.087	0.095
λ_L^{-1} (day)	36	30	17	11	10
τ	14	25	-	42	47

*The estimated precision of λ_L is ±5%, based on using different initial data and different segments of the same orbit.

The exact values τ vary somewhat with the offset criterion chosen, but the relative ordering $\tau_1 < \tau_2 < \tau_4 < \tau_5$ is always the same. This shows in fact longer persistence times for zonal sequences than for blocking sequences.

It is interesting to compare these results with the observational data in Figure 8 of Dole and Gordon (1983). They present the same type of statistics for regional anomalies of 500 mb geopotential fields from 15 winters of Northern Hemisphere atmospheric data (see Sections 6.1 and 6.6). The general character of the curves is very similar. Positive anomalies of a given duration occur more frequently than negative ones both in the PAC (North Pacific) region and in the ATL (North Atlantic) region, as defined by Dole and Gordon, with PAC positive being most persistent of all. The synoptic character of the flow across the continental margin for PAC positive and ATL negative anomalies is zonal, being blocked for PAC negative and ATL positive.

Thus one obtains reasonable qualitative agreement with atmospheric behavior. The actual values of τ for the present model are considerably higher than those reported from observations. This discrepancy is most likely due to the limited resolution and absence of baroclinic transition mechanisms between regimes in our model, and we shall return to this problem in Section 6.5.

Curves 1 and 2 deviate actually from a straight line by being convex, except near the origin, while 4 and 5 are rather concave up to very high persistences. More precisely, one can define separately the least-square slope of each curve for durations between 5 and 30 days, $\nu_j^{(1)}$, and between 35 and 60 days, $\nu_j^{(2)}$. The corresponding slopes satisfy the inequalities $\nu_j^{(1)} < \nu_j < \nu_j^{(2)}$ for blocked regimes, j = 1 and 2, and the opposite inequalities for zonal regimes, $\nu_j^{(1)} > \nu_j > \nu_j^{(2)}$ for j = 4 and 5. Thus for instance $\tau_2^{(1)} = 35$ days, $\tau_2 = 25$ days (see Table 6.1), and $\tau_2^{(2)} = 20$ days, where $\tau_j^{(k)} = 1/\nu_j^{(k)}$.

The curve for experiment 3, at the transition between zonal-flow dominated and blocked-flow dominated regimes in parameter space, is more complex. This is due to the fact that each of the two coexisting regimes actually has a different characteristic distribution of persistence times, with distinct means and hence distinct threshold values. Curve 3 is thus the weighted superposition of two separate curves (not shown), and no single, well defined value of τ_j exists.

The deterministic dynamics of the model under discussion thus produce waiting times for exit from a given flow regime, or durations of a persistent sequence, which have an approximately exponential distribution, i.e., approximately linear in semilogarithmic coordinates. For a pure, linear red-noise process (Eqs. (10.28, 10.29)), this distribution is exactly log-linear and its slope gives the probability of continued persistence of the given flow pattern, with the probability of exit being proportional to the absolute value of the slope, ν.

We suspect, therefore, that the local slope of the distribution in the present nonlinear, deterministic model is also proportional to the probability of exit from a given flow regime, for persistent sequences of given length. These considerations suggest introducing the concept of *regime predictability* of a flow pattern, in model and nature. This concept depends on the persistence properties of a finite region of phase space, be it an isolated attractor set, coexisting with other attractors, or a subset of an attractor which is only weakly connected to other portions of the attractor, and which is close to an unstable fixed point.

Regime predictability must be distinguished from the more familiar concept of *pointwise predictability*, which arises from the growth of small errors, at each point in physical space or in phase space, and at each moment in time. Pointwise predictability can be estimated by computing the mean rate of divergence of two trajectories starting close to each other in phase space. Predictability studies of the atmosphere by the three distinct methods of i) naturally-occurring analogs, ii) the statistical theory of turbulence and iii) numerical experiments with general circulation models (see Sections 4.5, 4.6 and 6.6) all show that small errors, on the average, grow exponentially. This fact is in agreement with the results of dynamical system theory.

The exponential growth of small errors in a system with chaotic dynamics, like those of Section 5.4 and of the present section, is given by the largest *characteristic exponent* λ_L. This number, also called the largest Liapunov exponent, provides a generalization to nonlinear systems of the familiar concept of instability exponent for a linear system (see Sections 4.4, 5.3 and 5.4).

Consider a system of n nonlinear ordinary differential equations, like (5.24), which we rewrite here for convenience as

$$\dot{x} = F(x), \tag{6.32}$$

The successive linearizations of (6.32) along a trajectory are given by the variational equation, or tangential differential system

$$\dot{y}(t) = DF(x(t))\, y(t), \tag{6.33}$$

where $DF = (\partial F_i/\partial x_j)$ is the Jacobian matrix, evaluated at $x = x(t)$. For a linear system with $F(x) = Ax$, the linearization (6.33) gives simply $DF = A =$ constant. The characteristic exponent λ_L generalizes therefore the concept of algebraically largest eigenvalue of A.

One defines

$$\lambda_L = \lim_{T\to\infty} \frac{1}{T} \ell n||y(T)||, \tag{6.34}$$

and it is possible to show under certain technical assumptions that the limit in (6.34) exists and is unique for almost all initial data $x(0)$, $y(0)$. These assumptions have to guarantee, in particular, the existence of a probability measure on the system's attractor, which is invariant under the phase-space flow governed by (6.32). The time mean in (6.34) is then equal to the corresponding ensemble mean with respect to this invariant measure.

It is currently believed that the situation described above prevails in the Navier-Stokes equations, at least in two space dimensions, and hence in the barotropic vorticity equation (6.23). Thus we should be able to calculate λ_L by (6.34) for the trajectories of (6.27). The results of this calculation are given in Table 6.1, for experiments 1 through 5.

The characteristic exponent increases monotonically from $\rho = \rho_1$ to $\rho = \rho_5$, in agreement with the observations about ρ-dependence of persistence made in discussing Figure 6.12. The inverse λ_L^{-1} measures the average e-folding time of small errors, and decreases by a factor of three from $\rho = \rho_1$ to $\rho = \rho_4$ or $\rho = \rho_5$.

It is clear from Table 6.1, by comparing the mean e-folding time with the e-folding time of small errors, λ_L^{-1}, that their order of magnitude, 0 (10 days), is the same, but their behavior is quite different: λ_L^{1} decreases with ρ, while τ increases with ρ. Actually, τ characterizes regime predictability, as defined earlier, while λ_L^{-1} characterizes pointwise predictability, as usual. Thus their comparable magnitude is somewhat fortuitous, and probably misleading in the analysis of atmospheric data. This assertion could be tested by computing the correlation between forecast skill verified at two days or less and skill at fifteen days or more in a number of advanced numerical weather prediction models. The former skill is presumably a measure of small error growth, i.e., of pointwise predictability, while the latter gives an indication of regime predictability.

In our model, λ_L^{-1} decreases with ρ and, moreover, is larger than τ for blocked flow evolutions, $\rho_1 \leq \rho \leq \rho_2$, and less than τ for zonal evolutions $\rho_4 \leq \rho \leq \rho_5$. Hence, persistent sequences carry considerably increased predictability in a zonal, but not in a blocked regime, when compared to all trajectory segments of the same length in the same regime. Since λ_L^{-1} is largest in blocking regimes, $\rho_1 \leq \rho \leq \rho_2$, it follows that in such a regime, all flow patterns, not only the ones which are blocked, lead to pattern evolutions which are more predictable on the average than those in a zonal regime. On the other hand, the zonal regimes, $\rho_4 \leq \rho \leq \rho_5$, are characterized by a distribution of predictability properties which is very inhomogeneous in phase space, i.e., depends very strongly on initial data. A zonal flow pattern in such a regime is much more likely to persist than a blocked pattern, while the reverse is less true in a blocking regime. Whether these results actually apply to large-scale midlatitude flows is still an open question, and we shall touch upon it in Section 6.5.

Predictability studies also indicate that error growth slows down and saturates, due to the atmosphere's bounded energy, which leads to the bounded volume in phase space of any attractor set which might characterize large-scale atmospheric flow (see Section 5.4). The fact that error growth, when averaged over all possible states accessible to the atmosphere, becomes progressively slower in time, suggests that nonlocal properties of the attractor always play a role. This role of the global dynamics can differ according to the flow regime under consideration, due to inhomogeneities of system behavior in phase space.

In particular, the duration of especially long persistent sequences will be influenced by "obstacles" to their leaving a preferred neighborhood, i.e., by the structure of basin boundaries for the attractor basin of one or more attractors present in the system's phase space at any given parameter values. Indeed, the numerically computed trajectories in our experiments are *pseudo-orbits* of the time-continuous sytem (6.27), and the numerical error at each time step can be considered as a realization of stochastic noise perturbing the orbits of (6.27) (see references to approximation results in dynamical system theory in Section 5.5).

To verify the effect of such numerical errors, and hence, more generally, of small modeling errors, on the persistence results discussed here, we have repeated experiment 5 with different initial data, as well as in the presence of additive random noise with a standard deviation of 10^5 times the machine accuracy, i.e., comparable to the truncation error of the scheme. The corresponding realizations of curve 5 in Figure 6.19 (not shown) have the same qualitative properties, and the ordinates at each duration differ by at most 2-3 percent from the realization shown in Figure 6.19.

The persistence results reported here are thus insensitive to pseudo-random errors in discretization, and one would hope the same might hold with respect to sufficiently small modeling errors in a complex, high-resolution numerical weather prediction model. Given the inevitable character of such errors, we recall that, when a portion of the boundary of an attractor basin of the time-continuous model system, e.g., (6.27), comes very close to the attractor itself, pseudo-orbits may leave the attractor, with mean waiting times for exit depending on the size of the basin and the structure of the boundary. Moreover, for any given mean exit time, the distribution of actual waiting times may be exponential or nearly so. This is in complete agreement with the results shown in Figure 6.19.

Basin boundaries are often made up, partly or entirely, of stable manifolds of certain fixed points or periodic solutions, which are themselves unstable. Such a situation was already discussed in connection with the attracting properties of Zonal 1 fixed points, so that the present description is entirely consistent with the interpretation of results given earlier. Here we emphasize the global properties of an attractor or of a large piece of an attractor, while before we stressed the connection between behavior on the attractor, on the one hand, and the presence of neighboring unstable fixed points or periodic solutions, on the other. Globally, a chaotic attractor can be generated or destroyed by separating from or colliding with a nonchaotic, simple attractor.

These concepts strongly suggest the following picture: for $\rho_1 \leq \rho \leq \rho_2$ there exists a single chaotic attractor of (6.27), partially co-existing with the Zonal 1 stable or unstable fixed points, but lying near the Blocked, unstable fixed points. This attractor carries flows which resemble synoptically blocking patterns and are on the average quiescent relative to the grand-ensemble behavior of all atmospheric flows. For $\rho_4 \leq \rho \leq \rho_5$ there also exists a single, chaotic attractor, carrying flows which resemble Zonal 1 or Zonal 2 stationary solutions, and are on the average more agitated. Near $\rho = \rho_3$ a crisis occurs, the two types of chaotic attractors exchange their stability, and pseudo-orbits can switch from the neighborhood of the one to the other.

The probability of exiting from the blocking-dominated regime increases with the length of residence time (Figure 6.19, Curves 1 and 2). For the zonal regime, this probability decreases with residence time (Figure 6.19, curves 4 and 5). In both cases, the fact that this probability is not constant, as it would be for a Markov process, suggests that the basin boundaries are rather complicated and might have fractional, rather than integer dimension, like the attractors or the destabilized "ghosts" of attractors themselves (compare discussion of Figure 5.12, and the "onion skin" structure of the attractor studied there).

At this point, we inquire whether the divergence of trajectories on the attractor is approximately constant in phase space. For this purpose, we consider the largest real part $\sigma_M(t)$ of the eigenvalues of $\mathcal{L}(t)$, where L is given by Eq. (6.30) as the instantaneous linearization of the equations of motion about a time-dependent solution $\Psi(t)$, rather than a stationary solution Ψ_s, cf. also Eq. (6.33). Thus σ_M is a local, rather than global measure of the divergence of trajectories, while the average $\bar{\sigma}_M$ of $\sigma_M(t)$ over a trajectory is at least λ_L, cf. Eq. (6.34). The fact that $\bar{\sigma}_M \geq \lambda_L$, rather than $\bar{\sigma}_M = \lambda_L$, is due to the variational solution $y(t)$ not being aligned at each point $x(t)$ with the eigenvector of $DF(x(t))$ which corresponds to $\sigma_M(t)$, so that $y(t)$ grows at most as fast as this eigenvector.

In Figure 6.20a,b, $\sigma_M(t)$ is plotted for experiments 2 and 4, respectively. It shows large variations over short time intervals when compared with other functionals of the flow field $\Psi(t;\rho)$ for the same values of the parameter ρ (compare Figures 6.14b,d and 6.17a,c, respectively). This large variability is due at least in part to the contribution to σ_M of rapidly varying, small-amplitude fluctuations in the spatially smallest scales of motion. Such fluctuations do not appear in $C(t)$, $D_i(t)$ or other "smooth" functionals, due to the small total energy of the small scales.

The average value $\bar{\sigma}$ of $\sigma_M(t)$ is approximately 0.07 day^{-1} for $\rho = \rho_2$ and 0.11 day^{-1} for $\rho = \rho_4$. In either case, λ_L is smaller than $\bar{\sigma}$, but of the same order. Occasionally, values of $\sigma_M(t)$ as large as 0.26 and 0.47 day^{-1}, respectively, obtain.

Negative values of $\sigma_M(t)$ obtain also, indicating that at such times all eigenvalues of $L(t)$ are in the left-half plane, and the trajectory $\Psi(t)$ at such points in phase space is attracting in all directions. Comparison of Figure 6.20 with 6.14b shows that negative values of $\sigma_M(t)$ precede quite systematically minima of $C(t)$, and large values of σ_M precede maxima of C. This lagged correlation between $\sigma_M(t)$ and $C(t)$ only exists at the onset of persistent or rapidly varying sequences; otherwise one type of sequence does not differ significantly in mean divergence rate from the other.

These observations confirm the previous picture of the relation between chaotic attractors and nearby unstable fixed points. Persistences are associated with gradual capture of the trajectory into a contracting phase-flow region near the stable manifold of the fixed point, and rapid transients with strong instabilities along the latter's unstable manifold, cf. Figure 6.12. This interpretation is consistent with the

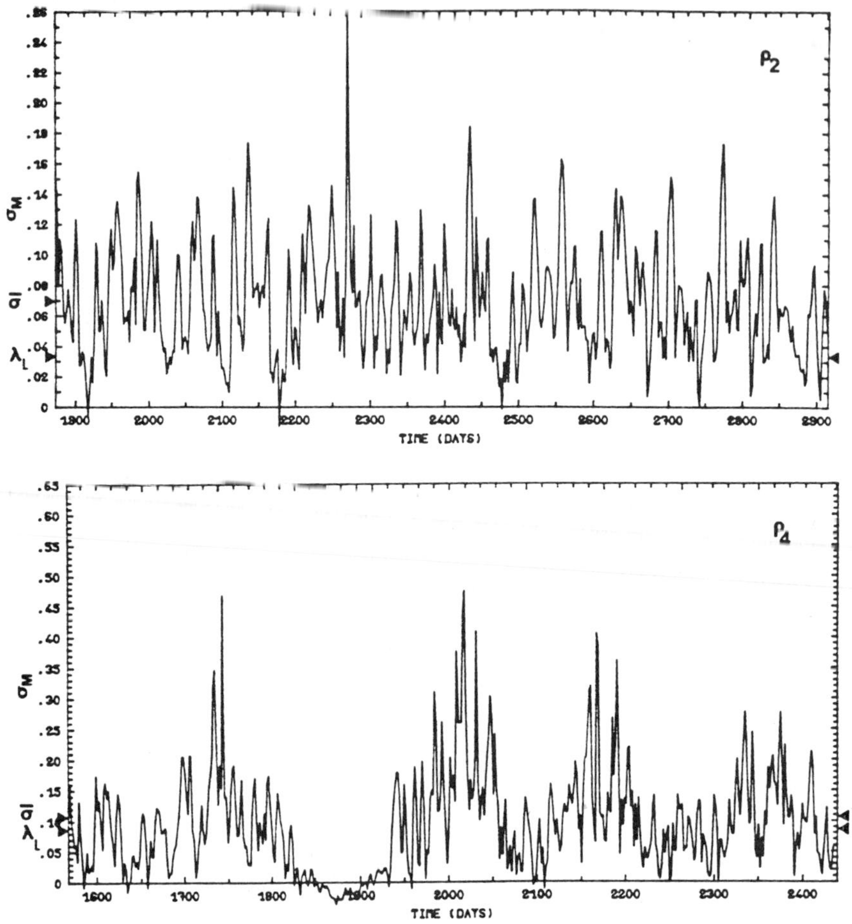

Fig. 6.20. Variations in time of the local rate of divergence of trajectories, σ_M; units of (days^{-1}) on the ordinate (from LG3).

inverted saw-tooth aspect of all curves which plot distance between a time-dependent solution and hyperbolic points nearby (Figure 6.17 here, and Figures 9 and 16 in LGs), or quantities correlated positively with such a distance (Figure 6.14): rapid growth of the distance occurs along $M^u(\Psi_s)$, followed by slow decrease as Ψ_s is approached again along $M^s(\Psi_s)$.

More generally, a point x_0 on an attractor set of (6.32) is called hyperbolic, even if it is not stationary, provided $L_0 = DF(x_0)$ does not contain zero in its spectrum $\Sigma(x_0)$. A rather complete theory exists for *hyperbolic attractors*, which contain only hyperbolic points x such that the spectra $\Sigma(x)$ are uniformly bounded away from zero. This is not the case for the attractors of (6.27) studied, as $\sigma_M(t) = \sigma_M(x(t))$ changes sign repeatedly. There are reasons to suspect that a change of sign of σ_M along a submanifold, i.e., a curve imbedded in the attractor, might correspond to a change in the dimension of the attractor itself along such a curve.

The phenomenon of approach to and quiescent behavior near a stable or otherwise slow manifold appears even more clearly in experiment 4, where the total correlation between $\sigma_M(t)$ (Figure 6.20b) and $C(t)$ (Figure 6.14d) is much larger. Here the minima of $\sigma_M(t)$ are all associated with those of $C(t)$, previously identified as persistent zonal sequences, and large divergence rates occur throughout the transient episodes. Hence contraction near the slow-manifold "ghost" of a Zonal 1 fixed point appears to be even stronger at times than along the actual stable manifold of the Blocked fixed point in experiment 2. This agrees with the funneling mechanism sketched in Figure 6.18. Long metastable transients like this one can occur also in connection with the disappearance and "ghostly survival" of a chaotic attractor, not only for the "ghost" of a fixed point.

When the contraction is particularly strong, increased predictability results well beyond that part of the trajectory where the contraction occurs. The components of the difference between two neighboring solutions which are transversal to the distinguished trajectory at the axis of the funnel (Figure 6.18) are so much reduced that they take a long time to grow again to an appreciable size after the contraction stops and the quasi-one-dimensional funnel starts widening again. This mechanism leads to recurring sequences of zonal flow as long as 525 simulated days for $L_R = \infty$.

Finally, it is interesting to mention another aspect of predictability associated with time intervals of rapid evolution. During such intervals, we have observed in all our experiments some reproducible sequences of events. These events are quite easily recognized by visual inspection of the time series of stream function fields. The identifiable events occur in the same sequence within each episode, but they do not last the same amount of time from one case to the other. The variation in relative duration makes such a recurrent sequence difficult to analyze by the usual objective methods.

These results suggest the following interpretation: trajectories tend to visit certain portions of the attractor in a preferred order, but dwell for variable intervals of time in each. One could thus distinguish between pattern predictability and phase predictability.

The situation is somewhat similar to numerical forecasting of frontal passages. The spatial structure of fully-developed, mature cyclone waves and associated warm and cold fronts is relatively well understood (Section 4.6), and hence the sequence of temperature changes and precipitation events at a given location along the trajectory of the traveling disturbance is predictable with high confidence. The exact time of occurrence of each event at the given location is, however, more difficult to determine. The only distinction is that, while a synoptic-scale sequence might last 24 h-72 h, the planetary-scale rapid-evolution sequences discussed here can last as long as 20 days.

6.5. Low-Frequency Atmospheric Variability

Observational evidence, both classical and recent, indicates that recurrent, persistent anomalies of large-scale flow from the "normal" circulation outlined in Section 4.1 occur in like locations and with like spatial patterns. The attempt to explain these anomalies, their flow patterns and the distribution of their durations has led us through the last two chapters in the study of changing and of multiple flow patterns, first in the rotating, differentially heated annulus, and then in the atmosphere itself. We shall now attempt to summarize and interpret the knowledge gained so far, and outline some gaps in this knowledge and ways to fill them. For additional approaches to the questions at hand we refer to Section 6.6.

We have shown in Section 6.4 that a nonlinear barotropic model of the global atmosphere, spectrally truncated to 25 spherical harmonics, possesses a large variety of behavior patterns, according to the values of the forcing and dissipation parameters. Wave-wave interactions among the nonzonal modes were shown to destabilize the stationary solutions obtained in quasi-linear models (Section 6.3), and lead to the existence of additional solutions, both stationary and nonstationary. Realistic-looking *blocked and zonal flow patterns* are obtained as coexisting, unstable stationary solutions. The synoptic realism of the blocked and zonal stationary solutions might be due to the predominantly barotropic nature of low-frequency variability in the atmosphere, to which we shall return in Section 6.6.

Recurrent, persistent sequences, zonal or blocked, are observed near the corresponding unstable stationary solutions in time-dependent integrations of the model. The persistence properties of solutions depend on the intensity of the forcing and on the Rossby radius of deformation. There is a demarcation in parameter space between a region where zonal flow is more persistent, and one where blocked flow is more persistent. In either case, there appears to be *no preferred time scale of persistence*, the number of episodes of blocked or zonal flow of given duration decreasing monotonically with the duration.

One can imagine that the forcing and dissipation in this barotropic model represent certain types of *boundary data* for the midlatitude, baroclinic atmosphere. These conditions, such as equatorial sea-surface temperature and pressure anomalies, may change from one season or year to another. As a result, episodes of blocked or zonal flow will occur predominantly at middle and high latitudes during the corresponding period, while transitions on shorter time scales between the two types of flow can still take place.

Such transitions are called *breaks* of a persistent pattern in classical, synoptic-statistical long-range forecasting (LRF). The timing of their occurrence is one of the major questions of LRF (Namias, 1982). The results of Section 6.4 indicate that there are two types of causes for breaks, external and internal.

External causes, such as sea-surface temperature anomalies (SSTAs), snow-cover anomalies or soil-moisture anomalies, will set up a certain predominant regime, in preference over another one. In the highly simplified model under discussion, such external anomalies correspond to a selection of the value of the forcing parameter ρ, say. For ρ small, the model possesses a single chaotic attractor, which carries predominantly blocked, quiescent flows. For ρ large, another chaotic attractor is the only one present, carrying predominantly zonal flows, which are more agitated on the average. Either attractor has a probability of exit from the predominant regime which is nearly constant; but this probability actually increases with length of persistent flow pattern for the blocked flows, while it decreases slightly with persistence length for zonal flows. These probabilities of exit, i.e., of breaks, can be computed from the global aspects of the model's deterministic, nonlinear dynamics, given the known properties of numerical and of modeling errors.

Internal, purely atmospheric causes for breaks have been discussed by the classical practitioners of LRF, but have not found a suitable place yet in the modern, dynamical and numerical studies of large-scale flows. The first steps in this direction are probably the articles of Egger and Schilling (1983) and of Reinhold and Pierrehumbert (1982), both of which introduce synoptic-scale waves as perturbations on the planetary-scale patterns whose persistence one wishes to forecast. These perturbations, whether deterministic or stochastic, destabilize the respective model's equilibria and lead to transitions between coexisting equilibria in finite time.

In the model of Section 6.4, internally-caused breaks correspond to the exchange of stability between two coexisting chaotic attractors, which obtain in an intermediate range of ρ values. In this range, the distribution of duration of persistent sequences results from a superposition of exit time distributions from either the zonal or the blocked attractor or ghost attractor. An equivalent situation can arise if two concentrations of invariant measure occur on a single attractor, separated by a region of low measure density.

When interpreting these results for LRF, it is important to realize that boundary data may also change slowly due to atmospheric effects and to other processes. Thus the distribution of exit times from a flow regime, i.e., of breaks, should not be considered as fixed on the slow time scale of a month or a season, but rather as slowly changing itself, due to atmosphere - biosphere - cryoshere - hydrosphere interactions (see Chapter 11 for a description of such interactions on even slower time scales).

To be more specific, we compare the results of Section 6.4 with certain observed characteristics of Northern Hemisphere atmospheric flow patterns over the two oceans, Pacific and Atlantic. This separate comparison is legitimate when considering the localizing effects of relatively high boundary-layer friction on large-scale flow, which produce distinct wave trains behind major orographic features, namely the Himalayas and the Rockies, rather than a zonally-periodic flow (Held, 1983; Kalnay-Rivas and Merkine, 1981). With such an interpretation, model results here suggest that the North Atlantic is, climatologically, in a blocking-dominated regime, since Dole and Gordon (1983) find that over the Atlantic blocking episodes of a certain length outnumber zonal-flow episodes of equal length. Over the Northeastern Pacific, the observed situation is reversed, longer zonal episodes being more likely. The latter fact has not received much notice, since the history of the subject (see Chapter 4) made practitioners expect mean zonal flow at all times.

The present results indicate that persistent zonal sequences are *a priori* neither more nor less likely than persistent sequences of blocking. Which ones occur more frequently depends simply on the boundary data. These boundary data change from one year to the other, most spectacularly so in the equatorial Pacific (Bjerknes, 1969; Namias, 1982; Rasmusson and Wallace, 1983). The climatologically exceptional situation there is associated with upper-ocean El Niño events (Cane, 1983).

In analyzing the impact of El Niño SSTAs on midlatitude atmospheric circulation, it has often been assumed that this impact is unique and relatively straightforward (compare also the situation for orbital perturbations and their climatic effects, cf. Sections 10.5 and 11.4). In fact, the El Niño event of 1976 was associated with the particularly long and strong blocking episode of the 1976-77 North American winter, while the 1982-83 El Niño event was associated with the exceptionally persistent and intense zonal circulation of that winter. More generally, the nine El Niño events during the forty-year period 1945-1984 were associated with six distinct seasonally-averaged temperature patterns over the United States (Namias and Cayan, 1984).

This variability in atmospheric response to the largest known boundary forcings appears to be more easily understood with the present results in mind. Changes in boundary data only select a preferred atmospheric regime, or the coexistence of two or more such regimes. Neither regime precludes long flow episodes of the opposite character, only renders them more unlikely. The persistence of either atmospheric regime will in turn affect and eventually modify boundary data, thus bringing about yet another change in flow regime (McWilliams and Gent, 1978; Philander *et al.*, 1984 ; Zebiak, 1985).

We have seen that the *relative persistence* of blocked and zonal episodes in the model of Section 6.4 corresponds roughly to observations. But the absolute value of these persistences is too large. There are two possible causes of this excessive persistence in the model: first, insufficient resolution, and second, lack of baroclinic processes. As far as the *resolution* is concerned, a number of numerical experiments were carried out by B. Legras with 100 real modes and by R. Benzi and F. Giorgi with approximately 250 modes. The experiments of Legras kept the same truncation at total wave number 9 as in Section 6.4, but removed the two symmetries of the flow fields, with respect to the equator and with respect to a rotation by 180° longitude. Those of Benzi and Giorgi used a higher truncation, at wave number 16, no symmetries, and a rigid lid ($L_R = \infty$).

Both sets of experiments showed roughly the same flow patterns and stability properties as those presented in Section 6.4. Moreover, the removal of sectorial symmetry permitted the appearance of blocked patterns over one ocean, while the other ocean exhibited zonal flow, in agreement with observations and with the discussion here. Statistics of persistence were insufficient due to increased computational cost, but indicated in general shorter persistences, as expected.

In general, given a finite-dimensional dynamical system obtained by spectral truncation from a set of partial differential equations governing fluid flow (Section 5.2), there are indications that the qualitative behavior of such a system might stabilize as the number of modes N retained in the truncation increases. Theoretical upper bounds on the number N sufficient to determine the qualitative long-term behavior of the Navier-Stokes equations in two space dimensions, and in certain three-dimensional cases, are reviewed in Barenblatt *et al.* (1983, Chapters 8 and 17). Numerical results for the two-dimensional, doubly-periodic case indicate stabilization of qualitative behavior at $N \simeq 50$ and of quantitative behavior, i.e., of the critical parameter values at which pitchfork and Hopf bifurcation occur, at $N \simeq 100$ (Franceschini *et al.*, 1984). Similar stabilization results, with $N \cong 100$, were obtained by Maschke and Saramito (1982a,b) for Rayleigh-Bénard convection (compare Sections 5.4 and 5.5) and for a simplified magneto-hydrodynamic flow of a confined plasma (compare Part III).

This order of magnitude for N is consistent with the approximate stabilization we observe for (6.23) at $25 \leq N \leq 250$ and with estimates of the actual number of degrees of freedom of large-scale atmospheric flow, namely $N = O(10^2 - 10^3)$. Such estimates, or educated guesses rather, can be based on various independent considerations. The number of variables necessary to describe the position, size, shape and motion of synoptic features on a weather map; the effective resolution of meteorological observations and of numerical models used in weather prediction and in general circulation studies, whether discretized spectrally or by finite differences; as well as the study of large-scale atmospheric analogs and of divergence of subsequent weather sequences all suggest a dimensionality of planetary and synoptic flows between 100 and 1000, loosely speaking. This number would appear to make the methods of investigation illustrated here for relatively low-order models feasible for rather complete, full-scale atmospheric models as well, within the computational means of the next decade.

Concerning the effect of baroclinic processes on persistence times, such processes were explicitly included in the model of Reinhold and Pierrehumbert (1982), which was, however, more severaly truncated, with only ten horizontal modes, and had zonal channel geometry. Their persistence times, like those of Section 6.4, were continuously distributed and had an excessively large mean value. This overestimate can be attributed to the severe truncation and constraining geometry of their model. On the other hand, they also noticed changes in slope of the distribution of exit times, occurring at lower persistences than those of Figure 6.19 and associated with the distinction in mean life time between baroclinic and barotropic processes. The changes in slope apparent in Figure 6.19 might be an indication of inherent distinctions between two or more low-frequency bands of barotropic dynamics.

In the context of our discussion, baroclinic processes affect not only the stability and persistence properties of long-lived, predominantly barotropic flow patterns. Baroclinic eddies also act in the atmosphere to transmit low-level orographic and thermal forcings to the upper levels, and to convert available potential into kinetic energy (Section 4.4). The way this transmission and conversion occurs in the presence of nonlinearities is interesting in its own right and may affect to some extent the equivalent-barotropic planetary flow patterns themselves, as well as their stability (Buzzi *et al.*, 1984; Itoh, 1985; Roads, 1982).

Ideally, the study of planetary flow regimes and of their persistence characteristics should be pursued with models of increasing complexity, including higher spatial resolution, baroclinic processes and explicit interactions with surface processes, such as changes in vegetation cover and in upper-ocean dynamics. In this pursuit, it is important to remember that the dynamic behavior of the relatively simple models presented in Chapters 5 and 6 could only be understood by exploring a large domain of parameter space around the physically most likely parameter values. The presence of solutions with certain attracting properties for some remote parameter values has important consequences for the persistence characteristics of solutions at the parameter values of interest. Hence, such explorations can often provide more physical insight and practical results than an increasingly detailed study of atmospheric flows restricted to a narrow domain of meticulously chosen parameters.

The message is to think globally, in both physical space and phase-parameter space. This does not mean neglect of local properties, in either space, but rather the opposite: spatial localization and suitable linearization are indispensable to an adequate description of global, nonlinear dynamics. The global and local points of view complement each other and only together allow for the study of low-frequency atmospheric variability.

6. Bibliographic Notes

Section 6.1. The idea of *weather types*, or multiple flow regimes, goes back to the end of the 19th and the early 20th century. A beautiful outline of the history of long-range forecasting (LRF) and of its scientific basis appears in Namias (1968). The interpretation of mean circulation in mid-latitudes being basically zonal, and hence every other flow regime being anomalous, is more recent. This interpretation is probably due, on the one hand, to the zonal regime being on the whole most prevalent, and on the other, to the long preoccupation from 1950 till 1980 with zonal flow and perturbations on it (Sections 4.5 and 4.6).

It would appear that blocked flow is the next most important mid-latitude flow regime. A number of descriptive papers appeared in the late 1940s and early 1950s, without being able to capture the attention of dynamic meteorologists, fruitfully engaged at that time in developing the theory of midlatitude storms reported here in Chapter 4. The most important papers of this number are those of Rex (1950a,b), in which criteria were given for the definition of spatial patterns and persistence of blocking events, as well as describing the climatic effects of these events.

The recent interest in blocking was stimulated by and produced in turn a few excellent observational papers, in which new, more extensive data sets and objective, high-speed methods of data analysis were applied to the difficult problem of flow-pattern recognition in space and time. A good perspective can be obtained by reading, among others, Austin (1980), Dole and Gordon (1983), Shukla and Mo (1983), and Wallace and Gutzler (1981). All of the papers just cited deal with Northern Hemisphere blocking. For the Southern Hemisphere, literature is much more scanty and we mention Trenberth and Mo (1985).

Transitions between zonally-dominant and pronounced meridional flow patterns were described in the context of an *index cycle* by Rossby *et al.*

(1939) and Namias (1950). These transitions are accompanied by a displacement in the latitudinal position of the subtropical jet (see also Sections 4.6, 5.3 and 6.4). The preference for certain geographic locations of both storm tracks and centers of action, i.e., of agitated and quiescent regions in space, respectively, was also emphasized in these papers.

To the articles about blocking in the atmosphere one should add those documenting blocking episodes in numerical experiments with large-scale models (Bengtsson, 1981; Halem *et al.*, 1982; Mechoso *et al.*, 1985; Miyakoda *et al.*, 1983). Such episodes were noticed in model output only after attention turned to them in the atmosphere and in simple model studies. But large models have the advantage over the atmosphere of being better documented and more easily studied, and over simple models of being more complete and detailed.

Section 6.2. Wave-wave interactions were studied extensively in the 1960s, first in connection with surface gravity waves in the ocean (Hasselmann, 1961; Phillips, 1966; Whitham, 1974, Sections 15.6 and 16.11), then for Rossby waves (Longuet-Higgins and Gill, 1967; Pedlosky, 1979, Section 3.26). The theory of resonant triad interactions, formulated originally for free waves, is weakly nonlinear, and thus not entirely compatible with fully-developed turbulence theories.

Statistical studies of turbulent flow over topography include Bretherton and Haidvogel (1976), Frederiksen and Sawford (1981) and Holloway (1978). The topography contains many wave numbers and multiple scales, and so do atmospheric motions. Clearly the effects studied in this section are a drastic simplification, and hence an exaggeration of what goes on in reality. The rich field of statistical theories of turbulence, and of the cascading exchanges of energy and enstrophy between wave numbers implied in two dimensional flow or in quasi-geostrophic turbulence can be entered via Charney (1973, Chapter XI), Pedlosky (1979, Section 3.27) or Sadourny (1985) (see also comments in Section 4.6 here).

We noted in the text that the numerical scheme used by Egger (1978) in his persistent-triad experiments (compare Isaacson and Keller, 1966, Section 8.3, for the analysis of Heun's scheme) is dissipative (see Richtmyer and Morton, 1967, Section 5.4). Without such numerical dissipation, his solutions could not stay bounded in time, due to the conservative character of the left-hand side of Eq. (6.1) and to the steady forcing provided by the right-hand side (see discussion of Eqs. (5.26-5.30)).

Resonant amplification of stationary Rossby waves in explaining blocking was also analyzed in a linear model by Tung and Lindzen (1979, and subsequent papers). Much of the effort in linear analyses has to go into studying the singular effect of critical lines, where the mean zonal wind of the basic flow is zero. As we saw in Section 6.4, such critical lines are poorly defined in monthly means and their actual lifetime is not longer than that of planetary flow features with meridional components.

A broad review, observational and theoretical, of standing waves in the midlatitude atmosphere is given in Hoskins and Pearce (1983). We shall only mention here that thermal forcing from the lower boundary is likely to play as large a role in the maintenance and low-frequency variability of quasi-stationary waves as topographic forcing (Kalnay and Mo, 1985). This complicates matters quite a bit, since thermal asymmetries change with season, while topographic asymmetries do not. The restriction to the study of topographic effects here is therefore only for the sake of simplicity.

Section 6.3. The article by Charney and DeVore (1979), on which this section is based, was followed by a study of baroclinic effects in Charney and Straus (1980) and by a comparison with observations in Charney *et al.* (1981). Among the numerous papers of other authors which contributed to this line of inquiry, we cite Pedlosky (1981b) for an analytic approach, Malguzzi and Speranza (1981) for a study of regional aspects, Davey (1981) and Källén (1984) for results in a rotating annulus and a spherical geometry, respectively.

Independently of the specific question of blocking as an alternative stable equilibrium for planetary flow, Vickroy and Dutton (1979) and Wiin-Nielsen (1979) pointed out the possible existence of multiple equilibria in simple, low-order models of barotropic, quasi-geostrophic flow. They also showed the existence of limit cycles in such a model. More specifically, Yoden (1985) studied the Hopf bifurcation by which the blocking solution in the six-component version of the Charney-DeVore model loses its stability to a periodic solution.

Section 6.4. Transition to chaotic behavior in fluids is reviewed by Ruelle (1985), mathematically, and by Libchaber (1985), experimentally, with many additional references. Among the possible routes from stationary or periodic to aperiodic flow (Eckmann, 1981; Gollub and Benson, 1980; Swinney and Gollub, 1981) the best known are: i) the direct bifurcation from a torus (quasi-periodic flow with two irrationally related frequencies) to a strange attractor (Ruelle and Takens, 1971), ii) intermittency related to a saddle-node bifurcation (Pomeau and Manneville, 1980), (iii) period-doubling cascade related to successive pitchfork bifurcations (Feigenbaum, 1978; Kadanoff, 1983), and iv) the saddle-focus homoclinic orbit route (Shilnikov, 1965; Arneodo *et al.*, 1982; Gaspard *et al.*, 1984).

All of these routes, or scenarios, are encountered at various points, or along various arcs of curve in the two-parameter plane of Figure 6.13 (see also Legras and Ghil, 1983, 1984, 1985). The point of this section, however, is *not* the transition to turbulent behavior, but rather the structure of phase space in the weakly turbulent regime itself. Successive bifurcations are important only to the extent that they help clarify this *inhomogeneous* structure. They do so by highlighting the role of unstable fixed points, and of their stable and unstable manifolds in guiding the phase-space flow and in creating concentrations of invariant measure on one or more coexisting attractors. Much of what is said in the text about this role, and about the mutual interactions of strange attractors and fixed points, follows Grebogi *et al.* (1983).

Pseudo-orbits and invariant measures on attractors (Bowen-Ruelle theory) are discussed by Katok (1981) and Ruelle (1981a,b). The multiplicative ergodic theorem underlying Liapunov exponents is reviewed by Ruelle (1985) and numerical aspects of their computation appear in Benettin *et al.* (1980) and Grebogi *et al.* (1983) (see also Section 5.5).

Concerning the question of spatial detail, and hence realism, of the flow patterns discussed here, a number of points arise. The actual resolution of globally interpolated observed fields (Bengtsson *et al.*, 1981) has not changed much over the last two decades: it still corresponds roughly to the linear span of spherical harmonics with $n \leq n_0 \cong 20$ (Ellsaesser, 1966; Savijärvi, 1984). Hence truncation at $n_0 = 9$ provides a modicum of realism in spatial detail.

As indicated already in Section 6.5, analysis along the lines presented here of a model with ten or more times as many variables as the one in Section 6.4 is easily within the means of electronic computers to be installed over the next decade. Still, brute-force increase of resolution might not be the best way to study the regional aspects of persistent anomalies. Localized, coherent structures imbedded in the large-scale flow could provide an appropriate basis for a better understanding of these aspects.

A fundamental problem in this approach is that *solitons* in one space dimension (Ablowitz and Segur, 1981; Warn and Brasnett, 1983) and *modons* in two (McWilliams, 1980; Malguzzi and Malanotte-Rizzoli, 1984) are solutions of a weakly nonlinear, *conservative* partial differential equation. The stability of solutions to completely integrable Hamiltonian problems with respect to *dissipative* perturbations is an open question even for ordinary differential equations (see also Section 12.7), although numerical evidence in the case of *modons* suggests some robustness to small enough perturbations, both conservative and dissipative. In any case, localized structures by themselves cannot provide a full explanation of persistent anomalies, any more than linear theories, local in phase space. In both situations, the interaction with the global, respectively with the mean flow, needs eventually to be taken into account, since this interaction occurs on a time scale comparable to or shorter than the lifetime of the anomaly.

Finally, the flow patterns of stationary solutions exhibited here are not strikingly more realistic than those of linearized large-scale models with carefully chosen mean flow and thermal forcing (Oopstegh and Van Der Pool, 1980; Held, 1983). The advantages of the approach used here are that: (i) no such constraints on model behavior as the imposition of both the largest scales and the detailed forcing are necessary to obtain this realism, (ii) transitions between distinct flow patterns are possible, and (iii) these transitions themselves exhibit many realistic features.

Section 6.5. The most important question not addressed in this chapter due to limitations of space is that of baroclinic processes and of their effect on low-frequency variability. While the structure of quasi-stationary patterns is to a large extent equivalent barotropic (Hoskins and Pearce, 1983, Chapters 2 and 3), baroclinic processes contribute strongly to the maintenance of these patterns and to transitions between flow regimes (Austin, 1980; Hoskins and Pearce, 1983, Chapter 7; Itoh, 1985). There seems to be an analogy between the interaction of quasi-geostrophic and inertia-gravity motions, on the one hand (see Chapter 3 and references there), and the interaction of barotropic and baroclinic processes, on the other (Held *et al.*, 1985). To stress this analogy with geostrophic adjustment, as the first interaction is called, one might refer to the second interaction as *barotropic adjustment* (Ghil, 1985a).

This adjustment of low-frequency atmospheric variability to an equivalent-barotropic structure is one of the most important questions in the theory and practice of long-range forecasting (LRF). Its understanding can greatly help in classifying from observations the different planetary flow regimes, and in computing the transition probabilities between these regimes from both observations and deterministic models of increasing complexity. The expected time for a break to occur, and what the most likely successor of the current flow pattern might be is the ultimate practical knowledge to be expected in LRF.

Finally, we shall mention another aspect of the global circulation to which some of the results of Section 6.4 might be relevant. Fluctuations in the atmosphere's zonal angular momentum by as much as 15 percent exist and exhibit a broad periodicity of 30-50 days (Hide *et al.*, 1980). While these fluctuations are apparent in tropical wind and cloud fields (Madden and Julian, 1971; Krishnamurti and Subrahmanyam, 1982), their origin must lie in midlatitude fluctuations of mountain drag (Ghil, 1986; Whysall and Hide, 1984).

This claim is supported by preliminary experiments in a rotating annulus with simplified bottom topography, in which vacillations with a period of 40 days approximately and with a large barotropic component have been observed (Li *et al.*, 1986). Thus the broad spectral peak associated with the transition to chaotic flow from the 40 day limit cycle in Section 6.4 would appear to be related to the 30-50 day oscillation in global angular momentum, as well as in the annulus. The Hopf bifurcation yielding the 40 day cycle at $\alpha^{-1} = 20$ days and $\rho \cong 0.123$ in Figure 6.13 might in turn be related to, but is not identical with the barotropic mode of linear instability observed for certain parameter values by Simmons *et al.* (1983).

PART III

DYNAMO THEORY

CHAPTER 7

MODELS OF GEOMAGNETISM: A SURVEY

We turn now to a problem of internal geophysical fluid dynamics connected with the explanation of the geomagnetic field. In the present chapter we shall examine some basic facts concerning geomagnetism and outline the main approaches that have been used to model it. Many of the ideas are applicable to planetary magnetism generally, and also to the solar magnetic cycle.

Observations of terrestrial magnetism can be traced back thousands of years, but its modern scientific study really begins with the book *De Magnete* of W. Gilbert, which appeared at the beginning of the 17th century. Another milestone was Gauss's application of potential theory to the analysis of the surface field, in 1830. Attempts to explain the observed structure of the field in terms of processes within the Earth's core began only in the early part of this century, and dramatic progress has been made in the last 25 years or so, although many questions remain unanswered.

In the present discussion we intend to emphasize two basic points: First, magnetism appears to be a general property of rotating bodies, and furthermore the geomagnetic field provides one of our few sensitive probes of the deep structure (especially the fluid structure) of the Earth. A model of the field is therefore of importance not so much as a prognostic tool, but rather as a test of our understanding of the internal dynamics of the core.

Second, the problems which will emerge can be thought of as "climate models" of the magnetic environment occupying a position roughly equivalent to climate models of the atmosphere with respect to its rapidly varying weather. The reason for this is that, while magnetism is a sensitive probe of the fluid core, the magnetic field observed at the Earth's surface is filtered both spatially and temporally. Usually one is concerned with time scales greater than 10 years, although magnetic maps must be revised in practice every few years. Time scales of the order of 10^3-10^6 yr. are characteristic of the most important processes in the models to be described below. Typical spatial scales of magnetic variability are ~1000 km, but actual core fields probably have a broad continuum of scales.

In Section 7.1, observations of the geomagnetic surface field, its spatial structure and variation in time are reviewed. *Aperiodic reversals of the* field's dominant *dipole* component are emphasized, and characteristic values of the geodynamo's physical parameters are listed.

Section 7.2 contains the full set of equations governing the problem: the fluid flow equations with rotation, coupled by the Lorentz force to the pre-Maxwell equations of the electric field. The likely role of *convection* in driving the dynamo-sustaining motion is stressed.

In Section 7.3 we show how the full, hydromagnetic problem has been partially decoupled into the *kinematic dynamo problem*, of magnetic field amplification by a given velocity field, and the problem of generation by convection of appropriate, dynamo-amplifying flow fields. The history of dynamo theory is outlined, and a tentative picture of *successive bifurcations* is drawn, which leads from a state of rest and no magnetic field to one of sustained motion and aperiodic reversals.

A simple laboratory device, a *shunted disc dynamo*, is studied in Section 7.4. It is shown to possess some of the major features of the natural system, and to exhibit aperiodic reversals.

Cowling's theorem states that no dynamo action is possible when coupling a flow field and a magnetic field symmetric about the same axis. The concept of *toroidal and poloidal* components of an axisymmetric vector field is introduced, and the theorem proved for a spherical core in Section 7.5.

In Section 7.6, we notice that the equation for the magnetic field B has the same form as that for vorticity in a fluid of constant density. The competition between magnetic flux line stretching and diffusion is studied. *"Pastry" dynamos* are obtained, where kinematic line stretching, folding and twisting in two and three dimensions can generate a field which overcomes the diffusion. Magnetic Reynolds numbers R_m are defined and their role in determining the intermittency of the field at the surface of the Earth and of the Sun is mentioned.

Finally, Section 7.7 contains a necessary condition for of magnetic energy in terms of a critical R_m. This condition emphasizes the role of sufficiently high fluid velocities in maintaining dynamo action. Bibliographic notes follow in Section 7.8.

7.1. General Features of the Earth's Magnetic Field

Spatial structure. For the present discussion, the Earth may be taken to be a spherical body consisting of three concentric layers, or shells. The region between the surface, $r = 6400$ km, to $r = 3500$ km is the *mantle*, the *fluid core* occupies the shell $1400 \text{ km} < r < 3500 \text{ km}$, and the sphere $r \leq 1400$ km comprises the *inner core*. Relative to the fluid core, both the mantle and the inner core may be viewed as composed of solid material, although creeping of the mantle is believed to occur and to be of significance in the theory of plate tectonics, as well as in ice-age dynamics (see Chapter 11).

This layered structure, which has been deduced from seismological evidence, is crucial to the theory of geomagnetism because of the fluid component. Larmor, in 1919, reasoned that any *rotating fluid* body which has a nonzero electrical conductivity might become magnetically active through the interaction between the velocity field within the body, and some ambient or seed magnetic field. Generally, such seed fields are always present on galactic scales.

The essential idea is that the process by which solid dynamo action is achieved in the laboratory and in industry, using rotating machinery, might be made to operate in a homogeneous fluid, provided the right motions can be excited. The difficulty with this idea, and the reason that it cannot be immediately decisive, is that the homogeneity of the material makes it difficult to see whether or not unwanted "short circuiting" would disrupt the process.

What are the principal features of the terrestrial magnetic field? The most basic property of the field is well known: the dominant field at the surface is roughly equivalent to that of a centered dipole with an inclination of about 11° relative to the axis of rotation. The remaining nondipole field, represents as much as 25% of the root-mean square (rms) surface field (Lowes, 1974).

The predominance of the dipole field at the surface could be misleading, however, since the mantle has a much lower electrical conductivity than the fluid core, and thus acts like an insulator. Magnetic fields generated within the fluid core decay through the mantle as potential multipoles, and this spatial filtering tends to emphasize the low-order components. Still, the time-dependent or *secular-variation* field, along with the nondipole components, contains the spatio-temporal fine structure by which one might hope to probe the fluid core.

One prominent and interesting feature of the secular field is a tendency for the global structure measured, for example, by contours of constant vertical or horizontal component, to drift from east to west as a fixed pattern, at a rate amounting to about 0.2° longitude per year. Within the fluid core, this angular speed yields a linear velocity of about 3×10^{-4} m/s. Such a figure is usually taken to represent a typical fluid velocity in the Earth's core, relative to the rotating frame of reference, although it could also be the speed of a hydromagnetic wave.

Other important data describe the spherical harmonic decomposition of the secular field at fixed time, as well as the time spectrum of the magnetic vector taken at fixed stations. The former, when extrapolated down to the fluid core, seems to indicate that the magnetic field contains significant contributions on many length scales. The complexity of the variation on the larger scales is comparable with that of the pattern of isobars of a typical global weather map (see Figures 4.2 and 6.1, which are schematic and averaged in time, respectively, for orientation).

Temporal variability and paleomagnetism. Turning now to the principal surface component, namely the main dipole field, the dipole moment is known to be time-dependent on the range of time scales indicated earlier. The intensity has decreased about 7% in the present century and the axis of the dipole is known to have wandered in its inclination and longitude over the last 1000 yr. More striking even is the paleomagnetic record over prehistoric times (Cox, 1969).

As sediments are accumulated at the bottom of the ocean, or else as material is extruded from mid-ocean ridges, the polarity of the dipole (as well as some indication of intensity) is preserved as a sort of "geomagnetic tape". Over the past decades these "tapes" have been obtained from locations all over the world, and "played" on instruments of increasing resolution. As a result, records such as that shown in Figure 7.1 exist, the black bars representing periods of "normal" (present-day) polarity, the white sections representing periods of reversed polarity.

We are dealing here with records over a significant fraction of the lifetime (10^9 yr) of the planet, so it is clear that the field is not a recent feature. It is found that these reversals have no simple history

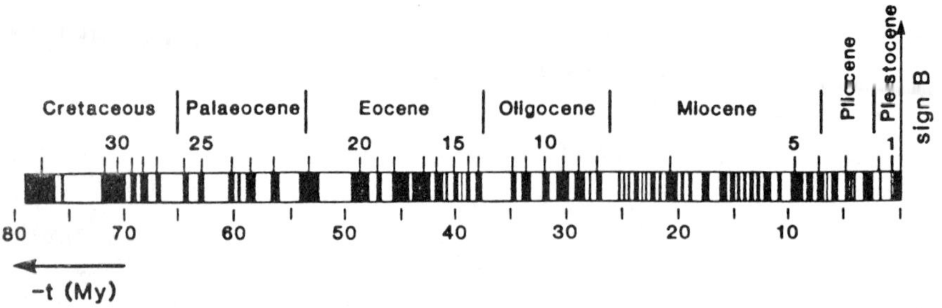

Fig. 7.1. Paleomagnetic polarity reversals. Dark sections show normal polarity, white sections are reversed (after Heirtzler *et al.*, 1968.)

and that the process appears to be random. In particular, Figure 7.1 shows relatively brief reversal *events* lasting $O(10^4\text{-}10^5$ yr), and longer intervals or *epochs* of the same polarity lasting $O(10^6\text{-}10^7$ yr). Of course as the resolution of observational techniques improves, we can expect more reversals to be found, altering somewhat the statistical distribution of the polarity intervals.

Records from various points on the Earth seem to show the same basic pattern, indicating that a reversal is indeed a global occurrence, and with refinement it may be possible to get information on the precise sequence of steps by which reversals take place. At present, it appears that the dipole field intensity first decreases to a fraction of its usual value over a few thousand years. The field direction then undergoes several swings up to 30° away from the axis of rotation, and subsequently reaches the reversed direction by an irregular excursion. Finally, the intensity of the field, now pointing in the new direction, increases to its usual absolute value. The global nature of the reversals has been very helpful in dating paleoclimatic and other geological records (see Section 11.1).

These data demonstrate a magnetically active planet over geophysical times. In Table 7.1 we have collected some of the current estimates of the various core parameters which might be involved in physical models of this behavior.

The value given in Table 7.1 for the electrical conductivity allows one to estimate the rate of decay of a magnetic field under the assumption

Table 7.1. Physical parameters of the geodynamo*

Symbol	Meaning	Units	Value
L	core radius	m	3.5×10^6
σ	electrical conductivity	$m^{-3}k^{-1}sq^2$=mho/m	3×10^5
η	magnetic diffusivity	m^2s^{-1}	3
τ	diffusion time	s	$4 \times 10^{12} = 10^5$ yr
π	fluid density	km^{-3}	10^4
χ	kinematic viscosity	m^2s^{-1}	10^{-6} (?)
Ω	angular velocity	s^{-1}	7.4×10^{-5}
T	core temperature	K = °Kelvin	4000
c_p	specific heat	$m^2s^{-1}\ K^{-1}$	670
γ	thermal conductivity	$mks^{-3}\ K^{-1}$	60
α	coef. of volume expansion	K^{-1}	5×10^{-6}
κ	thermal diffusivity	m^2s^{-1}	10^{-5}
g	acceleration of gravity	ms^{-2}	5
β	mean temperature gradient	$K\ m^{-1}$	2×10^{-3} (?)
B	magnetic field	$ks^{-1}q^{-1}$	10^{-2}= 100 gauss
V_A	Alfven speed	ms^{-1}	10^{-1}
U	speed	ms^{-1}	10^{-4}
Q	core heating rate	m^2ks^{-3}	$10^{12} - 10^{13}$

1 gauss = $10^{-4}\ kq^{-1}s^{-1}$

1 joule = 1 m^2ks^{-1} = 0.239 calorie

1 volt = 1 $m^2ks^{-1}q^{-1}$

1 ohm = 1 $m^2\ ks^{-1}q^{-2}$ = 1 mho^{-1}

μ = magnetic permeability = $4\pi \times 10^{-7}\ mkq^{-2}$

*Fundamentals units used for (m,k,s,q) = (meter, kilogram, second, coulomb), for (length, mass, time, charge). This is the rationalized MKSQ system of units (e.g., Sommerfeld, 1952).

that the fluid core is stationary. It is found that the present field would disappear in about 10^5 years, or in about 1/1000 of the time over which the paleomagnetic record indicates a magnetically active planet. No explanation other than the hypothesis of a core-based hydromagnetic dynamo has been able to account for this fact.

7.2. Equations of the Hydromagnetic Dynamo

The core material is thought to be an iron alloy which is essentially incompressible over the range of pressures encountered there, so most models of the dynamo take the velocity field to be divergence-free. The principal new feature of these models, compared to those in Part I, is the electromagnetic field present, and the coupling of this field with the fluid dynamics through the Lorentz force.

The equations of motion in the rotating frame are

$$\frac{du}{dt} + \frac{1}{\rho}\nabla p + 2\Omega \times u + \frac{1}{\rho} B \times J - \nu\nabla^2 u = cg, \tag{7.1}$$

$$\nabla \cdot u = 0. \tag{7.2}$$

In the momentum equation (7.1), $B(r,t)$ and $J(r,t)$ are respectively the magnetic field strength and the electrical current, and $\rho^{-1}B \times J$ is the Lorentz force. We have also included a body force cg expressing a possible inhomogeneity of the fluid density; the latter is expressed in the form $\rho c(r,t)$, where ρ is a constant reference value, as in (4.12a,d) and (5.31a).

We must now supplement (7.1), (7.2) with Maxwell's equations. For the terrestrial dynamo the time scales are all much larger than the travel time of light through the core, so it is possible to neglect the charge displacements associated with electromagnetic radiation. The resulting "pre-Maxwell" system takes the following form: if $E(r,t)$ is the electric field then we have

$$\nabla \times B = \mu J, \tag{7.3}$$

$$\nabla \times E = -\frac{\partial B}{\partial t}, \tag{7.4}$$

$$J = \sigma(E + u \times B), \tag{7.5}$$

$$\nabla \cdot B = 0, \tag{7.6}$$

$$\nabla \cdot E = q/\varepsilon. \tag{7.7}$$

Eq. (7.3) is Ampère's law, (7.4) Faraday's law and (7.5) Ohm's law with

$\sigma u \times B$ being the induction current. In the last equation, q is the charge density and ε the dielectric constant.

If we eliminate E from (7.3)-(7.6) we can obtain a single equation involving B and u in the form

$$\frac{dB}{dt} - B \cdot \nabla u - \eta \nabla^2 B = 0, \qquad \eta = (\mu\sigma)^{-1}. \tag{7.8}$$

This last equation is the key equation for understanding how motion of an electrical conductor can amplify and modify a magnetic field.

To close the system of equations it is necessary to provide an equation for the scalar field $c(r,t)$ in (7.1). Actually the inclusion of this buoyancy term has already prejudiced the model toward some kind of convection as a primitive source of energy. There is no widespread agreement on what energy source is most important, although the favored models are convectively-driven.

One method of driving convection is by thermal sources associated with radioactive decay. There is concern, however, that these thermal sources may be insufficient, and that a more likely energy source depends upon changes in composition of the core material associated with the growth of the solid inner core. The latter possibility is very attractive since it links the dynamo to the potential energy released as its internal structure evolves (see Section 9.6 below). In either case, any convection-driven model can probably be formulated as an equation for c of the form

$$\frac{dc}{dt} - \nabla \cdot D_c \nabla c = Q, \tag{7.9}$$

where D_c is a diffusion coefficient and Q is a source term (e.g. density of radioactive sources in the case of thermal convection).

The boundary conditions of the dynamo problem must reflect the fact that the mantle is a poor conductor, that the core-mantle and fluid core-inner core interfaces are surfaces of transition from fluid to solid, and that the exterior field decays to zero in a vacuum. At any surface where the properties change discontinuously the magnetic field as well as the tangential component of the electric field must be continuous. The velocity is usually (in the viscous case) made to vanish at the core-mantle interface. The appropriate boundary conditions on c will depend upon the details of the assumed convection model. Examples of specific boundary conditions will be given below.

7.3. Kinematic Versus Hydromagnetic Theory

Neither a complete analysis of the geomagnetic dynamo problem, nor direct numerical simulation using the full system of equations have been feasible, since fully three-dimensional models are needed, as will be seen presently. We wish therefore to outline first the main results of dynamo theory, as developed over the last half century, in order to understand how the problem has been divided into smaller manageable subproblems.

If pressure is eliminated from the problem by taking the curl of (7.1), then the resulting governing system can be written in the abbreviated form

$$\dot{Z} = F(Z;Q) \tag{7.10}$$

where $Z = (u,B,c)$ and in the function F we have suppressed all parameters except Q, the latter being taken as some measure of the intensity of the energy source. This is a forced, dissipative system, which has properties very similar to those of (5.26), although it has an infinite number of degrees of freedom.

It is clear from (7.8) that one class of solutions of (7.10), such as pure conduction, will have zero electromagnetic field, $Z = Z_0 = (u_0,0,c_0)$. Suppose, however, that such solutions are unstable to the electromagnetic field, in the sense that an infinitesimal seed field will always be amplified. Such a result bodes well for the dynamo hypothesis, since it indicates that we would not find general solutions of (7.10) ever settling into the subspace $Z = Z_0$; that is, the system remains magnetically active. In the presence of viscous and electrical dissipation, thermodynamic constraints will restrict solutions to a bounded region of phase space for any fixed value of Q. Examples (5.31-5.34) of Section 5.4, along with the general discussion of (5.26-5.30), make an aperiodic histograph such as Figure 7.1 at least understandable, if not unavoidable.

The amplification of small magnetic fields can be studied in models which neglect all coupling between u and B. This idea, of studying the effect of a *pre-existing* velocity field in the context of dynamo theory, defines the *kinematic dynamo problem*. If u in (7.8) is known, the problem for B becomes linear. The purpose of course is to learn what choices of u are likely to cause magnetic activity, and whether or not such preferred u are realizable as a solution Z_0 in a convective system.

In this respect, we are dealing with the inverse problem of deducing a velocity field from its effect on the magnetic field.

In the course of developing this kinematic theory, there have also appeared a number of "anti-dynamo" theorems, establishing classes of fields u which cannot succeed in amplifying a magnetic field. Some of the key steps in this work on kinematic dynamo theory are the following:

<u>1934</u> Cowling showed that a dynamo cannot operate in such a way that both magnetic and velocity fields are axisymmetric about the same axis. This was disappointing, since the axis of rotation is a natural axis of symmetry.

<u>1946</u> Elsasser studied interactions between nonsymmetric velocity and magnetic fields.

<u>1954</u> Bullard and Gellman presented suggestive, but not compelling numerical evidence for kinematic dynamo action in a sphere. Subsequent investigations strongly suggest that the core motions adopted by Bullard and Gelman are in fact not dynamo-generating. However this early study had the fortunate effect of stimulating others to take Larmor's proposal seriously.

<u>1955</u> Parker provided a physical argument to describe how an irregular small-scale motion of a fluid could amplify a large-scale magnetic field when the microscale process is averaged over space and time.

<u>1964</u> Braginsky considered nearly axisymmetric systems with high electrical conductivity using a formal expansion. In later papers he introduced nonthermal convection as an energy source.

<u>1966</u> Steenbeck, Krause and Rädler considered turbulent velocity fields, and by the use of smoothing methods derived mean-field electrodynamics for the dynamo problem.

One important aspect of these studies has been an understanding of how large-scale rotation of a body tends to favor velocity fields which are "good" from the standpoint of the kinematic dynamo problems. Recent work has returned to the role of convection in this regard, and there is an emerging picture of how magnetic activity can be acquired by a process of *successive bifurcations* of solutions of (7.10).

As we indicate in Figure 7.2, the onset of pure convection at a critical parameter value Q' is followed by a second bifurcation to magnetic activity at a critical value Q''; here the pure convection begins to amplify the magnetic field. The analysis of many magneto-convective systems shows that a sufficiently large magnetic field can in turn

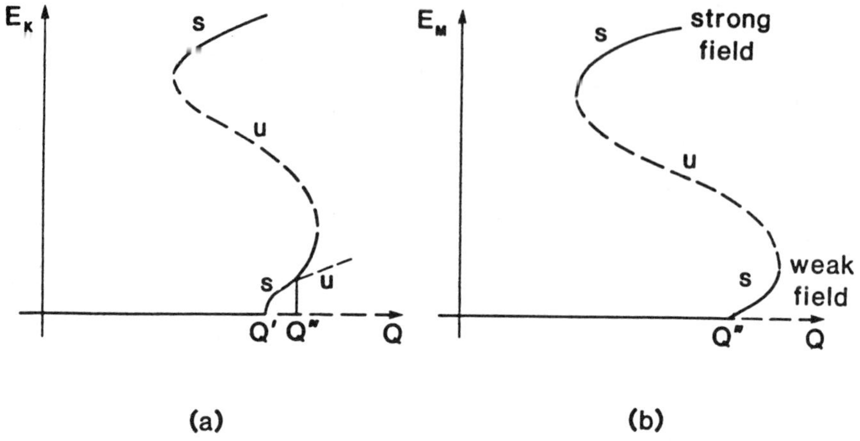

Fig. 7.2. Schematic diagram of successive bifurcations for the geomagnetic dynamo problem. E_K indicates total kinetic energy, E_M magnetic energy, $E_M = (1/2)\int B^2 dV$. Stable branches (s) are solid, unstable ones (u) are dashed.

destabilize the convection, with the result that branches may extend back to subcritical values of Q.

For values of Q which are sufficiently high, a globally attracting branch of solutions occurs. This stable branch might be a strange attractor, as in Sections 5.4 and 5.5, explaining the aperiodic appearance of paleomagnetic records like those of Figure 7.1.

Unfortunately, the subcritical bifurcation onto the upper branch of stable solutions is not easy to treat analytically, and this is a basic difficulty in modeling the Earth's dynamo. The second and third bifurcations involve, as we shall see in Section 7.5, truly three-dimensional, non-axisymmetric fields. The technical difficulties in the dynamo problem are thus analogous to the ones encountered in modeling the atmosphere's general circulation, while accounting for baroclinic instability, with its three-dimensional structure (see Section 4.1, Figure 4.5, Section 4.4, Figure 5.2, and Section 5.3).

7.4. A Disc Model

To schematize the basic interactions between the electromagnetic and the velocity field, it is helpful to model the process with a much simpler, but experimentally realizable system. A *shunted disc dynamo* is one such system; the usefulness of this particular example was emphasized by Malkus (1972) and was studied in detail by Robbins (1977). The mechanical system and its circuit diagram is sketched in Figure 7.3.

A conducting disc is rotated by a constant external torque T. The angular velocity of the disc, ω, is proportional to its angular momentum. A magnetic field normal to the disc will produce a radial electromotive force (emf) according to Faraday's law (7.4). The current I_2 flowing in the disc due to this emf is removed by a ring of collecting brushes and fed to an electric load consisting of a shunt and a coil.

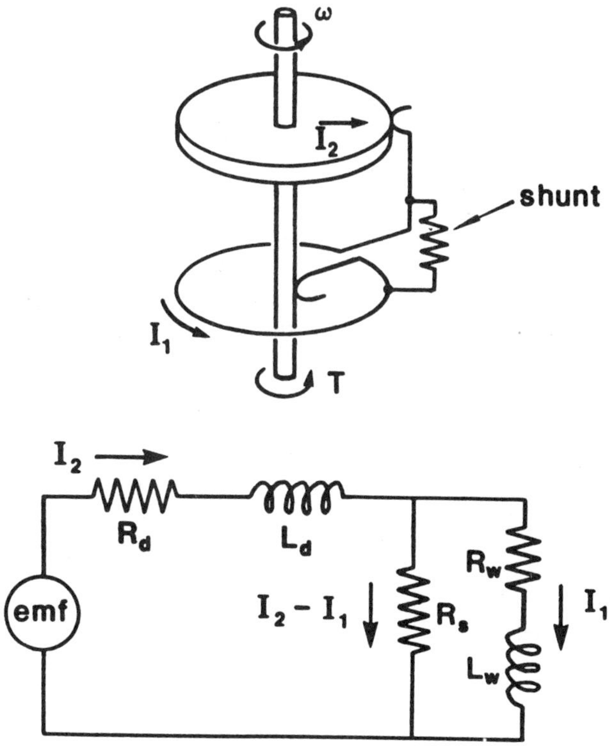

Fig. 7.3. Schematic diagram of the Malkus-Robbins disc dynamo.

The current I_1 through the coil produces a dipole-like magnetic field, approximately normal to the disc, according to Ampère's law (7.3). A reversal is said to occur when I_1 changes direction. The interaction of this magnetic field and the disc current I_2 induces a reactance opposing the motion, due to the presence of the Lorentz force in (7.1). The inductance L_W and resistance R_W of the brushes produce a phase difference between I_2 and I_1, which will favor reversals.

The equations for the currents and the angular velocity of the disc have the form

$$M\omega I_1 = L_d\dot{I}_2 + R_d I_2 + R_s(I_2 - I_1), \tag{7.11a}$$

$$(I_2 - I_1)R_s = L_W\dot{I}_1 + I_1 R_W, \tag{7.11b}$$

$$C\dot{\omega} = T + MI_1I_2 - \nu\omega, \tag{7.11c}$$

where $2\pi M$ is a mutual inductance, and $M\omega I_1$ is the emf. The transformations

$$[t,\omega,I_1,I_2] \to [\tau t, \delta(x - \frac{R_s}{M}), \beta y, \alpha z]; \tag{7.12a,b,c}$$

$$\tau \equiv \frac{L_d}{R_d R_s}, \quad \delta \equiv \frac{(R_W+R_s)(R_d+R_s)}{R_s M}; \tag{7.12d,e}$$

$$\alpha^2 \equiv \frac{R_W+R_s}{MR_s\nu\delta}; \quad \beta = \frac{R_s\alpha}{R_N+R_s}; \tag{7.12f,g}$$

yield the system

$$\dot{x} = R - yz - \nu x, \tag{7.13a}$$

$$\dot{z} = xy - z, \tag{7.13b}$$

$$\dot{y} = \sigma(x - y). \tag{7.13c}$$

It is interesting that the further substitutions $(x,y,z) \to (R/\nu - z, x, y)$ convert (7.13) into the Lorenz system (5.34). As a result, for certain values of the parameters, "magnetic reversals" occur in a deterministic aperiodic fashion, according to Figures 5.10 - 5.12.

To study the simplest example of a *kinematic* dynamo problem consider the case $R_s = \infty$, so that $I_1 = I_2 = I$, and take ω as a given quantity. Eqs. (7.11) reduce in this case to

$$L\dot{I} = (M\omega - R)I, \tag{7.14}$$

where $L = L_d + L_W$ and $R = R_d + R_W$. We see that the current will grow or decay exponentially depending upon whether ω is greater than or less than R/M. In other words, the *speed* with which a given velocity field is driven (here the mechanical system of disc and wires) is a stability parameter, with R/M an eigenvalue at which stationary fields of arbitrary amplitude may be maintained. This is a general property of kinematic dynamos involving a velocity field which is independent of time. That is, the dynamo effect is realized once the speed of the system exceeds a critical value.

7.5. Cowling's Theorem in a Spherical Core

Let us consider in more detail how the dynamo fails when axial symmetry is imposed. Consider a spherical core V_0 surrounded by an insulator layer, V_1. If both the velocity and the magnetic field are axisymmetric, with the same axis O_z, say, it is possible to represent them in the following form (cf. Figure 7.4):

$$u = u_T + u_P \equiv U\hat{\phi} + \nabla \times \psi\hat{\phi} \quad , \tag{7.15a}$$

$$\underset{\sim}{B} = B_T + B_P \equiv B\hat{\phi} + \nabla \times A\hat{\phi}, \tag{7.15b}$$

where $\hat{\phi}$ is the unit vector tangent to circles perpendicular to the z-axis. The components B_T and B_P are called the *toroidal* and *poloidal* parts of the field. They are the counterparts of zonal and meridional components in atmospheric dynamics (Part II).

Given u, the equations for A and B within the conductor follow from (7.8) expressed in cylindrical polar (z,r,ϕ) coordinates:

$$B_t + r\, u_P\cdot\nabla(B/r) - r\, B_P\cdot\nabla(U/r) = \eta(\nabla^2 - 1/r^2)B, \tag{7.16a}$$

$$A_t + r^{-1}u_P\cdot\nabla(rA) = \eta(\nabla^2 - 1/r^2)A. \tag{7.16b}$$

In the insulator, electric currents must vanish and we have

$$(\nabla^2 - 1/r^2)A = 0, \quad B = 0, \quad \text{in } V_1. \tag{7.17}$$

Also, continuity of the magnetic field at the boundary of the core, $R = (z^2+r^2) = R_0$, requires

$$B,\ A,\ \frac{\partial A}{\partial R} \quad \text{continuous on} \quad R = R_0. \tag{7.18}$$

Finally, to exclude sources at infinity we require that

$$A = O(R^{-2}), \quad R \to \infty, \tag{7.19}$$

this being an estimate consistent with a dipole-like decay of the poloidal vacuum field.

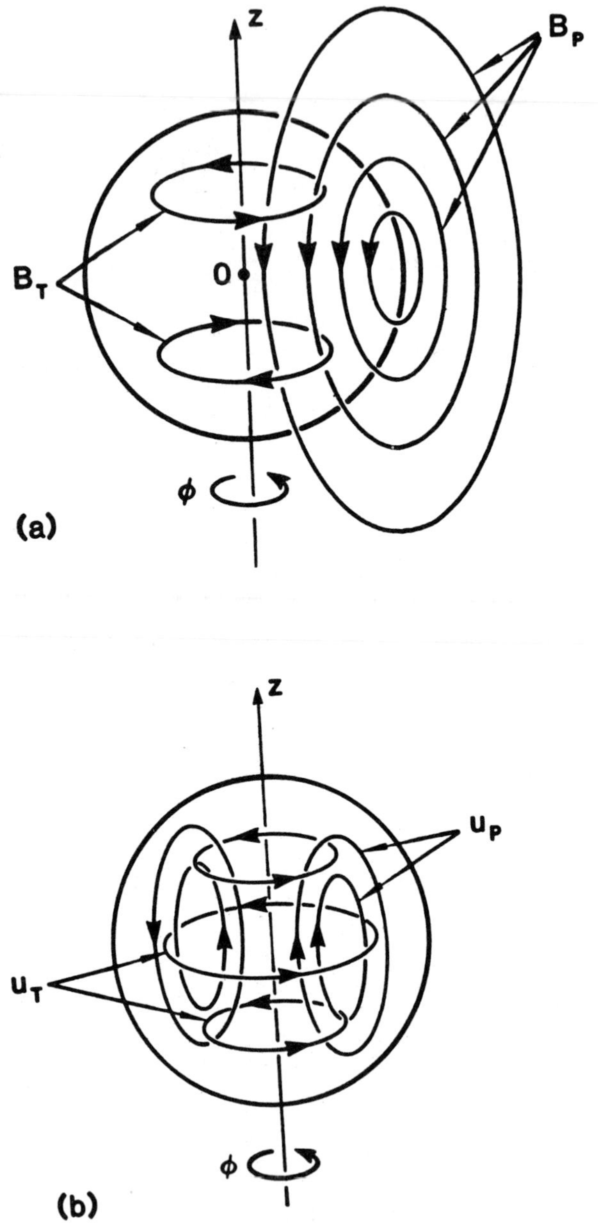

Fig. 7.4. Toroidal and poloidal components of: a) the magnetic field B; b) the velocity field u. Notice the symmetry of u_T about the equatorial plane, and the anti-symmetry of B_T.

Following Braginsky (1964), Cowling's theorem can be proved by multiplying (7.16a) by B/r^2 and (7.16b) by r^2A, and then integrating over V_0. The result of this calculation, after several applications of the divergence theorem is,

$$\frac{d}{dt}\int_{V_0}\left(\frac{B}{r}\right)^2 dV = \int_{V_0}\frac{B}{r}B_p\cdot\nabla(U/r)dV - \eta\int_{V_0}[\nabla(B/r)]^2 dV \tag{7.20a}$$

$$\frac{d}{dt}\int_{V_0}(rA)^2\,dV = -\eta\int_{V_0+V_1}[\nabla rA]^2\,dV \tag{7.20b}$$

Note that in the second equation a surface integral which arises has been expressed using (7.17) as a volume integral over V_1. Also we have assumed that u_p has vanishing normal component at the core boundary.

Now from (7.20b) we see that the poloidal component of the magnetic field must decay to zero. In (7.20a) the first term on the right suggests a source of toroidal field, and we shall indicate the mechanism below. However, this source is proportional to A, since $B_p = \nabla\times A\hat{\phi}$, and we have just seen that A tends to zero with time. Thus (7.20a) will imply the vanishing of B as well, and dynamo action fails.

This failure is a result of the symmetry imposed upon both the velocity and the magnetic field. Notice already the asymmetry built into the simple device of the previous section (Figure 7.3). In fact it can be shown that axisymmetric motions can act as dynamos, but that the induced field is then necessarily asymmetric (Moffatt, 1978, Ch. 6.).

7.6. The Dynamo Effect as Line Stretching

Rewriting (7.8) in the form

$$B_t + u\cdot\nabla B - \eta\nabla^2 B = B\cdot\nabla u, \tag{7.21}$$

we notice that this equation for B is identical to the equation for vorticity ω in a fluid of kinematic viscosity η and constant density $\rho \equiv 1$. It follows from Eqs. (1.22, 1.23) that, if $\eta = 0$, the lines of force of the magnetic field are material lines. From Section 1.4 it appears then that B can be amplified by the stretching of flux tubes. To be sure, in reality $\eta > 0$ and the diffusion of flux tends to oppose the enhancement of the field, to a degree which will depend upon the gradients of field strength. These two opposing effects are highlighted in the equation for the total magnetic energy.

Working again in the core-insulator domain $V_0 + V_1$, it is not difficult to show that

$$\frac{d}{dt}\int_{V_0+V_1} \frac{1}{2} B^2\, dV = \int_{V_0} u\cdot(B\times\nabla\times B)\, dV - \eta \int_{V_0} (\nabla\times B)^2 dV. \tag{7.22}$$

The first term on the right allows the possibility of field amplification by line stretching, while the second expresses the dissipation which is occurring by way of Joule heating. Since we are regarding u as given, the work being done to maintain the motion in the presence of the magnetic field is thus divided into a part which appears as a growth in magnetic energy, and a part which is lost from the system as heat. When a time mean is taken, assuming a mean magnetic energy exists, then mean rate of working exactly equals mean Joule dissipation.

is rather easy to see how field lines can be stretched by realistic core motions. For a spherical core, a velocity of the form $U(r,t)\hat{\phi}$ is the most general *geostrophic* motion allowed by rotation about the z-axis (compare Sections 1.5 and 2.3). In Figure 7.5 we indicate schematically how a line from the poloidal component B_P could be stretched out to produce a contribution to the toroidal part B_T. The process shown, called *differential rotation*, is described by (7.16a) with $u_P = 0$ and $\eta = 0$. It is more difficult to imagine how to continue the process so that the original poloidal component ends up being reinforced and maintained against dissipation, and it is this difficulty which leads to the analyses of kinematic induction considered in Chapter 8.

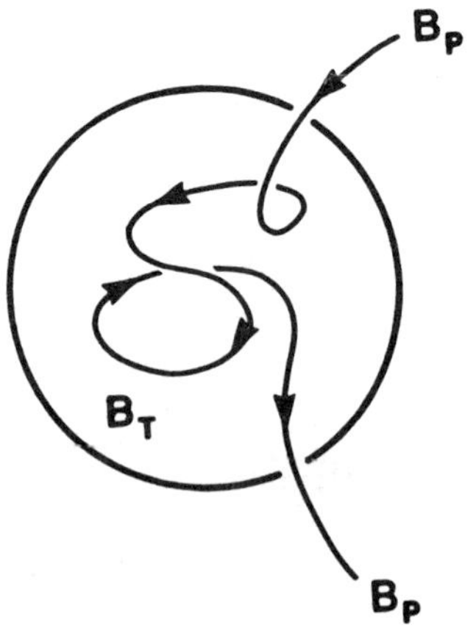

Fig. 7.5. Stretching and twisting of magnetic flux line.

As a geometric exercise, one can attempt to devise a dynamo theory based on prescribed fluid motions in three-dimensional space, chosen to do the necessary stretching of field lines without incurring too great a penalty in Joule heating. The two-dimensional process shown in Figure 7.6 is analogous to the folding of dough by a pastry chef.

In the sequence shown, if dissipation is neglected, the magnetic energy is increased by a factor of nine at each stage. Notice, however, that the field direction alternates between folds. One finds that, regardless of how small η may be, the dissipation which results will ultimately dominate the process, at least provided the power, and hence the speed of folding of the pastry chef is bounded! If dissipation connected with the fold is neglected, a Fourier series representation of the solution shows that, just after the n-th fold, the mean magnetic energy E_m satisfies (Childress, 1978)

$$E_m \le E_m(0)\ 9^n\ e^{-2\ ^2 9^n (\Delta t/\tau)}, \tag{7.23}$$

where $\tau = L_2^2/$, each fold is made instantaneoulsy and Δt is the time between foldings, while L_2 is indicated in the figure. This estimate implies the ultimate decay of E_m, irrespective of the value of τ.

A clever pastry chef could proceed differently and cut the dough after each stretching as in Figure 7.7a, but there is no way to do this to the magnetic field using smooth motions. Alvén (cf. Roberts, 1967, p. 96) has suggested a "twisted kink" mechanism in three dimensions. An important modification, due to Vainstein and Zeldovich (1972), involves a realignment of flux tubes, yielding the process shown in Figure 7.7b. Energy is quadrupled with a presumably small resistive penalty.

It should be clear from these examples and the form of (7.21) that the induction process created by a velocity field with a characteristic time T, length L, and speed U will involve the dimensionless parameters

$$R_{m_1} = UL/\eta, \quad R_{m_2} = L^2/\eta T, \tag{7.24}$$

which are characteristic *magnetic Reynolds numbers*. Roughly, a large magnetic Reynolds number implies that line stretching occurs without essential dissipation, that is, without diffusion of the field. This will be the case for large conductors, or for steady flow at high speed,

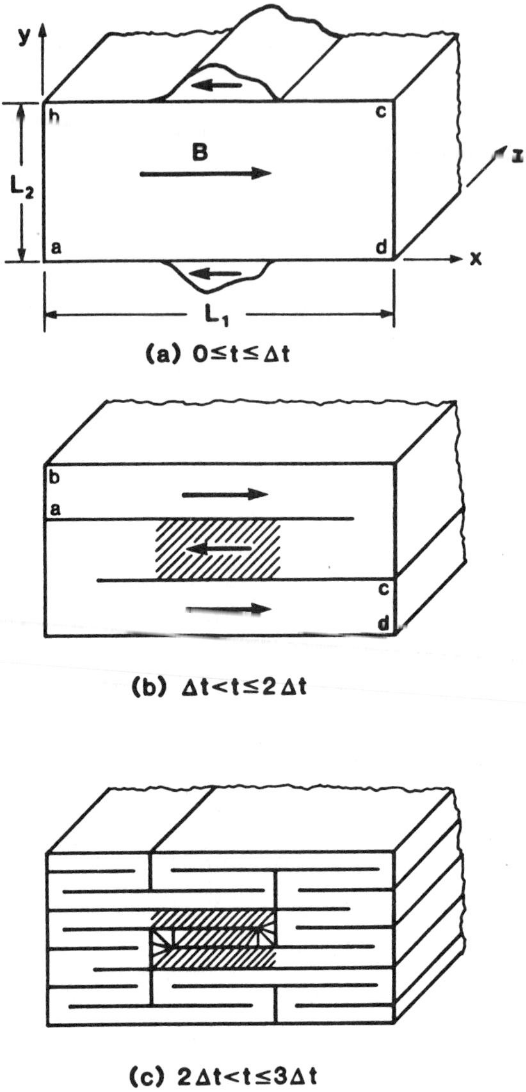

Fig. 7.6. Three steps of magnetic field folding by a pastry chef: shaded center indicates increasing gradients of B.

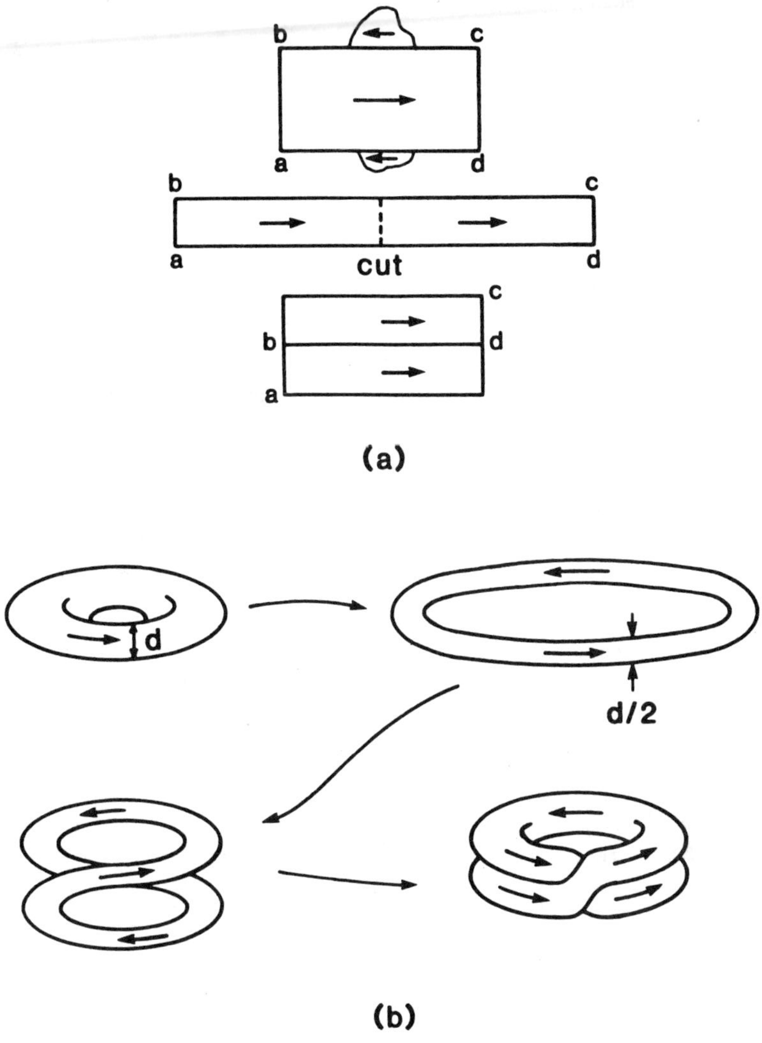

Fig. 7.7. Pastry dynamos: (a) in two dimensions, with cutting and reattachment; (b) in three dimensions, with smooth folding.

or for time-dependent motions with high frequency, since all of these properties imply a large R_m.

If L is core radius and U and η have the values given in Table 7.1, then R_{m1} has a value of about 100. This is a comfortably large figure for achieving a dynamo effect, as we shall see in the next section.

On the other hand, this very approximate value is not large enough to suggest that the limit of $R_{m1} = \infty$ (or $\eta = 0$) yields a reasonable kinematic description of the core magnetic field. If this were the case, the field would presumably be amplified locally by continual line stretching until a pattern of thin sheets and tubes of intense flux develops. It is difficult, for such highly intermittent fields, to construct a smoothly distributed heat source of convection, because of the complexity of the realizable geometries, and because the mollifying effect of field diffusion is confined to a relatively small domain.

For this reason, the only kinematic dynamo model which has successfully dealt with the case of large R_{m1} depends on the simultaneous realization of near axial symmetry of all fields (see Section 8.5). We note that recent observations of the surface solar magnetic field show a highly intermittent structure, a fact which cannot be avoided in modeling any dynamo action developed in the solar convection zone (see Section 8.6).

7.7. A Necessary Condition for a Dynamo Effect

We end this introduction to dynamo theory with a simple necessary condition for dynamo action in a bounded core V_0. Let u_m be the maximum speed of a time-independent velocity defined over the core. From (7.22) and the Cauchy-Schwarz inequality we have

$$\frac{d}{dt}\int_{V_0+V_1} \frac{1}{2} B^2 \, dV \leq u_m\left[\int_{V_0+V_1} B^2 dV \int_{V_0} (\nabla\times B)^2 dV\right]^{1/2} - \eta \int_{V_0} (\nabla\times B)^2 \, dV. \tag{7.25}$$

Now, for any magnetic field continuous at the core boundary and matching with a vacuum field in V_1, there is a constant L_0 such that

$$\int_{V_0} (\nabla\times B)^2 dV \geq L_0^{-2} \int_{V_0+V_1} B^2 \, dV \tag{7.26}$$

(e.g., Backus, 1958). In particular, if V_0 is a sphere of radius r_0, then $L_0 = r_0/\pi$.

Eqs. (7.25) and (7.26) imply that

$$\frac{d}{dt}\int_{V_0+V_1} \frac{1}{2} B^2 \, dV \le (u_m L_0 - \eta)\int_{V_0} (\nabla\times B)^2 \, dV. \tag{7.27}$$

Thus if $R_m^0 = u_m L_0/\eta$, the magnetic Reynolds number based upon the length L_0 is less than 1, the field decays and dynamo action is not possible.

This result again indicates the role of the fluid speed in achieving field amplification, but the estimate does not suggest how successful dynamos should be constructed. We note that, since (7.22) holds relative to any steadily rotating frame, R_m^0 must exceed 1 relative to *every* rotating frame in order for dynamo action to occur. Thus u_m is best thought of as an rms speed, the amount by which the velocity field differs from solid-body rotation.

7.8. Bibliographic Notes

General. Dynamo theory is rich in review articles. Among the earlier ones we mention the paper by Hide and Roberts (1961). More recently, Roberts (1971), Weiss (1971), Gubbins (1974), and Busse (1978) have discussed various aspects of the dynamo problem. The recent books by Moffatt (1978) and Parker (1979) are highly recommended. Stix (1981) has reviewed the status of dynamo theory as it applies to the sun. For concise yet readable accounts of the background MHD theory, see Cowling (1957) and Roberts (1967). Both of these monographs contain sections on dynamo theory. Some of the material in this and the following two chapters is adapted from Childress (1978). A recent addition to the literature is the proceedings of the 1980 Workshop on Dynamo Theory held in Budapest, Hungary, edited by A. M. Soward (1983).

Section 7.1. The numbers given in Table 7.1 represent values typical of various models and are not taken from any single source. For a recent summary of the status of estimates of core parameters see Loper and Roberts (1981).

Section 7.2. In equations (7.1)-(7.7) we adopt rationalized MKSQ units. For a discussion of alternative systems of units see Sommerfeld (1952).

Section 7.3. The Bullard-Gellman dynamo was re-examined by Gibson and Roberts (1969) and by Lilley (1970). These studies extended the truncation limit of the spherical harmonic expansion of the magnetic field used by Bullard and Gellman (1954). Although the results of the two computations did not agree completely, both confirmed in general terms a lack of convergence to an eigenvalue establishing stationary dynamo action. It is interesting that the motions considered here also fail to produce a dynamo when incorporated into Braginsky's kinematic theory, see Chapter 8. Suitable modification of the core motion, consistent with the demands of the kinematic theory, led to convergence of the numerical scheme and the onset of dynamo action, cf. Gubbins (1973).

Many of the important papers of Steenbeck, Krause and Rädler, originally in German, can be found in an English translation compiled by Roberts and Stix (1971).

Section 7.4. There is a considerable literature on various disc dynamo models, and there is current interest in them as realizations of physical systems with chaotic behavior. Bullard (1955) considered the single-disc dynamo in a geomagnetic context. An analysis of the corresponding two-disc coupled system, first studied by Rikitake, was given by Cook and Roberts (1970). For studies of some related models, motivated by the solar dynamo problem, see Jones, Weiss and Cattaneo (1985), and Weiss, Cattaneo and Jones (1984).

The exponential growth of I which is implied by (7.14) when ω exceeds its critical value can be shown to be in contradiction with analogous results for a fluid sphere. Moffatt (1979) has considered this paradox and shown that a modified disc geometry which more accurately represents the current field leads to realistic growth rates and behavior consistent with spherical models.

Section 7.7. It should be emphasized that the viewpoint taken in establishing necessary conditions for dynamo action, as well as in the derivation of "anti-dynamo" theorems, may turn out to be of great value in discussing dynamo action in realistic parameter ranges. For reasons which will be given in Chapter 9, gross bounds on dynamo activity may be all that one can hope to achieve by mathematical analysis within the parameter range ultimately reached in an isolated system. A number of workers have studied estimates for dynamo action which are more refined than (7.27), and so provide more information concerning the necessary structure of the velocity field. We mention the papers of Backus (1958), Busse (1975) and Proctor (1977b). As yet there has apparently been no systematic study of the possible use of variational techniques to derive approximating extremal fields along with bounds on the dynamo process.

CHAPTER 8

KINEMATIC DYNAMO THEORY

8.1. Introduction

Kinematic dynamo theory is concerned with solutions of the induction equation (7.8), which we rewrite here in the form

$$\frac{\partial B}{\partial t} - \eta\nabla^2 B = B \cdot \nabla u - u \cdot \nabla B, \tag{8.1a}$$

for a *given* velocity field $u(r,t)$. In its broadest sense, the theory seeks to discover, among some classes of admissible u, those functions which lead to "dynamo action", as discussed in Section 7.3. The precise meaning of dynamo action will generally depend upon the nature of the admissible functions u.

Realistic geodynamos involve a spherical or annular fluid core and incompressible, time-dependent flows. In this case a class of admissible motions might be defined as linear combinations of the form (see, for instance, Cowling, 1957)

$$u = \sum_{j=1}^{N} a_j \phi_j(r), \tag{8.1b}$$

where the ϕ_j are selected from a complete set of velocity modes determined by the domain, the boundary conditions on u, and the incompressibility condition.

If the a_j are stipulated to be independent of time, dynamo action can be defined by the existence of an eigenvalue λ with positive real part, such that

$$\eta\nabla^2 B^* + B^*\cdot\nabla u - u \cdot \nabla B^* = \lambda B^*, \tag{8.2a}$$

$$B = B^* e^{\lambda t}, \quad B^* = B^*(r). \tag{8.2b,c}$$

The implied growth of B for $\lambda > 0$ would correspond to the bifurcation point Q" in Figure 7.2. This eigenvalue problem has a rich history; a notable early numerical solution of this problem was undertaken by Bullard and Gellman (1954).

If the a_j in (8.1) are allowed to depend upon time, care must be taken to avoid a spurious dynamo action that might rely upon steadily increasing *kinetic* energy. Note that time-dependent kinematic dynamos can be analyzed most easily when the a_j are zero except for brief, regularly repeating intervals of activity. Backus (1958) utilized this approach to establish dynamo action in a spherical core.

Another important class of admissible u is suggested by the rotating components of laboratory dynamos. For example, let ϕ_j be the (discontinuous) flow equivalent to a solid sphere of radius a_j and center at r_j, rotating with unit angular velocity about an axis Ω_j. Two spheres of this kind, imbedded in a spherical core, can act as a time-independent dynamo (Herzenberg, 1958). Other such examples are reviewed by Roberts (1971) and Moffatt (1978).

All of these examples share an unusual feature: although the equation being solved is *linear* in B, the analysis of dynamo action necessarily involves the *nonlinear* dependence of the solution B upon the coefficients u and ∇u in (8.1). In the case of time-independent dynamos of the form (8.2) for example, the matter of interest is the dependence of $\mathrm{Re}(\lambda)$ upon the weights a_j. Since the point of the exercise is to prove dynamo action by velocity fields which might arise naturally from the dynamics of the core, the kinematic dynamo problem has many features of an *inverse problem* -- given that B grows in time or is maintained indefinitely, find u.

Recent efforts to "solve" the kinematic dynamo problem have relied on methods applicable to fields u of quite general structure, such as might be expected to occur in a turbulent core. These methods again depend upon the choice of admissible u; usually, they decide the question of dynamo action on the basis of solutions B which are functionally simpler than the field u. Indeed, a choice of u is often made which allows B to be computed by variants of the so-called *smoothing method*.

In Section 8.2 we introduce the smoothing method in the simplest possible context. After heuristic considerations for time-periodic and spatially small-scale velocity fields u, smoothing to first order of the effect of such fields on B is described. The magnetic field B is decomposed into a rough part B_R on the scale of u, and a smooth part B_S on a larger scale. The induction equation (8.1) is rewritten formally in terms of a projection P onto the smooth subspace of B-fields, and of its complement I-P, leading to an equation for B_S only, and one for B_R in terms of B_S.

Explicit model problems in one space dimension are solved to illustrate the general smoothing procedure, to first order and higher orders as well. It is shown how the procedure converges when the local magnetic Reynolds number $R_{ml}(\ell)$, based on the scale ℓ of u, is *small*.

The smoothing procedure is applied to spatially three-dimensional velocity fields, periodic in space and time, in Section 8.3. The singular character of the systematic expansion in small microscale Reynolds number $R_\ell = R_{ml}(\ell)$ is noted, and dynamo action shown to occur for large R_ℓ as well. A simple example exhibits clearly the α-*effect* by which the microscale B_R interacts with the microscale u to produce a nonvanishing mean effect on the macroscale induction equation for B_S. The connection between the α-effect and helicity in the resulting mean-field electrodynamics is pointed out.

The case of zero magnetic diffusivity, $\eta = 0$, is considered in Section 8.4. Examples indicate that an α-effect can still exist under these circumstances, which raises the question of finite α in the limit $\eta \to 0$ (Sec. 8.6).

The macroscale Reynolds number $R_L = R_{ml}(L)$ in the Earth's core, with L the core radius, is O(100). In Section 8.5 we consider a method, due to Braginsky (1964), for the study of mean-field dynamo action for *large* R_L and nearly axisymmetric fields u and B. The method is illustrated first for the advection of a scalar s by a velocity field which is nearly two-dimensional, with a small y-dependent part. A Lagrangian transformation of coordinates is introduced, depending on the smallness parameter ε, which reduces the field equation for the scalar s to one in "effective" variables, independent of y up to $O(\varepsilon^3)$.

Application to the dynamo problem proceeds by noticing that the previous transformation of coordinates leaves the non-diffusive part of the induction equation invariant. The diffusion term is multiplied by R_L^{-1} in the nondimensional form of the equation, and the method proceeds by taking $\varepsilon^2 = R_L^{-1}$ to combine the effects of small diffusivity and approximate axisymmetry. An α-type term is obtained in the final mean-field equations, which are compared with those of Section 8.3.

It was mentioned already in Section 5.5 that such generalized Lagrangian coordinate transformations have been extended to the study of large-scale atmospheric motions (Andrews and McIntyre, 1978; McIntyre, 1980). The problem of computing the mean effect of atmospheric waves on the axisymmetric general circulation is more difficult, since the vorticity ζ takes the place of the field B in (8.1) (Section 7.6), and ζ is not independent of u. Furthermore, the atmosphere is characterized by closed, cyclonic eddies, rather than by weak deviations from axisymmetry. Still, iteratively nonlinear modifications of the method might be useful, at least for an approximate description of certain features, accurate to first order in a suitable smallness parameter.

In Section 8.6, we consider small-diffusivity dynamos in which the advection time is much shorter than the diffusion time. This situation leads to a small effective Reynolds number $R_{m1}(\text{eff})$ and allows a special kind of smoothing to be applied. The limiting case of steady u is studied next in two geometries, and relevance to large-R, spatially-intermittent fields in the Sun's convection zone is discussed.

Spatial intermittency is taken up again in Section 8.7, along with a third scale of motion, that of the entire fluid core. A parameter is exhibited which distinguishes between smooth dynamo action, as encountered before, and a new "phase" of small-scale micro-dynamos. Bibliographic notes conclude the chapter in Section 8.8.

8.2. The Smoothing Method

This method can be illustrated heuristically in a special but important case, where u is periodic in time and is chosen to vary spatially on a scale ℓ much smaller than the core radius L. Let $B = B_0 + B'$ where B_0 varies on the scale L and B' varies on scale ℓ (and also, in general, L). Now u varies on scale ℓ, and has characteristic magnitude (say) u_0, so that the term $B_0 \cdot \nabla u$ in (8.1) can be estimated to have magnitude $|B_0|u_0/\ell$. Tentatively taking $|B'| << |B_0|$, we see from the various terms in (8.1) that B' must be generated by the dominant term above according to

$$\frac{\partial B'}{\partial t} - \eta\nabla^2 B' = L_S B' \cong B_0 \cdot \nabla u \tag{8.3}$$

where we use the subscript S to denote a "smooth operator". $B' = L_S^{-1}(B_0 \cdot \nabla u)$ can be estimated to have magnitude $|B_0|\ell u_0/\eta$ (since $\ell << L$) provided that the time-periodic variation of u is sufficiently slow. If the appropriate magnetic Reynolds number $R = R_{m1}(\ell) = U_0\ell/\eta$ is small, our Ansatz $|B'| << |B_0|$ is seen to be consistent with the solution.

On the other hand, extracting the smooth part of (8.1) by averaging over the small scales, we see that

$$\frac{\partial B_0}{\partial t} - \eta\nabla^2 B_0 = \nabla \times [\text{average of } (u \times B')] \sim |B_0| \frac{\ell U_0^2}{\eta L} . \tag{8.4}$$

Comparing the term on the right of (8.4) with the diffusion term on the left, we see that the two are comparable provided that

$$\frac{\ell L U_0^2}{\eta^2} \sim 1. \tag{8.5}$$

Since $\ell \ll L$, we can satisfy (8.5) even though $\ell U_0/\eta \ll 1$. Thus, even though we have restricted u to have a small *local* magnetic Reynolds number, the existence of two spatial scales yields an average dynamo effect (represented by the term on the right of (8.4)) which is comparable to the relatively small magnetic diffusion on the scale L. According to (8.5), it is the geometric mean of the two Reynolds numbers determined by the scales ℓ and L which is the deciding parameter.

We shall see below that, for certain choices of u, the term estimated on the right of (8.4) has the form $\nabla \times \alpha B_0$ where α is a pseudo-scalar constant. This term corresponds to an induced mean current αB_0, and the phenomenon has been termed the *alpha effect*. Additional "effects" of this nature occur as higher-order terms in the equation for the mean magnetic field when smoothing is carried out formally and recursively.

The essential features of the above procedure were first indicated by Parker (1955), although his arguments did not rely on a small local magnetic Reynolds number. The resulting dramatic simplification of the kinematic dynamo theory is achieved at the cost of a rather severe *structural* condition on the velocity field. In the example just considered, the dominant components of the velocity field are necessarily small-scale. When carried over to bounded domains such as the Earth's fluid core, this condition presupposes dominant motions on a scale small compared to core radius.

There is no reason, however, why a rotating, roughly spherical domain should not admit large-scale circulations, which might even dominate the small-scale flow. Fortunately, if it happens that the dominant large-scale flow has sufficient symmetry, a theory of almost-symmetric dynamos, due to Braginsky (1964), can be used. In its technical aspects Braginsky's method differs considerably from classical smoothing as outlined above. The small parameter is $R^{-1/2}$, and both magnetic and velocity fields in the core differ from axially symmetric fields by terms of this order. We postpone further discussion of Braginsky's approach to Section 8.5.

The most extensive developments of the smoothing technique as applied to dynamo theory are due to Krause, Rädler and Steenbeck. These authors deal with turbulent velocity fields as represented by stochastic functions of space and time endowed with well-defined characteristic scales of length and time. Large-scale components of the flow are also admitted in some calculations, where they contribute additional advection terms in the "smooth" differential operator L_S (cf. (8.3)). The resulting *mean-field electrodynamics* is in essence a theory of "effective" differential equations for an average magnetic field, and of their dependence on the parameters determining the statistics of the underlying turbulent velocity field.

Efforts to provide a rigorous justification for the smoothing approach have utilized deterministic fields which have relatively simple structure. For example, if the velocity field is periodic in space and time, it is straightforward to define a suitable vector space for the functions and operators appearing in the theory. In this way it is possible to study "smoothing to all orders" and examine the convergence of the series of approximations (Childress, 1970; Roberts, 1970). Because of the close correspondence between an analytical approach based upon velocity fields built up from periodic modes, and the spectral decomposition of one realization of a turbulent flow, we shall develop the theory below for the periodic case.

Smoothing to first order. Let us first indicate how the mean field equation (8.4) occurs as a first approximation in a sequence of smoothed equations. Since the magnetic field B is stretched by the velocity u, we expect B to acquire spatial and temporal scales which include all those present in u. Intuitively it is unlikely that scales much smaller than those present in u will be found in B since these would tend to be removed by diffusion. At the same time, it is possible that scales significantly *larger* than those of u might appear in B, and this is a key point which can be exploited, if true, by familiar asymptotic methods.

In this spirit, we can attempt to distinguish between "smooth" and "rough" fields, given that u is entirely "rough". The smooth part of B can be thought of as obtained under a projection P, and the rough part of B as obtained under I-P, which is also a projection. Recall that an operator is a projection if and only if $P^2 = P$.

We thus write, using subscripts R and S to refer to rough and smooth components,

$$B = B_S + B_R, \quad B_S = PB, \quad B_R = (I-P)B. \tag{8.6a,b,c}$$

Note that P is already referred to in this context as a "smoothing" operator although the precise meaning must await a definition of the underlying function space.

To indicate a formal approach to the solution of (8.1) utilizing smoothing, it is convenient to rewrite the equation in the form

$$LB \equiv L_S B + L_R B = 0, \quad L_S = \partial_t - \eta\nabla^2, \quad L_R = -\nabla \times (u \times (\cdot)), \tag{8.7a,b,c}$$

involving the operators L_S (cf. (8.3)) and L_R. In what follows, we shall assume that P and L_S commute when applied to admissible fields. If we then subtract the P-projection of (8.7a) from (8.7a) itself, this results in

$$(L_S - PL_S)B = -(L_R - PL_R)B, \tag{8.8a}$$

$$L_S B - L_S B_S = L_S B_R. \tag{8.8b}$$

Since the left-hand side of (8.8a) has zero P-projection, we need only assume that L_S has an inverse on "rough" functions, in order to solve (8.8) in the form

$$B_R = -L_S^{-1}(L_R - PL_R B) \equiv MB, \tag{8.9a}$$

or

$$B - B_S = MB. \tag{8.9b}$$

Further, if $(I-M)^{-1}$ is defined on the projected field, we have

$$B = (I - M)^{-1} B_S. \tag{8.10}$$

After substituting (8.10) into the second term in (8.7a) and multiplying the result by P we obtain

$$PL_S B + PL_R(I-M)^{-1}B_S = [L_S + PL_R(I-M)^{-1}]B_S = 0, \tag{8.11}$$

which is an equation for the smooth component B_S. It may be viewed as a compatibility condition for (8.7).

What has been accomplished by the derivation of (8.11)? The essential point is that this sequence of steps may be reversed: if one solves (8.11) as a linear equation for B_S, then the B defined from this B_S

by (8.10) will be a solution of the induction equation (8.7). Indeed suppose that (8.11) holds for some B_S. Then

$$LB = (L_S+L_R)(I-M)^{-1}B_S = (L_S+L_R+PL_R-PL_R)(I-M)^{-1}B_S$$
$$= L_S\{I + L_S^{-1}(L_R-PL_R)\}(I-M)^{-1}B_S + PL_R(I-M)^{-1}B_S$$

and using again (8.9a), as well as (8.11),

$$LB = L_S(I-M)(I-M)^{-1}B_S - L_SB_S = 0,$$

so we have indeed found a solution of the underlying equation (8.7).

This formal procedure replaces the solution of one linear problem by the solution of another, and it is not clear *a priori* what is gained by the effort. The advantages of starting with (8.11) will become obvious when we look at specific examples.

The justification for this approach rests upon an observation concerning the structure of the operator L_R. The operator L_R itself is not small in any appropriate sense relative to L_S, since the latter represents the effect of diffusion, while L_R describes advection and distortion of lines of force and must more than compensate diffusion to achieve a dynamo effect. Thus we cannot solve (8.10) perturbatively in any direct way (at least not while assuming the parameter η is fixed, cf. Section 8.7).

The operator M, however, as defined by (8.9), may well be a small operator. This operator involves L_S operating on the *rough* part of $L_R(\cdot)$, whereas (8.7), if written as $B = -L_S^{-1}L_RB$, involves L_S^{-1} on both smooth and rough components. It will be possible therefore to consider velocity fields for which $(I-M)^{-1}$ may be expanded in a rapidly converging Neumann series. Retaining only the terms I+M in this series gives what is known as *first-order smoothing*. In this approximate case, Eq. (8.11) for B_S reduces to

$$[L_S+PL_R-PL_RL_S^{-1}(L_R-PL_R)]\,B_S = 0, \tag{8.12a}$$

which, if $PL_RP = 0$, becomes

$$L_SB_S - PL_RL_S^{-1}L_RB_S = 0. \tag{8.12b}$$

We now may compare this first-order approximation with the heuristic result (8.4). We find that B', satisfying (8.3), is given by $B' = MB_0 = MB_S$, where B_S satisfies (8.4). Thus, the right-hand side of (8.4) is precisely $PL_R L_S^{-1} L_R B_0$.

A model problem. A simple example of the smoothing method can be given in a model which simulates many of the features of the full three-dimensional dynamo problem with periodic u. We consider one space dimension and allow both u and B to be complex-valued functions of x and t. In place of (8.1) we take

$$\partial_t B - \eta\partial_x^2 B = i\partial_x(uB^*), \qquad u = u_0 e^{i(kx+\omega t)}, \tag{8.13a,b}$$

where $-\infty \leq x \leq +\infty$ and ω, $k > 0$. Here the star denotes complex conjugation.

Given a real positive number $n < k/2$, define a vector space $V(n)$ to consist of functions B having the form

$$B = e^{inx+\sigma t} \sum_{m=-\infty}^{+\infty} a_m e^{im(kx+\omega t)}$$

$$+ e^{-inx+\sigma^* t} \sum_{m=-\infty}^{+\infty} b_m e^{im(kx+\omega t)}, \tag{8.14a}$$

with

$$\sum_{m=-\infty}^{+\infty} |a_m|^2 + |b_m|^2 < \infty, \tag{8.14b}$$

where σ is a complex constant, and define a projection P on $V(n)$ by

$$PB = B_S = (a_0 e^{inx+\sigma t} + b_0 e^{-inx+\sigma^* t}). \tag{8.15}$$

We also set

$$L_S = \partial_t - \eta\partial_x^2, \qquad L_R = -i\partial_x(u\cdot(\cdot)). \tag{8.16}$$

With the obvious definition of L_S^{-1}, we have for any B in $V(n)$,

$$\begin{aligned} MB = -u_0 e^{-inx+\sigma^* t} &\sum_{\substack{m=-\infty \\ m\neq 1}}^{+\infty} (a_m^*(k-km-n) \\ &\times \exp\{-i(1-m)(kx+\omega t)\}/[\sigma^*+i(1-m)\omega + \eta(k-km-n)^2] \\ -u_0\exp\{inx+\sigma t\} &\sum_{\substack{m=-\infty \\ m\neq 1}}^{+\infty} (b_m^*(k-km-n) \\ &\cdot\exp\{i(1-m)(kx+\omega t)\}/[(\sigma+i(1-m)\omega + \eta(k-km+n)^2]. \end{aligned} \tag{8.17}$$

If we define the norm of any such B by

$$||B|| = e^{\sigma_r t}\left[\sum_{m=-\infty}^{+\infty} |a_m|^2 + |b_m^2|\right]^{1/2}, \tag{8.18}$$

with $\sigma = \sigma_r + i\sigma_i$, then the norm of M is the least upper bound of all γ such that $||MB|| \le \gamma||B||$ for all B in V(n). Thus from (8.17) we have

$$||M|| \le \max_{\substack{0<n/k<1/2 \\ m \text{ integral} \neq 1}} \left\{\frac{u_0(k-km-n)}{\sigma+i(1-m)\omega + \eta(k-km-n)^2}\right\}. \tag{8.19}$$

From (8.19) we may estimate $||M||$ two ways:

(i) Set $\sigma = \omega = 0$,

(ii) Set $\sigma = 0$ and maximize with respect to k-km-n. We then obtain

$$||M|| \le \min[2u_0/k\eta,\ u_0/\sqrt{2\omega\eta}]. \tag{8.20}$$

Note that this last estimate exhibits the magnetic Reynolds numbers $R_{m1} = u_0/\eta k$ and $R_{m2} = \omega/\eta k^2$ (cf. (7.24)), with $||M||$ small if R_{m1} or $R_{m1}/(R_{m2})^{1/2}$ is small. These estimates carry over to the three-dimensional dynamo model of Section 8.3 below.

Our one-dimensional model problem is special in that the Neumann series $(I+M+M^2\ldots)B_S$ actually terminates with the term MB_S, so that *first-order smoothing is exact*: Indeed

$$\begin{aligned} B &= B_S + B_R \\ &= e^{inx+\sigma t} - u_0(k-n)(\sigma^*+i\omega+(k-n)^2\eta)^{-1}e^{-i(n-k)x+\sigma^* t+i\omega t} \end{aligned} \tag{8.21}$$

solves the equation exactly provided that

$$u_0^2 n(k-n) = (\sigma+n^2\eta)(\sigma^*+i\omega+(n-k)^2\eta). \tag{8.22}$$

Since n and k-n clearly yield the same values of σ_r, the condition $n < k/2$ introduced at the outset is not restrictive.

<u>Solutions for $\omega = 0$</u>. In this case of a steady velocity field, we have from (8.22) that $\sigma_i = 0$, and we can solve for σ_r:

$$\frac{\sigma_r}{ku_0} = -\frac{1}{2R}[N^2 + (N-1)^2] + \frac{1}{2}\sqrt{(1-2N)^2/R^2 + 4N(1-N)}, \tag{8.23}$$

where $R = R_{m1}$ and $N = n/k$. The variation of σ_r with respect to N for various R is indicated in Figure 8.1.

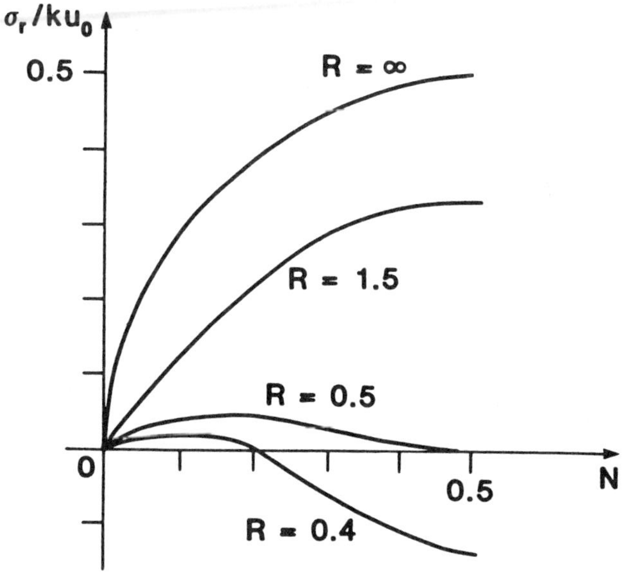

Fig. 8.1. Growth of the magnetic field B, measured by σ_r, as a function of magnetic Reynolds number R and truncation parameter $N = n/k$.

Smallness of M is insured by R being small, although no such condition is actually needed here since the series terminates. For small R, Eq. (8.23) may be expanded with $\tilde{N} = N/R^2$ fixed to obtain

$$\frac{\sigma_r}{ku_0R^3} \equiv \tilde{\sigma} = \tilde{N} - \tilde{N}^2 + O(R^2). \tag{8.24}$$

In particular growing modes $(\tilde{\sigma} > 0)$ occur in this limit when $\tilde{N}$ is sufficiently small.

We can also see in (8.23) an emerging multi-scale structure when R is small. Indeed, if ℓ is the wavelength of the velocity field, L the wavelength of the growing magnetic field, then $\ell/L \equiv \varepsilon \sim R^2$; note that in this ordering a magnetic Reynolds number based upon L is large, since $R_{m1}(L) = (1/\varepsilon)R \sim \varepsilon^{-1/2}$. In fact the length scale relative to which the magnetic Reynolds number is of order unity is the geometric mean $(\ell L)^{1/2}$ of these two lengths.

From (8.21) it is evident that B_R is of order R when (8.24) is valid. Consequently, the structure of the velocity and magnetic fields can, in the limit of small R, be sketched in the manner of Figure 8.2.

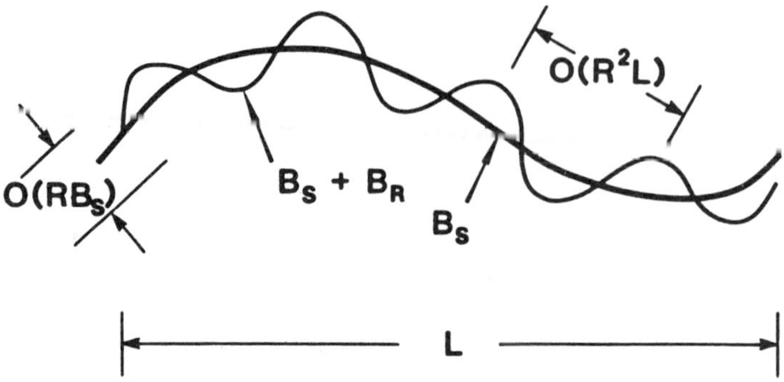

Fig. 8.2. Spatial structure of the magnetic field $B = B_S + B_R$ in the smoothing approach, with R small.

That is, a relatively small-scale velocity field, having small local magnetic Reynolds number $R = R_{ml}(\ell)$, operates to amplify a magnetic field consisting of a dominant smooth part and a rough, small-scale component which is smaller by a factor R.

Multi-scale expansion. The single-mode projection just treated suggests how the procedure of solving (8.13) asymptotically can be carried out using multi-scale methods. The scaling reflected in (8.24) suggests that the variable $\xi = R^2 x$ be defined, along with an expansion which takes the dimensionless form

$$B = \sum_{j=0}^{\infty} R^j B_j(x,\xi,\tau). \tag{8.25a}$$

$$u = e^{ix}, \quad \tau = R^3 \eta\, t, \tag{8.25b,c}$$

Restricting ourselves to the case $\omega = 0$ and $k = 1$ the equations become

$$\partial_t B - R^{-1}\, \partial_x^2 B = i\partial_x(uB^*). \tag{8.25d}$$

The partial ∂_x in (8.21) is now to be replaced by $(\partial_x)_\xi + R^2(\partial_\xi)_x$.

With these substitutions in (8.12), we may collect terms of the same order in R to obtain

$$\partial_x^2 B_0 = 0,$$

$$\partial_x^2 B_1 = -i\partial_x(e^{ix}B_0^*),$$

$$\partial_x^2 B_2 = -2\partial_x\partial_\xi B_0 - i\partial_x(e^{ix}B_1^*), \tag{8.26a,b,c}$$

$$\partial_x^2 B_3 = -2\partial_x\partial_\xi B_1 - i\partial_x(e^{ix}B_2^*) - i\partial_\xi(e^{ix}B_0^*),$$

$$\partial_x^2 B_4 = -2\partial_x\partial_\xi B_2 - i\partial_x(e^{ix}B_3^*) - i\partial_\xi(e^{ix}B_1^*) + \partial_\tau B_0 - \partial_\xi^2 B_0, \text{ etc.} \tag{8.26d,e}$$

Thus $B_0 = B_0(\xi,\tau)$, $B_1 = -e^{ix}B_0^*$, $B_2 = 0$ and $B_3 = -ie^{ix}\partial_\xi B_0^*$. Assuming that B_4 is periodic with respect to x with period 2π, the last of these equations integrates to give

$$\partial_\tau B_0 - \partial_\xi^2 B_0 + i\partial_\xi B_0 = 0, \tag{8.27}$$

which is the differential equation associated with the terms exhibited in (8.24).

According to (8.27), the microscale velocity field has introduced the term $i\partial_\xi B_0$ into the equation for the dominant part of the smooth magnetic field. While the total magnetic field also involves a small microscale part, the complete solution is generated by solving for the smooth field. It turns out that this model problem accurately reflects the structure of the solution of the exact three-dimensional induction equation when microscale velocity fields with small local magnetic Reynolds numbers are utilized, as we shall see in Section 8.3 below.

Before turning to the latter problem we note that the necessary condition for dynamo action, discussed in Section 7.7, can also be derived in the present model, with $L_0 = n^{-1}$. Setting $\sigma_r = 0$ in (8.23) there results

$$\left(\frac{u_0 L_0}{\eta}\right)^2 = \frac{k}{n} - 1, \quad \frac{n}{k} < 1/2. \tag{8.28}$$

Thus values of u_0L_0/η greater than but arbitrarily close to 1 produce dynamo action.

8.3. Application to The Dynamo Problem

The preceding model mimics very well the exact kinematic dynamo problem. The main adjustments needed in passing to the full, three-dimensional problem are technical details concerning the class of velocity fields and the exact nature of the differential operators.

We first consider the class of solenoidal velocity fields which are periodic in space and time,

$$u(x+2\pi\ell(m_1e_1 + m_2e_2 + m_3e_3),\ t + \frac{2\pi}{\omega} m_4) = u(x,t),$$
$$\nabla \cdot u = 0, \tag{8.29}$$

where (e_i) is a basis in $\mathbb{R}^3$ and the m_i are arbitrary integers. For simplicity we take u to consist of a finite complex linear combination of modes $\exp i[k\cdot x + \omega t]$ where, in terms of the dual basis (e^j), $e^j\cdot e_i = \delta_{ij}$, we have

$$k = \ell^{-1} \sum_{j=1}^{3} m_j e^j. \tag{8.30}$$

The methods used to study these periodic fields are similar to Bloch wave analysis in solid-state physics (e.g., Brillouin, 1953, pp. 139-143). However, our approach carries over to more general cases, for example, square-integrable periodic velocity, or to suitable classes of stochastic functions, where the smooth projection is onto a statistical mean. In the latter class of problems, the present theory has come to be referred to as mean-field electrodynamics.

With (8.29), (8.30) we introduce the wave number vector n and growth rate σ of the smooth field, and so create the (complex) vector space $V(n)$ spanned by basis elements of the form

$$E = \exp[i(k+n)\cdot x + (im\omega + \sigma)t], \quad m \text{ integer}. \tag{8.31}$$

There remains to define the smooth projection P. By analogy with (8.15), P could be defined as

$$PE = 0 \text{ if } k \neq 0 \text{ or } m \neq 0,$$
$$= E \text{ if } k = 0 \text{ and } m = 0.$$

It is simpler and more useful in other contexts to think of P as a suitable space-time mean.

If n and σ are not in the set of k and ω occurring in V, we define

$$PB = e^{in\cdot x+\sigma t}\langle e^{-in\cdot x-\sigma t}B\rangle, \tag{8.32}$$

where $\langle\cdot\rangle$ represents the space-time mean. With the definition of norm through mean-square amplitudes, the estimate (8.20) carries over to the present theory and the smoothed equation (8.11) takes the form

$$L_S B_S \equiv (\partial_t - \eta\nabla^2)B_S = -PL_R(I-M)^{-1}B_S = -\nabla \times \sum_{j=1}^{\infty} A_j\cdot B_S, \tag{8.33a}$$

$$A_j(n,\sigma) = -PL_R M^j. \tag{8.33b}$$

It can be shown (Childress, 1969) that the matrices A_j have the property that

$$A_j^{T*}(n,\sigma) = (-1)^j A_j(n,\sigma^*), \tag{8.34}$$

where T here denotes transpose. For modes with real σ, A_j is therefore Hermitian or anti-Hermitian, depending upon whether *j* is even or odd. Furthermore, $A_j(0,0)$ is a real matrix. This result is important, since $A_1(0,0)$ determines the form of (8.33) in the limit of small microscale Reynolds number, provided it does not vanish.

More explicitly, if we take ℓ and ℓ^2/η as units of length and time and go over to a dimensionless formulation, the equation analogous to (8.25d) or (8.27) has the differential form

$$(\partial_t - R^{-1}\nabla^2)B_S = \nabla \times \alpha \cdot B_S, \tag{8.35}$$

where α is a constant symmetric matrix. Indeed, in the dimensionless formulation the matrices $A_j(0,0)$ can be written $A_j^0 R^j$ where A_j^0 is constant and independent of R; we then define $\alpha = A_1^0$. Of course regarded as a differential equation for B_s, the full compatibility equation (8.33) is of infinite order, so that strictly speaking (8.35) is obtained by a singular limit for small R.

If we now ask for what choices of α equation (8.35) possesses growing Fourier modes as solutions, the answer turns out to be surprisingly simple: G. O. Roberts (1970) has observed that such modes exist, for sufficiently small wavenumbers *n*, provided that α is nonsingular. A necessary and sufficient condition, weaker than the preceding one, is that the matrix adjoint of α have at least one positive eigenvalue.

Since the matrix α is obtained by first-order smoothing, it is a quadratic function of the velocity. If, for example, u(x,t) is taken to be any element of the Hilbert space of mean-square integrable periodic solenoidal fields, the scalar stability condition $\mathrm{Det}(\alpha) = 0$ determines a set of motions of measure zero. This set is a hypersurface of nonzero codimension in the infinite-dimensional space of Fourier amplitudes. In this sense, *almost all periodic fields operate as dynamos*, a point emphasized by G. O. Roberts.

The preceding results were obtained for small R, but it is possible to show that growing magnetic modes with small *n* exist for almost all values of R for almost all space-integrable solenoidal fields *u*. This is established by noting that the full series $\sum A_j(0,0)$ is an analytic function of R, and by using the properties of the A_j noted above.

An example. The canonical example of a spatially periodic, steady dynamo can be deduced by looking for fields which yield $\alpha = I$. We thus obtain $u = (\sin y + \cos z, \sin z + \cos x, \sin x + \cos y)$. Applying the two-scale procedure introduced in the one dimensional model, we can use this particular u and $\xi = R^2 x$, $t = R^3 \tau$, to rederive (8.35) as follows:

$$B = B_0(\xi,\tau) + R\, B_1(x,\xi,\tau) + \cdots \tag{8.36a}$$

$$-\frac{1}{R}\nabla_x^2 B_1 = -\nabla \times (u \times B_0) = -B_0 \cdot \nabla u. \tag{8.36b}$$

Hence

$$B_1 = R\, B_0 \cdot \nabla(\sin y + \cos z, \ldots, \ldots) = R(B_{02}\cos y - B_{03}\sin z, \ldots, \ldots)$$

and

$$(\partial_\tau - \nabla_\xi^2) B_0 = \nabla_\xi \times \langle u \times B_1 \rangle = \nabla_\xi \times B_0. \tag{8.36c}$$

The α-effect. We have exhibited here the so-called "alpha effect" of mean-field electrodynamics, first mentioned following Eq. (8.5). The last calculation shows that the dominant microscale magnetic field interacts with the microscale velocity field to produce a nonvanishing mean effect on the induction equation. This "smoothed" equation has therefore a class of solutions fundamentally different from the unsmoothed, exact equation.

In particular, the "anti-dynamo" theorems do not apply, and we may expect that geometries having an appealing and simplifying degree of symmetry can be allowed. The most useful example are dynamos which have axisymmetric mean fields. It is important to keep in mind that the full magnetic field involves the rough component, and the smooth problem being solved is a sort of "generator" of the full field, as exhibited above more precisely for the periodic case.

The *physical basis* of the α-effect can best be understood by considering only part of the velocity field. We observe in (8.36) that each constituent "wave", e.g., $(0, \cos x, \sin x)$ contributes to a single entry along the diagonal of the matrix $\alpha = I$, with no interaction between waves on the average. We can therefore concentrate on "one-third" of the motion.

More generally, let us restore dimensions and time dependence and set $u = U(0, \sin \xi, \cos(\xi - \phi))$, $\xi = kx + \omega t$. Then

$$\frac{\partial}{\partial t} B_R - \eta \nabla^2 B_R \simeq B_S \cdot \nabla u \tag{8.37a}$$

and therefore, with $\tan \psi = \eta k^2/\omega$,

$$B_R \simeq (B_S \cdot i)k(\eta^2k^4 + \omega^2)^{-1/2}u(\xi+\psi). \tag{8.37b}$$

The "sources" of B_R on the right of (8.37a) are shifted in phase by $\pi/2$ relative to u, since the distortion of magnetic field results from the shear of the flow. For stationary sinusoidal sources, the inversion of the diffusion (now Laplace) operator only alters the amplitude, but for moving sources there is an auxiliary shift, and the combination of the two gives the phase shift ψ in (8.37b). When $\phi = \pi/2$, u and B_R are circularly polarized so that $u \times g$ is proportional to $\sin \psi$. For arbitrary ϕ, the expression for the (dimensionless) α reduces to

$$\alpha_{11} = -\eta k^2(\eta^2k^4 + \omega^2)^{-1/2}\sin \phi \sin \psi = -\sin \phi \sin^2\psi. \tag{8.38}$$

Note that, through the ψ-dependence, α vanishes as $\eta \to 0$, i.e., fully developed propagating waves cannot regenerate the magnetic field in a perfect conductor (Krause and Roberts, 1973; Moffatt, 1976). Nevertheless, from conditions analogous to those given following (8.19), we find that first-order smoothing is valid for motions periodic in t with zero mean provided that $Uk \ll (\eta^2k^4 + \omega^2)^{1/2}$ and $|\alpha_{11}| \ll \eta k/U$. Both of these inequalities may be satisfied for small Uk/ω, independently of the value of Uk/η.

It is curious that the *regeneration* of the field, as well as its decay, should be due to the finiteness of resistivity. It should be borne in mind that the present argument deals with the amplification of the "large-scale" magnetic field, while the process depicted in Figure 7.7b does not. We may conclude that, in the present case, dissipation is needed for reconnection and smoothing on the smaller scales, and that this process produces small-scale magnetic fields capable of interacting favorably with the velocity field, resulting in a nonzero mean current density.

The specific results given above are special in two ways. First, there are many ways in which a conductor can flow so as to introduce several scales of space and time, and we have not as yet considered any examples involving temporally intermittent processes, typical of fully turbulent flow. Rapid, jerky motions followed by periods of rest can produce intense local distortions of a mean magnetic field, and since this occurs quickly the effective local magnetic Reynolds number R_{m2} will be quite large. If the field distortions are also on a microscale, they will decay rapidly during the rest periods, so the end effect can be accessible to a smoothing procedure of a somewhat different kind.

Indeed, Parker foresaw in 1955 the use of smoothing in the dynamo problem by imagining the effect of small regions and intervals of "cyclonic activity". He then showed, by appealing to an essentially high magnetic Reynolds number analysis, that the result should be the α-effect in the functional form described by (8.35). For periodic movement this approach requires $R_{m1}^2 \ll R_{m2}$. However (apart from the almost symmetric cases considered in Section 8.5) there is really no completely satisfying formal theory that can be reasonably referred to as "smoothing at large R_{m1} with arbitrary R_{m2}", and this is an important gap in the mathematical analysis of the kinematic dynamo problem.

Our aim in Chapter 9 will be to consider the consequences of the α-effect in the construction of models of the geodynamo. It is important, therefore, to know whether or not velocity fields of the kind which give rise to a nonzero α will actually be produced in a contained volume of rotating fluid when subjected to some energy source.

Because of the global dynamics of a contained, rotating fluid, a nonzero α will vary with position. In the context of microscale-macroscale spatial smoothing, there is no reason why u should not vary on both scales, so that α would then also vary on the larger scale.

It is found in many problems that the kinds of microscale motions which are rather easily produced in a rotating fluid have a nonzero macroscale *helicity* H, where (cf. Moffatt, 1978, Ch. 10)

$$H = \langle u \cdot \nabla \times u \rangle. \tag{8.39}$$

In example (8.36) above, $\alpha = I$ and $\nabla \times u = -u$. This example is therewith a *Beltrami field*, in which vorticity and velocity are everywhere parallel, and $H = -1$. Bounded rotating fluids having zero helicity everywhere initially tend, when suitably driven (e.g., by temperature differences through a convective instability) to acquire helicity on the macroscale, although the net helicity over the domain may remain zero. In this case we can think of rotation as "polarizing" the velocity into regions of positive and negative helicity. In the process, however, an α-effect tends to be created throughout the conductor.

Finally, it should be pointed out that the dynamical reaction of the growing magnetic field on the velocity field which produces it can have a significant effect on the structure of the flow. As a result, the separation of spatial scales exploited above may disappear in the process (see Chapter 9).

8.4. The α-Effect in a Perfect Conductor

The preceding calculations have depended very much on the diffusion of magnetic flux. It is of interest to see what happens when an α-effect is computed in its absence. We consider here a simple flow for which this can be done explicitly.

The motion takes place in the unit square $|x| \leq 1$, $|y| \leq 1$. The conductor is rigid except within the unit circle $r^2 = x^2 + y^2 < 1$. If (r,θ,z) are cylindrical polar coordinates, the velocity field (u_r,u_θ,u_z) within $r < 1$ is defined by

$$u_r = 0, \quad u_\theta = r\omega(r), \quad u_z = W(r), \quad \omega(1) = W(1) = 0. \tag{8.40}$$

The governing equations, with

$$(B_r,B_\theta,B_z) = \left(-\frac{1}{r}\frac{\partial A}{\partial\theta}, \frac{\partial A}{\partial r}, B\right), \tag{8.41}$$

and A,B functions of r, θ, and t are obtained by setting $\eta = 0$ in (7.8) and substituting (8.41), which yields

$$\frac{\partial A}{\partial t} + \omega\frac{\partial A}{\partial\theta} = 0, \tag{8.42a}$$

$$\frac{\partial B}{\partial t} + \omega\frac{\partial B}{\partial\theta} + \frac{W'}{r}\frac{\partial A}{\partial\theta} = 0. \tag{8.42b}$$

We suppose that $A(r,\theta,0) = -r\sin\theta$, $B(r,\theta,0) = 0$, corresponding to an initial field parallel to the x-axis. Since $\eta = 0$, $\partial A/\partial t$, $\partial B/\partial t = 0$ in the rigid conductor, and therefore $A(1,\theta,t) = \sin\theta$. Thus

$$A(r,\theta,t) = -r\sin(\theta - \omega(r)t), \tag{8.43a}$$

$$B(r,\theta,t) = tW'(r)\cos(\theta - \omega(r)t) \tag{8.43b}$$

over the moving conductor, with a discontinuity of B at $r = 1$.

The average of the z-component of $u \times B$ over the unit disc then gives a value for a, the diagonal component of the α matrix corresponding to the z-direction, equal to

$$\begin{aligned} a &= \frac{1}{\pi}\int_0^{2\pi}\int_0^1 r(u_\theta\cos\theta\, B - B_\theta W\cos\theta - B_r W\sin\theta)dr \\ &= \int_0^1 (tr^2\omega\, W'\cos\omega t + W'r^2\sin\omega t)dr. \end{aligned} \tag{8.44}$$

A representative choice for functions ω, W is

$$\omega = 1 - r, \quad W = 1 - r^2, \tag{8.45}$$

in which case (8.44) yields

$$a(t) = -24t^{-3} - 12t^{-3} \cos t + 36t^{-4} \sin t. \tag{8.46}$$

This function is negative and decays like t^{-3} for large t. The negative sign of a is interesting in view of a general property of mean-field dynamo theory. As we noted following equation (8.39), the simplest examples of periodic dynamo action show that flows with fixed positive helicity produce negative components of α. It is clear that if W' in (8.44) is chosen to be zero except at a value of r where $d(r^2\omega)/dr \neq 0$, then $a(t)$ can exhibit a growing oscillation, at least initially.

This range of behavior can be understood in terms of line stretching (cf. Section 7.6). The apparent value of $a(t)$ measures x-directed current, and by Stokes' theorem this is indicative of the presence of field components parallel to the yz-plane. These components are created by simultaneously lifting and twisting a field line (see Figure 8.3). As the twisting continues, the x-current tends to average to zero, a result of the rapid oscillation of the y and z components of the field, leading to the decay of $a(t)$.

This example highlights the importance of randomness, finite η, or both, in the generation of a significant α-effect. We might imagine a random pattern of square regions in the xy-plane within which motions of this kind occur over a certain time interval. At the end of this interval, we repeat the process with the same initial condition as before. In this way an ensemble average can be constructed without ever considering

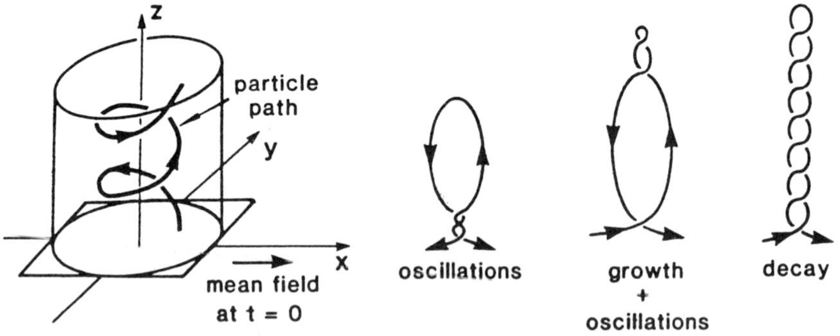

Fig. 8.3. Simple illustration of α-effect in a perfect conductor. (a) The velocity field; (b) the time evolution of mean x-directed current.

the limit of infinite time, and the average of a, defined in this way, can have a non-negligible value since the motions are never operating long enough to cause decay.

In the Earth's core, the small electrical conductivity could, for sufficiently long periods of rest, produce the same effect. Certainly finite conductivity completely alters the induction problem for large times, and it has been suggested (see Section 8.6) that for many three-dimensional motions the resulting α satisfies

$$\lim_{\eta\to 0} (\lim_{t\to\infty} \alpha) \neq 0. \tag{8.47}$$

The above calculation indicates that in many cases the reversal of the two limits in (8.47) yields zero, a result that is not entirely unexpected since it is known that in general the two limits do not commute.

8.5. Almost Symmetric Dynamos

In 1964 Braginsky published the first of a now classical series of papers dealing with the kinematic dynamo problem in a spherical core. His method treats the induction equation,

$$B_t + u\cdot\nabla B - B\cdot\nabla u - R^{-1}\nabla^2 B = 0 \tag{8.48}$$

in nondimensional form, under the assumption that the macroscale magnetic Reynolds number $R = R_{m1}(L)$ is large compared to 1. Estimates of this R in the Earth's core vary, but if the unit of length L is the core radius, a value of R on the order to 100 is reasonable (see also Section 7.6).

Braginsky also noted that the near-symmetry of the Earth's magnetic field suggests that departures of solutions of (8.48) from exact axial symmetry, departures which, by Cowling's theorem (Section 7.5), are inevitable in a working kinematic dynamo, might nevertheless be small. The key point is the *simultaneous* use of the limit $R \to \infty$ and of the assumption of near axial symmetry for *both* magnetic and velocity fields. We might therefore describe Braginsky's theory as one of "asymptotic symmetry breaking".

The result of the theory is, as in the smoothing methods of Sections 8.2 and 8.3, a modified induction equation which avoids the force of Cowling's theorem. We shall find that the modification can again be expressed as an α-effect. In the present case, however, the smoothed fields are actually axisymmetric. In the general context of smoothing, we can view Braginsky's procedure as one of averaging with respect to longitude (zonal averaging in the terminology of Part II).

The original derivation of the modified system by Braginsky (1964a) utilizes a systematic expansion in $1/R$. This expansion becomes quite lengthy, since it is necessary to proceed to higher order terms than the ones that actually appear in the final equations. These equations are obtained by a compatibility argument applied to an inhomogeneous higher-order calculation.

In an important series of papers, Soward (1971a,b, 1972) reexamined and clarified the essential mathematical content of Braginsky's method, and we shall follow Soward's approach here (see also Moffatt 1978, Ch. 8). Soward notes that Braginsky, in effect, exploits an invariance property of the induction equation in a perfect conductor ($R \to \infty$). He shows how the invariance may be used constructively when the fields are almost symmetric. To understand the basis of the method it is helpful to begin by considering a problem much simpler than the dynamo problem, namely advection of a scalar quantity by an almost symmetric velocity field.

<u>Almost symmetric advection</u>. Consider Cartesian coordinates $\tilde{\underset{\sim}{x}} = (\tilde{x},\tilde{y},\tilde{z})$, a solenoidal velocity field $\tilde{u}(\tilde{x},\tilde{y},\tilde{z},t)$, and a scalar field $\tilde{s}(\tilde{x},\tilde{y},\tilde{z},t)$ satisfying

$$\tilde{s}_t + \tilde{u} \cdot \tilde{\nabla}\tilde{s} = 0. \tag{8.49}$$

We use the tilde to associate $\tilde{u}$ and $\tilde{s}$ with particle paths which have a fluctuating component in addition to a symmetric part. We shall define a symmetric field as one which is independent of $\tilde{y}$, and take $\tilde{u}$ to be close to such a field. A velocity field having this structure is appropriate to a version of Braginsky's problem where the symmetric field is independent of one Cartesian coordinate instead of being independent of an angular coordinate.

We introduce a small parameter ε and postulate that $\tilde{u}$ has the form

$$u(\tilde{x},\tilde{y},\tilde{z},t) = (0,U(\tilde{x},\tilde{z},t),0) + \varepsilon u'(x,y,z) + \varepsilon^2(\psi_{\tilde{z}},0,-\psi_{\tilde{x}}), \tag{8.50}$$

where $\psi = \psi(\tilde{x},\tilde{z},t)$. The first term on the right of (8.50) is analogous to an axisymmetric toroidal component (cf. (7.15)), the second is a term contributing the fluctuation, and the third is analogous to a small axisymmetric poloidal component.

Equation (8.49) can be "solved" formally by introducing the Lagrangian coordinates $\tilde{x} = \tilde{X}(a,t)$, since the equation states that $\tilde{s}$ is a material invariant:

$$\tilde{s}(\tilde{X}(a,t),t) = \tilde{s}(a,0). \tag{8.51}$$

Here $\tilde{x}$ stands, as usual in our convention, for $(\tilde{x},\tilde{y},\tilde{z})$, and similarly $\tilde{X}$ also stands for $(\tilde{X},\tilde{Y},\tilde{Z})$. But the "solution" (8.51) does not reveal the structure of the field $\tilde{s}$ and may be difficult to obtain explicitly. We shall therefore examine a procedure which reaches a solution in a different way by solving a modified *symmetric* equation.

Consider the same problem in new variables (without tildes):

$$s_t + u \cdot \nabla s = 0. \tag{8.52}$$

If $\tilde{s}$ and s have the same initial values, then the equation

$$\tilde{s}(\tilde{X}(a,t),t) = s(X(a,t),t) \tag{8.53}$$

describes how the function transforms in passing from one set of variables to the other. This transformation can be represented by a time-dependent function $\tilde{x}(x,t)$, so that

$$\tilde{x}(X(a,t),\ t) = \tilde{X}(a,t). \tag{8.54}$$

From (8.54) we see that

$$\tilde{u} = \frac{d}{dt}\tilde{X} = \left(\frac{\partial \tilde{x}}{\partial t}\right)_x + u \cdot \nabla\tilde{x}(x,t), \tag{8.55}$$

which tells us how the velocity fields $\tilde{u}$ and u are related under the transformation $\tilde{x}(x,t)$.

The idea is that u will be a simpler velocity field than $\tilde{u}$, that is, the fluctuations are removed, at least partially, by transformation to the new variables. There is therefore a close connection between $\tilde{x}(x,t)$ and the asymmetric part u' in (8.50). We can then use (8.53) to introduce the function $\tilde{s}$ into (8.52), when $\tilde{s}$ is expressed as a function of x and t in the form $\tilde{s}(\tilde{x}(x,t),t)$. Since this new equation for s has a simpler velocity field, we can anticipate an easier task in solving it, by averaging in y if necessary.

Since we are dealing with divergence-free velocity fields, we want all transformations of space to preserve volume. It is therefore convenient to suppose that the transformation $\tilde{x} \overset{f}{\to} x$ occurs through a solenoidal motion. Following Moffatt (1978), we introduce for this purpose a solenoidal motion $v(x)$ and consider

$$f\colon\ x \to \tilde{x} = e^{tv(x)\cdot\nabla}x \tag{8.56}$$

where the operator $e^{tv(x)\cdot\nabla}$ takes a function of x into the same function of $\tilde{x}$. Thus

$$\begin{aligned}(\frac{\partial\tilde{x}}{\partial t})_x &= v(x) + tv\cdot\nabla v(x) + \frac{t^2}{2}(v\cdot\nabla)^2 v(x) + \ldots \\ &= \exp\{tv(x)\cdot\nabla\}v(x) = v(\tilde{x}).\end{aligned} \tag{8.57}$$

Letting the flow be "turned on" for a very short time t_0 and setting $t_0 v(x)\cdot\nabla = \varepsilon$, we obtain a class of volume preserving transformations in the form $f(x) = \tilde{x}$,

$$\tilde{x} = x + \varepsilon\xi + \frac{\varepsilon^2}{2}\xi \cdot \nabla\xi + \frac{1}{3!}\varepsilon^3(\xi \cdot \nabla)^2\xi + \ldots, \tag{8.58}$$

with $\nabla\cdot\xi = 0$. This can be regarded as an instantaneous map, since the use of t above is unrelated to the actual evolution of the fields. Hence ξ depends now on x, y, z and t. We have assumed that ξ is independent of ε, but (8.58) also applies to any solenoidal function $\xi(x,t,\varepsilon)$, so there is considerable flexibility in this class of mappings with respect to the exact form of the expansion in powers of ε.

Let us try to choose a mapping which maps $\tilde{u}(\tilde{x},t)$ given by (8.55) and (8.58) into

$$\begin{aligned}u(x,y,z,t) &= (0,\ U(x,z,t)(1+O(\varepsilon^2)),\ 0) + \varepsilon^2(\psi_{ez}, 0, -\psi_{ex}) \\ &\quad + \varepsilon^2 u''(x,y,z,t) + O(\varepsilon^3),\end{aligned} \tag{8.59}$$

where $\psi_e = \psi_e(x,z,t)$. That is, we want the mapping to suppress the fluctuations of order ε. We assume that the term u' in (8.50) has zero mean in the second coordinate, and that the same is true for u'' in (8.59). We write this as a projection P onto a field which is independent of the second coordinate:

$$Pu' \equiv \lim_{L\to\infty}\frac{1}{2L}\int_{-L}^{+L} u'\, d\tilde{y} = 0, \quad Pu'' = 0. \tag{8.60a,b}$$

The following calculation compares (8.50) and (8.59) when the mapping has the form (8.58). The comparison proceeds in two steps. First, $O(\varepsilon)$ terms yield

$$\xi\cdot\nabla U(x,z,t)\hat{j} + u'(x,y,t,z) = \xi_t + U\xi_y, \tag{8.61}$$

where $\hat{j} = (0,1,0)$. Second, P applied to the x and z components of $O(\varepsilon^2)$ terms gives

$$P[\xi\cdot\nabla u'(x,y,z,t)] + (\psi_z(x,z,t),0,-\psi_x(x,z,t))$$
$$= P[\partial_t+U\partial_y)(\tfrac{1}{2}\,\xi\cdot\nabla\xi)] + (\psi_{e_z}(x,z,t),0,\psi_{e_x}(x,z,t)). \tag{8.62}$$

From (8.62) we obtain an equation connecting u' and ξ. If u' were given and ξ were allowed to be a function of ε, this would actually be an equation for $\xi(x,t,0)$. We can think of ξ as given and of (8.61) as defining the asymmetric component of velocity. In any case we have

$$u' = D\xi - \xi\cdot\nabla(0,U,0), \tag{8.63}$$

where $D = \partial_t + U\partial_y$.

Equation (8.63) tells us how the symmetric "poloidal" component of velocity transforms. We note that

$$\xi\cdot\nabla D\xi - D(\tfrac{1}{2}\,\xi\cdot\nabla\xi) = \tfrac{1}{2}\,\xi\cdot\nabla D\xi \tag{8.64a}$$

$$-\tfrac{1}{2}\,D\xi\cdot\nabla\xi + \tfrac{1}{2}\,\xi\cdot\nabla U\xi_y = -\tfrac{1}{2}\,\nabla\times(\xi\times D\xi), \tag{8.64b}$$

and therefore we have

$$\psi_e = \psi + \tfrac{1}{2}\,P[(\xi\times D\xi)\cdot\hat{j}]. \tag{8.65}$$

The subscript "e" refers to an "effective" variable, in the sense that it determines an effective circulation which includes a drift due to the asymmetric component.

To tie these observations together we now change slightly our point of view and simply regard (8.52) as an equation to be solved with (8.59) defining the velocity. Projecting (8.52) according to (8.60) and substituting (8.59) we obtain

$$\frac{\partial Ps}{\partial t} + \varepsilon^2\,\frac{\partial(Ps,\psi_e)}{\partial(x,z)} + \varepsilon^2\nabla\cdot P(u''s) = O(\varepsilon^3). \tag{8.66}$$

Assume now all projected functions of time are actually projected functions of $\varepsilon^2 t = \tau$. Then a family of solutions of (8.66) has the form

$$s = s_0(x,z,\tau) + O(\varepsilon^2), \tag{8.67}$$

where $s_0(x,z,\tau)$ solves

$$\frac{\partial s_0}{\partial \tau} + \frac{\partial(s_0,\psi_e)}{\partial(x,z)} = 0. \tag{8.68}$$

Given a solution of (8.68), the desired field $\tilde{s}(x,y,z,t)$ is obtained from (8.53) and (8.58),

$$\tilde{s}(x,y,z,t) = s_0(x,z,\tau) - \varepsilon\xi \cdot \nabla s_0(x,z,\tau) + O(\varepsilon^2). \tag{8.69}$$

Hence this method accounts for certain averaged effects of advection by a complicated velocity field. It resembles the smoothing method in some respects, since the desired solution is generated from the symmetric field as in (8.69), while the equation for the "smooth" field again involves the averaged effects of asymmetric components of the velocity.

As an example of the method, consider the velocity field given by (8.50) with

$$u' = (\cos ky,\ 0,\ \sin ky). \tag{8.70}$$

Using (8.70) in (8.63) we may solve for ξ to obtain:

$$\xi = \frac{1}{kU^2}\,(U \sin ky,\ -U_x \cos ky - U_z \sin ky,\ -U \cos ky). \tag{8.71}$$

Thus from (8.65)

$$\psi_e = \psi - \frac{1}{2kU}\,. \tag{8.72}$$

We could obtain (8.72) directly from a regular representation if we recognize (cf. (8.69)) that $\tilde{s} = s_0(x,z,\tau) + \varepsilon s' + \dots$. Then substitution into (8.49), (8.50) gives, for the motion (8.70),

$$s' = \frac{1}{kU}\left(-\sin ky\,\frac{\partial s_0}{\partial x} + \cos ky\,\frac{\partial s_0}{\partial z}\right), \tag{8.73}$$

$$\frac{\partial s_0}{\partial \tau} + \frac{\partial(s_0,\psi)}{\partial(x,z)} + P(u'\cdot\nabla s') = 0. \tag{8.74}$$

Now (8.70) and (8.73) give

$$\begin{aligned} P(u'\cdot\nabla s') &= \frac{1}{2}\frac{\partial}{\partial x}\left(\frac{1}{kU}\frac{\partial s_0}{\partial z}\right) - \frac{1}{2}\frac{\partial}{\partial z}\left(\frac{1}{kU}\frac{\partial s_0}{\partial x}\right) \\ &= \frac{\partial(s_0, -(2kU)^{-1})}{\partial(x,z)}, \end{aligned} \tag{8.75}$$

and we obtain (8.72).

A direct expansion as in (8.73 - 8.75) was used by Braginsky (1964a,b) for the dynamo problem. As can be appreciated from the model problem above, this method does not reveal the underlying reason for the appearance of the effective variable ψ_e, which depends on the cancellation of mixed partials of s_0 in (8.75). The mystery deepens for the dynamo equations, when effective variables are found in *both* velocity and magnetic field variables of the advection term.

Soward's approach, on the other hand, explains the effective velocity field for the above model as resulting from the map which "smooths out" the $O(\varepsilon)$ perturbation in $\tilde{u}$. The averaging operation P, when applied to (8.52), commutes with $u\cdot\nabla$ to $O(\varepsilon^2)$, so that Ps becomes the new dependent variable, as in (8.66). The calculations are also capable in principle of providing an equation for Ps to any order in ε, although explicit results are usually sought only for the dominant terms (cf. (8.68)).

Application to the dynamo problem. The construction above suggests that similar procedures apply to the induction equation (8.48), provided that we consider velocity fields which involve small asymmetric components. There are, however, two important differences. First, the connection between $B(x,t)$ and $\tilde{B}(\tilde{x},t)$, defined analogously to that between $s(x,t)$ and $\tilde{s}(\tilde{x},t)$, now involves the Jacobian of the Lagrangian coordinate function (see (8.78) below). Second, the equation we want to solve includes a diffusion term, which apparently cannot remain invariant under the transformation f. Consequently, the transformed equation will not have the same structure as the one we start with. The modified equation will incorporate the crucial α-effect.

It is essential here to utilize the assumed largeness of R to make the terms introduced by small asymmetry and by diffusion of comparable order. These considerations suggest that the parameter ε used above should be ordered relative to R in a precise way. The quadratic dependence of (8.65) on ξ and the fact that diffusion is $O(R^{-1})$ suggest that ε be taken of order $R^{-1/2}$.

For simplicity of exposition, we again consider a symmetric field as independent of y and give Braginsky's results for this case. The corresponding system in cylindrical polar coordinates will be exhibited in Chapter 9, cf. (9.2) - (9.6).

The Lagrangian transformation f, again defined by (8.58) in terms of ξ, leaves the reduced, non-diffusive equation

$$B_t + u\cdot\nabla B - B\cdot\nabla u = 0 \tag{8.77}$$

invariant, provided that the magnetic fields are connected by

$$\tilde{B}_i(\tilde{x},t) = B_j(x,t)\,\frac{\partial\tilde{x}_i}{\partial x_j}\,, \tag{8.78a}$$

$$\tilde{u}_i(\tilde{x},t) = \frac{\partial\tilde{x}_i}{\partial t} + u_j(x,t)\,\frac{\partial\tilde{x}_i}{\partial x_j}\,. \tag{8.78b}$$

Since diffusion is actually present, we must see how

$$\tilde{B}_t + \tilde{u}\cdot\tilde{\nabla}\tilde{B} - \tilde{B}\cdot\nabla\tilde{u} - R^{-1}\tilde{\nabla}^2\tilde{B} = 0 \tag{8.79}$$

transforms under f^{1}. We shall need the identity

$$\varepsilon_{ijk}\ \partial x_i/\partial\tilde{x}_\ell\ \partial x_j/\partial\tilde{x}_m\ \partial x_k/\partial\tilde{x}_n = \varepsilon_{\ell mn}, \tag{8.80}$$

where the summation convention is used and ε_{ijk} is a tensor with unit components, skew-symmetric in all pairs of indices. This identity expresses the fact that $\mathrm{Det}(\partial x_i/\partial\tilde{x}_j) = 1$.

Using (8.80),

$$\begin{aligned}\tilde{\nabla}^2\tilde{B} &= -\tilde{\nabla}\times(\tilde{\nabla}\times\tilde{B}) = \varepsilon_{ijk}\partial x_m/\partial\tilde{x}_j\ \partial/\partial x_m(\tilde{\nabla}\times\tilde{B})_k \\ &= -\varepsilon_{\ell mn}\ \partial\tilde{x}_k/\partial x_n\ \partial/\partial x_m\ (\tilde{\nabla}\times\tilde{B})_k.\end{aligned} \tag{8.81}$$

We also have

$$\begin{aligned}-&\varepsilon_{\ell mn}\ \partial\tilde{x}_k/\partial x_n\ \partial/\partial x_m\ (\tilde{\nabla}\times\tilde{B})_k \\ &= -\varepsilon_{\ell mn}\ \partial\tilde{x}_k/\partial x_n\ \partial/\partial x_m[\varepsilon_{kpq}\ \partial x_s/\partial\tilde{x}_p\partial/\partial x_s(B_r\ \partial\tilde{x}_q/\partial x_r)] \\ &= -\varepsilon_{\ell mn}\ \partial/\partial x_m[\partial\tilde{x}_k/\partial x_n\ \varepsilon_{kpq}(\partial x_s/\partial\tilde{x}_p\ \partial\tilde{x}_q/\partial x_r\ \partial B_r/\partial x_s \\ &\qquad - B_r\ \partial\tilde{x}_q/\partial x_s\ \partial/\partial x_r\ (\partial x_s/\partial\tilde{x}_p))] \\ &= \varepsilon_{\ell mn}\ \partial/\partial x_m[\varepsilon_{nro}\ \partial x_o/\partial\tilde{x}_p\ \partial x_s/\partial\tilde{x}_p\ \partial B_r/\partial x_s \\ &\qquad + \varepsilon_{sno}\ B_r\ \partial x_o/\partial\tilde{x}_p\ \partial/\partial x_r\ (\partial x_s/\partial\tilde{x}_p)],\end{aligned} \tag{8.82}$$

where we have used (8.81) and also

$$\varepsilon_{\ell mn}\ \partial^2\tilde{x}_k/\partial x_m\partial x_n = 0, \tag{8.83a}$$

$$\begin{aligned}&\partial x_s/\partial\tilde{x}_p\ \partial/\partial x_r\ (\partial\tilde{x}_q/\partial x_s) \\ &\qquad = -\partial\tilde{x}_q/\partial x_s\ \partial/\partial x_r\ (\partial x_s/\partial\tilde{x}_p).\end{aligned} \tag{8.83b}$$

Since the other terms of the equation are invariant, we have from (8.82)

$$B_t + u\cdot\nabla B - B\cdot\nabla u - R^{-1}\nabla^2 B = R^{-1}\nabla\times E, \tag{8.84}$$

where the vector field E has the form

$$E_i = \beta_{ijk}(x,t)\,\frac{\partial B_j}{\partial x_k} + \alpha_{ij}B_j, \tag{8.85}$$

$$\beta_{ijk} = \varepsilon_{ijq}\left(\frac{\partial x_k}{\partial \tilde{x}_p}\frac{\partial x_q}{\partial \tilde{x}_p} - \delta_{kq}\right), \tag{8.86}$$

$$\alpha_{ij} = \varepsilon_{i\ell m}\frac{\partial x_\ell}{\partial \tilde{x}_p}\frac{\partial}{\partial x_j}\left(\frac{\partial x_m}{\partial \tilde{x}_p}\right). \tag{8.87}$$

Thus the terms contributed by diffusion involve both B and its first derivatives.

Taking $\varepsilon^2 = R^{-1}$, we wish to consider velocity and magnetic fields which have the forms

$$P\tilde{u} = (0,\ U(1+O(\varepsilon^2)),\ 0) + \varepsilon^2(\psi_{\tilde{z}},0,-\psi_{\tilde{x}}) + O(\varepsilon^3), \tag{8.88a}$$

$$(I - P)\tilde{u} = \varepsilon u'(\tilde{x},\tilde{y},\tilde{z},t) + O(\varepsilon^2), \tag{8.88b}$$

$$P\tilde{B} = (0,\ B(1+O(\varepsilon^2)),0) + \varepsilon^2(A_{\tilde{z}},0,-A_{\tilde{x}}), \tag{8.89a}$$

$$(I - P)\tilde{B} = \varepsilon b'(\tilde{x},\tilde{y},\tilde{z},t) + O(\varepsilon^2). \tag{8.89b}$$

Here ψ and A as well as U and B are functions of x, z, and $\tau = \varepsilon^2 t$. Our aim is to choose ξ so that u and B have the following forms:

$$Pu = (0,\ U(x,z,t),0) + \varepsilon^2(\psi_{ez},0,-\psi_{ex}) + O(\varepsilon^3), \tag{8.90a}$$

$$(I - P)u = O(\varepsilon^2), \tag{8.90b}$$

$$PB = (0,B,0) + \varepsilon^2(A_{ez},0,-A_{ex}) + O(\varepsilon^3), \tag{8.91a}$$

$$(I - P)B = O(\varepsilon^2). \tag{8.91b}$$

As in the scalar advection model, $O(\varepsilon)$ contributions yield

$$u'(x,t) = D\xi - \xi\cdot\nabla U\hat{j}, \tag{8.92}$$

$$b'(x,t) = B\frac{\partial\xi}{\partial y} - \xi\cdot\nabla B\hat{j}. \tag{8.93}$$

With the terms of order ε^2 in (8.78), the connections between fields may be established and the following equations are obtained when diffusive contributions are included:

$$\frac{\partial B}{\partial\tau} + \frac{\partial(B,\psi_e)}{\partial(x,z)} - \frac{\partial(U,A_e)}{\partial(x,z)} - \nabla^2 B = 0, \tag{8.94a}$$

$$\frac{\partial A_e}{\partial\tau} + \frac{\partial(A_e,\psi_e)}{\partial(x,z)} - \nabla^2 A_e = \alpha B. \tag{8.94b}$$

Here the effective variable ψ_e is given by (8.65), while A_e has the form

$$A_e = A + \frac{1}{2} P[(\xi \times \frac{\partial \xi}{\partial y}) \cdot \hat{j}]B. \tag{8.95}$$

The expression for α is found to be, once (8.58) is used in (8.87),

$$\alpha = \varepsilon_{2k\ell} P\left[\frac{\partial \xi_k}{\partial x_m} \frac{\partial^2 \xi_\ell}{\partial x_m \partial x_2}\right] = 2P[\nabla \xi_3 \cdot \nabla \frac{\partial \xi_1}{\partial y}]. \tag{8.96}$$

From (8.94), we observe that the problem is reduced to studying symmetric equations with one additional term, describing an α-effect creating mean y-current proportional to the mean y-field. Exactly the same system, but with no effective variables exhibited, would be obtained by expanding in R^{-1} the smoothed equation (8.35), when fields are postulated to have the ordering of (8.90) and (8.91). It is gratifying that two rather different methods yield similar conclusions, and there are perhaps deep reasons for this which are not understood as yet.

Braginsky's results suggest that the ordering in powers of $R^{-1/2}$ is the "natural" one for geodynamo modeling. It turns out that the first asymmetric magnetic field in spherical geometry, which is nonzero outside the fluid core, is a term of order $R^{-3/2}$. Since the mean poloidal field is $O(R^{-1})$ according to (8.91a), this model is consistent with the observation that the dominant field is an aligned dipole.

Equation (8.96) indicates that the essential feature of a nonzero α can be achieved quite easily. For example, we may take ξ_1 and ξ_3 to have a sinusoidal variation with respect to y, and be 90° out of phase. This simple example highlights again the importance of helical waves in the generation of a magnetic field.

8.6. Small Diffusivity Aspects of Kinematic Dynamos

We have commented on the difficulty of carrying out smoothing in certain low resistivity limits, and will now focus on some aspects of this problem. Recall first from (7.24) and (8.20) that two distinct magnetic Reynolds numbers, R_{m1} and R_{m2}, can be defined in terms of a characteristic advection time $T = 2\pi/\omega$, a characteristic length $L = 2\pi/k$ and a characteristic velocity $U = u_0$, namely $R_{m1} = u_0/\eta k$ and $R_{m2} = \omega/\eta k^2$. In (8.23) and until now, $R = R_{m1}$ was used. If $R_{m1} >> 1$, the physical meaning of this ordering will depend upon the time scale of motion T. If R_{m2} is also large, we are dealing with motions whose advection is short compared to their diffusion time, and we have observed that no particular problem with smoothing to all orders arises, provided that $R_{m1}/(R_{m2})^{1/2} = u_0/\sqrt{\omega\eta} << 1$. We illustrate now this situation of *fast advection.*

Small effective R_{m1}. If the above ordering is satisfied, then the time derivative dominates the diffusion term in the equation for B_R, (8.8b). To first order we then have, for a time-periodic u with characteristic frequency ω_0 and characteristic amplitude u_0,

$$B_R \simeq \nabla \times (\hat{u} \times B_S), \tag{8.97}$$

where $\partial\hat{u}/\partial t = u$. Thus the dimensionless partial differential form of the mean-field equation (8.35) becomes, to first order in $R = R_{m1}$,

$$\frac{\partial B_S}{\partial t} - \frac{1}{R}\nabla^2 B_S = \langle \nabla \times [u \times \nabla \times (\hat{u} \times B_S)] \rangle \equiv F, \tag{8.98}$$

where the average $\langle\ \rangle$, in contradistinction to (8.32), is now with respect to *time alone*. The right-hand side of (8.98) can be manipulated as follows:

$$F_i = \langle \varepsilon_{ijk} \frac{\partial}{\partial x_j} \varepsilon_{k\ell m} u_\ell \left[B_{S_n} \frac{\partial \hat{u}_m}{\partial x_n} - u_n \frac{\partial B_{Sm}}{\partial x_n} \right] \rangle$$

$$= \langle \frac{\partial}{\partial x_j} \left[u_i B_{Sn} \frac{\partial \hat{u}_j}{\partial x_r} - u_i \hat{u}_n B_{Sn} \frac{\partial \hat{u}_i}{\partial x_n} + u_j \hat{u}_n \frac{\partial B_{Si}}{\partial x_n} \right] \rangle$$

$$= \langle B_{Sn} \frac{\partial}{\partial x_n} \hat{u}_j \frac{\partial u_i}{\partial x_j} + u_j \frac{\partial \hat{u}_n}{\partial x_j} \frac{\partial B_{Si}}{\partial x_n} \rangle$$

$$= \langle B_{Sn} \frac{\partial}{\partial x_n} \hat{u}_j \frac{\partial u_i}{\partial x_j} - \hat{u}_j \frac{\partial u_n}{\partial x_j} \frac{\partial B_{Si}}{\partial x_n} \rangle$$

or

$$F = \omega^{-1} \nabla \times (v \times B_S), \tag{8.99a}$$

where

$$v = \omega \langle \hat{u} \cdot \nabla u \rangle . \tag{8.99b}$$

Note that we have treated B_S as slowly varying, since $\hat{u}$ is formally $O(\omega^{-1})$, with $\omega = \omega_0/u_0 k$. Thus, (8.98) may be written,

$$\frac{\partial B_S}{\partial t} - \frac{1}{R}\nabla^2 B_S \simeq \omega^{-1} \nabla \times (v \times B_S). \tag{8.100}$$

The effect of the rapid oscillation averages out over time to produce an effective advection by the non-oscillatory field v, in the manner of the secondary motion frequently encountered in nonlinear time-dependent hydrodynamics. By a change of time scale, (8.100) can be reduced to a new non-oscillatory problem with effective $R_{m1}^{(eff)}$ equal to $R_{m1}/\omega = u_0^2/\eta\, \omega_0 \ll 1$, provided that originally $R_{m1}/(R_{m2})^{1/2} \ll 1$ as stipulated.

The multi-scale smoothing of Section 8.2 can be applied to the spatial variation in this latter problem, thereby completing the averaging over the scales of u. We are dealing, it should be remarked, with a case of smoothing not equivalent to first-order smoothing. This can be seen from the fact that, when first-order smoothing is applied formally to (8.100), the resulting α, if nonzero, is homogeneous of degree four in the components of u, while in true first-order smoothing, cf. (8.35), it is quadratic in u.

As an example, suppose that

$$u = u_S \sin \omega t + u_c \cos \omega t. \tag{8.101}$$

Then (8.99b) yields

$$v = \frac{1}{2} [u_C \cdot \nabla u_S - u_S \cdot \nabla u_C]. \tag{8.102}$$

If

$$u_S = (\cos z, \sin z, -\sin y), \tag{8.103a}$$

$$u_C = (\cos z, \sin z, \sin y), \tag{8.103b}$$

then

$$v = 2(-\psi, \psi_z, -\psi_y), \quad \psi = \sin y \sin z. \tag{8.104a,b}$$

The flow field (8.104) is a well known example of a cellular flow, independent of one coordinate, producing an α matrix with two nonzero elements at second and third positions along the diagonal. In fact (8.104) is equivalent under translation, rotation, and a change of scale, to (sin y + cos z, sin z, cos y), which is "two-thirds" of the periodic dynamo (sin y + cos z, sin z + cos x, sin x + cos y) discussed in the example (8.36). (See also Roberts, 1972.)

The important point here is the reliance once again on small $R_{m1}^{(eff)}$. Because of this, *on the average*, the magnetic field is advected as if diffusivity were not near zero. The oscillation produces a rapidly changing field, which *does* respond as if the diffusivity were zero. The effective mean motion of the fluid however, is sufficiently slow to drastically reduce $R_{m1}^{(eff)}$.

This somewhat artificial, rapidly oscillating, large-R_{m1} dynamo highlights the role of multiple time scales in studying small diffusivity dynamos. Still, we must ultimately accept that a steady or quasi-steady motion can also distort greatly the magnetic field, producing regions of intense field over relatively small domains such as thin sheets and tubes.

There seems to be no theory at present where smoothing is valid asymptotically, and which encompasses this physical situation.

The most difficult case to treat is that of $R_{ml} >> 1$ in the presence of a *steady* velocity field u. No satisfactory kinematic dynamo theory exists in this limit, except in Braginsky's case of weakly asymmetric flows. Nevertheless, we may attempt to calculate the α-matrix resulting from a formal application of first-order smoothing, or rather the asymptotic behavior of α as $R_{ml} \to \infty$. This amounts only to evaluating a spatial average of $u \times B_R$, where B_R corresponds to a *uniform* B_S.

The α-effect for steady flows. The approach suggested above was taken by Childress (1979) for two velocity fields. The first is two-dimensional and of the form

$$u = (\psi_y, -\psi_x, w(\psi)), \tag{8.105a}$$

$$-\nabla^2\psi > 0 \quad \text{in} \quad D, \quad \psi = 0 \quad \text{on} \quad \partial D, \tag{8.105b}$$

where D is a rectangle determining the spatially periodic function ψ everywhere, and ∂D is its boundary. Note that (8.104) is a flow of this kind, for $D = \{0 < y, z < \pi\}$. The second motion is axisymmetric and has the cylindrical polar form

$$(u_z, u_r, u_\phi) = \left(r^{-1}\frac{\partial\psi}{\partial r}, -r^{-1}\frac{\partial\psi}{\partial z}, r^{-1}w(\psi)\right), \tag{8.105c}$$

$$-\left(\nabla^2 - \frac{2}{r}\frac{\partial}{\partial r}\right)\psi > 0 \quad \text{in} \quad D, \quad \psi = 0 \quad \text{on} \quad \partial D, \tag{8.105d}$$

with D now a cylinder, $D = \{0 < z < 1, 0 < r < r_0\}$. Both flows are characterized by a function ψ with a single extremum (maximum) of ψ in the interior of D, and by helical flow lines.

Numerical calculations of time-dependent induction of B by motions of type (8.105a,b) were performed by Weiss (1968), who studied in detail how flux is gradually expelled from the center of the eddy toward the boundary, eventually leading, for large $R \equiv R_{ml}$, to a layer of thickness $O(R^{-1/2})$ near ∂D where most of the flux is concentrated. In analogous axisymmetric calculations, Galloway, Proctor, and Weiss (1978) found that certain motions of type (8.105c,d) lead to concentration of flux into a tube of diameter $R^{-1/2}$ along the axis of the cylinder D. The calculation of α for large R requires, in each case, that one exploit the special geometry of these boundary layers.

For the two-dimensional problem, with periodicity conditions on the magnetic field at ∂D, there results $\alpha(R) = \alpha_0 R^{-1/2}$ where α_0 is a constant symmetric matrix which can be computed numerically. This result is consistent with the numerical values of $\alpha(R)$ computed by Roberts (1972). It is possibly representative of contributions from free, compressed sheets of flux for arbitrary steady flow in three dimensions.

For the second, axisymmetric motion (8.105c,d) it is natural to expect the central flux rope to play a prominent role in developing an α-effect. Surprisingly, this turns out not to be the case, at least for boundary conditions on the magnetic field as adopted by Childress (1979). If $B = (\,^{-1}\frac{\partial a}{\partial r}, -r^{-1}\frac{\partial a}{\partial z}, rb)$, then these conditions are

$$rb \to 0 \quad \text{as} \quad r \to 0, \quad 0 \leq z \leq 1; \tag{8.106a}$$

$$b = 0, \quad r = r_0, \quad 0 \leq z \leq 1; \tag{8.106b}$$

$$\frac{\partial a}{\partial z} = \frac{\partial b}{\partial z} = 0, \quad 0 < r < r_0, \quad z = 0,1; \tag{8.106c}$$

$$a = r_0^2/2, \quad r = r_0, \quad 0 \leq z \leq 1. \tag{8.106d}$$

Note that (8.106d) insures that a mean field of unit strength parallel to the axis of symmetry permeates the flow.

It is found that the component α_{33} of the α-matrix, the only nonzero entry in the present example, tends to a constant as $R \to \infty$, and that its generally nonzero value comes from contributions to $\langle u \times B_R \rangle$ due to the two boundary layers at top and bottom, $z = 0,1$. To see how these two boundary layers fit into the cylinder D, consider Figure 8.4. We indicate there that an inner flux tube of diameter $\sim R^{-1/2}$ merges with a "compressed" flux tube of diameter $\sim R^{-1/4}$ to provide the initial profile of a boundary layer extending across the top from A to B, of thickness $\sim R^{-1/2}$.

The dominant contribution to α_{33} occurs in the boundary layers along the top AB and the bottom CD of the cylinder. This can be seen from

$$\alpha_{33} = 2r_0^{-2}\int_0^1\int_0^{r_0} r(u_r B_\phi - u_\phi B_r)\,dr\,dz. \tag{8.107}$$

In each such layer $B_\phi \sim R^{1/2}$, $u_r \sim 1$, but the layer thickness is $R^{-1/2}$, hence contributions of order one are possible, and can be easily computed by recasting the boundary layer analysis as an equivalent problem in one-dimensional heat conduction.

Stix (1980) has remarked that the most obvious application of dynamo calculations at large R would be to the magnetic cycle of the Sun, but the preceding estimate of α_{33} must be viewed with caution for several

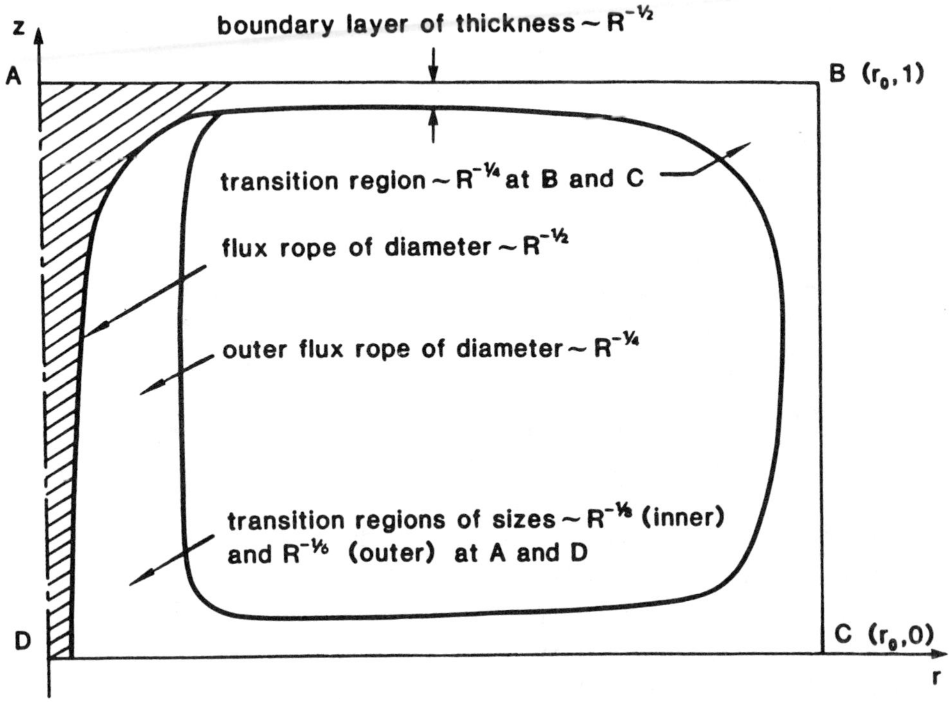

Fig. 8.4. Structure of the magnetic field $B=B(a,b)$ induced by the flow (8.105c,d) with boundary conditions (8.106). The structure of a and b can be represented as shown: An intense flux rope of diameter $\sim R^{-1/2}$ develops from D to A along the axis, turns through a transition region of size $R^{-1/3}$ and spreads into a boundary layer $A \to B \to C$ of thickness $\sim R^{-1/2}$. Transition regions of size $\sim R^{-1/4}$ occur at B and C. At D the boundary layer passes through a $R^{-1/6}$ transition and emerges as an outer flux tube of diameter $\sim R^{-1/4}$. This outer tube also goes through a $R^{-1/6}$ transition at A, and it, along with the $R^{-1/3}$ wide contribution from the inner flux tube, determines the profile of the emerging boundary layer.

reasons. First, in treating only the kinematic model, we have excluded the important feedback from magnetic forces which are developed in the Sun's photosphere, and which have been studied numerically for axisymmetric thermal convection by Galloway *et al.* (1978). Second, the estimate of α_{33} is affected by changes in boundary conditions. A drop from $O(1)$ to $O(R^{-1/2})$ occurs if the top and bottom boundary-layer contributions are not present, and this might occur if, say, b in (8.106) were required to vanish at $z = 0,1$. Finally, it is not clear how to imbed the axisymmetric geometry into a real three-dimensional field so that it might be justifiably regarded as a "canonical" structure for evaluating α. It is worth mentioning that an estimate α of order unity would probably be much too *large* to be compatible with other data (Stix, 1980).

These somewhat limited results underline the need for theoretical studies of the alpha-effect mechanism when the magnetic field is highly compressed and distorted and the flow field fully three dimensional. There is no *a priori* reason why some sort of smoothing procedure should not be applicable, even though the mean magnetic field is actually small compared to maximum fields. Nevertheless, the existing theories of smoothing fail because there is no apparent convergence of the series of terms representing B_R. It is perhaps not surprising that the analytical tools fall short of dealing with cases where the "mean field" is best regarded as a ficticious average of small, intense fields. As Stix (1980) has remarked, it might be preferable to adopt a statistical viewpoint, not only to deal with a turbulent velocity field, but also to treat random ensembles of small magnetic structures. The mean or expected field might be small or even zero, but first-order statistics would nonetheless provide the crucial information concerning field geometry. We shall adopt in fact this viewpoint for another purpose in Section 8.7.

A final point concerns the common features of three-dimensional flows which might be identified as contributing to the α-effect in the limit $R \to \infty$ under steady or quasi-steady conditions. From Figure 8.4 and (8.105c,d) it is tempting to surmise that flow lines of steady motions connecting two stagnation points, that is the heteroclinic orbits of the flow, could play a crucial role as locations of the most intense flux ropes. The stable and unstable manifolds emanating from these critical points might then locate the $R^{-1/2}$ boundary layers where the α-effect is developed. It is possible that these conjectures could be tested analytically in certain simple cases such as the periodic Beltrami flow $(\sin y + \cos z, \sin z + \cos x, \sin x + \cos y)$ (cf. Section 8.8).

8.7. Multi-Scaling and Breakdown of Smoothing

We shall now consider the smoothing of kinematic dynamos whose velocity fields involve two or more spatial scales. Such a situation might arise if a velocity field u has a small-scale component producing an α-effect under first-order smoothing, together with a modulation of the amplitudes of the small eddies over a second, distinct scale of length.

A third, generally distinct scale is introduced by the boundary conditions on the mean magnetic field. This problem was studied by Kraichnan (1976; see also Moffatt, 1978, p. 176) who showed that an auxiliary smoothing procedure leads to an equation for the mean field involving a possibly *negative* effective diffusivity. This argument, developed for turbulent flow, would therefore suggest that the possibility of formulating a well-posed problem for the mean magnetic field is questionable for certain multi-scale velocity fields.

A related question was studied by Childress (1983a), in the context of *steady* induction by a velocity field with high spatial intermittency. Here the paradox raised by Kraichnan can be studied quite explicitly while allowing the intermittency to be random, through the use of an approximate closure of the hierarchy which arises when one attempts to compute α. Motion is assumed to occur only within well separated domains. It is found that, for a class of motions determined by a parameter K, α exists only provided $K < 1$. It is suggested that when $K > 1$ a "change of phase" of the magnetic field occurs, wherein the inductive process comes to be dominated by "microdynamos" operating on the scale of pairs of active domains. That is to say, the process of collective growth of large-scale fields is replaced by local growth of small-scale fields, when K increases through 1.

We describe now briefly the examples treated by Childress (1983a). The velocity u is nonzero only within identical nonoverlapping spheres s_i of radius r_0 and center r_i, $i = 1,\ldots,N$. Within each sphere a velocity field having a small scale relative to r_0 is prescribed. This field is chosen so as to insure that an isolated sphere cannot itself act autonomously as a dynamo. On the other hand, the velocity fields u_i should allow an arbitrary α-matrix to be constructed using large numbers of spheres s_i. As Childress (1970) has shown, one possible choice of u_i within s_i has the form

$$u_i = U\nabla \times \psi\left[e_i' \cos\left(\frac{e_i\cdot r}{\ell}\right) + e_i'' \sin\left(\frac{e_i\cdot r}{\ell}\right)\right], \quad |r-r_i| \le r_0, \tag{8.108}$$

where (e_i, e_i', e_i'') is a right-handed triple of orthonormal vectors, U a constant, ℓ the small length scale and ψ is an infinitely differentiable function which is zero on the boundary of each sphere and is unity in its interior, except within a distance $\delta \ll 1$ of the boundary. The periodic interior field (8.108) produces an α locally with a single nonzero entry on the diagonal. We assume $u \equiv 0$ exterior to the s_i, and that

$r_0^3 N << V_s$, V_s being the volume over which the spheres are distributed.

Given this choice of spatially intermittent flow, a statistical problem can be formulated by noting that the interaction of the velocity domains via a magnetic field can be studied in a *far-field* approximation. Induction within each isolated domain s_i in the presence of a uniform ambient magnetic field B_0 yields a magnetic field B_i whose expansion at infinity is of the form

$$B_i = B_0 + \frac{1}{4\pi\eta} \nabla \times \left(\frac{A_i B_0}{|r-r_0|} \right) + O(|r-r_i|^3), \tag{8.109}$$

where $(A_i)_{jk} = a_0(e_i)_j(e_i)_k$, $a_0 = \eta^{-1} U^2 \ell(\frac{4}{3}\pi r_0^3)$. The interaction of two such fields, B_1 and B_2 say, allows *pairs* of domains to develop dynamo action provided that the matrix M defined by

$$MB = \frac{-R \times [A_2 \cdot (R \times (A_1 \cdot B))]}{(4\pi\eta)^2 |R|^6} \tag{8.110}$$

with $R = r_2 - r_1$, has an eigenvalue unity.

Under simple assumptions for the statistics of the domains, it is found that the expected value of M is just

$$\langle M \rangle = \frac{a_0^2 n}{6\pi\eta^2 b} I, \tag{8.111}$$

with $n = N/V_s$ the number of spheres per unit volume. Here b is a cut-off distance for the domain-pair distribution function. The parameter K suggested by (8.111) is

$$K = \frac{2}{9} \frac{a_0^2 n}{\pi b \eta^2} (1 - \frac{4}{9} \pi b^3 n). \tag{8.112}$$

The closure method used above shows that α fails to exist when $K > 1$ precisely because the density of domain pairs, linked interactively by dynamo action, is finite (see Figure 8.5).

We remark that these results are limited to steady velocity fields, and it is not known if a similar phase transition can arise in the presence of temporal variations. Because of the many time scales apparent in the paleomagnetic record of geodynamo polarity (Section 7.1), a general analysis of spatio-temporal transition to microdynamo activity, for a more realistic class of motions, is likely to be very fruitful.

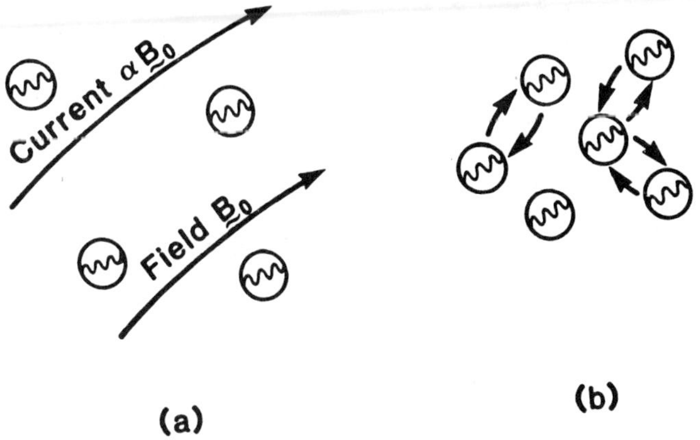

Fig. 8.5. Phase change from large-scale to small-scale dynamo action. (a) K < 1: Domains interact collectively to produce an effective α equal to a multiple of the identity. (b) K > 1: Domains interact pairwise, local growth of field occurs.

8.8. Bibliographic Notes

Sections 8.1, 8.2. Smoothing methods have a long history and essentially the same algebraic technique has appeared in a variety of forms. There is a formal equivalence, for example, between the operator statements outlined here and expansions generated by the perturbation theory of abstract linear operators, see Friedman (1956) and Friedrichs (1965). For stochastic velocity fields, the usual procedure is to include a dependence of our exact operator L (cf. (8.7)) on a random variable v. Ensemble averaging over v then replaces the projection P. For a discussion of smoothing from this viewpoint see the appendix to Keller (1969). A review of these methods is given by Nayfeh (1973).

Section 8.3. Further development and applications of smoothing to kinematic dynamo theory, including numerical studies of specific spherical models, may be found in the general references given at the end of Chapter 7. See, in particular, Moffatt (1978, Ch. 9), Roberts and Stix (1971, 1972), Roberts (1972), and Stix (1973).

Section 8.6. A detailed discussion of the asymptotic theory for the two-dimensional cellular flow (cf. (8.105a)) has recently been given by Anufriyev and Fishman (1982). Their results substantiate the structure summarized in Figure 8.4, include a complete numerical solution for the field structure, and an analysis of passive non-uniformities near the corners of the cell. They compute α_0 (defined below (8.105b)) and obtain a value smaller by a factor of approximately 0.2 than the value obtained by Childress (1979). Comparison with the calculated results of Roberts (1972) would, using this revised value of α_0, be favorable up to $R = 64$; it would also indicate, however, that the asymptotic value of $R^{1/2}\alpha$ is approaches at magnetic Reynolds numbers of this order.

The Beltrami field $u = (B \sin y + C \cos z, C \sin z + A \cos x, A \sin x + B \cos y)$ has, apart from its emergence in dynamo theory, a recent interesting history as a rotational solution of Euler's equations. Arnold (1965) gave this flow as an example of steady-state, inviscid motion without a simple topological classification of particle orbits. Hénon (1966) studied the flow numerically for certain values of A, B, C and found complicated behavior of certain orbits. Suppose, for example, we consider the position of a moving fluid particle in $z \bmod 2\pi$. In general this projection, or Poincaré map, will exhibit isolated points and sets of points dense on closed curves. Hénon also found apparently space-filling, wandering orbits, for some choices of the constants.

It is of interest to try to understand the complicated geometry of this Arnold-Hénon flow and its bearing on dynamo-action at large R. Recent studies by Childress and Soward (1985) for the symmetric case $A = B = C = 1$ suggest that the axisymmetric rope-sheet structure indicated in Figure 8.4 is actually typical of the behavior of *steady* flows near intense flux tubes: the dominant contributions to the alpha-effect occurs in these examples as the flux tube is pulled out into a two-dimensional flux sheet. Tentative estimates suggest that α is $O(1)$ in the limit of infinite R.

<u>Section 8.7</u>. Recent work by Moffatt (presented to the American Physical Society, Washington, 1982) on dynamo action by velocity fields with many length scales utilized a renormalization group approach, and thus bears on the possibility of breakdown of smoothing at large R. Other aspects of high-R turbulent dynamos are discussed by Moffatt in Soward (1983), pp. 3-16.

The mechanism shown in Figure 7.7b is one example of a dynamo which might produce exponential growth of magnetic field on one of the length scales of the velocity field. Such "fast" dynamos, whatever their form might be, provide mechanisms for breakdown of smoothing. Recently Arnold and Korkina (1983) have published numerical results for kinematic dynamo action by the Arnold-Hénon flow with $A = B = C = 1$ up to magnetic Reynolds numbers of about 20. Their results show exponentially growing spatially-periodic modes when R lies between 8.93 and 17.54, the period being that of the velocity field. Hence in this range breakdown of smoothing occurs in the sense used in this section.

It is worth noting that dynamical processes would presumably limit the growth of small-scale magnetic structures in real hydromagnetic systems, so there remains the possibility that smoothing as a practical technique can be applied at moderate and even large R. The question of smoothing at large values of R is particularly tantalizing since the results of Arnold and Korkina (1983) imply stability for a certain interval beyond the upper limit 17.54 of the instability window.

In connection with the analysis of dynamo action at large R, we note the paper by Arnold *et al.* (1981), which treats dynamo action on a Riemannian manifold. Since the structure is put into the manifold, the flow can be simple and is taken to be unidirectional. The result is that magnetic field lines are stretched exponentially. The diffusion at large R cannot arrest the process and so dynamo action occurs.

Readers interested in the large-R limit as it arises in astrophysical models should consult the recent monograph of Zeldovich *et al.* (1983). The importance of topological constraints and magnetic line reconnection upon the growth of magnetic energy in the limit of infinite R is emphasized. Insofar as the dynamo remains kinematic, properties of the underlying fluid flow, regarded as a map on three-dimensional space which are essentially of a geometrical character, are likely to play important roles in the construction of models appropriate to the large-R dynamo theory.

CHAPTER 9

THE HYDRODYNAMIC BASIS FOR GEOMAGNETISM

9.1. Introduction

We turn now to the study of those aspects of the hydrodynamics of rotating, electrically conducting fluids, which can be expected to play some role in establishing the geodynamo. Current thinking favors convection within the spherical fluid shell between mantle and inner core as a mechanism apt to create the appropriate flow field u on which the magnetic field B can grow and be maintained against dissipative loss in u and B.

For the present qualitative purposes, it is immaterial whether the convection has a thermal origin, and extends throughout the fluid core, or is gravitationally driven by the release of a relatively light component of a composite fluid near the inner boundary (see Section 9.6 below); in either case, the density is determined by a scalar field c, cf. Eq. (7.9). Quantitatively, as Braginsky (1964b) has remarked, the difference between the efficiency of a thermal and that of a nonthermal convective "engine" is significant, and may provide a strong case for the gravitationally driven dynamo.

Let us consider the equations for the full, hydromagnetic dynamo, Eqs. (7.1)-(7.9), in a spherical domain, and suppose that a suitable Rayleigh number Ra is defined, cf. (5.32), based on the source term Q in (7.9) and boundary conditions on c. If convective instability is studied in such a system with the magnetic field equal to zero (point Q' in Figure 7.2), linear theory indicates that many details of the problem are unimportant in the limit of large Taylor number T (compare Eq. (5.2)). It is usually found that the critical Rayleigh number Ra_c is proportional to $T^{2/3}$ as $T \to \infty$ (Chandrasekhar, 1961).

If this calculation is carried out with a given, nonzero magnetic field threading the fluid, it is found that this estimate is altered, and the smallest Ra_c occurs when $M^2 \sim T^{1/2}$, in which case $Ra_c \sim T^{1/2}$ (Eltayeb and Roberts, 1970; Eltayeb, 1972); here M is the Hartmann number, defined by $M^2 = B_0^2L^2/\mu\rho\nu\eta$ (see Table 7.1). According to these estimates for large T, the presence of a magnetic field tends to destabilize the system to convective motion (point Q'' in Figure 7.2).

Intuitively, it stands to reason that the kinetic energy of convective flow increases with $Ra\text{-}Ra_c$. If this flow produces an α-effect by one of the mechanisms discussed in Chapter 8, it can be argued that the system would acquire and sustain the strongest accessible magnetic field. Indeed, once motion begins, the α-effect suffices to amplify a seed field in accordance with kinematic dynamo theory. The growing field will tend in turn to lower the critical Rayleigh number, Ra_c and thereby, for fixed T and Ra, will increase $Ra\text{-}Ra_c$. This increases the α-effect, thereby further enhancing the excitation of the magnetic field, and the process continues until nonlinear saturation effects take over and Ra_c begins to increase with field amplitude, which happens when $M^2 \sim T^{1/2}$.

This order-of-magnitude relation may be tested for the Earth using the numerical values given in Table 7.1, since the questionable value for the fluid viscosity cancels out of the ratio $M^2/T^{1/2}$. For a core field of 100 gauss we obtain a ratio of about 5. This reasonable estimate is encouraging and suggests that the prevailing dynamical balance in the core is indeed between Coriolis forces and magnetic forces. Hence magnetic pressure may play the same role here as the hydrostatic pressure in Part II.

In this chapter, we shall follow a few approaches to the study of interacting hydrodynamic, magnetic and thermodynamic fields. In Section 9.2, the *weak-field approximation* is discussed. The velocity field is still independently deduced, but only to first order, and the magnetic field is allowed to act back on it and produce equilibration of both fields at a small, but finite amplitude. The shortcomings of this approach are outlined, especially its reliance on unrealistic scale separation. The difficulties in analyzing the complete interaction of axisymmetric, large-scale fields and the microscale, turbulent fields leads to macrodynamic theories.

In Section 9.3, the *macrodynamic model equations* are formulated. They involve a prescribed α-effect, acting on the poloidal component A of the magnetic field, as in Chapter 8, and on the toroidal component B as well. This secondary α-effect, with strength α^*, is found to play an important role in establishing dynamo action. Taylor's constraint, a diagnostic relation which has to be satisfied by A, B and the toroidal velocity component ω at all times, is derived for macrodynamic models. It is both analogous to and connected with the geostrophic relation (1.36, 2.28). Finally, it is demonstrated that no dynamo action can be sustained by such a model in the presence of a constant density field and with $\alpha^* = 0$, if $A/B = O(R_m^{-1}) \ll 1$.

In Section 9.4, we study a number of *minimal systems* exhibiting finite-amplitude, steady or sustained oscillatory dynamo action. Malkus and Proctor showed that a stable, nonzero equilibrium occurs in a spherical dynamo if $A/B = O(1)$ and $\alpha^* \neq 0$. A minimal system of three ordinary differential equations (ODEs) for A, B and ω captures some of the features of their model. In particular, it has a nontrivial, stable steady state, which is approached, under certain conditions, by damped oscillations.

Another approach to sustained dynamo action, due to Braginsky, is to allow density variations in space and time within a macrodynamic model. This introduces a toroidal "thermal wind", while the poloidal field B is mostly parallel to the axis of rotation, the z-axis. We next explore a system of five ODEs, augmenting the previous one by equations for A', B', the field components at a different point in, or in a different region of, the core. In this system $c \neq 0$ is a stratification parameter, and its solution include a stable Malkus-Proctor equilibrium, as well as an equilibrium of the Braginsky "model-Z" type. The latter is only stable on a short time scale, suggesting that a more complicated attractor set, such as a limit cycle or strange attractor, might replace it on longer time scales.

This possibility is examined further in Section 9.5. The coupling between microscale motions and the macrodynamics is modeled, again in the simplest possible way, by allowing for two stratification variables, a mean C and a microscale c, within a system of five ODEs for A, B, C, c and velocity u. The equation for u reflects the effect of the magnetic energy $E_m = A^2 + B^2$ on the amplitude of convection. The qualitative bifurcation picture in Figure 7.2 can be followed for this system. Aside from steady convection with and without dynamo action, stable periodic solutions of varying complexity occur. Various similarities between the behavior of these solutions and the "geomagnetic tape" of Figure 7.1, on the one hand, and with the succession of "blocked" and "zonal" episodes in some solutions of Section 6.4, on the other, are pointed out.

In Section 9.6, we return to energy balance considerations and the likely power source of the geodynamo. The rate of change of kinetic and potential energy in terms of compressional, precessional and tidal work and Joule heating is derived from the equations of motion. The dynamo is shown schematically to be driven by a convective heat engine, with the heat generated inside the solid inner core and the fluid outer core, and eliminated at the core-mantle interface. The efficiency η of this heat engine is estimated in terms of the temperature T_0 of the interface and the highest temperature in the fluid core, T_{max}, with $\eta \leq (T_{max}/T_0) - 1$.

For an estimate of $10^{-2} \leq \eta \leq 10^{-1}$, it is clear that thermal energy sources within the core are insufficient to achieve the estimated magnitude of the geomagnetic field in the core. This suggests a gravitational source of convection, with a phase change at the surface of the inner core producing light material floating up through the liquid core. This power source seems to be insufficient in order to drive the geodynamo. The chapter concludes with bibliographic notes in Section 9.7.

9.2. Weak-Field Models and Constant α

Weak-field dynamics and scale separation. In order to investigate the growth of magnetic fields from small seed fields, it is of interest to consider first so-called *weak-field* dynamical models, wherein the velocity field can, to first approximation, be deduced independently. It was suggested by Childress and Soward (1972) that thermal convection between isothermal, infinite parallel planes presents an interesting example of small-scale motions capable of exciting magnetic fields, provided that the system rotates rapidly with respect to an axis perpendicular to the fluid layer. This Rayleigh-Bénard system was studied in detail, for weak magnetic fields, by Soward (1974). Soward demonstrates that a dynamo is established at the weak-field level, and that the equilibration involves a sensitive reaction of the magnetic field on the energy distribution among the convective modes allowed by lateral boundary conditions. For an infinite fluid layer, real-field equilibration in Soward's dynamo depends crucially on the degeneracy of the structure of allowable convective fields occurring at the onset of thermal instability.

Soward's model can be thought of as a rough approximation to the convection which might occur in a thin spherical fluid shell in the region near the North or South Pole (Figure 9.1a). A more reasonable approximation to convection in a spherical shell, based upon the form which convection takes at the onset of instability in such a shell, was suggested as

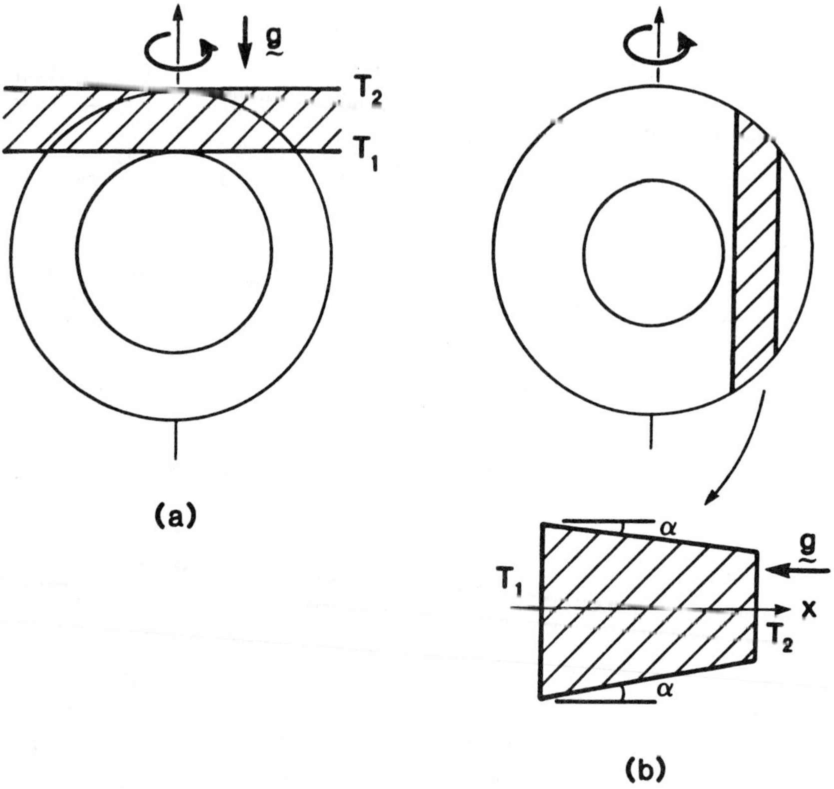

Fig. 9.1. Weak-field models; (a) Soward's convective dynamo as a tangent-plane model; g denotes gravity, T temperature. (b) Busse's annulus model.

a model for the geodynamo by Busse (1975b, 1976). As we indicate in Figure 9.1b, Busse isolates a portion of the shell where vertical convection cells are established, and models in a wedge geometry the inclined sidewalls as well as the local direction of gravity and mean heat flux. Busse's analysis, again carried out in the weak-field regime, demonstrates local equilibration of the dynamo. The equilibrating mechanism is different from Soward's, since Busse's convective field has a predetermined structure. This structure seems to correspond more closely to the geometry of the geodynamo.

The restriction to the weak-field regime is a crucial element of both models. For the Rayleigh-Bénard dynamo above, Fautrelle and Childress (1982) find that there is no solution analogous to Soward's once the initial magnetic field is sufficiently large. The system is eventually destabilized by the magnetic field and no equilibration is observed.

Another peculiarity of the weak-field models is the absence of any large-scale motion resembling the dominant velocity field in Braginsky's model (Section 8.5). As we shall see below, such a flow occurs naturally as the geostrophic component of the velocity field for dynamically self-consistent models, and it is clearly an ingredient in many dynamo mechanisms; such flows might be expected to arise through the build-up of appreciable magnetic torques as the field grows. For a sphere, the form taken by this geostrophic component was given in Section 7.6. More generally, geostrophic flows in rotating containers are treated in Greenspan (1968, Sec. 2.6).

A final and extremely troublesome difficulty with weak-field models is their reliance on smoothing theory (Section 8.2). If the system is unstable to virtual increases in the strength of the magnetic field, and if the result is to enhance the rise of magnetic energy, the only realistic brake on the growth is through magnetic stresses operating on the scale of the core. This will tend to eliminate the small-scale components of the motion (Eltayeb, 1972), and lead to the breakdown of the asymptotic validity of the smoothing in a first approximation. The conclusion looms large that natural evolution of the magnetic field may well be toward a state in which the kind of scale separation needed to make the notion of an α-effect useful is no longer relevant.

We should note in this connection the recent extensive numerical study by Gilman and Millar (1981) on self-consistent dynamo action in a spherical shell, which they compare with observations of the solar dynamo. These numerical results agree with certain features of observed solar activity. They do not agree with the numerical values of α which are needed in kinematic dynamos based on smoothing (Section 8.3) to achieve a good match with observation. Interestingly enough, one reason for this discrepancy is the relative weakness in the numerical simulation of the so-called "ω-effect", i.e., of the stretching of poloidal field into toroidal field (Figure 7.5). The nonlinear, interactive α-effect computed numerically more than compensates for this, since it is much larger than that required by the linear, kinematic models. The compensation can occur since, roughly speaking, dynamo action requires only that the product of the intensities of the α- and ω-effects be sufficiently large.

Wave-mean flow interaction and constant α. These difficulties with the calculation and interpretation of α apply to methods based on scale separation. Braginsky's asymptotic symmetry breaking theory could still be utilized in principle. Nevertheless, it is difficult to implement this theory too, because the asymmetric motions responsible for the α-effect, which can be thought of as waves riding on the axisymmetric flow, are certainly coupled to the symmetric field. A sufficiently strong field could tend, therefore, to suppress the α-effect.

This wave-mean flow interdependence, however, is difficult to calculate explicitly (see also remarks on Lagrangian-mean methods in Section 5.5), and recent modeling of the hydrodynamic dynamo has focused on systems with prescribed α. Such an approach leads to a "macrodynamic theory" of spherical dynamos based upon a system satisfied by the *symmetric* fields, which is equivalent to (7.1)-(7.7) together with the modified induction equation

$$\frac{dB}{dt} - \eta\nabla^2 B = B\cdot\nabla u + \nabla \times (\alpha \cdot B), \tag{9.1}$$

with $\alpha(r,t)$ a given symmetric matrix.

In Braginsky's expansion in a large magnetic Reynolds number R for the macroscale, the α-effect appears only in Eq. (8.94b) for the poloidal magnetic field. Here, we shall retain the more general form in order to exhibit the secondary α-effect, with strength α^*, in the equation for the toroidal field, see Eq. (9.3) below. In the following sections we study some aspects of macrodynamic models in a spherical core.

9.3. Macrodynamic Model Equations

Formulation. The governing equations in this approach, when expressed in cylindrical polar coordinates (z,r,ϕ), with $R = (r^2+z^2)^{1/2}$, take the dimensional form:

$$\frac{\partial A}{\partial t} + r^{-1}u_p\cdot\nabla(rA) = \eta(\nabla^2 - r^{-2})A + \alpha B, \tag{9.2}$$

$$\frac{\partial B}{\partial t} + ru_p\cdot\nabla(r^{-1}B) = \eta(\nabla^2 - r^{-2})B + [\nabla\omega\times\nabla(rA)]_\phi + [\nabla\times\alpha^*(\nabla\times A\hat{\phi})]_\phi, \tag{9.3}$$

$$\rho\,\frac{du}{dt} - \rho\nu\nabla^2 u + \frac{1}{\mu}\,\underset{\sim}{B} \times \nabla \times \underset{\sim}{B} + \nabla p + 2\Omega\rho\hat{z} \times u = \rho c g, \tag{9.4}$$

$$\frac{\partial c}{\partial t} + u_p \cdot \nabla c - \nabla \cdot D_c \nabla c = Q, \tag{9.5}$$

$$u = r\omega\hat{\phi} + u_p, \quad u_p = \nabla \times \psi\hat{\phi}, \quad \underset{\sim}{B} = B\hat{\phi} + \nabla\times A\hat{\phi}. \tag{9.6a,b,c}$$

Here $(\hat{\ })$ denotes a unit vector in the given coordinate direction, A is the vector potential for the poloidal field, B is the toroidal field component, and all quantities are independent of ϕ.

In a spherical core $0 \le R \le R_0$ bounded by an insulator, we have

$$u = 0, \quad B = 0, \quad c = 0, \quad (\nabla^2 - r^{-2})A = 0, \quad \text{for} \quad R > R_0, \tag{9.7}$$

and the conditions

$$B = 0 \quad \text{and} \quad A, \frac{\partial A}{\partial R} \quad \text{continuous on} \quad R = R_0. \tag{9.8}$$

We shall assume that $c = 0$ on $R = R_0$. The boundary condition on u deserves comment, since we have retained a nonzero viscosity and for small $E \equiv T^{-1/2}$ a thin viscous boundary layer, the Ekman boundary layer, will be present. As a result, the apparent boundary condition at $R = R_0$ (now regarded as the outer edge of the boundary layer) for the normal component of velocity becomes (Roberts and Soward, 1972)

$$u_p \cdot \hat{R} = -\frac{1}{2}(\nu/\Omega)^{1/2} \frac{1}{\sin\theta}\frac{\partial}{\partial\theta} \frac{\omega(R_0\cos\theta,\ R_0\sin\theta, t)\sin\theta}{|\cos\theta|^{1/2}}, \tag{9.9}$$

where $\cos\theta = \hat{z}\cdot\hat{R}$. We have assumed here that u_p is very small compared to the toroidal component $r\omega$. Since E is small, (9.9) will determine whether or not a zero-slip condition is approximately met, once u_p is ordered relative to E. If the poloidal velocity field u_p is of order larger than $E^{1/2}$, then $u_p \cdot \hat{R} \cong 0$ and the boundary condition is unaffected by the presence of the Ekman layer.

The α-effect has been included here in a form which assigns the scalar α to the toroidal component of the induced current, and a second scalar α^* to the poloidal component. We have referred to the latter as a "secondary" α-effect, since it would not appear using Braginsky's ordering of the magnetic field components. For α^* to appear formally in the expansion of Section 8.5, it would have to be greater than α by a factor R_m^2, where $R_m = UR_0/\eta$ and U is a characteristic magnitude of $r\omega$. This is rather unlikely and α^* accounts to zeroth order for the otherwise neglected wave-mean field interactions.

<u>Taylor's constraint</u>. We turn to an important dynamical constraint discovered by J. B. Taylor (1963), who studied the dynamical balance which results when Lorentz and Coriolis forces are dominant and of comparable magnitude. Let us define

$$\mathbf{F} = -\frac{d\mathbf{u}}{dt} + \nu\nabla^2\mathbf{u} + c\mathbf{g}, \tag{9.10}$$

so that $\mathbf{F} = (F_z, F_r, F_\phi)$ equals the sum of the Lorentz, Coriolis and pressure gradient forces. Component-wise, Eq. (9.4) can be written, accordingly, as

$$\frac{1}{\rho}\frac{\partial p}{\partial z} + (\mu\rho)^{-1}\left[B_r\left(\frac{\partial B_r}{\partial z} - \frac{\partial B_z}{\partial r}\right) + B\frac{\partial B}{\partial z}\right] = F_z, \tag{9.11a}$$

$$\frac{1}{\rho}\frac{\partial p}{\partial r} + (\mu\rho)^{-1}\left[B\left(\frac{1}{r}\frac{\partial rB}{\partial r}\right) - B_z\left[\frac{\partial B_r}{\partial z} - \frac{\partial B_z}{\partial r}\right]\right] - 2\Omega r\omega = F_r, \tag{9.11b}$$

$$-(\mu\rho)^{-1}\left[B_r\left(\frac{1}{r}\frac{\partial rb}{\partial r}\right) + B_z\frac{\partial B}{\partial z}\right] + 2\Omega u_r = F_\phi, \tag{9.11c}$$

where $\mathbf{B} = (B_z, B_r, B_\phi)$.

Suppose that we integrate F_ϕ over the cylindrical surface $r = r_0$, $-z_0 \le z \le +z_0$, $z_0 = (R_0^2 - z_0^2)^{1/2}$, $0 \le \phi \le 2\pi$ (see Figure 9.2). Using the formulae for B_z and B_r in an axisymmetric situation,

$$B_z = \frac{1}{r}\frac{\partial rA}{\partial r}, \quad B_r = -\frac{\partial A}{\partial z} \tag{9.12a,b}$$

in (9.11c), and the fact that $B = 0$ on the boundary, integration by parts yields the following expression for the terms involving the magnetic field:

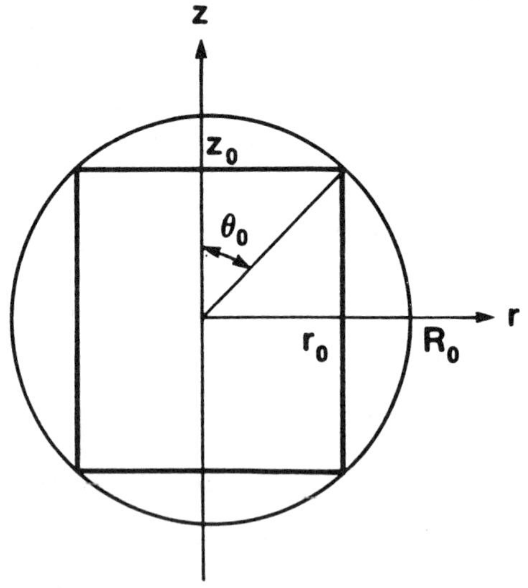

Fig. 9.2. Cylindrical integration surface for Eqs. (9.11 - 9.14).

$$2\pi(\mu\rho)^{-1}\int_{-z_0}^{z_0}\left[\frac{\partial A}{\partial z}\frac{\partial rB}{\partial r}+\frac{\partial}{\partial r}\left(r\,\frac{\partial A}{\partial z}\right)B\right]_{r=r_0}dz$$
$$= 2\pi(\mu\rho)^{-1}\left[\frac{1}{r}\frac{d}{dr}\,r^2\int_{-z_0}^{z_0}B\,\frac{\partial A}{\partial z}\,dz\right]_{r=r_0}. \qquad (9.13)$$

The integral of u_r over this cylindrical surface must equal minus the sum of mass fluxes into the two spherical caps determined by the intersection of the cylinder with the core boundary, because u is a solenoidal flow within the core. The mass flux can then be evaluated using the boundary condition (9.9). We thus obtain the following expression from (9.11c):

$$2\pi r_0\int_{-z_0}^{+z_0}F_\phi(z,r_0,t)\,dz$$
$$= 2\pi(\mu\rho)^{-1}\left[r^{-1}\frac{d}{dr}\,r^2\int_{-z_0}^{z_0}B\,\frac{\partial A}{\partial z}\,dz\right]_{r=r_0} \qquad (9.14)$$
$$+\ 2\pi(\Omega\nu)^{1/2}r_0^2(\cos\theta_0)^{-1/2}[\omega(z_0,r_0)+\omega(-z_0,r_0)],$$

where $\theta_0 = \tan^{-1}(r_0/z_0)$.

Given the left-hand side, this is a functional constraint that must be met by A, B and ω at each instant of time. It expresses the physical fact that all torques acting on the control cylinder must add to zero. If F_ϕ and ν were both zero, then (9.14) would express the vanishing of net electromagnetic torque and thus impose a direct constraint on the functional forms of A and B.

This vanishing of the ϕ-component of magnetic torque on each cylindrical surface, which in the present axisymmetric case reduces, cf. (9.14), to

$$\left[r^{-1}\frac{d}{dr}\,r^2\int_{-z_0}^{+z_0}B\,\frac{\partial A}{\partial z}\,dz\right]_{r=r_0} = 0, \qquad (9.15)$$

is customarily referred to as *Taylor's constraint*. Note that the last expression, if the integral is regular at $r = 0$, requires that

$$\int_{-z_0}^{z_0}B\,\frac{\partial A}{\partial z}\,dz = 0,$$

which expresses the functional implication of the Taylor constraint under the condition of axial symmetry.

We remark that essentially the same analysis can be carried out for non-axisymmetric $\underset{\sim}{B}$ and with the same result: if $F_\phi = 0$ and the fluid is inviscid, then $[\underset{\sim}{B} \times (\underset{\sim}{\nabla} \times \underset{\sim}{B})]_\phi$ has zero integral over each cylindrical surface $r = r_0$ contained in the core. We write this now as

$$<M[\underset{\sim}{B} \wedge \underset{\sim}{\nabla} \wedge \underset{\sim}{B}]_\phi (r_0, t) = 0>.$$

The fact that the constraint is imposed at each instant of time allows the last equation to be differentiated with respect to time, to obtain

$$<M[\underset{\sim}{B}_t \times \underset{\sim}{\nabla} \times \underset{\sim}{B}]_\phi + M[\underset{\sim}{B} \times \underset{\sim}{\nabla} \times \underset{\sim}{B}_t]_\phi = 0.$$

By use of the induction equation (in either its primitive form or, following smoothing, in the form (9.1)), to eliminate $\underset{\sim}{B}_t$, an equation for the velocity field in terms of $\underset{\sim}{B}$ and its spatial derivatives will result. If, at the instant considered, the magnetic field is *known,* then $\underset{\sim}{u}$ is known up to its geostrophic component.

In principle, this component is determined therefore by the Taylor condition (9.15). Note that the variable count is consistent here: the constraint applies to all $r = r_0 < R_0$ and t, thereby determining a geostrophic velocity depending on r and t. The above procedure was used by Taylor (1963) to determine the geostrophic flow when inertial terms are neglected and the momentum equations cease to be predictive.

This method relies on the neglect of all terms in (9.14) except the term on the right. An essential question is whether or not such an approximation is valid, even when E, as well as inertial terms, are nominally small. The question is a basic one in current attempts to construct hydromagnetic dynamo models and Section 9.4 is devoted to consideration of the alternatives which have been examined to date.

<u>A minimal system</u>. We consider next the simplest macrodynamical model which might operate qualitatively like the geodynamo. Suppose that $c = 0$, that $\alpha^* = 0$, and that inertial and viscous terms in (9.4) are negligible, although we retain the effect of viscosity through the boundary condition (9.9). We also adopt Braginsky's ordering $A/B \sim 1/R_m \ll 1$, which will allow some simplification of the expression for the Lorentz force in (9.11).

Physically, these approximations leave one possible mechanism for driving the dynamo. Whatever the flow field responsible for the α-effect, some force distribution must be present to sustain the flow and the work done by these forces has to balance the system losses. Regarding the underlying flow as a microscale motion, the process thus envisaged is

$$\text{microscale work} \rightarrow \alpha \rightarrow \text{macroscale } A, B, u \rightarrow \text{macroscale dissipation}.$$

We shall show, however, that in this simple form the idea fails and the dynamo must decay, irrespective of the prescribed structure of α. To see this we proceed as in the proof of Cowling's theorem in Section 7.5 and multiply (9.3) by B/r^2, then integrate over the core. In view of the boundary condition on B and the vanishing of α^*, we have immediately

$$\int \frac{B}{r} u_p \cdot \nabla(r^{-1}B)\,dV = \frac{1}{2}\int \nabla\cdot(u_p B r^{-2})\,dV = 0, \tag{9.16a}$$

and

$$\eta\int Br^{-2}(\nabla^2 - r^{-2})B\,dV = \eta\int [\nabla(Br^{-1})]^2 dV, \tag{9.16b}$$

where the volume integration is over $R \leq R_0$.

To study the term involving ω in (9.3), we first use the assumed dominance of B over A in (9.11a) and the vanishing of F from our assumptions, to obtain

$$p + \frac{B^2}{2\mu} = \pi(r,t) \tag{9.17}$$

where π is an arbitrary function. Combining this with (9.11b) we then have

$$2\rho\Omega r\omega = \frac{\partial\pi}{\partial r} + \frac{B^2}{\mu r}\,. \tag{9.18}$$

The second term on the right contributes nothing to the integral, since

$$\begin{aligned}\frac{1}{2\Omega\mu\rho}\int\left(\frac{B}{r^2}\times\nabla rA\right)_\phi dV &= \frac{1}{3\Omega\mu\rho}\iint\left(\nabla\left(\frac{B}{r}\right)^3\times\nabla rA\right)_\phi dz\,dr\\ &= \frac{1}{3\Omega\mu\rho}\iint\left[\frac{\partial}{\partial z}\left(\frac{B}{r}\right)^3\frac{\partial}{\partial r}rA - \frac{\partial}{\partial r}\left(\frac{B}{r}\right)^3\frac{\partial}{\partial z}rA\right]dz\,dr\\ &= \frac{1}{3\Omega\mu\rho}\int\frac{\partial}{\partial z}\left(\frac{B^3}{r^3}\frac{\partial rA}{\partial r}\right) - \frac{\partial}{\partial r}\left(\frac{B^3}{r^3}\frac{\partial rA}{\partial z}\right)dz\,dr = 0.\end{aligned}$$

The remaining contribution to the integral comes from $\pi(r,t)$, and we must evaluate the volume integral of the second term on the right of (9.3) with

$$\omega = \frac{1}{2\Omega r\rho}\frac{\partial\pi}{\partial r}\,. \tag{9.19}$$

Hence ω here is entirely geostrophic. Writing $\omega = \zeta(r,t)$, we obtain

$$\begin{aligned}I &\equiv \int\frac{B}{r^2}[\nabla\zeta\times\nabla(rA)]_\phi\,dV\\ &= -2\pi\iint\frac{\partial\zeta}{\partial r}B\frac{\partial A}{\partial z}\,dz\,dr.\end{aligned} \tag{9.20}$$

Defining

$$m(r,t) = \int_{-(R_0^2-r_0^2)^{1/2}}^{(R_0^2-r_0^2)^{1/2}} B \frac{\partial A}{\partial z}\, dz, \tag{9.21}$$

there results from the use of (9.14) with $F_\phi = 0$ and the end-point behavior of ζ and m,

$$\begin{aligned} I &= -2\pi \int_0^{\pi/2} \zeta \frac{\partial m}{\partial \theta}\, d\theta \\ &= -2\pi k^{-1} \int_0^{\pi/2} \left(2 \frac{(\cos\theta)^{1/2}}{\sin^2\theta} m + \frac{(\cos\theta)^{-1/2}}{\sin\theta} \frac{\partial m}{\partial \theta}\right) \cdot \frac{\partial m}{\partial \theta}\, d\theta \\ &= -2\pi k^{-1} \int_0^{\pi/2} \left\{\frac{(\cos\theta)^{-1/2}}{\sin\theta} \left[\left(\frac{\partial m}{\partial \theta}\right)^2 + \frac{1}{2} m^2\right] + 4 \frac{(\cos\theta)^{3/2}}{\sin^3\theta} m^2\right\} d\theta < 0, \end{aligned} \tag{9.22}$$

where $k = 2\rho\mu(\Omega\nu)^{1/2}R_0^2$. From (9.16) and (9.22) we therefore have

$$\frac{\partial}{\partial t} \int \frac{B^2}{r^2}\, dV < -\int \left(\nabla \frac{B}{r}\right)^2 dV \tag{9.23}$$

and the dynamo has to decay.

The geostrophic component $r\zeta$ of the velocity occurs here as an arbitrary function of integration of the dynamical equations, as noted earlier. The present discussion, which follows Braginsky (1975), generalizes Taylor's analysis to include the effect of Ekmann suction. It is not surprising that this additional dissipative effect does nothing to prevent decay of the magnetic field under the conditions set out above.

9.4. The Simplest Hydromagnetic Models

The Malkus-Proctor models. The minimal system of the preceding section must be altered in some way if dynamo action is to occur. Malkus and Proctor (1975) and Proctor (1977) proposed restoring the secondary α-effect and also retaining A and B as fields of possibly comparable magnitude. The former assumption seems to be the key to obtaining dynamo action in their studies, and their numerical simulations of spherical dynamos based upon this idea indicated successful equilibration to a state of nonzero magnetic energy.

If the amplitude of the magnetic field is measured in units of $(\Omega\eta\mu\rho)^{1/2}$ and denoted by ε, an ordering equivalent to $M^2/T^{1/2} \sim \varepsilon^2$, the Malkus-Proctor approach depends crucially upon the parameter $\varepsilon^2 T^{1/2} = \varepsilon^2 E^{-1/2}$ when both ε and E are small. Only if this parameter is large does the problem become essentially inviscid. Then the induced field is observed to evolve toward a state which appears to satisfy approximately the Taylor condition. The time evolution toward this equilibrated state was found to involve weakly damped oscillations.

It is possible to reproduce some features of these models with a highly truncated system of ordinary differential equations (Section 5.2). The system

$$\dot{A} + A = \alpha B, \tag{9.24a}$$

$$\dot{B} + B = \omega A + \alpha^* A, \tag{9.24b}$$

$$E_M \dot{\omega} + E^{1/2}\omega = -AB \tag{9.24c}$$

is an extremely simple dimensionless analog of (9.2)-(9.4). Here α and α^* are constants, and $E_M = \eta/L^2$ is a measure of the ratio of inertial force to Lorentz force (in these units of the magnetic field).

To study the inviscid limit of system (9.24) we set $\omega^* = \omega + \alpha^*$ and define ε by $\varepsilon^2 = E^{1/2}\alpha^*$, in which case (9.24b,c) become

$$\dot{B} + B = \omega^* A, \tag{9.25a}$$

$$E_M \dot{\omega}^* + E^{1/2}\omega^* = \varepsilon^2 - AB. \tag{9.25b}$$

Letting E_M and E tend to zero, we obtain

$$\varepsilon^2 - AB = 0, \tag{9.26}$$

which is the analog of the Taylor condition. To use it we differentiate (9.26) with respect to time and employ (9.24a) and (9.25a) to eliminate time derivatives:

$$B\dot{A} + A\dot{B} = B(\alpha B - A) + A(\omega^* A - B) = 0,$$

yielding

$$A^2\omega^* = 2AB - \alpha B^2. \tag{9.27}$$

The solution evolves from any initial condition consistent with (9.26) according to

$$\dot{A} + A = \frac{\alpha\varepsilon^2}{A}, \quad B = \frac{\varepsilon^2}{A}, \quad \omega^* = \frac{2\varepsilon^2}{A^2} - \frac{\alpha\varepsilon^4}{A^4}. \tag{9.28a,b,c}$$

Clearly the solutions of this model tend to a stable equilibrium given by (9.28b,c) with

$$A = \varepsilon\alpha^{1/2}. \tag{9.28d}$$

To study the possibility of fast oscillations about the field evolving to (9.28b,c,d), we assume E_M and E are small and of comparable order, and perturb as follows:

$$(A,B) = (A_0(t),B_0(t)) + E^{1/2}(\delta A(t,\tau),\delta B(t,\tau)) + o(E^{1/2}), \tag{9.29a}$$

$$\omega^* = \omega_0^*(t) + \delta\omega^*(t,\tau), \tag{9.29b}$$

where A_0, B_0, and ω_0^* satisfy (9.28) and $\tau = t/E^{1/2}$. Expanding (9.24a), (9.25a), and (9.25b) to orders δ, δ, and $E^{1/2}\delta$ respectively we obtain

$$\frac{\partial\delta A}{\partial\tau} = 0, \qquad \frac{\partial\delta B}{\partial\tau} = \delta\omega^* A_0,$$

$$\frac{E_M}{E} \frac{\partial\delta\omega^*}{\partial\tau} + \delta\omega^* = -\omega AB_0 - \delta BA_0,$$

and therefore

$$\frac{E_M}{E}\frac{\partial^2\delta\omega^*}{\partial\tau^2} + \frac{\partial\delta\omega^*}{\partial\tau} + A_0^2\delta\omega^* = 0. \tag{9.31}$$

Damped oscillations occur if $4E_M A_0^2 > E$. Their period in units of t is

$$T = \frac{4\pi E_M}{\sqrt{4A_0^2E_M - E}}. \tag{9.32}$$

As a final property of this simple ODE system, note that if $\alpha^* = 0$, then solutions decay to zero, since (9.24b,c) yield

$$\frac{d}{dt}\left(\frac{1}{2}B^2 + \frac{1}{2}E_M\omega^2\right) = -B^2 - E^{1/2}\omega^2 \tag{9.33}$$

and (9.24a) accordingly yields that $A \to 0$ as well.

Braginsky's Model Z. Another, perhaps physically more interesting simple hydromagnetic system is obtained by restoring the Archimedean force. If we regard c as a given function of space and time, but otherwise retain all features of the minimal system, then (9.18) is replaced by

$$2\rho\Omega r\omega = \pi_r + \frac{B^2}{\mu r} - \int_0^z [\nabla c\times g]_\phi \, dz. \tag{9.34}$$

The new term represents the "thermal wind" created by the stratification of the fluid (cf. Section 4.2). This term makes a generally nonzero contribution to the volume integral (9.23) which defeated the minimal system of Section 9.3.

The approach we now consider, due to Braginsky (1975, 1976), also differs from the preceding theory in its use of the Taylor condition. This condition can be written in the form (cf. (9.14) and (9.21))

$$\frac{1}{r^2}\frac{\partial r^2}{\partial r}\, m(r,t) = f, \tag{9.35}$$

where Taylor's assumptions mean that f is small.

Rather than equating the term on the left to zero, Braginsky argues that this term is *a priori* of small order and in equilibrium with f. From (9.14) it follows that f involves the geostrophic component of velocity and its derivatives with respect to r in a linear fashion. Thus, given B, we can invert (9.35) to obtain the form of the geostrophic velocity. In other words, the expression becomes an equation for the geostrophic component of velocity rather than a condition on the magnetic field (from which the velocity can be inferred afterwards using the induction equation). It is argued that the balance (9.35) is achieved by the smallness of B_r over the core, i.e., because the poloidal field is approximately parallel to the z-axis; hence the model's name.

Various considerations suggest that this state of affairs is consistent with the governing equations and observational evidence. Furthermore, if both c and α^* are nonzero, the system might in fact exhibit multiple equilibria, one class corresponding to the Malkus-Proctor solutions, another to Model Z. We consider now an enlargement of the system (9.24) of ordinary differential equations, which allows additional equilibria analogous to the solutions envisaged in Model Z. Introducing an auxiliary pair of variables (A',B') we adopt the following system of five equations for A,B,A',B',ω.

$$\dot{A} + A = \alpha B, \tag{9.36a}$$

$$\dot{B} + B = \omega(A-A') + cA, \tag{9.36b}$$

$$\dot{A}' + A' = \beta B', \tag{9.36c}$$

$$\dot{B}' + B' = k\omega A', \tag{9.36d}$$

$$E_m\dot{\omega} + E^{1/2}\omega = -(A-A')B - \chi A'B'. \tag{9.36e}$$

Here α, β, and c are positive constants; for reasons which will be given below, k and χ are negative constants. We also take, for simplicity, $E_m = 0$ in the following analysis. The thermal wind is represented by the term cA in (9.36b).

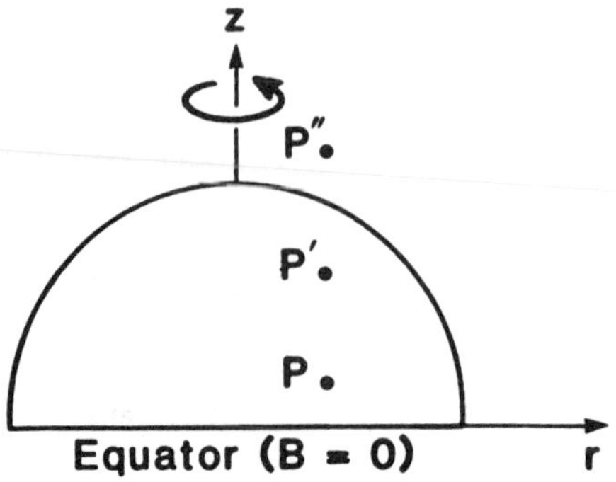

Fig. 9.3. Conceptual model for system (9.36). All fields vanish at the exterior point P".

The variables (A,B) and (A',B') can be viewed as values of a spatial field at two distinct points, P and P' respectively, as indicated in Figure 9.3. This use of an augmented system of ordinary differential equations to simulate spatial variations was suggested by the work of Jones (1980). The terms (A-A') in (9.36b) and (A'-0) in (9.36d) indicate that the derivative $\partial A/\partial z$ in (9.21) is represented by the differences of field values between two points. In (9.36b) the difference is between P and P', while in (9.36d) it is between point P' and infinity, A" being zero.

We note that (9.36) reduces to (9.24) when $A' = B' = 0$ and c is identified with α^*. The nonzero equilibrium of this reduced system is easily seen to be stable to the growth of A', B' provided that

$$\frac{\beta k}{\alpha}(1 - \alpha c) < 1. \tag{9.37}$$

There is, in addition, the possibility of a nonzero equilibrium with $A = B = 0$, but from (9.36c,d,e) we see that this exists only if $\chi k < 0$, and we will assume instead that $k\chi > 0$. If $\alpha c > 1$ and (9.37) is satisfied, the Malkus-Proctor equilibrium (9.28) remains the only stable equilibrium of the enlarged system, with $A' = B' = 0$.

There are, however, equilibria with both (A,B) and (A',B') nonzero, provided that $\alpha c > 1$ and that (9.37) does not hold. These are given by

$$A_e = \pm E^{1/4}\{(k\beta/\alpha)^2(\alpha c-1)-\chi k[1+(k\beta/\alpha)(\alpha c-1)]^2\}^{-1/2}, \tag{9.38a}$$

$$A_e' = A_e[1+(k\beta/\alpha)(\alpha c-1)], \quad B_e = \alpha^{-1}A_e, \quad B_e' = \beta^{-1}A_e'. \tag{9.38b,c,d}$$

Thus (9.38) and the Malkus-Proctor equilibrium can coexist, and we are interested in the possible stability of (9.38).

The limiting solution treated by Braginsky in a spherical geometry is based upon the assumed smallness of $E^{1/2}$, which is linked to the consequent smallness of $\partial A/\partial z$. Note that if $A_e - A_e'$ is to be small in (9.38), then $(k\beta/\alpha)(\alpha c - 1)$ is small and so, by (9.37), the Malkus-Proctor equilibria are stable in the present model.

To discuss an analog of Braginsky's limiting solution for (9.36), we set $\delta = E^{1/6} << 1$ and introduce order-one variables, marked by tilde, as follows:

$$\tilde{\omega} = \omega\delta^2, \quad \tilde{A} = \frac{A-A'}{\delta}, \quad \tilde{k} = k\delta^{-2}, \quad \tilde{\kappa} = \kappa\delta^{-1}, \quad \tilde{c} = c\delta. \tag{9.39}$$

Since $A \cong A'$, (9.36a,c) imply that $\beta B' \cong \alpha B$. Therefore (9.36b,c,d,e) reduce to the following limits as $\delta \to 0$ in tilde variables:

$$\tilde{\omega}\tilde{A} + \tilde{c}A = 0, \tag{9.40a}$$

$$\dot{A} + A = \alpha B, \tag{9.40b}$$

$$\dot{B} + B = (\beta\tilde{k}/\alpha)\tilde{\omega}A, \tag{9.40c}$$

$$\tilde{\omega} = -\tilde{A}B - (\tilde{\chi}\alpha/\beta)AB. \tag{9.40d}$$

Using (9.40a,d) we obtain

$$\tilde{\omega}^2 = cAB - \tilde{\omega}(\tilde{\chi}\alpha/\beta)AB \tag{9.41}$$

or

$$\tilde{\omega} = \frac{1}{2}\,[-(\tilde{\chi}\alpha/\beta)AB \pm \sqrt{(\alpha\tilde{\kappa}AB/\beta)^2+4cAB}. \tag{9.42}$$

Since $\tilde{\omega}\tilde{k} > 0$ at equilibrium and $\alpha, \beta, \tilde{c}, \tilde{\chi}\tilde{k} > 0$, the sign to be chosen in (9.42) must be that of $\tilde{k}$.

To test stability of the equilibrium for the system (9.40) we write

$$(A, B, \tilde{\omega}) = (A_e, \alpha^{-1}A_e, 1/\beta\tilde{k}) + \text{perturbations}, \tag{9.43a}$$

where A_e is given by

$$A_e^{-2} = (\tilde{k}\beta/\alpha)^2\alpha\tilde{c} - \tilde{\chi}\tilde{k}. \tag{9.43b}$$

The perturbation, if proportional to $e^{\sigma t}$, is then found to have growth rate σ satisfying $(\sigma+2)(\sigma-\Gamma) = 0$, where $\Gamma = (2+Ae^2\tilde{\chi}\tilde{k})^{-1}$, and so the equilibrium is unstable.

As Braginsky (1976) has pointed out, there is a natural dynamical time scale for Model Z which is shorter by a factor δ than that implicit in (9.40). To examine stability of our model equilibrium on this short time scale, we define $\tilde{t} = t\delta$ as a new order 1 time variable. Again, taking the limit $\delta \to 0$ in tilde variables we obtain

$$\frac{dA}{d\tilde{t}} \simeq 0, \quad \frac{dA'}{d\tilde{t}} \simeq 0, \quad \frac{dB'}{d\tilde{t}} \simeq 0, \tag{9.44a}$$

$$\frac{d\tilde{A}}{d\tilde{t}} \simeq \alpha B - \beta B', \tag{9.44b}$$

$$\frac{dB}{d\tilde{t}} \simeq \tilde{\omega}\tilde{A} + \tilde{c}A. \tag{9.44c}$$

If use is made of (9.40d) in (9.44c), the perturbation equations imply

$$\left[\frac{d^2}{d\tilde{t}^2} + (\tilde{c}\beta\tilde{k}A_e)^2 \frac{d}{d\tilde{t}} - (\tilde{c}\beta\tilde{k}A_e^2 + \alpha/\beta\tilde{k})\right](B-B_e) \simeq 0 \tag{9.45}$$

for small deviations $B-B_e$. From (9.45) it follows that the equilibrium is stable on this short time scale provided that $\tilde{k}$ (and hence $\tilde{\chi}$) is negative.

The instability of the Z-type equilibrium in system (9.36) on longer time scales is somewhat disappointing. But there are a number of aspects of Braginsky's Model Z which cannot be well represented by an elementary system of differential equations, e.g., there must exist a thin current layer at the core-mantle interface. Therefore it is not possible, based on such a system, to make a definitive statement about the actual stability of this proposed solution in the full system of partial differential equations.

It is also possible that modifications of (9.36), or a parameter range different from the small-δ limit, could lead to stability of these equilibria. Even more interesting would be cases where no stable equilibria exist, but there is a periodic or a strange attractor (see also Sections 6.3, 6.4 and 7.4). In contradistinction to Section 6.4, unsteady solutions of (9.36) would not be close to Model-Z equilibria.

9.5. Coupling of Macrodynamics to Microscale

The preceding discussion has focused on the balance of forces in a dynamo with given macroscale parameters α and c. In fact both α and c are internally generated. At the next level of approximation, it is helpful to visualize this coupling in terms of a convective motion on a spatial microscale, although in reality such a scale separation may be only marginally valid. Even if we assume such a scale separation, the magnetic field generated within the system will affect α and c through magnetohydrodynamic coupling with the microscale. These effects, which are completely neglected by macrodynamic modeling, are probably quite important in the time evolution of the magnetic field.

Continuing in the spirit of this chapter, we shall model this coupling using a system of ordinary differential equations (Fautrelle and Childress, 1982). We again use (9.24a,b) with $\alpha^* = C$,

$$\dot{A} + \mu A = \alpha B, \tag{9.45a}$$

$$\dot{B} + \eta B = \omega A + CA, \tag{9.45b}$$

but now introduce the parameters μ, η on the left-hand sides.

The variable C is capitalized in order to emphasize its new interpretation, as a mean density perturbation associated with a field of convection. The latter will involve a microscale density perturbation c and a microscale velocity u. The coupling between C, c, and u is assumed to take the following form:

$$\dot{C} + C = uc, \tag{9.45c}$$

$$\dot{c} + \chi c = Ku(1-kC). \tag{9.45d}$$

Note that (9.45c) models the influence of convective transport of material, while (9.45d) expresses the connection between the microscale fields when the mean density gradient is $K(1-kC)$.

To complete the system we need equations for α, u, and ω. Adopting (9.24c) with $E_M = 0$, we have

$$\omega = -E^{-1/2}AB. \tag{9.45e}$$

As for α, we recall its derivation in Section 8.3, which suggests that in general it should depend quadratically on the microscale velocity. Since we also expect the convection to be suppressed when the magnetic energy density is large, we assume

$$\alpha = \frac{u^2}{1+\lambda E_m}, \quad E_m = A^2 + B^2. \tag{9.45f}$$

The final equation, for u, must reflect the important role of the magnetic field in enhancing convection (Figure 7.2). We assume

$$\varepsilon\dot{u} = u[Ra(1-kC) - f(E_m) - u^2], \tag{9.45g}$$

where Ra is a Rayleigh number. The function $f(E_m)$ is chosen to reflect the dependence of critical Rayleigh number for convective instability in a rotating system upon the magnetic field strength, as described in Sections 7.3 and 9.1. We take

$$f = E_m(E_m - E_1)(E_m - E_2), \quad 0 < E_1 < E_2. \tag{9.45h}$$

The global structure of solutions to system (9.45) is difficult to treat analytically. But it is rather easy to see how the bifurcation to convection with zero magnetic field (point Q' in Figure 7.2), then to convection with weak magnetic field (point Q" in the figure) occur.

If $A = B = 0$, the equilibrium is given by

$$a_e^2 = \frac{\chi}{2kK}\left[-1 + \left(1 + \frac{4RakK}{\chi}\right)^{1/2}\right], \tag{9.46a}$$

$$C_e = Ku_e^4/\chi Ra, \tag{9.46b}$$

$$c_o = Ku_e^3/\chi Ra. \tag{9.46c}$$

It is then obvious from (9.45a,b) that the magnetic field begins to grow when $C_c u_e^2$ exceeds $\mu\eta$, or when

$$u_e^2 > (\mu\eta\chi Ra/K)^{1/3}. \tag{9.47}$$

Since the left-hand side of (9.47) involves $Ra^{1/2}$, this inequality is met for sufficiently large Ra, $Ra > Ra_c$ say.

We thus see that convection without dynamo action occurs when $0 < Ra < Ra_c$ and convection with weak magnetic field occurs when Ra slightly exceeds Ra_c. The values $Ra = (0, Ra_c)$ are thus analogous to the values $Q = (Q', Q")$ in Figure 7.2.

We show some numerical results for the above system in Figure 9.4. The time scale in the figure, cf. (9.45), is the characteristic diffusion time of the mean variable. In these computations we set $\eta = \mu = 0.5$, $k = K = 0.2$, $\varepsilon = \lambda = E^{1/2} = 1$, $E_1 = 0.1$, $\chi = 10^{-4}$, and consider various values of Ra and E_2. Due to the highly idealized nature of the system, no attempt is made here to choose numerical values representative of the Earth's core.

Notice the occurrence of periodic solutions with simple structure in Figure 9.4a, while in Figure 9.4b the solution still appears to be periodic, but has more complex structure, involving more than one characteristic time scale. Solution variability on the shorter and longer time scale is reminiscent of polarity reversal "events" and "epochs" of the

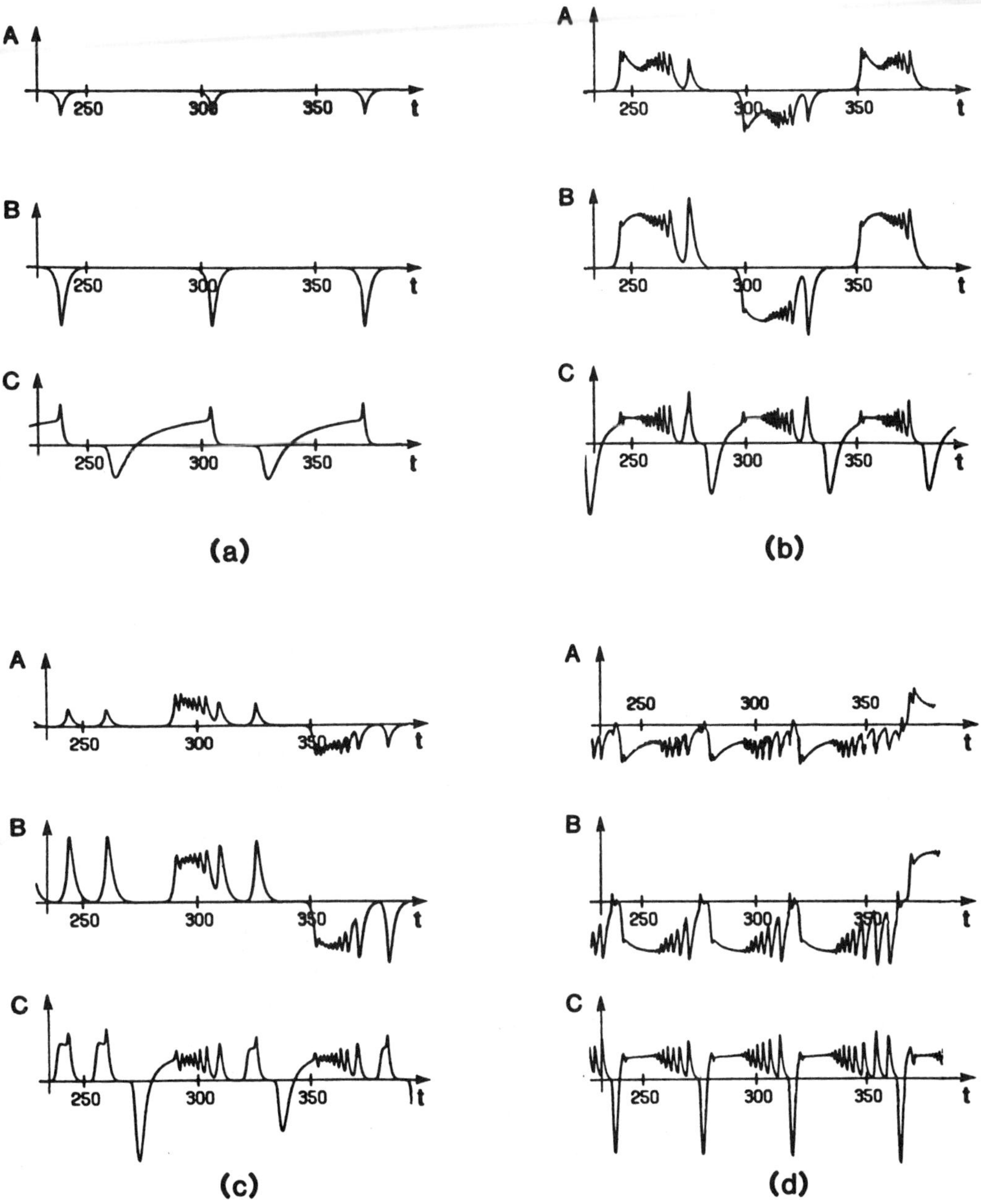

Fig. 9.4. Calculated values of A, B, and C for the model (9.45). (a) $E_2 = 3$, Ra = 1; (b) $E_2 = 4$, Ra = 2; (c) $E_2 = 3$, Ra = 2; (d) $E_2 = 3$, Ra = 7.

paleomagnetic record, respectively (Section 7.1).

The solutions shown in Figures 9.4c and 9.4d indicate a random pattern of reversals, suggestive of the data in Figure 7.1. The increasing amplitude of oscillations before reversal is also in qualitative agreement with the detailed sequence of geomagnetic field changes as outlined in Section 7.1.

Figure 9.4c shows two characteristic types of behavior or "signatures" of the same solution. One consists of an isolated pulse, or a brief succession of such pulses, the other is a much broader signal, with rapid oscillations superimposed on a nearly constant mean. A clear analogy exists with the succession of "blocked" and "zonal" episodes, respectively, in Figures 6.14a-e.

9.6. Thermal or Gravitational Source of Convection

We shall try to clarify here the distinction which must be drawn between thermal and nonthermal convection as a possible source of power to sustain a hydromagnetic dynamo. At the same time we recall that the precession of the axis of rotation of the Earth introduces another possible power source. As indicated in Section 7.2, recent estimates of the power available from thermal and precessional driving suggest that neither is sufficient to account for the observed magnetic field.

Such estimates have led to proposals for a nonthermal, physico-chemical power source. This source is both plausible from the viewpoint of core chemistry, and apparently more than adequate in terms of power availability.

Energy balance of the geodynamo. We study the energy balance for the fluid occupying a spherical shell $r_1 < r < r_2$, where r_1 is the radius of the solid inner core, r_2 the radius of the core-mantle interface, and the region $r > r_2$ is taken as a nonconductor. Since it is likely that Joule heating predominates over viscous dissipation in the fluid core, we shall consider an inviscid fluid, but allow for the possible time dependence of the angular rotation rate Ω. Relative to a rotating frame, the momentum and continuity equations are (see Sections 1.3 and 7.2)

$$\rho \frac{du}{dt} + 2\rho\Omega\times u + \rho\dot{\Omega} \times r + \nabla p + B\times J = \rho\nabla\phi, \tag{9.48}$$

$$\frac{d\rho}{dt} + \rho\nabla\cdot u = 0. \tag{9.49}$$

To compute the mechanical energy balance, assume $u\cdot n = 0$ on the boundary of the fluid core, take the dot product of (9.48) with u, and integrate over the core. The result, after making use of the divergence theorem and (9.49), has the form (since ϕ is independent of time)

$$\partial_t(E_K+E_P) = W_c + W_p - \int_{core} u\cdot(B\times J)\ dr, \tag{9.50}$$

where

$$E_K = \text{core kinetic energy} = \frac{1}{2}\int_{core} \rho u^2 dr, \tag{9.51a}$$

$$E_P = \text{core potential energy} = -\int_{core} \rho\phi\ dr \tag{9.51b}$$

$$W_c = \text{compressional work} = \int_{core} \rho\nabla\cdot u\ dr, \tag{9.51c}$$

$$W_p = \text{precessional and tidal work}$$

$$= -\int_{core} \rho[\frac{\partial\phi}{\partial t} + u\cdot(\dot{\Omega}\times r)]dr. \tag{9.51d}$$

On the other hand, if E_M denotes the magnetic energy of the system as a whole (the integral of $B^2/2\mu$ over all space), (7.22) gives

$$\partial_t E_M = \int_{core} u\cdot B\times J\ dr - Q_{Joule}, \tag{9.52a}$$

$$Q_{Joule} = \frac{1}{\sigma}\int_{core} J^2 dr. \tag{9.52b}$$

Eliminating the common term in (9.51) and (9.52) we obtain

$$\partial_t(E_K+E_P+E_M) = W_c + W_p - Q_{Joule}. \tag{9.53}$$

We now turn to the energy equation, under the assumption that there is a source of heat at a rate q_{source} over the fluid core. Introducing also an internal energy density e, temperature T, and specific entropy s, we may write the energy equation in the form

$$\begin{aligned}\rho T\frac{ds}{dt} &= \rho\frac{dc}{dt} - \frac{p}{\rho}\frac{d\rho}{dt} = \frac{\partial\rho e}{\partial t} + \nabla\cdot(\rho e u) + p\nabla\cdot u \\ &= \nabla\cdot(\lambda\nabla T) + q_{source} + \frac{1}{\sigma}J^2,\end{aligned} \tag{9.54}$$

where λ is the thermal conductivity. In terms of temperature, another form of (9.54) is

$$\rho c_p\frac{dT}{dt} = \alpha T\frac{dp}{dt} + \nabla\cdot(\lambda\nabla T) + q_{source} + \frac{1}{\sigma}J^2. \tag{9.55}$$

If E_I and Q_{source} are the integrals of ρe and q_{source} over the core, (9.54) yields

$$\partial_t E_I = -W_c + Q_{source} + Q_{Joule} + \int_{core\ boundary} \lambda n \cdot \nabla T \, dA. \qquad (9.56)$$

Combining (9.53) and (9.56), and writing $E = E_K + E_P + E_M + E_I$, the basic energy balance is expressed as follows:

$$\partial_t E = W_p + Q_{source} + \int_{core\ boundary} \lambda n \cdot \nabla T \, dA. \qquad (9.57)$$

Taking the time average $\langle\cdot\rangle$ of the terms in (9.53) and (9.57), we obtain balance relations as indicated in Figure 9.5. We regard the convecting core as consisting of a heat engine powering the dynamo. The two work terms $\langle W_p \rangle$ and $\langle W_c \rangle$ drive the dynamo, but the two systems are coupled since the Joule heat is returned to the engine. It is precisely this point which distinguishes thermal from nonthermal convection, although the efficiencies of various convective systems depend upon details not revealed in the balance equations.

Efficiency of heat exchanges in the core. To measure the efficiency of the thermal engine in Figure 9.5, we assume $W_p = 0$ and take the efficiency to be the ratio η of the Joule heating to the heat supplied to the system. The total heat supplied, Q_0, equals Q_{source} plus the heat Q_b supplied to the fluid core from the inner core, so that $\eta = Q_{Joule}/Q_0$.

Following Backus (1975) and Hewitt *et al.* (1975), an expression for η can be obtained by taking the convection to consist of many Carnot cycles: a fluid parcel (label i) absorbs heat Q within the core at a temperature

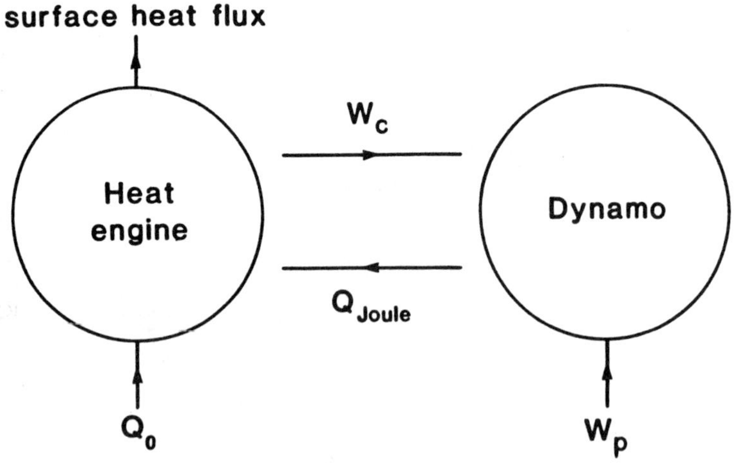

Fig. 9.5. Schematic energy balance for the geodynamo.

T_i and gives it up at the core-mantle interface, at a temperature T_0 say. Since entropy does not decrease in the process,

$$\sum_i \frac{Q_i}{T_i} \leq \frac{Q_0}{T_0} . \tag{9.58}$$

If $Q_i = Q_{Joule_i} + Q_{0_i}$ is the decomposition of each Q_i into parts belonging to Q_{Joule} and Q_0, and if T_{max} is the maximum of the T_i, then (9.58) implies

$$\frac{Q_0 + Q_{Joule}}{T_{max}} \leq \frac{Q_0}{T_0} ,$$

or

$$\eta \leq \frac{T_{max}}{T_0} - 1. \tag{9.59}$$

T_{max} is not known with certainty, but estimates of the efficiency of the thermal engine lie between 1/100 and 1/10. This is a rather low figure, which is not consistent with the estimated magnitude of the core field and estimates of Q_0.

For a Boussinesq theory of convection it is possible to give a sharper bound on $\langle Q_{Joule} \rangle$. If the heat sources are concentrated uniformly over a spherical fluid core, so that $Q_0 = Q_{source}$, it can be shown that

$$\langle Q_{Joule} \rangle \leq \frac{\beta}{5} \langle Q_0 \rangle . \tag{9.60}$$

Here β is the ratio of the (outer) radius of the spherical core to the temperature scale height $c_p/\alpha g$, and thus $\eta \leq \beta/5$. The catch is this: Boussinesq theory is valid only under certain parameter restrictions, one of which is that $\beta \ll 1$. If the latter is met, then Joule heating may be neglected in the energy equation, but thermal efficiency is necessarily low (Backus, 1975; Hewitt *et al.*, 1975).

Of course, for nonthermal Boussinesq convection these constraints are not the same. The meaning of β in (9.60) has nothing to do with the heat balance, and we only need density changes to be small. Low efficiencies are also found to be associated with precessional driving, cf. Rochester *et al.* (1975), and so the most plausible possible mechanism for sustaining the dynamo would seem at present to be nonthermal convection.

The gravitationally powered dynamo. In a recent series of papers, Loper and Roberts (1978, 1980, 1981) have examined in detail a model of convection based upon changes in the composition of the core material (see also Soward, 1983, Ch. 6). The proposal that such a process might be relevant to dynamo theory was first made by Braginsky (1964b), and more recently has been studied by Gubbins (1977). A key feature of this model is the linking of the power supply to the release of gravitational potential energy associated with the growth of the solid inner core. Although thermal effects are shown to be relatively weak in the dynamics of the outer regions of the fluid core, heat flux from the core is closely associated with the process of solidification at the boundary of the inner core. This phase change, in effect, produces relatively light material, whose floating up releases gravitational potential energy.

Principally because of the different physical properties of the light fraction of the iron alloy invoked in the process, compared with the temperature field in a thermally-driven dynamo, the gravitationally powered dynamo is considerably more efficient and estimates of available power appear to be quite sufficient to account for the observed magnetic field. Further developments of this model are awaited with great interest.

9.7. Bibliographic Notes

General. Portions of this chapter are based upon Childress (1977; 1978, Lectures 6 and 10; 1982b) and Moffatt (1978, Ch. 12). See also the recent review paper on convection-driven dynamos by Soward, in Soward (1983), pp. 237-244.

Sections 9.1, 9.2. We have not, in this chapter, taken up the general problem of obtaining bounds on the performance of hydromagnetic dynamos, analogous to estimates of heat flux in classical thermal convection. One such result follows easily from (9.60), (9.52), and (7.26), leading to an estimate of magnetic energy realized by a given heat input (see Childress, 1977). Other results along these lines are given by Kennett (1974), Proctor (1979), and Soward (1980). In view of the serious analytical obstacles encountered in strong-field theory, the bounding approach probably deserves further scrutiny.

We have not dealt explicitly with numerical simulation of hydromagnetic dynamos in a sphere or spherical shell from the primitive equations, although such calculations are now possible. Cuong and Busse (1981) study the dynamo produced in a thick spherical shell, and compare the results with the annulus model. They find that dipolar fields are preferred for sufficiently large Rayleigh numbers, but that the sign of the dominant helicity is opposite to that of the annulus model. Recently Gilman (1983) has extended the computations of Gilman and Miller (1981) for dynamo action in the solar convective zone. By reducing the viscosity and thermal diffusivity by a factor of 10, the fraction of the kinetic energy contained in the large-scale differential rotation (which creates the "ω-effect") is increased from 30% to 80%. There remains a poleward migration of fields in disagreement with solar observations. It is suggested that the effects of compressibility may be important in setting up a more realistic, equatorward migration pattern.

Section 9.4. Soward and Jones (1983) have considered a macrodynamic model in which both α-effects, as well as a geostrophically controlled ω-effect operate. The system is driven by a prescribed body force. Both Lorentz forces and Ekman suction contribute to determine the geostrophic motion. The interesting new results of the paper concern the appearance of states which are essentially viscosity-independent and which satisfy Taylor's constraint. These solutions, which are obtained by nonlinear bifurcation analysis in a slab geometry, thus establish analytically the existence of the Taylor states suggested by the numerical results of Proctor (1977a). It would be very interesting to see whether or not the approach used by Soward and Jones (1983) could be applied to Braginsky's Model Z.

PART IV

THEORETICAL CLIMATE DYNAMICS

CHAPTER 10

RADIATION BALANCE AND EQUILIBRIUM MODELS

Climate dynamics is a relatively new member of the family of geophysical sciences. Descriptive climatology goes back, of course, at least to the ancient Greeks, who realized the importance of the Sun's mean zenith angle in determining the climate of a given latitude belt, as well as that of land-sea distribution in determining the regional, zonally asymmetric characteristics of climate. The general human perception of climate change is also preserved in numerous written records throughout history, starting with the floods described in the epic of Gilgamesh and in the Bible.

Only recently did the possibility of global climate monitoring present itself to the geophysical community, through ground-based observational networks and space-borne instrumentation. This increase in quantitative, detailed knowledge of the Earth's current climate was accompanied by the development of elaborate geochemical and micropaleonotological methods for sounding the planet's climatic past.

Observational information about present, spatial detail, and about past, temporal detail were accompanied in the 1960s by an increase of computing power used in the processing of climatic data, as well as in the modeling and simulation of the seasonally varying general circulation, as outlined in Sections 4.1 and 4.6. The knowledge thus accumulated led to an increase of insight which was distilled in simple models, in an attempt to analyze the basic ingredients of climatic mechanisms and processes.

In the following three chapters, we describe a few simple models, and try to convey the flavor of the new, theoretical climate dynamics. As in every area of the exact sciences, the fundamental ideas suggested by simple models have to be tested by further observations and detailed simulations of the phenomena under study. We hope that this description of preliminary, theoretical results will stimulate the comparisons and verfications required to develop the theory further.

10.1. Radiation Budget of the Earth

The major characteristics of a physico-chemical system, such as the climatic system, are given by its energy budget. The climatic system's energy budget is dominated by the short-wave radiation, R_i, coming in from the Sun, and the long-wave radiation, R_0, escaping back into space. The approximate balance between R_i and R_0 determines the mean temperature of the system. The distribution of radiative energy within the system, in height, latitude and longitude, determines to a large extent the distribution of climatic variables, such as temperature, throughout the system.

We consider in this section first the global radiation balance, then its variation with latitude. In Section 10.2, we study a spatially zero-dimensional (0-D) model of radiation balance, with mean global temperature as the only variable. The dependence of the solar radiation's reflection on temperature, the so-called ice-albedo feedback, and the dependence of infrared absorption on temperature, the greenhouse effect, are discussed. Stationary solutions of this model and their linear stability are investigated.

The necessary physical concepts involved in horizontal heat exchanges within the system are introduced next. A model for zonally-averaged, latitudinally-dependent sea-level temperature is formulated. The analysis of equilbrium solutions and of their linear stability is carried out for this slightly more complex, spatially one-dimensional (1-D) model.

In Section 10.3, nonlinear stability of the 0-D model's multiple equilibria is discussed, using a variational principle. The geometric picture of attractor basins in the system's phase space thus obtained is helpful in analyzing stochastic perturbations of the equilibria for the 0-D model. Finite residence times obtain near each equilibrium. The expected length of the residence time near a given equilibrium depends on the global structure of the attractor basin about that equilibrium.

In Section 10.4 we consider some simple modifications of energy-balance models which lead to nonstationary solutions. First, an albedo dependence on cloud amount is introduced into the previous, spatially 1-D model. This dependence increases the number of the model's equilibrium solutions, while the stability properties of the pre-existing equilibria are left unchanged. Next, periodic forcing and stochastic perturbations are introduced. Their combined effect produces a considerable amplification of the deterministic forcing due to the random noise.

Finally, a delay is introduced into the 1-D model. It represents rather crudely a time lag in the ice-albedo feedback due to the slow response of continental ice masses to regional temperature changes. The additional degrees of freedom associated with this delay lead, in the absence of forced variations or noise, to periodic and aperiodic solutions. The aperiodic solutions' Fourier spectra show a qualitative resemblance to those of many climatic time series, on scales from hundreds to millions of years (see for instance, Figs. 11.1 and 11.3): a few broad peaks superimposed on a continuous background, with spectral density increasing as frequency decreases.

In Section 10.5, references are given for the precursor and companion disciplines of climate dynamics, descriptive and physical climatology. Recent articles on rocket and satellite measurements of radiative fluxes are mentioned.

Articles important in the development of energy-balance models, and reviews of the literature, are cited next. The question of the field of motion in models without explicit velocity variables is touched upon. The study of climate sensitivity using equilibrium and nonequilibrium models is briefly reviewed.

References for the calculus of variations, global analysis and stochastic differential equations follow, along with appropriate articles on applications to climate dynamics. Finally, there are references for delay-differential equations and spectral analysis.

Global radiation budget. The discussion of radiation budgets in this section is based on London and Sasamori (1971; to be referred to as LS throughout). Figure 10.1 shows the annually and globally averaged radiation budget for the earth-atmosphere system.

The method of calculation used in LS made the budget components relatively insensitive to the actual value of the solar radiative flux normally incident at the top of the atmosphere, along a straight line connecting the Earth and the Sun. This value, the so-called solar constant, S.C., is currently believed to be 1370 Wm^{-2}, to within $\pm 2Wm^{-2}$. The quantity $Q_0 = (1/4)$ S.C. is the value of the incoming solar radiative

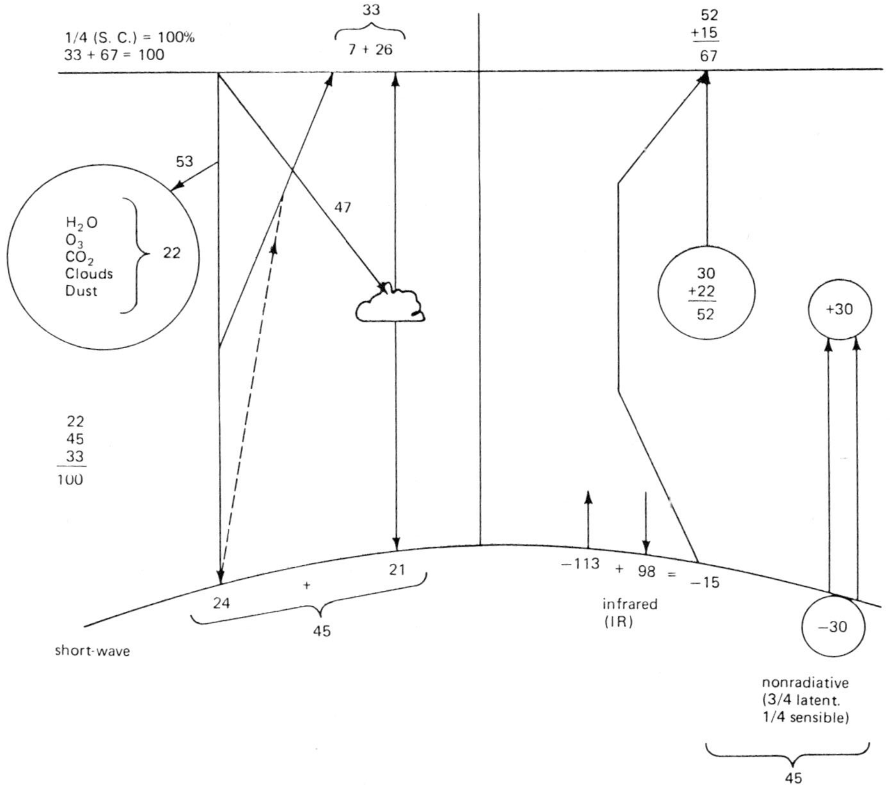

Fig. 10.1. Globally and annually averaged radiation budget of the Earth (after Ghil, 1981a).

flux averaged over the year and over the surface of the Earth. The factor 1/4 results from the Earth's sphericity: the surface area over which the incoming radiation has to be averaged in $4\pi a^2$, a being the radius of the Earth, while the total amount of solar radiation intercepted by the atmosphere is $\pi a^2 Q_0$ per unit time.

The radiation coming in from the Sun is dominated by short-wave components due to the Sun's very high temperature. This radiation is either absorbed by the atmosphere (22%), or transmitted to and absorbed by the ground (45%), or reflected back to space (33%). The short-wave radiation balance appears on the left of Figure 10.1.

A particularly important role in both the absorption and reflection of solar radiation is played by clouds: they cover on the average 50% of the earth's surface. The average reflectivity of the earth-atmosphere system is called the planetary *albedo* and it has an approximate value of 0.33. This value includes the contribution of clouds. The radiation reflected at the earth's surface also contributes to the total short-wave radiation returned to space at the top of the atmosphere. The surface albedo, i.e., the amount of radiation reflected as a fraction of the radiation received ($0.45\ Q_0$) at the surface, is on the average considerably less than 0.33, but it varies widely with location, depending on the nature of the surface.

The short-wave radiation absorbed by the atmosphere throughout its depth (22%), and by the lithosphere and hydrosphere at the surface (45%), heats these parts of the climatic system. They cool off by emitting long-wave, infrared (IR) radiation. The net IR emission at the surface is 15% and the emission by the atmosphere out to space is 52%; together they total 67%. Thus the radiative flux of the system to outer space is made up of 33% short-wave and 67% long-wave radiation. The IR radiation balance appears at center-right of Figure 10.1.

To equilibrate the radiative tranaactions at the ground and throughout the atmosphere, we have to include the nonradiative fluxes on the right of Figure 10.1. Sensible heat (7%) and latent heat (23%) flow from the surface into the atmosphere. The combined nonradiative flux of 30%, together with 15% emitted at the surface as net IR radiation, balances the 45% received at the surface in short-wave flux. Together with the 22% solar radiation absorbed by the atmosphere, this 30% makes up the 52% radiated by the atmosphere to space.

The numerical values given in Figure 10.1 are continually being revised due to new data provided by operational and by experimental meteorological satellites. The purpose of the present discussion is merely to understand qualitatively the different radiative and nonradiative fluxes involved in the earth-atmosphere's global energy budget, and not to provide the most accurate or most recent numbers.

An important consequence of this discussion is that, within the accuracy of the available data and calculations, the climatic system is close to radiative equilibrium, when considering global and anually-averaged quantities. This equilibrium determines global temperature, which makes the Earth a relatively pleasant place to live, on the average. A very simple model for this mean temperature will be dealt with in the following section.

Local radiative imbalances make certain zones more pleasant to live in than others. This is our concern in the remainder of the present section.

Local imbalances and meridional fluxes. It was already clear to the ancient Greeks that the inclination (κλισις) of the Sun, which changes with latitude, was the most important factor in determining the climate (κλιμα) of a zone, or latitude belt. In Figure 10.2 we show the latitudinal distribution of annually and zonally averaged radiation, absorbed and emitted, for the earth-atmosphere system.

The absorbed flux, R_i, which averages 0.67 Q_0, falls off sharply from the Equator towards the poles, with the mean annual zenith angle. The emitted flux, R_0, which has approximately the same average as R_i, originates mostly within the atmosphere (0.52 Q_0 on the average), rather than at the surface (0.15 Q_0). Free air temperature, at mid-troposphere say, is observed to be a much weaker function of latitude than the Sun's zenith angle. Hence the outgoing radiation R_0, emitted according to the Stefan-Bolzmann law, is much more constant from Equator to Pole than the absorbed one. This results in a local excess of mean-annual, radiative energy in the tropics (33°N to 37°S, approximately), and in a deficit at higher latitudes. As a consequence, the excess has to be carried off from low to high latitudes, by the atmosphere and by the oceans.

The average annual energy transport required by local energy-balance considerations is a horizontal flux from one latitude belt to the next, F, which is taken positive northward. If we assume an interannually stationary climate, the local excess of energy is the divergence of this flux,

$$R_i(\phi) - R_0(\phi) = \frac{1}{a \cos \phi} \frac{\partial F}{\partial \phi}, \tag{10.1}$$

where ϕ is latitude.

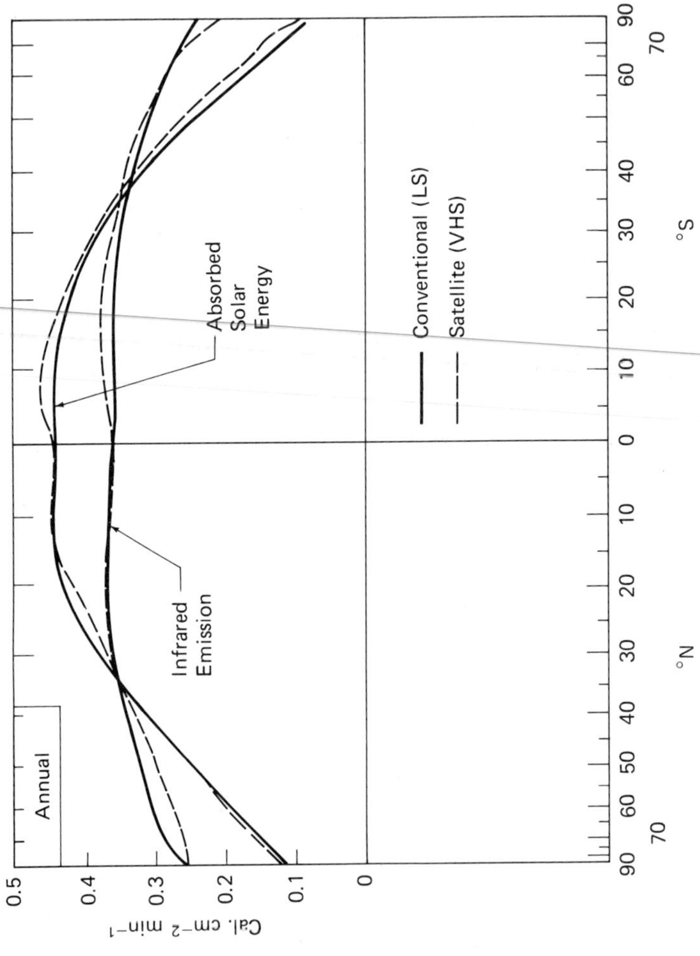

Figure 10.2. Zonally and annually averaged radiation budget of the Earth (after Ghil, 1981a).

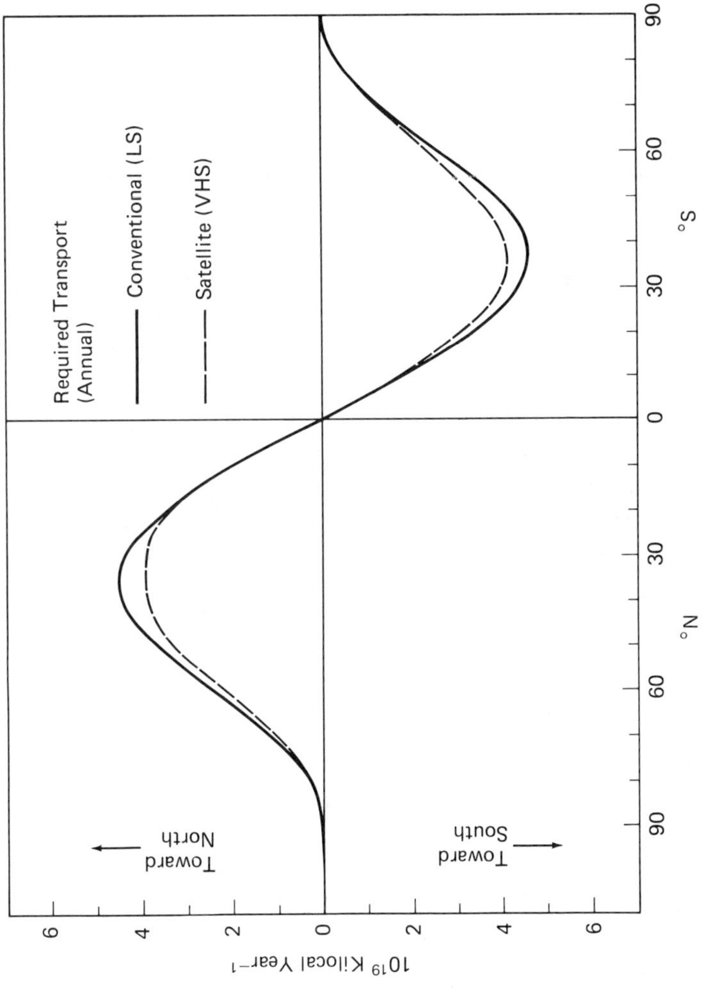

Fig. 10.3. Meridional heat fluxes balancing the radiation budget of the previous figure (after Ghil, 1981a).

Numerical integration, according to Eq. (10.1), of the difference between the two solid or the two dashed curves in Figure 10.2 yields the flux values shown in Figure 10.3. The level of current observational uncertainty is reflected by the difference between the solid curves and the dashed curves in either figure; the respective curves are based on data sets obtained from different observing systems about fifteen years apart. It becomes clear, therefore, that the problem of climatic change, i.e., of the difference on various time scales between the corresponding averages of the two sides of Eq. (10.1), is difficult: climatic change involves the small difference of two large, poorly observed and poorly understood quantities.

The energy transport in Figure 10.3, assuming stationarity, achieves the temperature distribution which resulted in the curve for R_0 in Figure 10.2. In fact both the latitudinal distribution of R_i and R_0 depend on properties internal to the system: cloud distribution, surface albedo, temperature. To understand better these interdependencies and the climatic features they determine, we turn now to the formulation and analysis of some models.

10.2. Energy-Balance Models (EBMs): Multiple Equilibria

The variable perceived most widely as defining climate is temperature, T. It is also most important in determining the components of the radiation balance. We start therefore with the simplest model of climate, one for the annually averaged temperature of the earth-atmosphere system, $\overline{T}$.

<u>A model for global temperature</u>. The equation governing the model is:

$$c \frac{d\overline{T}}{dt} = Q_0\{1 - \alpha(\overline{T})\} - \sigma g(\overline{T})\overline{T}^4. \tag{10.2}$$

It expresses the approximate radiation balance between absorbed radiation

$$R_i = Q_0\{1 - \alpha(\overline{T})\}, \tag{10.3a}$$

and emitted radiation,

$$R_0 = \sigma g(\overline{T})\overline{T}^4. \tag{10.3b}$$

Q_0 is the global mean solar radiative input, α the planetary albedo, σ the Stefan-Boltzmann constant and g the grayness of the system, i.e., its deviation from black-body radiative emission, $\sigma\overline{T}^4$; α and g are functions of $\overline{T}$. Any slight imbalance between R_i and R_0 leads to a change in the temperature of the system, at the rate $d\overline{T}/dt$, t being time; c is the heat capacity of the system, whose heat storage is $c\overline{T}$. This type of model derives from the work of Budyko (1969) and Sellers (1969), and has been investigated extensively since.

In principle, the dependence of α on $\overline{T}$ should express the change in both cloud and surface albedo with $\overline{T}$. But the dependence of cloud albedo on $\overline{T}$ is still not well understood, not even to the extent of knowing whether, or under which circumstances, it increases or decreases with $\overline{T}$. On the other hand, surface albedo varies most strongly with the presence or absence of snow and ice, on land or sea. Hence we take α to decrease linearly with $\overline{T}$, as ice cover decreases over part of the Earth, and to be constant for all $\overline{T}$ for which the Earth would be either entirely ice covered, $\overline{T} < T_\ell$, or ice free, $\overline{T} > T_u$. The mechanism so described is the *ice-albedo feedback*.

The dependence of emissivity σg on $\overline{T}$ has to express the *greenhouse effect*, i.e., the process by which outgoing IR radiation is partly absorbed by trace gases in the atmosphere, and by clouds. We choose

$$g(\overline{T}) = 1 - m \tanh (\overline{T}/T_0)^6, \qquad (10.3c)$$

with $m = 0.5$, for 50 percent cloud cover, and $T_0^{-6} = 1.9 \times 10^{-15} K^{-6}$, determined empirically by Sellers (1969). This expression indicates that, as temperature increases, g decreases, so that the greenhouse effect becomes stronger. The shapes of $\alpha(\overline{T})$ and $g(\overline{T})$ are reflected in the graphs of R_i and R_0 in Figure 10.4.

Stationary solutions and stability to perturbations. The intersections of $R_i(\overline{T})$ and $R_0(\overline{T})$, or those of $R_i - R_0$ with the $\overline{T}$-axis, determine the steady states, or *equilibria*, of model (10.2). We notice in Figure 10.4 that, for α and g as we have taken them, there are three stationary solutions of Eq. (10.2): T_1, T_2 and T_3. Hence, even as simple a model as the present one shows the possible existence of more than one climate, if we are willing to interpret the equilibria of the model as steady-state climates of the Earth; T_1 would represent the present climate, and T_2, T_3 would be colder climates, possibly ice ages.

In the real world, small deviations from a given temperature regime always appear, due to a multitude of mechanisms not included in the model. It is important, therefore, to investigate the stability of the model's solutions to small perturbations. Assume for instance that at time

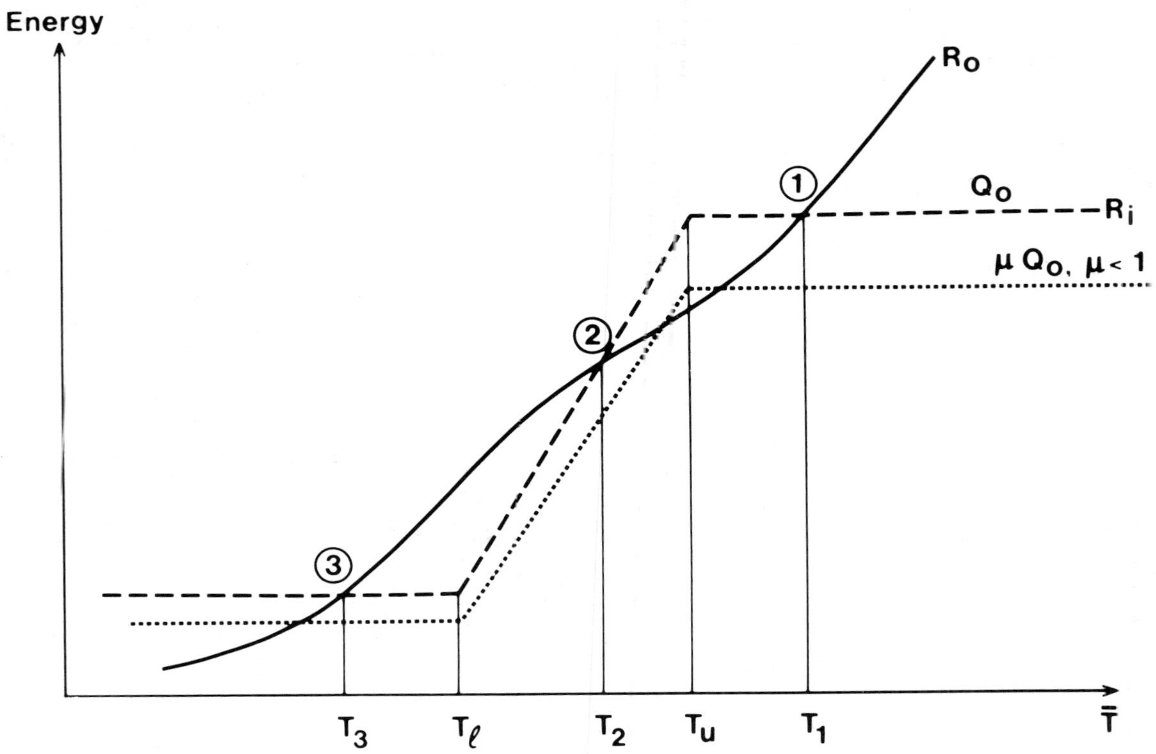

Fig. 10.4. Outgoing radiation R_0 (solid) and absorbed incoming radiation (dashed) for a global climate model (from Ghil, 1981a). Notice the three stationary solutions $\bar{T} = T_1, T_2, T_3$.

$t = 0$, $\overline{T}(0) = T_1 + \theta_0$. We are interested in knowing how the deviation $\theta(t)$ of $\overline{T}$ from T_1, which is initially equal to θ_0, evolves in time.

Assuming that $\theta(0) = \theta_0$ is suitably small, it suffices at first to consider a linearized equation for $\theta(t)$, obtained by expanding (10.2) in $\theta = T_1 - \overline{T}$:

$$c\frac{d}{dt}(T_1 + \theta(t)) = R_i(T_1 + \theta) - R_0(T_1 + \theta)$$
$$= R_i(T_1) - R_0(T_1) + \left\{\left.\frac{dR_i}{d\overline{T}}\right|_{\overline{T}=T_1} - \left.\frac{dR_0}{d\overline{T}}\right|_{\overline{T}=T_1}\right\}\theta + \cdots .$$

Using $dT_1/dt = 0 = R_i(T_1) - R_0(T_1)$ and neglecting the terms of higher order in θ indicated by the dots, the expansion above yields

$$\frac{d\theta}{dt} = \lambda_1\theta, \tag{10.4a}$$

where

$$\lambda_1 = -(Q_0\alpha_1 + \sigma g_1)/c, \tag{10.4b}$$

$$\alpha_1 = \left.\frac{d\alpha}{d\overline{T}}\right|_{\overline{T}=T_1}, \qquad g_1 = 4g(T_1)T_1^3 + \left.\frac{dg}{d\overline{T}}\right|_{\overline{T}=T_1} \cdot T_1^4. \tag{10.4c,d}$$

This is a linear ordinary differential equation (ODE) for the deviation $\theta(t)$ of $\overline{T}(t)$ from the equilibrium value T_1.

The solution of Eq. (10.4) is

$$\theta(t) = \theta_0 e^{\lambda_1 t}. \tag{10.5}$$

Hence θ will grow exponentially if $\lambda_1 > 0$ and decay to zero with time if $\lambda_1 < 0$. Thus T_1 is linearly *stable* if $\lambda_1 < 0$, and *unstable* if $\lambda_1 > 0$. A similar analysis holds for T_2 and T_3.

In Figure 10.4, an equilibrium T_j, $j = 1,2,3$, can be seen to be stable if the corresponding slope of $R_i(\overline{T}) - R_0(\overline{T})$ is negative at $\overline{T} = T_j$, and unstable if it is positive. It follows that T_1 and T_3 are stable, while T_2 is unstable. Indeed, near T_1, we have $d\overline{T}/dt < 0$ when $\overline{T} > T_1$, causing a decrease of $\overline{T}$ with time, while $d\overline{T}/dt > 0$ when $\overline{T} < T$, causing an increase. In other words, the energy balance (10.2) tries to restore $\overline{T}$ to its equilibrium value T_1; the same situation occurs for T_3. Near T_2, quite the opposite happens: once $\overline{T} > T_2$ it will increase further, while $\overline{T}(0) < T_2$ will lead to further decrease of $\overline{T}$, away from T_2.

This stability discussion suggests that it is reasonable to identify T_1 with the present climate; T_2, however, is not a good candidate for a colder climate, such as an ice-age, since it is unstable and could never have persisted for any length of time. The equilibrium T_3 corresponds to a completely ice-covered earth, referred to colloquially as the "deep freeze" solution. While the model shows it to be stable, it has never been observed in the paleoclimatological record of the past; we hope it will not occur in the future either.

Structural stability. It remains to be seen how the existence of one or more solutions for Eq. (10.2), and their stability, can be affected by changes in the model's parameters. The most important among these is the value of the solar input, Q_0. It could change as a result of variation in the Sun's energy output, in its distance from the Earth, or equivalently, in the atmosphere's optical properties. Within our simple model, either one of these changes can be expressed by replacing Q_0 in (10,2, 10.3a) with μQ_0, and taking the insolation parameter $\mu \neq 1$. Such a change will leave R_0 as it is in Figure 10.4, but the graph for μR_i, $\mu \neq 1$, will be either above or below the curve representing R_i.

In Figure 10.4, we have drawn the situation for a certain value of μ less than 1, i.e., solar input smaller than that corresponding to present-day conditions. Clearly, T_1 and T_2 in this new situation lie closer to T_u, while T_3 lies more to the left of its value for $\mu = 1$. If μ were to decrease further, a situation would obtain in which the graph of μR_i just touches that of R_0, at $T_2 = T_u = T_1$. Let the corresponding value of μ be μ_c. For $\mu < \mu_c$, the solutions T_1 and T_2 of Eq. (10.2) disappear altogether, and T_3 is the only solution left.

Recalling now that T_1 stands for the present climate within the model, it follows that while the solar input μQ_0 decreases, the Earth's climate cools off slowly at first, as T_1 moves to the left. Then, as μ crosses the value μ_c, the temperature of the system would have to decrease dramatically to that of a completely ice-covered earth, T_3. It is this type of result that attracted attention to the original models of Budyko and of Sellers, as well as the possibility of human activities reducing μ in the future, more than natural events had done in the past.

In the opposite situation, of μ increasing, the curve μR_i moves up. Then T_1 moves to the right, while T_2 and T_3 approach T_ℓ. As μ increases through and beyond a critical value μ_d, T_2 and T_3 coalesce, then disappear, while T_1 keeps increasing. This situation is therefore less dramatic, although even a gradual increase of T_1 might eventually lead to a climate much less desirable than the present one.

It should be noticed that for any value of μ, the analysis of the stability for the model's equilibria can be carried out as it was for $\mu = 1$. Indeed, Eqs. (10.4) become

$$\frac{d\theta}{dt} = \lambda_j(\mu)\theta, \tag{10.6a}$$

$$\lambda_j(\mu) = -\{\mu Q_0 \alpha_j(\mu) + \sigma g_j(\mu)\}/c, \tag{10.6b}$$

$$\alpha_j(\mu) = \left.\frac{d\alpha}{d\overline{T}}\right|_{\overline{T}=T_j(\mu)}, \tag{10.6c}$$

$$g_j(\mu) = 4g(T_j(\mu))T_j^3(\mu) + \left.\frac{dg}{d\overline{T}}\right|_{\overline{T}=T_j(\mu)} \cdot T_j^4(\mu), \tag{10.6d}$$

where $T_j(\mu)$, $j = 1,2,3$, is the position of solution T_j as μ increases or decreases. Clearly, each T_j is a continuous function of μ near $\mu = 1$, and therefore $\alpha_j(\mu)$ and $g_j(\mu)$ are continuous, so that λ_j depends continuously on μ near $\mu = 1$.

In particular, the sign of λ_j will not change near $\mu = 1$. The sign of λ_j is, in fact, the same as that of the slope of the curve $R_i(\overline{T};\mu) - R_0(\overline{T};\mu)$ in Figure 10.4, whether the curve moves with μ or not. This suggests, and careful evaluation of Eqs. (10.6b,c,d) confirms, that the sign of $\lambda_j(\mu)$ stays the same until the solution $T_j(\mu)$ coalesces with another solution, or until it disappears. Thus, T_1 is stable to small perturbations θ_0 in $\overline{T}$ for $\mu > \mu_c$, T_3 is stable for $\mu < \mu_d$, and T_2 is unstable for $\mu_c < \mu < \mu_d$. To summarize, T_1 and T_3 are stable, and T_2 is unstable, over the entire μ-interval for which they exist.

We have just considered the way in which the number of solutions and their *internal* stability to small perturbations in initial conditions changes when one of the model's parameters is changed. It is customary to call these properties the model's *structural* or external stability. A more precise definition will be given below, in connection with Figure 10.6.

Horizontal heat transport. We have seen in Section 10.1 that the radiation balance of the earth-atmosphere system changes with latitude ϕ (Figure 10.2). This latitudinal dependence of $R_i - R_0$ gives rise to and is maintained by zonally averaged heat fluxes, $F(\phi)$ (Figure 10.3). In the first part of this section, a spatially zero-dimensional (0-D) model was formulated for the globally averaged radiation balance, corresponding to Figure 10.1. In a similar way, we wish to formulate now a model for the latitudinally-dependent energy balance of Figures 10.2 and 10.3. The results obtained with the 0-D model (10.2) will guide us in the study of the following, spatially one-dimensional (1-D) model.

To formulate such a model, we need to consider horizontal heat transfers, in the same way in which the discussion of Figure 10.1 concerned itself with vertical heat transfers. There are three kinds of heat transfer: radiative, conductive and convective. In *radiative* transfer, energy passes between the medium's molecules by electromagnetic radiation due to photon emission. Radiative transfer plays an important role in the vertical distribution of temperature, which will not be discussed here. It is negligible in horizontal heat transport, compared to the following modes of heat transfer, which are more efficient for horizontal motions.

In *conduction*, energy passes from one molecule to another by thermal agitation. It is transported, therefore, from parts of the system with higher temperature to those where temperature is lower, without the medium's motion contributing to the transport. The corresponding heat flux F_c can be taken in many cases as being proportional to the local temperature difference, or gradient,

$$F_c = -k_c \nabla T. \tag{10.7}$$

In Eq. (10.7) ∇ is the gradient operator (see Eq. (1.2)), and k_c is the so-called conduction, or diffusion, coefficient; depending on the system, k_c can be constant, a function of position, of T itself or of other quantities.

In heat transfer, (10.7) is often called Fourier's law. In the diffusion of trace constituents, a similar formula is also valid, and called Fick's law. Notice that for a system in *thermodynamic equilibrium*, $T \equiv$ const. or $\nabla T \equiv 0$. When we speak of an equilibrium or equilibrium solution, we only mean a steady state of the system or a stationary solution of its governing equation.

Conduction is the main form of heat transport in solid media, such as the lithosphere. In a solid, in the absence of internal sources of energy, temperature will change in time only due to the convergence or divergence of the conductive heat flux at a given location,

$$c \frac{\partial T}{\partial t} = -\nabla \cdot F_c = \nabla \cdot k_c \nabla T; \tag{10.8}$$

here $\nabla\cdot$ is the divergence operator.

In fluid systems, the medium often is in motion, and thermal energy of the molecules is carried along with this motion. This type of heat transfer is called in general in fluid dynamics *convective* (see also Chapters 4, 5, 7 and 9). In meteorology and oceanography, the term convection is mostly reserved for transfer by vertical, small-scale motions of the fluid, while large-scale, horizontal motions are said to *advect* heat. The advective heat flux, $F_a = -vT$, modifies the local temperature T of the fluid according to the equation

$$c\,\frac{\partial T}{\partial t} = -\nabla\cdot F_a = \nabla\cdot vT, \tag{10.9}$$

v being the velocity of the fluid which carries internal energy at the temperature T with it. For large-scale, slow geophysical motions $\nabla\cdot v \cong 0$ (see Chapter 3) and $-\nabla\cdot F_a \cong v\cdot\nabla T$.

It is clear from (10.9) that atmospheric and oceanic dynamics play an important role in the climatic system's local energy balance and temperature distribution T via the velocity field v. To determine v at the same time as T is a considerably more complicated task than computing T alone. Also, more is known about T than about v for climates different from the present one.

When considering only the largest, planetary scales of the temperature field, $O(10^4$ km), and time scales longer than months and years, it is reasonable to attempt to eliminate the velocity field from our considerations, by using the so-called eddy diffusive approximation,

$$-\nabla\cdot F_a \cong \nabla\cdot k_e\nabla T. \tag{10.10}$$

In Eq. (10.10), k_e is an *eddy diffusion* coefficient. This type of approximation is used in many areas of fluid dynamics, with k_e being usually much larger than k_c. It often gives acceptable qualitative results when the most interesting time and space scales of T are much larger than the spatial extent and life span of the fluid motions which have the highest velocities. This is the case when studying planetary-scale, long-term climate change, given the typical scales of atmospheric and oceanic eddies, $O(10^2 - 10^3$ km) and O(weeks-months), respectively (see Chapter 4). Notice that, like k_c, k_e can be a function of position, T itself, and furthermore of ∇T. In the sequel, the zonally-averaged, meridional heat fluxes will be expressed by combining Eqs. (10.7-10.10) in

$$-\nabla\cdot F = \nabla\cdot k\nabla T; \tag{10.11}$$

here $F = F_c + F_a$, $k = k_c + k_e$, and k is taken to be a function of latitude ϕ alone, $k = k(\phi)$.

A one-dimensional model for surface temperature. The energy balance of a zonal slice of the climatic system, located at latitude ϕ and extending to the top of the atmosphere and down to a prescribed depth in the oceans and the continents, is governed by the equation

$$C(\phi)\frac{\partial T}{\partial t} = \frac{1}{\cos\phi}\frac{\partial}{\partial\phi}[k(\phi)\cos\phi]\frac{\partial T}{\partial\phi} + \mu Q_0 S(\phi)\{1-\alpha(\phi,T)\} - \sigma g(\phi,T)T^4. \tag{10.12}$$

This equation summarizes our discussion of predominantly radiative, vertical heat fluxes and of the predominently advective, horizontal heat fluxes. Here T is the sea-level temperature of the air, $T = T(\phi,t)$.

In comparing (10.12) with (10.2), we notice that R_i and R_0 have become functions of ϕ and T, rather than of the single, global variable $\overline{T}$. Model (10.2) was spatially zero-dimensional (0-D) and hence governed by a nonlinear ODE. Model (10.12) is one-dimensional (1-D), the spatial dimension taken into account is latitude, and it is governed by a partial differential equation (PDE). The ratio $\partial T/\partial t$ is the rate of change of T in time at location ϕ, while $\partial T/\partial\phi$ is the rate of change of T in latitude at time t; they are related to each other by (10.12).

The zonal slice at ϕ exchanges energy $(R_i - R_0)$ with outer space and energy $(\nabla\cdot F)$ with adjacent slices; as a result its temperature changes at the rate $\partial T/\partial t$, inversely proportional to its heat capacity $C = C(\phi)$. The function $S(\phi)$ gives the distribution of solar radiation incident at the top of the atmosphere with latitude; its average is one, so that Q_0 is the same constant as in Eq. (10.2). The first term on the right expresses horizontal flux divergence $\nabla\cdot F$ in the meridional direction, according to (10.11).

Before proceeding with an analysis of solutions to (10.12), it is interesting to compare this model with other 1-D energy-balance models (EBMs). The original models of Budyko and of Sellers were one-dimensional and we shall set them up side by side in Table 10.1.

Here

$$\overline{T}(t) = \int_0^{\pi/2} T(\phi,t)\cos\phi\, d\phi, \tag{10.13}$$

α_M and α_m are albedo values assigned to ice-covered and ice-free surfaces, respectively, and T_s is the assumed temperature of the "ice margin"; notice that the Sellers formulation allows for a partially ice-free zone, $T_\ell < T < T_u$, rather than requiring a sharp ice margin. The

Table 10.1. Comparison of Budyko's and Sellers' models.

Heat Flux	Budyko	Sellers
$R_i = Q(1 - \alpha(T))$ Absorbed solar radiation, as a function of ice-albedo feedback	Step-function albedo $\alpha = \begin{cases} \alpha_M, & T < T_s, \\ \alpha_m, & T \geq T_s, \end{cases}$ $\alpha_M > \alpha_m$, $T_\ell \leq T_s \leq T_u$	Ramp-function albedo $\alpha = \begin{cases} \alpha_M, & T < T_\ell \\ \alpha_M - \frac{T-T_\ell}{T_u-T_\ell}(\alpha_M - \alpha_m) & T_\ell \leq T < T_u, \\ \alpha_m, & T_u \leq T \end{cases}$,
R_0 Outgoing IR radiation	Linear, empirical $A + BT$	Stefan-Boltzmann law with greenhouse effect $\sigma T^4\{1 - m \tanh(T^6/T_0^6)\}$
$\nabla \cdot F$ Horizontal flux divergence	Newtonian cooling $\kappa\{T(\phi) - \overline{T}\}$	Eddy-diffusive $\nabla \cdot \{k(\phi)\nabla T(\phi)\}$

constants A, B and κ in Budyko's formulation of IR flux and horizontal heat flux, like the other constants in Table 10.1, are determined by matching the computed fluxes to current climatic data. The explicit dependence of α and g in (10.12) on latitude ϕ is rather weak and was introduced by Sellers to better match the available data.

Held and Suarez (1974) pointed out that, in spite of its empirical simplicity, Newtonian cooling as used by Budyko led to certain counter-intuitive results, being compatible with infinitely many positions of the presumed ice margin. In the context of a simple, zonally-averaged EBM, the ice margin $\phi_s(t)$ of a solution is defined by

$$T(\phi_s, t) = T_s. \tag{10.14a}$$

If temperature is a monotone function of latitude from pole to equator, and

$$\partial T/\partial\phi \neq 0 \quad \text{at} \quad \phi = \phi_s, \tag{10.14b}$$

the ice margin thus defined is unique, in each hemisphere, for such a solution.

The indeterminacy of ice margin position in Budyko's model comes from the discontinuous character of its solutions. It can be removed by adding diffusion, which renders solutions of the modified model continuously differentiable. The 1-D EBM formulation which became most widely used is thus the one in heavy outline in the Table, which combines Budyko's step-function albedo and linear IR flux with Sellers' eddy-diffusive heat flux, taking moreover a spatially constant diffusivity (North, 1975a). Such a formulation can easily be solved analytically, by expanding $T(\phi,t)$ in Legendre polynomials with argument $y = \sin\phi$ and with time-dependent coefficients, and matching the two linear problems for the ice-free and the ice-covered portion of the Earth across the "ice margin". The analytical simplicity of North's formulation, along with the striking results of Budyko and Sellers, led to the great popularity of EBMs in the literature of the late 1970s.

We pursue here the slightly more complicated model (10.12), with ramp-function albedo and greenhouse effect according to Sellers. The main reason for doing so is to illustrate how careful numerical methods, combined with mathematically informed qualitative reasoning, can give the same detailed information about solution behavior as analytical methods. Such a combination will prove very fruitful for the subsequent material in this chapter and in the next two chapters, for which no exact analytic methods are available.

The analysis of the 1-D model (10.12) (Ghil, 1975, 1976) will follow the steps for the 0-D model (10.2) (Ghil, 1981a). Results will be compared at the end of this section with those of other 1-D formulations.

Stationary solutions. Eq. (10.12) is to be solved subject to boundary conditions of zero heat flux at pole and equator

$$\frac{\partial T}{\partial \phi} = 0 \quad \text{at} \quad \phi = 0, \pi/2. \tag{10.15a,b}$$

The condition at the pole is natural, while that at the equator is equivalent to assuming the symmetry of the two hemispheres; such a symmetry assumption is reasonable for the simplicity of the model under consideration (compare also Section 6.4). The appropriate initial condition is

$$T(\phi,0) = \overset{\circ}{T}(\phi). \tag{10.15c}$$

We start by considering present solar input conditions, $\mu = 1$. One expects that, if (10.2) is a reasonable first approximation to (10.12), Eq. (10.12) will also have steady-state solutions. This turns out to be so, and can be proven to be the case under rather general assumptions on the functions k, α and g in (10.12).

Computation of stationary solutions, using numerical methods of high, controlled accuracy (compare Sections 5.3 and 5.5), yields the three equilibria $T_1(\phi)$, $T_2(\phi)$ and $T_3(\phi)$ in Figure 10.5. The model solution $T_1(\phi)$ closely matches the data for the present climate (open circles in the figure). The "deep freeze" $T_3(\phi)$ has a mean temperature $\overline{T}_3$ about 100K below that of the present, $\overline{T}_1$, and a much smaller pole-to-equator temperature difference

$$\Delta T(t) = T(0,t) - T(\pi/2,t). \tag{10.16}$$

For the present climate, the midlatitude surface temperature difference $T_1(\phi = 30°N) - T_1(\phi = 60°N)$ is much larger than the tropical difference $T_1(\phi = 0) - T_1(\phi = 30°N)$. As stated at the end of Section 5.1, this is in fluid-dynamical agreement with the existence of a Hadley circulation in the tropical atmosphere and of a Rossby circulation in midlatitudes. The smaller-scale aspects of these circulations provide in turn eddy diffusion coefficients $k(\phi)$ which are much larger on the average in the tropics than in mid-latitudes (see Ghil, 1976, Figure 1b), rendering the thermodynamic aspects of the atmospheric general circulation consistent with its fluid-dynamic aspects.

It would be nice if these values of $k(\phi)$ could be derived, at least approximately, from the stability properties of the large-scale atmospheric and oceanic flow in the tropics and in mid-latitudes, respectively, and from the corresponding energy and enstrophy cascades (see Sections 4.6 and 6.6). In the absence of a satisfactory derivation from first principles it remains to note that larger values for the combined atmospheric and oceanic contribution to $k(\phi)$ in low latitudes are empirically necessary to obtain the flatness of the temperature profiles $T_1(\phi)$ there.

In order to compare the solutions of (10.12) with those of (10.2) first, we notice that both $T_1(\phi)$ and $T_2(\phi)$ straddle T_ℓ, the perennial snow-line temperature, as well as T_u, the snow-absence temperature. It should be remembered that (10.12) is still a model for annually averaged temperature; hence it is customary to take $T_\ell = -10°C \cong 263K$ and $T_u = 10°C \cong 283K$.

In model (10.2), one had $T_3 < T_\ell < T_2 < T_u < T_1$. This is in fact true of the mean equilibrium temperatures here, i.e.,

$$\overline{T}_3 < T_\ell < \overline{T}_2 < T_u < \overline{T}_1; \tag{10.17a}$$

actually $T_3(\phi)$ lies entirely below T_ℓ, $T_3(\phi) < T_\ell$. From the results

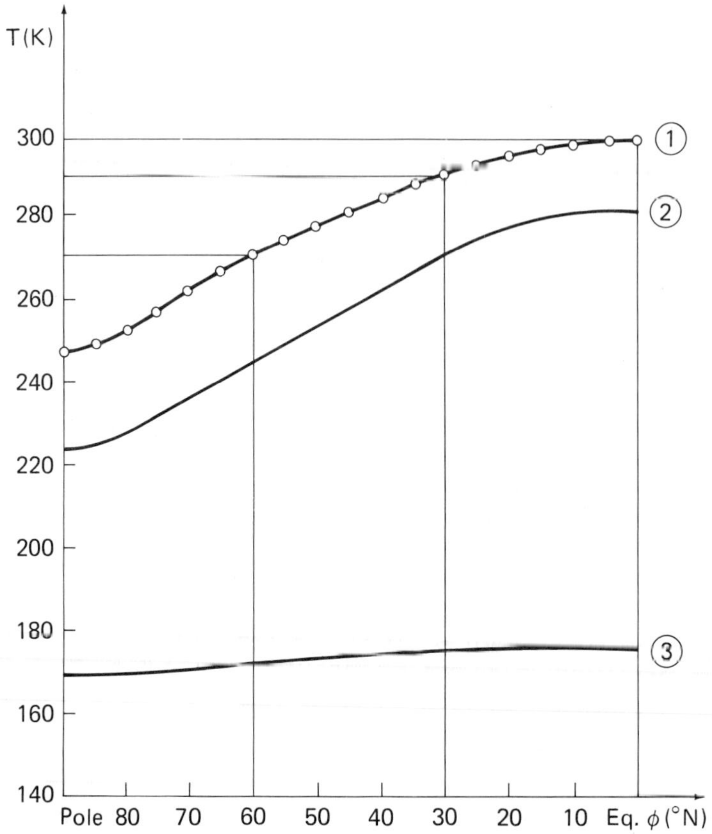

Fig. 10.5. Latitude dependence of the three equilibrium solutions of a zonally-averaged climate model (after Ghil, 1975). Open circles indicate observed temperature.

for (10.2) one might have been tempted to guess that

$$T_{\ell} < T_2(\phi) < T_u < T_1(\phi) \tag{10.17b}$$

also holds for all latitudes ϕ. This is however not the case, and Eq. (10.17b) is false, while (10.17a) is true.

The elementary discussion above illustrates both the usefulness and dangers of simple models. Eq. (10.2) led us to conjecture that (10.12) would have three equilibria, which proved correct. It also suggested the chain of inequalities (10.17a), which does hold for (10.12). However, model (10.12), in accordance with reality, does not satisfy (10.17b); the latter would have been an unrealistic interpretation of the results for model (10.2).

Internal stability. We are interested now in the stability to small perturbations of the equilibrium solutions to Eq. (10.12). Assume that, for some reason, at $t = 0$ the temperature in a latitude band would be slightly different from the "present climate", i.e.,

$$T(\phi,0) = T_1(\phi) + \varepsilon\theta_0(\phi), \tag{10.18}$$

for ε small and $\theta_0(\phi)$ arbitrary, but bounded. We would like to know how the temperature difference $\theta(\phi,t;\varepsilon) = T(\phi,t;\varepsilon) - T_1(\phi)$, initially equal to $\theta(\phi,0;\varepsilon) = \varepsilon\theta_0(\phi)$ evolves in time.

A linear PDE for $\theta(\phi,t)$ can be derived from (10.12) in a way similar to that in which the linear ODE (10.4) was derived from (10.2). Substitution of $T(\phi,t;\varepsilon) = T_1(\phi) + \theta(\phi,t;\varepsilon)$ into (10.12), and differentiation with respect to ε yields, at $\varepsilon = 0$,

$$\theta_t = -L_1\theta, \tag{10.19}$$

with $(\ )_t = \partial(\ )/\partial t$. Here L_1 is a linear operator obtained from the right-hand side of (10.12),

$$L_1\theta = -\frac{1}{r(x)}\{p(x)\theta_x\}_x + q(x)\theta, \tag{10.20a}$$

with x normalized co-latitude, $x = (\pi/2 - \phi)/(\pi/2)$, so that $x = 0$ at the pole and $x = 1$ at the equator, and $(\ )_x = \partial(\ )/\partial x$. The functions r, p and q are given by

$$r(x) = C(x)\sin\frac{\pi x}{2}, \tag{10.20b}$$

$$p(x) = \frac{4}{\pi^2}\sin\frac{\pi x}{2}\,k(x), \tag{10.20c}$$

$$q(x) = \{\mu\, Q_0 S(x)\, h(x,T_1) - d(x,T_1(x))\}/C(x), \tag{10.20d}$$

and

$$d(x,T) = c(x)\frac{\partial}{\partial T}\sigma g(T)T^4, \tag{10.20e}$$

where $c(x)$ contains the explicit x-dependence of $g(x,T)$, while $h(x,T_1)$ is a piecewise constant function equal to zero when $\alpha(x,T_1(x))$ is constant, and equal to the T-slope of α when T_1 is along the ramp portion

$$T_\ell < T_1(x) < T_u. \tag{10.20f}$$

All this, although lengthy, is quite straightforward and rather general. Similar linearizations L_j, $j = 2,3$, are obtained about $T_2(x)$ and $T_3(x)$. For convenience, we drop the subscript $j = 1$ from L_j in Eq. (10.19).

The role of the coefficient λ_1 in (10.4) will now be played by the eigenvalues $\lambda^{(k)}$ of L, defined by

$$L\psi_{(k)}(x) = \lambda^{(k)}\psi_{(k)}(x). \tag{10.21a}$$

It is again possible to show, under rather general assumptions on the form of $k(x)$, $\alpha(x,T)$ and $g(x,T)$, that the eigenvalues $\lambda^{(k)}$ are all discrete, real and tend to infinity,

$$\lambda^{(k)} \to +\infty \quad \text{as} \quad k \to \infty. \tag{10.21b}$$

It follows that solutions to (10.19) can be expanded in the eigenfunctions $\psi_{(k)}(x)$ of L,

$$\theta(x,t) = \sum_0^\infty a_k \, e^{-\lambda^{(k)}t}\psi_{(k)}(x), \tag{10.22}$$

which is the 1-D form of (10.5). If $k(\phi) \equiv$ const., $g(T)$ is chosen so that gT^4 may be a linear function of T, and α is piecewise constant, then $\psi_{(k)}(y)$, with $y = \sin\phi$, are just the Legendre polynomials used by North.

From (10.21, 10.22) it follows that at most a finite number of $\lambda^{(k)}$ will be negative, and that a stationary solution $T_j(x)$ will be stable if and only if no $\lambda_j^{(k)}$ is negative. Neutral stability corresponds to a zero eigenvalue, $\lambda_j^{(0)} = 0$, strict (or asymptotic) stability to $\lambda^{(0)} > 0$. The technical aspects of all these statements are discussed in Ghil (1976).

At this point, the linear stability problem for stationary solutions of (10.12) has been reduced to determining the sign of $\lambda_j^{(0)}$ for each solution $T_j(x)$. It turns out that this is easily done by a single numerical integration of the linear ODE

$$L_j w(x) = 0, \tag{10.23a}$$

with initial conditions

$$w_x(0) = 0, \quad w(0) = 1. \tag{10.23b}$$

If $w(x)$ is nonnegative for $0 \le x \le 1$, then $\lambda_j^{(0)}$ has the same sign as $w_x(1)$, so that $T_j(x)$ is stable according to whether w_x at $x = 1$ is positive or not. This result is a simple extension to singular boundary conditions at the pole of comparison and oscillation theorems for regular Sturm-Liouville problems, and we shall not dwell here on its proof (given in Ghil, 1975, 1976).

For (10.12), the numerical integration of (10.23) shows, as expected from (10.4), that the "present climate" $T_1(x)$ and the "deep freeze"

$T_3(x)$ are stable, while $T_2(x)$ is unstable; in fact, for the latter exactly one eigenvalue, $\lambda_2^{(0)}$, is negative.

Structural stability. With the results for $\mu = 1$ in hand, we can turn to the study of stationary solutions and their internal stability for all values of the insolation parameter μ. Figure 10.6 shows the average temperature $\overline{T}$ of stationary solutions as a function of μ. The solutions $T_j(x;\mu)$, $j = 1,2,3$, and their averages were found numerically for all μ shown, as in Figure 10.5 for $\mu = 1$. For the 0-D model (10.2), such a graph of $\overline{T} = \overline{T}(\mu)$ would contain all the information required, since $\overline{T}$ determines a solution completely.

For a 1-D model, infinitely many numbers $a_k(\mu)$ are required to determine a stationary solution, cf. (10.22); one can choose for instance $\overline{T}(\mu) = a_1(\mu)$, $\Delta T(\mu) = a_2(\mu)$, etc. Still, Figure 10.6 is very informative: for $\mu = 1$ there are the three solutions, represented by $\overline{T}_1$, $\overline{T}_2$, $\overline{T}_3$.

As μ decreases, climate branch A, which passes through the point $(\mu = 1, \overline{T} = \overline{T}_1)$, and branch B, passing through $(1, \overline{T}_2)$, approach each other and merge, at a point $(\mu_c, \overline{T}_c)$. For $\mu < \mu_c$, only the "deep freeze" branch C exists. For $\mu > 1$, branches C and B approach each other, merging at $(\mu_d, \overline{T}_d)$. Beyond $\mu = \mu_d$, only the "warm climate" branch A exists.

The dependence of solution number on μ is thus quite analogous to (10.2), as discussed in connection with Figure 10.4. From Figure 10.4 and Eqs. (10.4, 10.5), we were also able to infer, using the T-slope of $R_i(\overline{T};\mu) - R_0(\overline{T};\mu)$, that all solutions of (10.2) corresponding to branches A and C were stable, while branch B in the 0-D case is unstable. There is no simple counterpart to this geometric argument in the general 1-D case.

We have seen in the 0-D case that $\lambda_j(\mu)$, $j = 1,2,3$, were continuous functions of μ near $\mu = 1$, cf. Eqs. (10.6). This continuity provided an alternative argument for the stability persistence of $T_j(\mu)$. It is this argument which can be extended in all generality to the 1-D case.

In fact, we noticed that for $T_2(x;\mu)$, $\mu = 1$, exactly one eigenvalue, $\lambda_2^{(0)}$, was negative. It turns out that $\lambda_j^{(k)}(\mu)$ are all continuous functions of μ, for each *k*. The function $\lambda_j^{(0)}(\mu)$ changes sign where two branches, A and B or B and C, coalesce, i.e., $\mu = \mu_c$ and $\mu = \mu_d$, respectively, while all other $\lambda_j^{(k)}(\mu)$, $k \geq 1$, stay positive

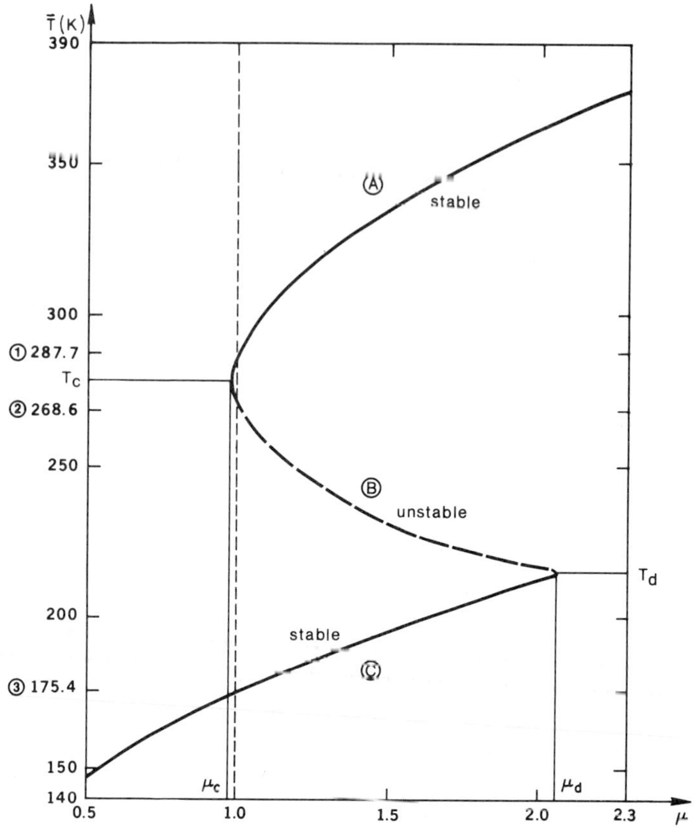

Figure 10.6. Dependence of solutions on fractional change in insolation for the model in the previous figure (after Ghil, 1975).

for all j and all μ. This mathematical analysis confirms the physical intuition, embodied in Figure 10.4, that a stationary climate branch $T(x;\mu)$ is stable if $\overline{T}$ is an increasing function of insolation μ along the branch, and unstable if $\overline{T}(\mu)$ is decreasing.

Figure 10.6 can thus represent exactly the dependence of stationary solutions of Eq. (10.12) on the parameter μ, in spite of the phase space of $T(x,t)$ requiring, cf. (10.22), an infinite number of dimensions to represent nonstationary solutions with arbitrary initial data (10.15c). The surprising usefulness of this figure is not fortuitous: it is due to the fact that Eq. (10.12) is *structurally stable*, i.e., that the number and stability of its stationary solutions does not change for arbitrary small changes in the equation, thus allowing more than a simple change in μ.

A consequence of this structural stability is that the values of μ for which the number and internal stability of stationary solutions can change are isolated. These isolated values, μ_c and μ_d in Figure 10.6, are called critical values, and the appropriate points in phase-parameter space are called *bifurcation points*. Both bifurcations in the figure are of the *saddle-node* type, in which two branches of stationary solutions coalesce, with a single real eigenvalue changing sign from one branch to the other (compare also Sections 5.3 and 6.3, especially the bifurcation diagrams in Figures 5.8, 6.5 and 6.18). For this type of bifurcation, the dependence of the stationary solutions on the bifurcation parameter is parabolic near the bifurcation point, e.g., $\overline{T}_1(\mu) - T_c \sim (\mu-\mu_c)^2 \sim \overline{T}_2(\mu) - T_c$, with the same constant of proportionality for both branches, but a different constant at (μ_d,T_d).

Concluding remarks on 1-D EBMs. The independent analyses of 1-D EBMs by Ghil (1975, 1976), Held and Suarez (1974), North (1975a,b) and other investigators gave the following picture of these models:

(a) Multiple stationary solutions exist for fixed solar input; among these solutions one is close to the present climate and one represents a completely ice-covered Earth;

(b) The "present climate" and the "deep freeze" are stable to internal perturbations;

(c) All stationary solutions depend continuously on the insolation parameter μ, with various solution branches merging at certain critical values of μ.

In fact, North (1975b) showed that the diffusively-modified Budyko model with constant k (outlined in Table 10.1) and the original Budyko model with Newtonian cooling are equivalent if exactly two even Legendre polynomials are retained in the expansion of the problem (cf. (10.22)). In a two-mode approximation, various other forms of 1-D EBMs are also equivalent.

Unfortunately, the number of solution branches changes in North's (1975a,b) model if an additional Legendre polynomial is retained in the expansion: the three-mode model has five solution branches, rather than three, as we found here. The diffusive Budyko model does actually have five solutions, for all truncations higher than three, as shown already by Held and Suarez (1974).

The two additional solutions, related to the so-called "small ice-cap" instability, are an artifact of the step-function formulation of albedo: they disappear when a ramp function is introduced, as in Sellers

(1969) and here (Held *et al.*, 1981, Appendix B). Furthermore, the two-mode step-function formulation leads to a "present climate" branch which is entirely ice free (cf. North, 1975b, Fig. 2), like the 0-D model (10.2), in contradiction to current climatic data.

In spite of such technical differences, the fundamental agreement between the results of different EBMs led to verification of these results with more detailed models. Clearly, the "present climate" branch $T_1(x;\mu)$ alone is of interest in studying climatic change on the interannual and decadal time scale. Thus, the most important feature of EBMs which had to be verified by general circulation models (GCMs) is the dependence of solutions on solar input near present conditions.

A great deal of effort went into "tuning" EBMs on the one hand, and GCMs on the other, to yield the same "sensitivity" of climate to insolation variations, $d\overline{T}/d\mu$ at $\mu = 1$, $T = T_1(x)$. Agreement now is satisfactory, with a sensitivity of approximately 1K change in global temperature for one percent change in solar input.

Aside from the derivative of $\overline{T}(\mu)$ at $\mu = 1$, the shape of the curve $\overline{T}_1(\mu)$ for $\mu \geq \mu_c$ is of interest. This shape, cf. Figure 10.6 and discussion above, is parabolic, with vertical slope at the bifurcation point $\mu = \mu_c$, $\overline{T}_1(\mu_c) = \overline{T}_2(\mu_c) = T_c$. A parabolic dependence, with slope increasing as μ decreases, was obtained in the GCM experiments of Wetherald and Manabe (1975, Figure 5). This non-tuned, qualitative result offers rather striking confirmation to the simple story that EBMs are telling about climate.

10.3. Nonlinear Stability and Stochastic Perturbations

In the previous section, we discussed stability for small or, more precisely, infinitesimal perturbations in the initial data of an evolution equation for the global climate. The early growth of such perturbations could be analyzed by a linearized form of the equation governing the actual solution.

In this section, we shall examine what happens if the initial perturbations are of finite and possibly large size. Furthermore, perturbations which occur not only initially, but during the entire evolution of the solution will be considered. A stochastic model for such perturbations is introduced and their effect on the solution in the long run is studied.

A variational principle. For simplicity, we illustrate again the basic ideas with the 0-D model analyzed at the beginning of Section 10.2. It will be rewritten for convenience as

$$\dot{T} = f(T), \tag{10.24a}$$

where we identify T with $\overline{T}$; $\dot{T}$ is then dT/dt, and

$$f(T) = \frac{Q_0}{c}\{1 - \alpha(T)\} - \frac{\sigma}{c} g(T)\, T^4. \tag{10.24b}$$

We know that, for a reasonable range of parameter values, $f(T)$ has three roots, $T = T_1, T_2, T_3$, which are steady-state solutions of Eq. (10.24).

The question now is: what will be the solution $T(t)$ of (10.24) for initial data far away from an equilibrium solution? This can be easily settled in the 0-D case at hand by considering $f(T)$ as the derivative of another function, $-F(T)$. Then F is just the indefinite integral of $-f$,

$$F(T) = -\int^T f(s)\, ds, \tag{10.25}$$

or one can fix an arbitrary initial point of the integration, such as $T = T_1$, the "present climate", so that $F(T_1) = 0$.

A qualitative picture of the pseudo-potential $F(T)$ for $f = R_i - R_0$ of Figure 10.4 is given in Figure 10.7. Where $f(T)$ is negative, $F(T)$ increases, and where $f(T)$ is zero, $F(T)$ has an extremum. Thus T_1 and T_3 are minima, with $F(T_3) < F(T_1) = 0$, and $F(T_2)$ is a maximum. We shall justify the use of the term pseudo-potential below (see Eq. (10.38)).

With (10.25) Eq. (10.24) becomes

$$\dot{T} = -\frac{dF}{dT}. \tag{10.26}$$

Multiplying by $\dot{T}$ yields

$$\frac{dF(T(t))}{dt} = -\dot{T}^2 \leq 0, \tag{10.27}$$

for any solution $T(t)$ of (10.24).

Given an arbitrary initial datum $T(0) = T_0$, the motion of the solution $T(t)$ can be represented in Figure 10.7 as motion along the curve $F = F(T)$, starting at the point $(T_0, F(T_0))$. The motion of the solution point $(T(t), F(t))$, according to (10.27), will always be towards lower values of F. The solution point can remain at rest only at the extrema of F, where $\dot{T} = 0$.

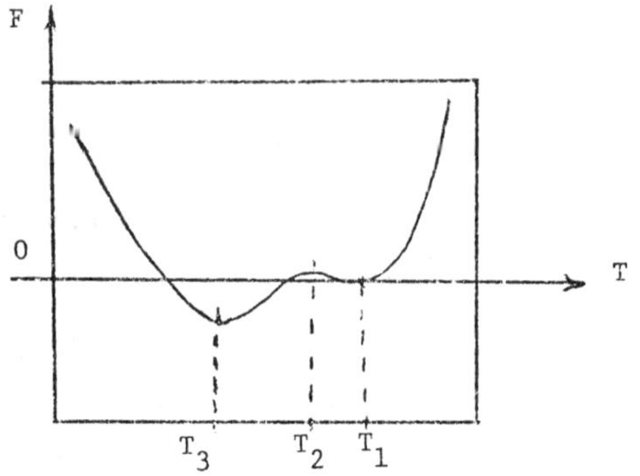

Fig. 10.7. The pseudo-potential $F(T)$ for an EBM with three equilibria (courtesy of C. Nicolis).

In fact, for any initial point to the right of T_2, $T_2 < T_0 < +\infty$, the solution starting at that point will "flow" into T_1. For all points to the left of T_2, solutions will flow into T_3. Thus, the eventual fate of all solutions $T(t)$ in this simple model is to end up either at $T = T_1$ or at $T = T_3$. The speed of approach is given by the slope $F' = dF/dT$ at any given point. The solutions will actually reach T_1 or T_3 only after an infinite length of time, since the "final approach" is governed approximately by the linear equation (10.4), with exponential solutions (10.5).

Stochastic perturbations and Brownian motion. The previous discussion settles the question of finite-amplitude initial perturbations for the 0-D model (10.24). What happens if the perturbation, instead of occurring at just one moment in time, occurs repeatedly, nudging the solution from where it is again and again? This is the case of Brownian motion, where a large particle in a suspension is subject to repeated collisions by smaller particles.

Such a situation, while more realistic than the "single collision" case discussed before, is also harder to analyze. A model is needed for the multiplicity of small nudgings, or collisions. It was formulated in the case of Brownian motion by Einstein (1905).

The simplest derivation, for our purposes, of Einstein's main result proceeds as follows. Let us consider the motion of a particle along a straight line, subject to linear friction and a random force. The governing equation is the so-called *Langevin equation*

$$\dot{u} = -\lambda u + \eta(t), \tag{10.28}$$

where $u = u(t)$ is the velocity of a particle with unit mass, λ the constant friction coefficient, and $\eta(t)$ represents the collisions with the smaller particles. The idea is that the collisions are (i) random and (ii) have much shorter correlation times than the deterministic relaxation time given by $1/\lambda$.

The Einstein model for $\eta(t)$ is essentially that of *white noise,*

$$E\{\eta(t;\omega)\} = 0, \tag{10.29a}$$

$$E\{\eta(t;\omega)\eta(t+s;\omega)\} = \sigma^2\delta(s). \tag{10.29b}$$

Here E is the expectation operator, or ensemble average, over individual realizations of the particle's history, labeled by ω. The variance σ^2 of the collisions is related to the thermal agitation of the small particles, and is fixed. The short correlation times are reflected by taking $\delta(s)$ in (10.29b) to be the Dirac δ-function. Properties (10.29a,b) define the *random process* called white noise and are sufficient, subject to the technical assumption of Gaussianity, to obtain the most important information concerning the motion of the particle.

Due to the presence of the random forcing in (10.28), u itself becomes a random process, $u = u(t;\omega)$. The mean $\overline{u}(t) = Eu(t;\omega)$ of this process is given by applying E to (10.28) to yield (cf. (10.29a))

$$\dot{\overline{u}} = -\lambda\overline{u},$$

with solution

$$\overline{u}(t) = \overline{u}_0 e^{-\lambda t}. \tag{10.30a}$$

Here $\overline{u}_0 = Eu(0;\omega)$ is the mean of the ensemble of initial data. Clearly

$$\overline{u}(t) \to 0 \quad \text{as} \quad t \to +\infty, \tag{10.30b}$$

independently of $\overline{u}_0$. The fact that the random force $\eta(t;\omega)$ has zero mean, together with the linear damping, thus imply that the velocity also has to have zero mean, eventually.

The important result, however, concerns the variance of the fluctuations of the velocity around its deterministic value. Let the initial value of the velocity u_0 be known exactly, so that

$$u(t;\omega) = u_0 e^{-\lambda t} + e^{-\lambda t}\int_0^t e^{\lambda s}\eta(s;\omega)\,ds. \tag{10.31a}$$

We are interested in

$$E(u(t;\omega) - u_0 e^{-\lambda t})^2 = e^{-2\lambda t}\int_0^t\int_0^t e^{\lambda(r+s)}E\eta(r;\omega)\eta(s;\omega)\,dr\,ds.$$

Making the change of variables $r+s = \rho$ and $r-s = \theta$ allows one to use (10.29b), which says that the integrand is concentrated along the "diagonal" $r-s = 0$. This yields, using also (10.30a),

$$\overline{u^2}(t) - u_0^2 e^{-2\lambda t} = \frac{e^{-2\lambda t}}{2} \int_0^{2t} e^{\lambda\rho} d\rho \int_{-t}^{t} \sigma^2 \delta(\theta) d\theta. \tag{10.31b}$$

The second integral on the right-hand side is numerically equal to σ^2, but has dimensions of $(\text{length})^2 \times (\text{time})^{-3}$, while σ^2 is $(\text{acceleration})^2$. In order to make the distinction, it is convenient to let

$$\tau_* = \int_{-\infty}^{+\infty} \sigma^2 \delta(s) ds. \tag{10.32a}$$

One thus obtains

$$\overline{u^2} - u_0^2 e^{-2\lambda t} = \frac{\tau_*}{2\lambda} (1 - e^{-2\lambda t}), \tag{10.32b}$$

and in the limit of large time, $t >> 1/\lambda$,

$$\overline{u^2} = \tau_*/2\lambda. \tag{10.32c}$$

If one considers the initial velocity to be random, but uncorrelated with the forcing, the variance of fluctuations about the *mean* velocity (10.30a) can be computed. It suffices to replace u_0 by $\overline{u_0}$ in the derivation above, and the same result (10.32) is obtained. From the point of view of Brownian motion proper, the importance of (10.32) was that the macroscopically observable variance of the velocity fluctuations of the large, suspended particles could be related to the temperature of the liquid and that this relation could be verified experimentally.

The striking fact, more generally, is that the *displacement* of the particle $x(t;\omega)$, given by $\dot{x} = u(t;\omega)$, has a variance $\overline{x^2}(t)$ which, for times much longer than the deterministic relaxation time $1/\lambda$, is given by

$$\overline{x^2}(t) = \frac{\tau_*}{\lambda^2} t. \tag{10.33}$$

This result is derived by means entirely analogous to (10.30) and (10.32). The linear growth in time of variance is somewhat counterintuitive; one might have expected the *standard deviation* $(\overline{(x^2}(t))^{1/2}$ of the displacement to be linear in time, since the standard deviation of velocity is constant.

<u>Stochastic differential equations and exit time</u>. The experimental verificationof Einstein's results (10.32, 10.33) led to the realization

that other situations in the sciences can be described in a simplified manner by interactions on two time scales: a slow time associated with relatively simple interactions, and a fast one with much greater complexity. It is then convenient to model the slow interactions between a few degrees of freedom deterministically, and the fast interactions involving many more degrees of freedom stochastically.

A general model for such situations is a stochastic differential equation

$$du = f(u)dt + \sigma\, d\beta. \tag{10.34}$$

Here $\beta(t;\omega)$ is an abstract, normalized "Brownian motion", or *Wiener process*, which is characterized by the properties

(i) $\beta(0;\omega) = 0$ with probability one, (10.35a)

(ii) $\beta(t;\omega)$ is normally distributed with mean zero and variance σ_t^2, and (10.35b)

(iii) $\beta(t;\omega)$ has increments which are probabilistically independent of each other and stationary, i.e., independent of time. (10.35c)

From (10.35) it follows in particular that

$$E\beta(t;\omega) = 0, \tag{10.36a}$$

and

$$\sigma_t^2 \equiv E\beta^2(t;\omega) = ct; \tag{10.36b}$$

here c is a positive constant, which is taken equal to one for normalization.

For $f(u) = -\lambda u$, Eq. (10.34) becomes (10.28), if we set $\eta = d\beta/dt$. In fact, the individual path of any realization of β is almost nowhere differentiable, although it is frequently said that white noise is the "derivative" of a Wiener process, in the same way that the Dirac δ-function is the "derivative" of the Heaviside step function (see Eqs. (11.20, 11.21)). Writing (10.34) in terms of differentials, rather than derivatives, also reminds us that $d\beta^2$ is proportional to dt, cf. (10.33) and (10.36b), rather than being negligibly small. For a presentation which is both rigorous and readable, Jazwinski (1970, Ch. 3-4) is a good reference.

We have now a reasonable description of repeated perturbations due to phenomena on smaller spatial and shorter time scales than the global temperature evolution of Eq. (10.24). We now rewrite this equation as

$$dT = f(T)dt + \sigma\, d\beta. \tag{10.37}$$

What are then the consequences of such repeated perturbations on the stability of the equilibria, T_1 and T_3?

Returning to Figure 10.7, we saw that in the deterministic case all solutions "flowed" into one of the two minima. For the Brownian particle of Eq. (10.28), a pseudo-potential is also immediately obtained,

$$G(u) = \lambda u^2/2; \tag{10.38}$$

it has a single minimum at $u = 0$.

As a matter of fact, this justifies our use of the term pseudo-potential, rather than potential, for the function $F(T)$ in Eq. (10.25) and for $G(u)$ above; the usual potential function in mechanics depends on position, rather than on velocity. Mathematically speaking, it appears in a second-order ODE for the displacement, rather than in a first-order equation, like (10.24) or (10.28). In particular, a solution point does not come necessarily to rest at the bottom of a true potential well, as it does here, but can keep moving beyond the minimum, if it arrives with nonzero velocity. A "first-order" potential like $F(T)$ or $G(u)$ above is also called kinetic potential in thermodynamics.

The Brownian particle can thus be viewed as subject to a deterministic flow which is directed towards $u = 0$. The random perturbations, however, keep it "diffusing" against this flow, in both the physical and mathematical sense, with the result that its variance is asymptotically constant and nonzero. This variance is the result of the competition between the deterministic flow $-\lambda u = -G'(u)$ and the random forcing.

As a result of the sustained nonzero variance of $u(t;\omega)$, there is a nonzero chance that u^2 will be above any preset value after some finite time. In other words, the small collisions keep pumping kinetic energy into the Brownian particle, energy which is dissipated by the viscosity. The ensemble mean kinetic energy $\overline{u^2}$ is constant, but its instantaneous value for any realization can become arbitrarily large, provided all collisions in that realization are in the same direction over a sufficiently long time interval. The probability for this to occur becomes smaller and smaller as the required time interval becomes longer and longer, but it does not become exactly zero for any finite value of the kinetic energy. This is a consequence of the Gaussian white noise assumption made about the small collisons.

In the case of the double-well pseudo-potential $F(T)$ of Figure 10.7, something interesting can happen due to the finite variance of the temperature in either well. Clearly one can approximate the shape of either well by a parabola, like (10.38), and obtain a result like (10.32) for $\overline{T^2}$, with $\lambda_{1,3} = F''(T_{1,3})$. The question then arises how long will it take for $F(T(t))$ to exceed the "potential barrier" $\Delta F = F(T_2) - F(T_1)$ or $\Delta F = F(T_2) - F(T_3)$, depending on the well the solution lies in at the initial time.

This waiting time or residence time or *exit time* θ is itself a random variable, whose expected value we shall denote by $\overline{\theta} = E\theta$. More precisely, $\overline{\theta}$ is a function of the initial point T_0 at $t = 0$, and of the end points of the interval out of which we are waiting for the point to exit. Thus $\overline{\theta} = \overline{\theta}(T_0;T_2,+\infty)$ if the climate point lies originally to the right of T_2, and we are waiting for it to cross over into the lower-temperature well, since $F(T) \to \infty$ as $T \to +\infty$.

It turns out that for any $T_0 \geq T_2 + \varepsilon$, with $\varepsilon > 0$ small, $\overline{\theta}$ is nearly constant and thus independent of T_0. This is due to the fact that almost surely exit does not occur due to one single, large increment of $\beta(t;\omega)$, but due to an accumulation of small "pushes". Only near T_2 does the probability for a sufficient accumulation of such pushes vary rapidly with distance from T_2.

We shall only state the asymptotic formula, in the limit of small noise $\sigma^2 \ll \Delta F$, of the mean exit time from an asymmetric well,

$$\overline{\theta}(T_1;T_2,+\infty) \simeq \frac{\pi}{\{-F''(T_1)F''(T_2)\}^{1/2}} \exp \frac{F(T_2)-F(T_1)}{\sigma^2/2} . \tag{10.39}$$

For relatively simple derivations of this formula, which is due to Kramers, we refer to Chandrasekhar (1943), Nicolis and Nicolis (1981), and Sutera (1981).

In the case of the low-temperature well, T_1 is replaced by T_3 on both sides of (10.39). The striking feature of (10.39) is the exponential dependence on the height of the barrier ΔF, and on the noise level σ^2. The dependence on ΔF gives a precise meaning to the intuitive feeling that the deeper the well, the more stable the equilibrium at its bottom.

The sensitivity of the dependence of $\overline{\theta}$ on the noise level creates difficulties in applying the double-well model to climate theory. Indeed, the total high-frequency variance of heat content σ^2 in the climatic system depends on the time scales which we agree to call short, compared to the relaxation time $|1/f'(T)| = |1/F''(T)|$, and is hard to estimate accurately. Indications are that this variance is sufficiently small to make (10.39) a good approximation to the expected exit time. On the other hand, σ^2 is also sufficiently small for an exit from the neighborhood of the "present climate" T_1 to be extremely unlikely during the lifetime of the Earth: the present climate in this model is stochastically stable. A modification of the model which allows for exit in time $O(10^5 a)$, with $1a = 1$ year, will be studied in the next section.

Considerations of *stochastic stability*, like the ones above, are as important as those of stability to initial perturbations and structural stability. The theory of the latter two, as presented in Section 10.2, can be carried out with relative ease for models which are considerably more complex than the single, scalar ODE (10.24). The analysis of stochastic stability, on the other hand, becomes very rapidly more complicated as the number of degrees of freedom in the deterministic problem increases beyond two. The study of stochastic perturbations for systems of increasing complexity is an active field of research and one which should play a growing role in climate theory.

10.4. Modified EBMs: Periodic and Chaotic Solutions

One might suspect that there is more to climate evolution than steady states and random fluctuations about them. In the next chapter, we shall describe the complex structure of this evolution, and in Chapter 12 we shall explore some models including other physical processes in addition to energy balance, which lead in a natural way to solutions with a rich temporal structure.

In this section, we present merely some simple modifications of EBMs which also exhibit nonstationary solutions. The point of departure is the 1-D model (10.12). First, we shall consider a modification of the albedo dependence on temperature, and its effect on the number of solutions, their relative position, and their stability. A 1-D version of the variational principle in the previous section will be used to study the solutions' nonlinear stability.

Next, we consider the joint effect of periodic forcing and stochastic perturbations for a 0-D version of the model above. The interaction of small forcing and small noise can lead, for certain parameter values, to large excursions of the solution from the neighborhood of one equilibrium to another.

Finally, we consider the effect of delays on the 1-D modified model introduced in this section. Periodic and aperiodic solutions are studied, along with their Fourier spectra.

Cloud effect on albedo and stationary solution structure. In Eq. (10.12), $\alpha = \alpha(\phi,T)$ at constant ϕ was taken as a Sellers-type ramp function, cf. Table 10.1. We mentioned in the discussion of ice-albedo feedback after Eq. (10.3) that the dependence of α on cloud amount is equally important, but hard to formulate in terms of global temperature $\overline{T}$. It is possible to do a slightly better job in terms of the zonally-averaged surface temperature $T(\phi,t)$.

The idea of the albedo mofication which we discuss is familiar from descriptive climatology and synoptic meteorology: slightly equatorward of an ice margin, one observes preferentially the passage of midlatitude storms. Such a storm track is accompanied by high cloudiness, and hence increased albedo, as well as by mean temperatures which are higher than over the ice. The ensuing picture of albedo dependence on temperature is given in Figure 10.8. At the same time, any explicit dependence of α on latitude ϕ is removed, so that $\alpha = \alpha(T)$ exclusively, although $T = T(\phi,t)$. We shall rewrite Eq. (10.12), with the albedo function of Figure 10.8, as

$$C(\phi)T_t = \frac{1}{\cos\phi}\frac{\partial}{\partial\phi}[k(\phi)\cos\phi\, T_\phi] + \mu Q_0 S(\phi)[1 - \alpha(T)] - \sigma g(\phi,T)T^4, \tag{10.40}$$

where $T_t = \partial T/\partial t$ and $T_\phi = \partial T/\partial\phi$.

As a result of the change in the albedo function, the model's stationary solution structure changes. Five steady states, rather than the three steady states of Figure 10.5, become possible solutions. Their dependence on the insolation parameter μ is shown in Figure 10.9.

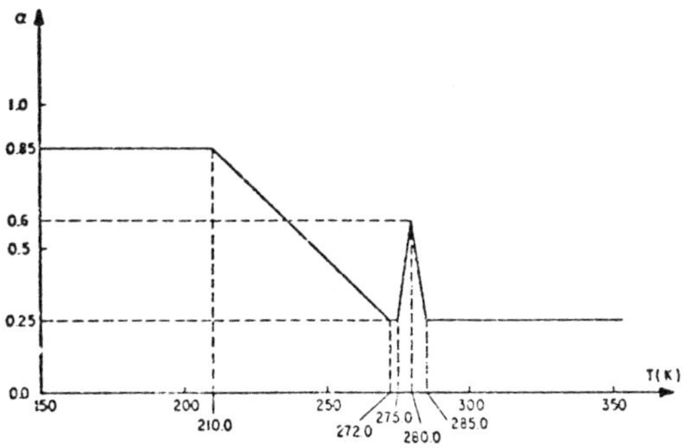

Fig. 10.8. The dependence of albedo α on temperature T as a result of both ice-albedo feedback and increased cloudiness near the ice margin (from Bhattacharya *et al.*, 1982).

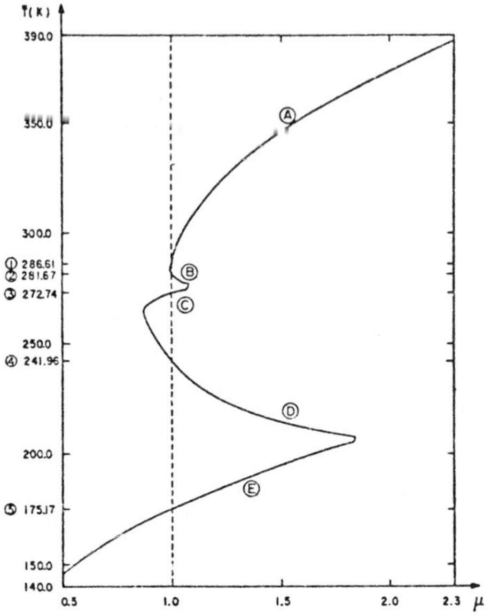

Fig. 10.9. Bifurcation diagram of a 1-D EBM with cloud-dependent albedo (from Bhattacharya *et al.*, 1982).

The stability of the five solutions to initial perturbations alternates: branches A, C and E are stable, B and D are unstable. This agrees with the heuristic connection between the slope of the $\overline{T} = \overline{T}(\mu)$ curve at a given solution, and the internal stability of that solution, as stated in the discussion of Figure 10.6. A comparison of the two figures shows that branch A here corresponds to the "present climate", branch E to the "deep freeze", and branches B and C are introduced by the "cloudiness kink" of Figure 10.8.

The stable branch C of this model is a somewhat more plausible candidate for a quasi-stationary ice age than either of branch B or branch C of the 1-D model in Section 10.2, branch B there being unstable and C totally ice covered. For present insolation conditions, $\mu = 1$, branch C here lies O(10K) below the present climate, rather than O(100K), which was the case for $\overline{T}_3$ of model (10.12).

<u>Variational principle and nonlinear stability</u>. The nonlinear stability of solutions for fixed μ can be studied by an extension of the variational principle (10.25-10.27). The nonlinear *functional* $J\{T(x);\mu\}$ which takes the place of $F(T)$ here is given by

$$J\{T(x);\mu\} = \int_0^1 \{\frac{1}{2}\, p(x) T_x^2 - r(x) G(x,T(x);\mu\} dx, \tag{10.41a}$$

with x again being normalized co-latitude, $T_x = \partial T/\partial x$ and the functions $p(x)$ and $r(x)$ defined in Eqs. (10.20b,c). The function $G(x,T;\mu)$ in the integrand of (10.41a) is the indefinite integral of the net radiative flux $R_i - R_0$ in (10.40),

$$G(x,T;\mu) = \int^T \{\mu Q_0 S(x)[1 - \alpha(\Theta)] - \sigma g(x,\Theta)\Theta^4\} d\Theta, \tag{10.41b}$$

where the integration is at fixed x with respect to the dummy variable Θ, and $\alpha(T)$ is given by Figure 10.8.

The stationary solutions of the model are the extrema of the functional J. In fact, one can rewrite (10.40), by analogy with (10.26), as

$$r(x) C(x) T_t = -\delta J\{T;\mu\}/\delta T. \tag{10.42a}$$

Here the right-hand side is the functional derivative of J with respect to variations in the temperature distribution $T(x)$, defined by

$$\int_0^1 \frac{\delta J}{\delta T}\, \Theta(x)\, dx = \lim_{\varepsilon \to 0} \frac{J\{T(x)+\varepsilon\Theta(x)\} - J\{T(x)\}}{\varepsilon}, \tag{10.42b}$$

for arbitrary $\Theta(x)$. This takes the place of the scalar derivative dF/dT in (10.26). An extremum, or critical point, $T_*(x)$ of J is given by

$$\delta J\{T_*;\mu\}/\delta T = 0.$$

To obtain a geometric picture of the functional $J\{T(x);1\}$, one has to expand $T(x)$ in a set of basis functions,

$$T = \sum_0^\infty T_n \phi_n(x).$$

A convenient orthonormal set $\{\phi_n\}$ are the Legendre polynomials $\phi_n(x) = P_n(\cos(\pi x/2))$. Using the coefficients T_n as coordinate axes, one can plot $J\{\Sigma\, T_n\, P_n(x);1\}$ as the height h of the hyper-surface $h = J\{T\}$ given by (10.41).

Figure 10.10 shows such a plot for a truncated representation of $T(x)$, with $n = 0,2$; the normalized coordinates are I_n/Q_0, where $I_n = A + BT$ (compare Table 10.1). The plot is valid for a model similar to (10.12), having three stationary solutions, marked as I, II and III in the figure.

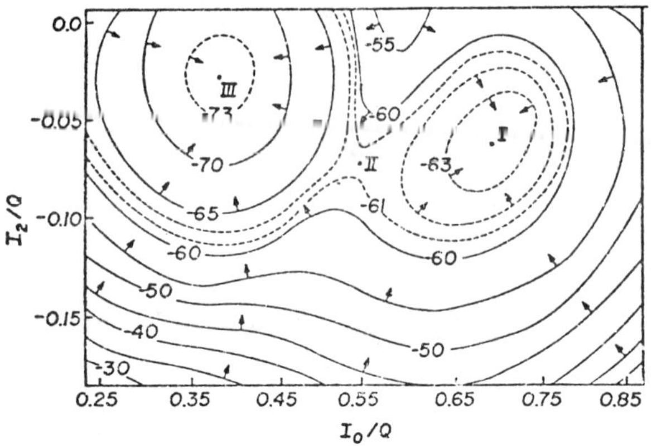

Fig. 10.10. Contour plot of a two-mode approximation to the functional $J\{T(x);1\}$ (after North *et al.*, 1979).

For $C(x) = C$ constant, as in the model of Figure 10.10, and an orthonormal basis like the Legendre polynomials, Eq. (10.42) can be rewritten as a decoupled system of ODEs for the evolution of the coefficients $T_n = T_n(t)$,

$$C\,\dot{T}_n = -\partial J/\partial T_n. \tag{10.43}$$

Figure 10.10 shows by little arrows the flow field on the right-hand side of (10.43); it is the gradient, or steepest descent, of the "topography" of the functional.

The solution will thus move along the gradient $-\delta J/\delta T = -(J_1, J_2, \ldots, J_n, \ldots)$, where $J_n = \partial J/\partial T_n$. This gradient is perpendicular to the contour surfaces $J\{T\} = h$ in a multi-dimensional plot analogous to Figure 10.10. The fact that the solution will always move "downslope" is guaranteed by the analogue of (10.27), namely

$$\begin{aligned} \frac{dJ\{\Sigma\, T_n(t)P_n(x)\}}{dt} &= \sum \frac{\partial J}{\partial T_n}\dot{T}_n \\ &= -\sum C\,\dot{T}_n^2 \le 0, \end{aligned} \tag{10.44}$$

since $C > 0$.

In the slightly more general and more realistic case that the heat capacity $C(x)$ is not constant with latitude, as in (10.40), the variational formulation (10.42) cannot be reduced to the diagonal form (10.43) by Legendre polynomials. The appropriate result, though, can still be obtained, by applying the chain rule and integration by parts to (10.41), subject to boundary conditions (10.15). One obtains

$$\begin{aligned}\frac{dJ\{T(x,t)\}}{dt} &= \int_0^1 \{p(x)T_xT_{xt} \\ &\quad -r(x)[\mu Q_0S(x)(1-\alpha(T)) - \sigma g(x,T)T^4]\,T_t\}\,dx \\ &= -\int_0^1 \{(pT_x)_x\,T_t + r\,\mu Q_0S(1-\alpha) - \sigma gT^4]T_t\}dx \\ &= -\int_0^1 r(x)C(x)T_t^2\,dx \le 0,\end{aligned} \tag{10.45}$$

since $r(x)C(x) > 0$ over the interval of integration, $0 < x < 1$.

In this case, the flow of solution points is not necessarily perpendicular to the "contour lines" of the functional. Still it is always downslope, cf. (10.45), at an angle with strictly positive direction cosine from the inward pointing normal. Again, solutions come to rest at the minima of the pseudo-potential $J\{T\}$, while other extremal points, i.e., maxima and saddles, are nonlinearly unstable.

Notice that I and III in Figure 10.10, like T_1 and T_3 in Figure 10.7, are local minima, and thus nonlinearly stable. Point II is a saddle, rather than a local maximum, as T_2 was. In more than two dimensions, a *saddle* is a local minimum in a certain number of directions, equal to or greater than one, and a local maximum in all the remaining directions, numbering not less than one.

The number of directions of stability and instability is given by the number of negative and positive eigenvalues of the second variation of J, $\delta^2J/\delta T^2$, evaluated at the saddle. For the simplified case of Eq. (10.43), this second variation is just the symmetric bilinear form defined by the *Hessian*, or matrix of second derivatives $(\partial J_k/\partial T_\ell)$, with $k,\ell = 0,1,2,\ldots$.

The saddles $T_2(x)$ and $T_4(x)$ in model (10.40) have each a single unstable direction. Indeed, the linearization $L_j(x)$ of the right-hand side in (10.40), which is analogous to (10.20), has a single positive eigenvalue along branch B and branch D, $j = 2$ and $j = 4$ respectively, all other eigenvalues being negative.

In general, a critical point T_*, $\delta J\{T_*\}/\delta T = 0$, is *nondegenerate* if its Hessian, which is always a linear operator, has no zero eigenvalue. Such nondegenerate critical points are almost always isolated. The *Morse lemma* states that in the neighborhood of an isolated nondegenerate point, one can always find local coordinates which will make the functional be a quadratic near that point. Thus, nondegenerate extrema are true minima, true maxima or "true" saddles, according to our geometric intuition.

Moreover, under very general circumstances, a saddle will lie between two minima. In one dimension, cf. Figure 10.7, a "saddle" is just a maximum, and the result is obvious for any continuously differentiable function $F(T)$. In two dimensions, as in Figure 10.10, the result is known as the *mountain pass lemma*.

Consider a closed valley, with its bottom at I and a contour line of height $h_0 > J_I$ surrounding it. Outside this contour there is another point, such as III, with $J_{III} < h_0$. All paths Γ from I to III must cross the h_0-contour around I. Thus, for each such path, $\max_{T\in\Gamma} J\{T\} \geq h_0$. One can show, for continuously differentiable J, that the number h_* defined by $\inf_{\Gamma} \max_{T\in\Gamma} J\{T\} = h_* \geq h_0$ is a stationary value of J, and hence corresponds to a "pass" or saddle.

This result can be applied to the functional (10.41), which has only isolated critical points. From Figure 10.9 we see that for different values of μ there are either two minima of $J\{T;\mu\}$ and one saddle, or three minima and two saddles, or one minimum and no saddle. Likewise, for the model (10.12) and Figure 10.6, or the model of Figure 10.10 and Eq. (10.43), one of the two latter situations arises, according to the value of μ. Each of the saddles above has exactly one direction of instability, all other directions being stable. More details and references on the calculus of variations and global analysis will be found in Section 10.5.

Periodic forcing and quasi-equilibrium response. The preceding stability discussion suggests that slow changes in model parameters for an EBM will only lead to slow adjustments of the stable equilibria to the new parameter values. Thus, if $T_* = T_*(x;\mu)$ is a minimum of the pseudo-potential $J\{T(x);\mu\}$ for a given, fixed value of the insolation parameter μ, and μ undergoes a variation $\mu(t)$ which is slow compared to the relaxation time at T_*, then $T(x,t) = T_*(x;\mu(t))$ is an excellent approximation to the actual solution of (10.40) with $\mu = \mu(t)$. Various numerical experiments have confirmed that the so-called *climate sensitivity* $\partial\overline{T}_*/\partial\mu$ for an EBM is a good guide to the actual temperature variations resulting from slow and moderately large changes in μ (North *et al.*, 1981; compare also Section 10.2).

Over the last few millions of years of the Earth's history, changes in μ due to solar variability and to variations in the Earth's orbit around the Sun have certainly not exceeded a small fraction of one percent. On the other hand, global temperature changes over the same period were possibly as large as 10K. Given the usually accepted sensitivity value for $\partial \overline{T}_1/\partial\mu$ of roughly 1K for one percent, EBMs as discussed so far do not seem to give a satisfactory answer to the problem of paleotemperature variations.

In the remainder of this section, we shall consider two simple modifications of the EBM (10.40) which do lead to large temperature variations. The first such modification exhibits solutions which visit alternately the neighborhood of $T_1(x)$, the warm "present climate", and of $T_3(x)$, the cold "ice age" climate 10K below. The second modification has solutions which "live" near $T_1(x)$, but still have sustained fluctuations of 2K in amplitude.

Stochastic resonance. In Section 10.3 we saw that sustained, random perturbations due to phenomena on smaller spatial scales and shorter time scales can be modeled as additive white noise. This leads to exit from a finite-height potential well defined by the model's slow, large-scale, deterministic part, in finite time, with probability one. The mean exit time $\overline{\theta}$ is a sensitive, exponential function of the noise level σ^2 and of the height of the potential barrier ΔF between two wells. This sensitive dependence suggests that relatively small changes in ΔF can have an important effect on the long-term behavior of the stochastically perturbed solution.

For the investigation of this effect, our point of departure is a highly simplified, 0-D version of model (10.40), with two potential wells 10K apart. Ignoring the third, completely ice-covered potential well, and all the other details, we consider the model problem for $T = \overline{T}(t)$

$$dT = (R/c)\{\gamma(1-T/T_1)(1-T/T_2)(1-T/T_3)(A \cos \omega t + 1) + A \cos \omega t\}\, dt + \sigma\, d\beta, \tag{10.46}$$

where the insolation parameter μ was taken to vary periodically,

$$\mu = 1 + A \cos \omega t,$$

with period $2\pi/\omega = 100{,}000\text{a} = 100$ ka and amplitude $A = O(10^{-3})$, while $T_3 < T_2 < T_1$ as in Figure 10.7, but $T_1 - T_3 = O(10\text{K})$ now. The constants R and c are characteristic values of the infrared radiation $R_0(T)$ and the heat capacity $C(x)$, respectively, and γ is a tuning parameter. For details we refer to Benzi *et al.* (1982).

A similar 0-D model was derived by Nicolis (1982) directly from (10.2-10.3), without specific reference to the distance between the stable equilibria. In the absence of forcing, $A = 0$, and of stochastic perturbations, $\sigma = 0$, the solutions of (10.46) are $T = T_1$, T_2 and T_3, with T_1 and T_3 nonlinearly stable, and T_2 unstable, according to the discussion of Figure 10.7.

For constant forcing, $\mu = 1$, and $\sigma \neq 0$, the situation is as described in Section 10.3: the Kramers formula (10.39) holds, provided σ^2 is small enough compared to ΔF, yielding a mean exit time much longer than the deterministic relaxation times $1/F''(T_1)$ and $1/F''(T_3)$. If $\sigma = 0$ and $A \neq 0$, but the period of the forcing $2\pi/\omega$ is much longer than the relaxation time of either stable equilibrium, we saw that a solution starting near a stable equilibrium will simply follow the slight, hardly perceptible shift in position of the potential minimum due to the change in $\mu(t)$. This is illustrated in Figure 10.11.

What happens if both $A \neq 0$ and $\sigma \neq 0$? Assuming σ^2 is still small and $2\pi/\omega$ large, one would expect the transition from T_3 to T_1 in case (a) of Figure 10.11 to be much more probable than that from T_1 to T_3. Likewise, in case (c) the transition from T_1 to T_3 is much more probable. One feels that a situation could arise, for appropriate choices of period ω, noise level σ, and change in height of barrier ΔF, where the solution point would jump almost every time the bottom of the potential well it lies in is "up", and almost never while the well bottom is "down".

Numerical simulations indicate that almost regular jumps occur for a certain range of noise levels (Figure 10.12a). If the noise is too small, the mean exit time $\overline{\theta}$ will be longer than the half-period π/ω during which the potential barrier for a certain well is down, missing the jump. If the noise is too large, $\overline{\theta}$ will be shorter than π/ω even for the barrier being up, leading to frequent and irregular jumps from either well. The situation of nearly regular jumps for intermediate noise levels could be called *stochastic resonance*.

The condition for "resonance", i.e., for nearly regular jumps with an approximate period of $2\pi/\omega$ is, with respect to T_1 say,

$$\overline{\theta}_c(T_1) \ll \pi/\omega \ll \overline{\theta}_a(T_1). \tag{10.47a}$$

For fixed, small σ^2, and fixed, long π/ω, one can look to the Kramers formula (10.39) for guidance on the potential barrier lowering necessary to achieve (10.47a). Dropping the middle term π/ω in the comparison

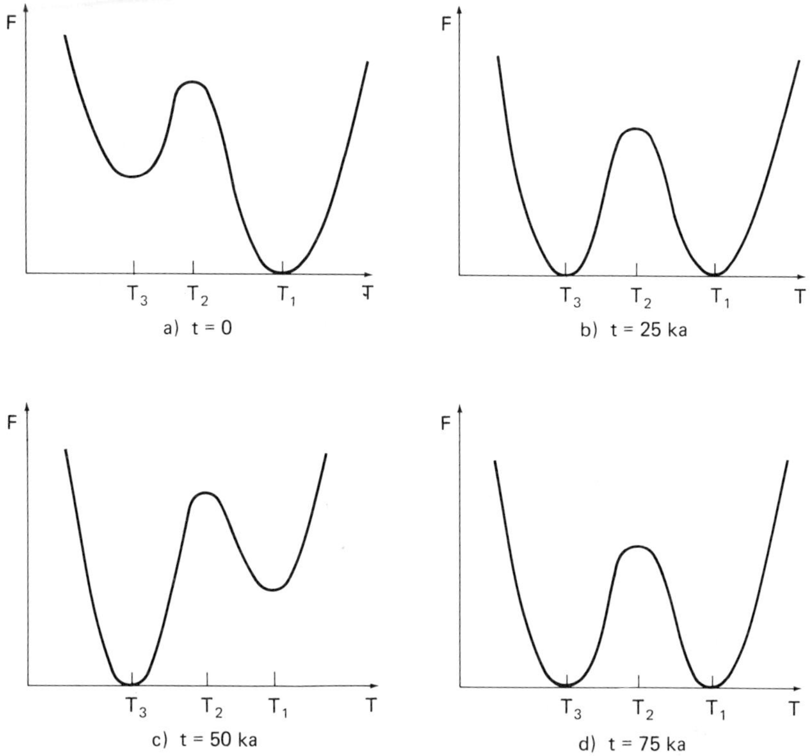

Fig. 10.11. Periodic changes in the potential $F(T;\mu(t))$ for $F'(T) = -f(T)$ given by Eq. (10.46); $T_1 \cong 290K$, $T_3 \cong 280$ K (after Benzi *et al.*, 1982).

above, the Kramers formula yields the condition

$$\exp \frac{(\Delta_{12}F)_c}{\sigma^2/2} << \exp \frac{(\Delta_{12}F)_a}{\sigma^2/2}$$

where $\Delta_{12}F = F(T_2) - T(T_1)$, assuming the product of pseudo-potential curvatures $F''(T_1)F''(T_2)$ is not too different from situation (a) to situation (c) in Figure 10.11. Thus (10.47a) implies

$$\exp\frac{\Delta(\Delta F)}{\sigma^2/2} >> 1, \qquad (10.47b)$$

with $\Delta(\Delta F) = (\Delta_{12}F)_a - (\Delta_{12}F)_c$.

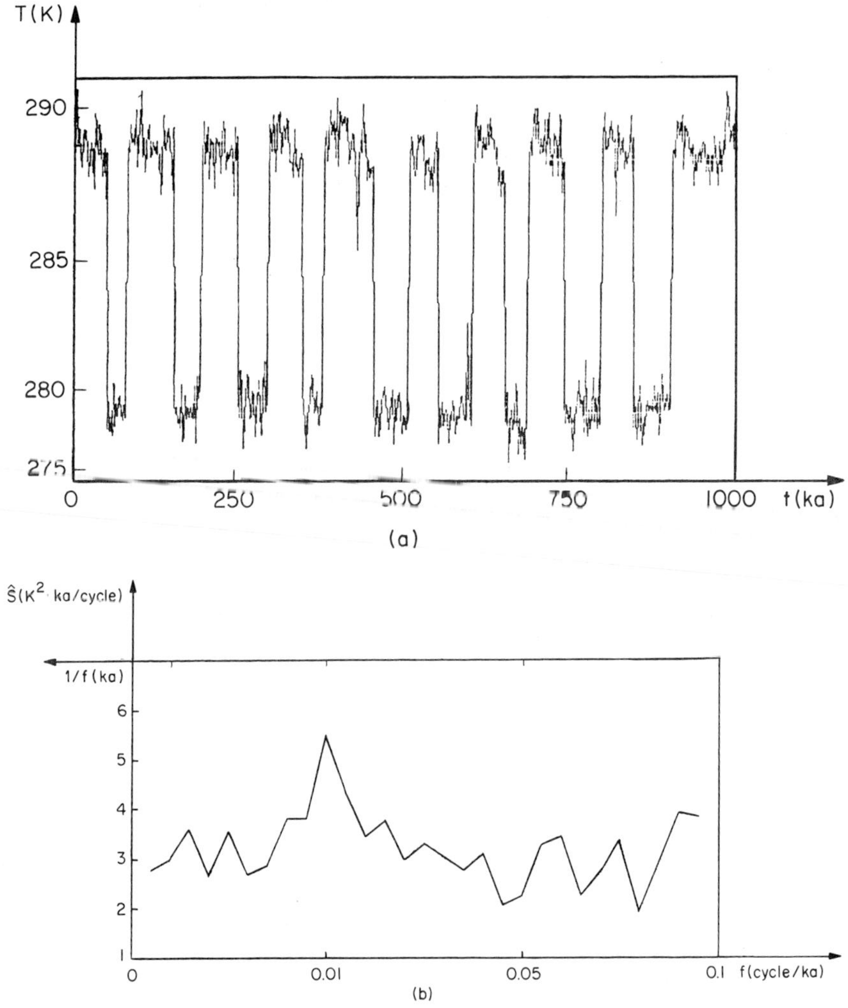

Fig. 10.12. Periodic jumps between two potential wells, under the combined effect of periodic forcing and stochastic perturbations (after Benzi *et al.*, 1983). a) Time series of $T(t)$; b) power spectrum of $T(t)$ above.

Recall that $\sigma^2 << (\Delta_{12}F)_c$ is necessary for (10.39) to be a valid approximation of the shorter exit time $\overline{\theta}_c$, so that the potential in Figure 10.11, with $\Delta(\Delta F) \geq (\Delta_{12}F)_c$, clearly satisfies (10.47b). Hence, with $\Delta(\Delta F)$ given as above, one can find an admissible range of noise levels σ^2 for which the resonance condition (10.47a) provides two-sided inequalities. The power spectrum of the nearly regular jumps in Figure 10.12a is shown in Figure 10.12b.

A broad peak at ω is visible in Figure 10.12b. Similar spectra for σ^2 smaller or larger than given by (10.47a) merely show a quadratic, red-noise like decay of the continuous background. These red-noise spectra result from the jumps in $T(t)$ being completely irregular, as they would be in the absence of periodic forcing. The jumps in this case have exponentially distributed waiting time (compare Figure 6.19 and discussion there) and a variance roughly equal to the mean. Towards the end of this section, more will be said about the harmonic analysis of stationary time series in general, and about red noise in particular.

The relatively small irregularities of jumps in Figure 10.12a suggest that a reduction in variance occurs near resonance. Figure 10.13 show that indeed the ratio R of variance of exit time $E(\theta - \overline{\theta})^2$ to its squared mean $\overline{\theta}^2$ decreases drastically for a small range of σ-values. This range of values is analogous to the resonance band width of a deterministic oscillator.

The phenomenon of stochastic resonance, as discussed here, represents an interesting application of nonlinear analysis and stochastic ideas to climate theory. Continuous records of climatic change available on different time scales (see, for instance, Figure 11.2) show no evidence for the combination of bistable equilibria and of symmetric, sharp jump transitions between them which are associated with this phenomenon in its simplest form analyzed so far. Still, extending the study of stochastic perturbations to nonlinear systems with *nonstationary* solutions should help explain the complexity of climatic behavior on all time scales. In this sense, it represents an important step on the road to ultimate modeling and understanding of climatic interactions, just like the study of deterministic multiple equilibria in the first two sections of this chapter.

<u>The effect of time lags</u>. We return now to the full, 1-D deterministic model (10.40), and will study the effect that the delay in ice-albedo feedback due to slow growth and decay of ice cover has on model solutions, in the absence of any periodic forcing. To illustrate the potential importance of such a delay effect, we consider first the simplest possible problem where it occurs.

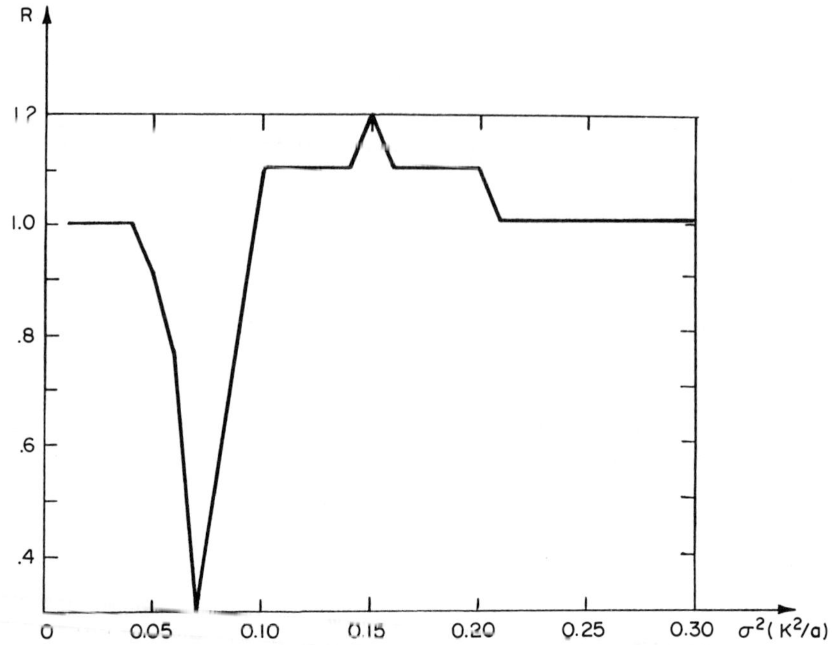

Fig. 10.13. Ratio R between the variance of transition times $E(\theta-\overline{\theta})^2$ and their squared mean $\overline{\theta}^2$ for the model of Figure 10.11, as a function of noise level σ (after Benzi *et al.*, 1983).

The prototype of decay to a stable equilibrium is the equation

$$\dot{u} = -u,$$

with solution

$$u(t) = u_0 e^{-t}.$$

Introducing a delay τ,

$$\dot{u}(t) = -u(t - \tau) \tag{10.48a}$$

gives, for $\tau = \pi/2$, a family of oscillatory solutions

$$u(t) = u_0 \cos t + v_0 \sin t, \tag{10.48b}$$

where u_0 and v_0 are arbitrary. Similar solutions exist for $\tau = k\pi/2$, where k is any integer. One way of understanding the effect of the delay τ is to introduce the auxiliary dependent variable v,

$$v(t) = -u(t - \pi/2); \tag{10.49}$$

the same solution as (10.48b) is obtained from the system of two ODEs

$$\dot{u} = v, \tag{10.50a}$$

$$\dot{v} = -u, \tag{10.50b}$$

subject to (10.49) and initial conditions

$$u(0) = u_0, \quad v(0) = v_0. \tag{10.50c,d}$$

The single *delay-differential equation* (DDE) (10.48a) is thus equivalent to the system of two ODEs with no delays (10.50). For references to the literature of DDEs and the variational interpretation of (10.48) and (10.50), see Bhattacharya *et al*. (1982) and Section 10.5.

Recall now that $\alpha(T)$ in (10.2) and (10.12) was meant to represent the heightened albedo of snow and ice, which were assumed to cover land and water instantaneously as temperature dropped. In talking about time scales longer than a year and as long as many millenia, one realizes that considerable accretions of ice take a long time to form, stabilize and extend, as well as withdraw. Likewise, the additional mean climatic cloud effect included in (10.40) was attached to the long-term ice margin, and thus would suffer the same time lag in its latitudinal displacement as the ice itself.

The way this delay is introduced into Eq. (10.40) is the following. The albedo α will be computed as a function of past, as well as present temperatures,

$$\alpha = \alpha^*(T^*;\tau,s_0), \tag{10.51a}$$

where

$$T^*(t) = \int_0^\infty w(s)\ T(t-s)ds; \tag{10.51b}$$

the weight given to past temperatures is

$$w(s) = \begin{cases} Ae^{-(\tau-s)^2/2s_0^2}, & 0 \le s < 2\tau, \\ 0 & , \quad 2\tau \le s, \end{cases} \tag{10.51c}$$

with A chosen so that $T(t) \equiv 1$ implies $T^*(t) \equiv 1$. The *distributed delay* of (10.51b,c) (Bhattacharya *et al.*, 1982) is more realistic than the *discrete delay* considered for simplicity in (10.48) and used by Ghil and Bhattacharya (1979). The largest weight is given to the temperature lagged by τ, and s_0 is the half-width of the weight distribution.

Available observations do not permit establishing values of τ and s_0 *a priori* with any certainty. The point of the investigation will be to see how the behavior of solutions changes as τ and s_0 vary within reasonable limits.

An appropriate time scale for this investigation is provided by the relaxation time to the "present climate" of model (10.40) with $\tau = 0$. Choosing the heat capacity $C(\phi)$ to correspond to the ocean volume present at each latitude ϕ, this relaxation time, or *e*-folding time, is equal to 2060 a in the model. We consider therefore delays of hundreds and thousands of years, like those involved in the build-up of continental ice sheets during ice ages.

The stationary solutions of (10.40) with $\alpha(T)$ replaced by $\alpha^*(T^*)$ are the same as before. Their stability, however, for $\tau > 0$ and $s_0 \geq 0$ is different, as suggested by the simple example (10.48).

Consider first the case of a discrete delay, $s_0 = 0$, where $w(s) = \delta(s-\tau)$. If the problem were linear, we would expect, according to (10.48), periodic solutions to arise at certain discrete values of τ, and only at those values, and have arbitrary amplitude. For the nonlinear Eq. (10.40) we expect instead, as seen for instance in Section 5.3, the stationary solution $T_1(x)$ to be stable up to a certain value of τ, and then to transmit its stability to a periodic solution, which starts out with zero amplitude and has increasing amplitude as τ increases. In the discrete-delay version of the model (Ghil and Bhattacharya, 1979), such behavior was actually observed. The general properties of stability transfer from a stationary to a growing periodic solution will be discussed in Section 12.2.

For s_0 comparable to τ or larger, we expect $T_1(x)$ to keep its stability, no matter what the value of τ: if the influence of past temperatures is averaged out over an entire cycle, incipient oscillations will be damped out. Thus, for a given τ, there exists a critical value of s_0, called $\sigma_c = \sigma_c(\tau)$, such that for $s_0 > \sigma_c$ there are no sustained oscillations. Figure 10.14 shows this critical half-width σ_c as a function of τ, scaled on the vertical axis to the left of the figure.

Sustained oscillations about $T_1(x)$ exist and are stable to initial perturbations for $\tau \geq \tau_0 \simeq 475$ yr. The peak-to-peak amplitude A of the oscillations in mean global temperature $\overline{T}(t)$ depends mostly on τ and very little on s_0 for $s_0 < \sigma_c$. The amplitude is also shown in Figure 10.14, scaled on the axis to the right. Between $\tau \simeq 1000$yr and $\tau \simeq 3000$ yr it is approximately constant and equal to 1K. The temporal average

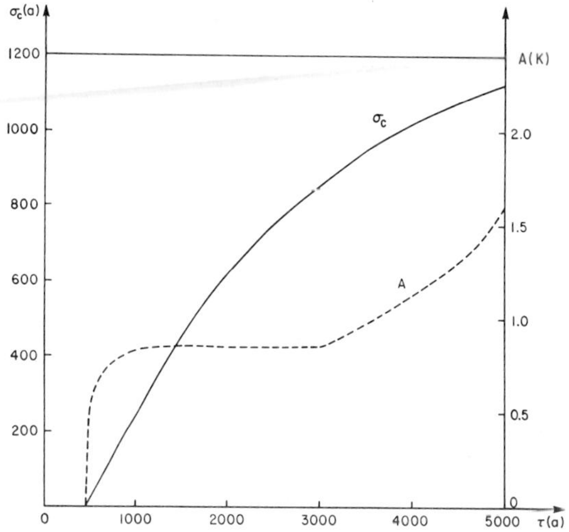

Fig. 10.14. Dependence of sustained model oscillations on the lag parameters τ and s_0 (after Bhattacharya *et al.*, 1982). No oscillations occur for $s_0 \geq \sigma_c$ (solid curve). The amplitude of the oscillations for $s_0 < \sigma_c$ is A (dashed curve).

over one period of the global temperature $\overline{T}(t)$ is independent of τ and equals precisely $\overline{T}_1$.

The shape of the oscillations in global temperature $\overline{T}(t)$ is shown in Figures 10.15a-f. For low τ, 500 a $\leq \tau \leq$ 1000 a, a nearly sinusoidal oscillation of period about 2τ is evident (Figures 10.15a-c). It is modulated in amplitude with periods of the order of $10\ \tau$ - $100\ \tau$.

At $\tau \simeq 1500$ a (Figure 10.15d) the shape of the "carrier-wave" oscillation becomes square, and prominent spikes appear at $\tau = 5000$ a and $\tau = 10000$ a (Figures 10.15e-f). The modulation is now completely irregular.

To describe more completely the transition from the periodic, regular behavior of Figures 10.15a-c to the aperiodic, irregular solution behavior of Figures 10.15e-f, it is useful to consider the power spectra of the solutions. Given the Fourier transform $\hat{T}(f)$ of the real, stationary time series $\overline{T}(t)$,

$$\hat{T}(f) = \frac{1}{2\pi}\int_{-\infty}^{\infty} \overline{T}(t)e^{-ift}\, dt, \tag{10.52a}$$

the spectral density $\hat{S}(f)$ of its power is an even function of frequency f defined by

$$\hat{S}(f) = \hat{T}(f)\ \hat{T}^*(f), \tag{10.52b}$$

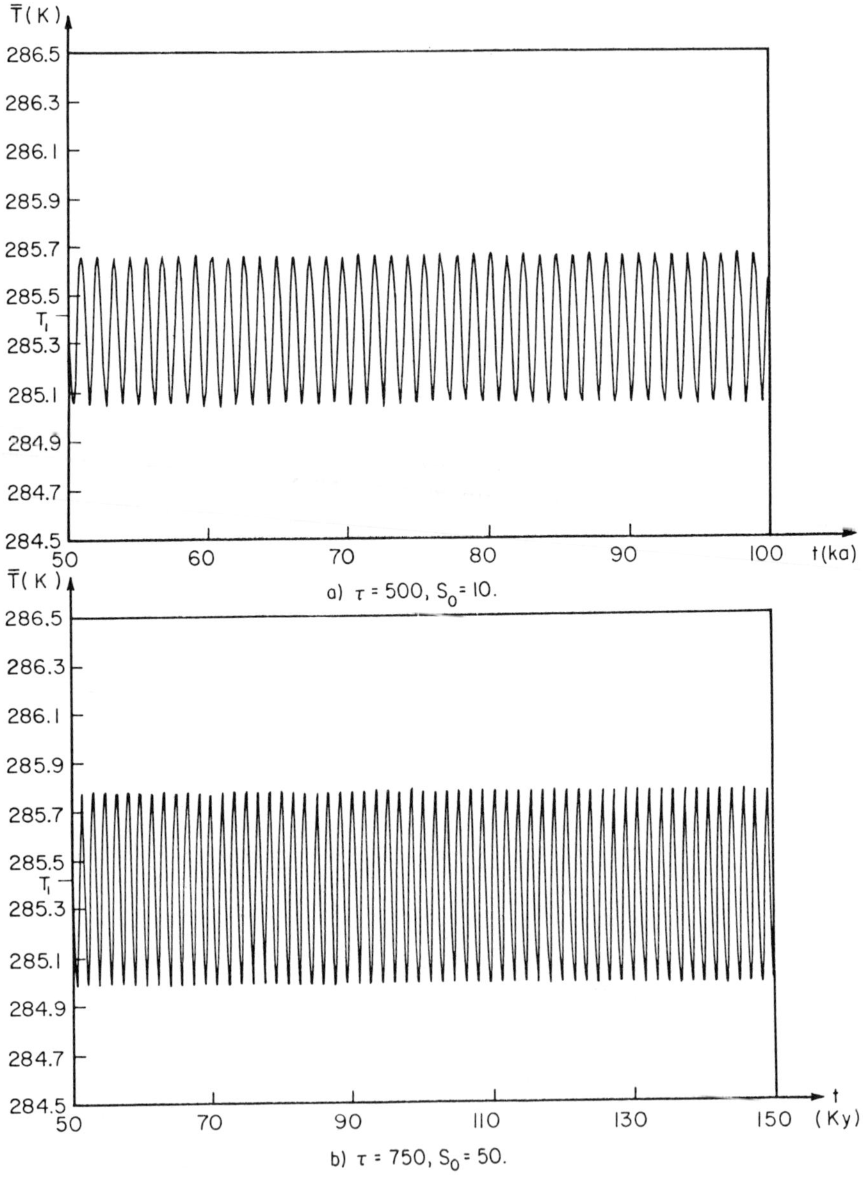

Fig. 10.15. Time evolution of the globally averaged temperature $\overline{T}(t)$ (after Bhattacharya *et al.*, 1982).

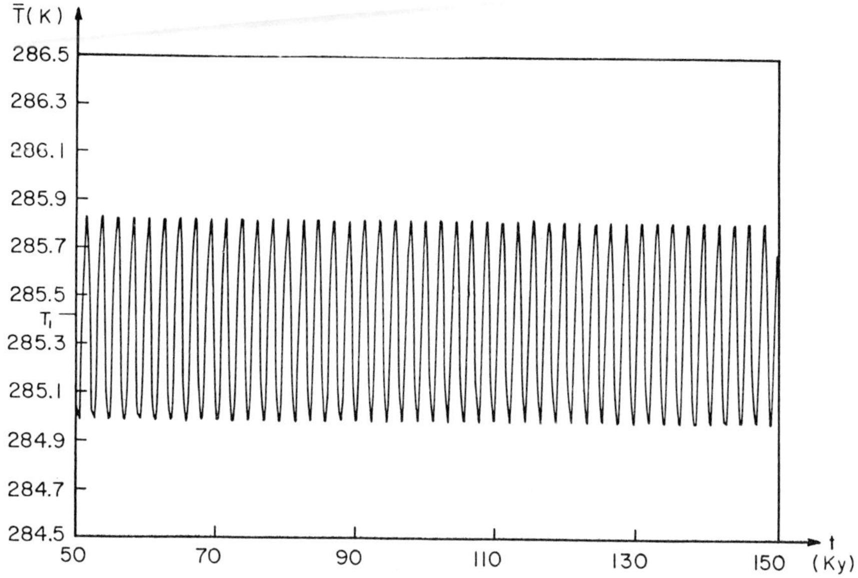

c) $\tau = 1000, S_0 = 100.$

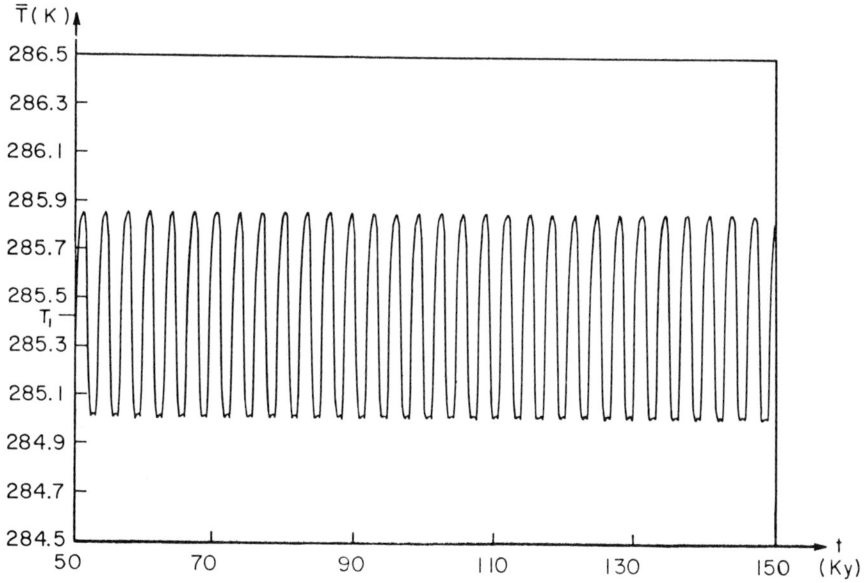

d) $\tau = 1500, S_0 = 100$

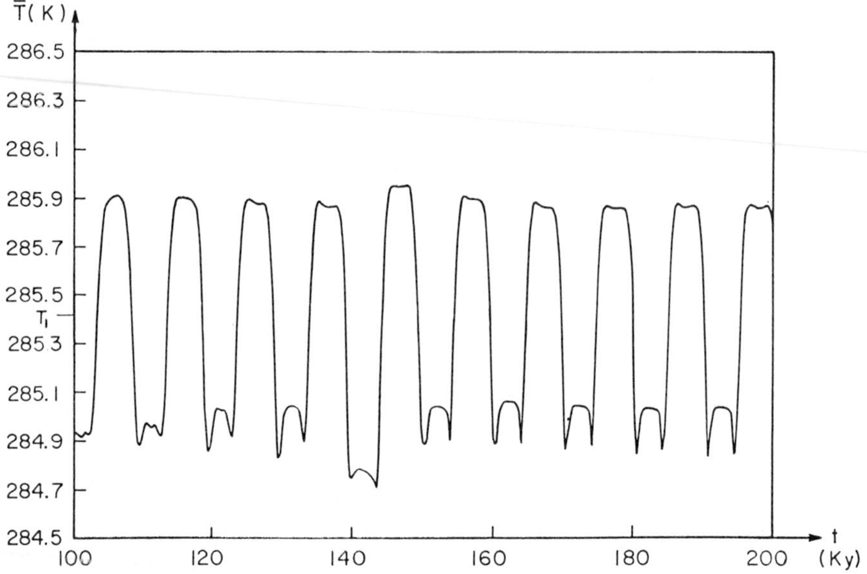

e) $\tau = 5000$, $S_0 = 500$.

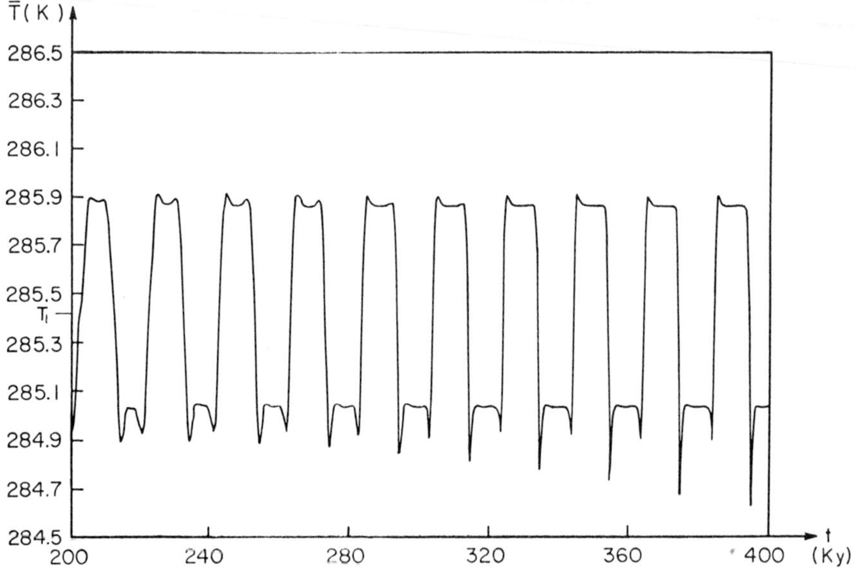

f) $\tau = 10000$, $S_0 = 500$.

where $\hat{T}^*$ is the complex conjugate of $\hat{T}$. By Parseval's theorem

$$\int_{-\infty}^{\infty} \overline{T}^2(t)\, dt = 2 \int_0^{\infty} \hat{S}(f)\, df. \tag{10.52c}$$

The power spectrum $\hat{S}(f)$ can be computed from the lagged correlations $R(s)$ of $\overline{T}(t)$,

$$R(s) = \lim_{L\to\infty} \frac{1}{2L} \int_{-L}^{L} \overline{T}(t)\, \overline{T}(t+s)\, dt, \tag{10.53a}$$

using the *Wiener-Khinchin formula*

$$\hat{S}(f) = \frac{1}{2\pi} \int_{-\infty}^{\infty} R(s)\, e^{-ifs} ds. \tag{10.53b}$$

In practice, due to finite length L of the time series and sampling problems, a Bartlett spectral window is used (see Appendix B in Bhattacharya *et al.*, 1982, for details and further references). The power spectra of $\overline{T}(t)$ in Figure 10.15a-f, as shown in Figure 10.16a-f, respectively.

In Figures 10.16a-c, a sharp peak at $f_0 \simeq 1/2\tau$ dominates the spectrum. One, two or three sharp harmonics at $2f_0$, $3f_0$, $5f_0$ and $6f_0$ also rise above the background in each panel, to the right of the fundamental f_0. In Figures 10.16c-d the peaks broaden and their number increases further. Finally, in Figures 10.16e-f the distinction between peaks and background disappears. The background in the last panel has an exact (-2) slope in the figure's log-log coordinates.

Thus transition from regular to irregular behavior in the time domain corresponds to a transition from sharp peaks to a continuous background in the spectral domain. Similar transitions occur in fluid dynamical laboratory experiments (Figure 10.17), and in many other physical phenomena (compare Sections 5.3-5.5 and 6.4-6.6). What is the significance of the (-2) log-log slope?

We return for a moment to the *Langevin equation* (10.28), as a model for Brownian motion, and compute the power spectrum of its solution ensemble. One could use the solutions' (10.31a) lagged correlations, similar to (10.32), and the Wiener-Khinchin formula (10.53) to do so.

A quicker and more enlightening way is to compute the spectral density directly, by Fourier transforming the equation. Recall the differentiation rule for the Fourier transform (10.52a), namely,

$$\hat{\dot{u}} = if\,\hat{u},$$

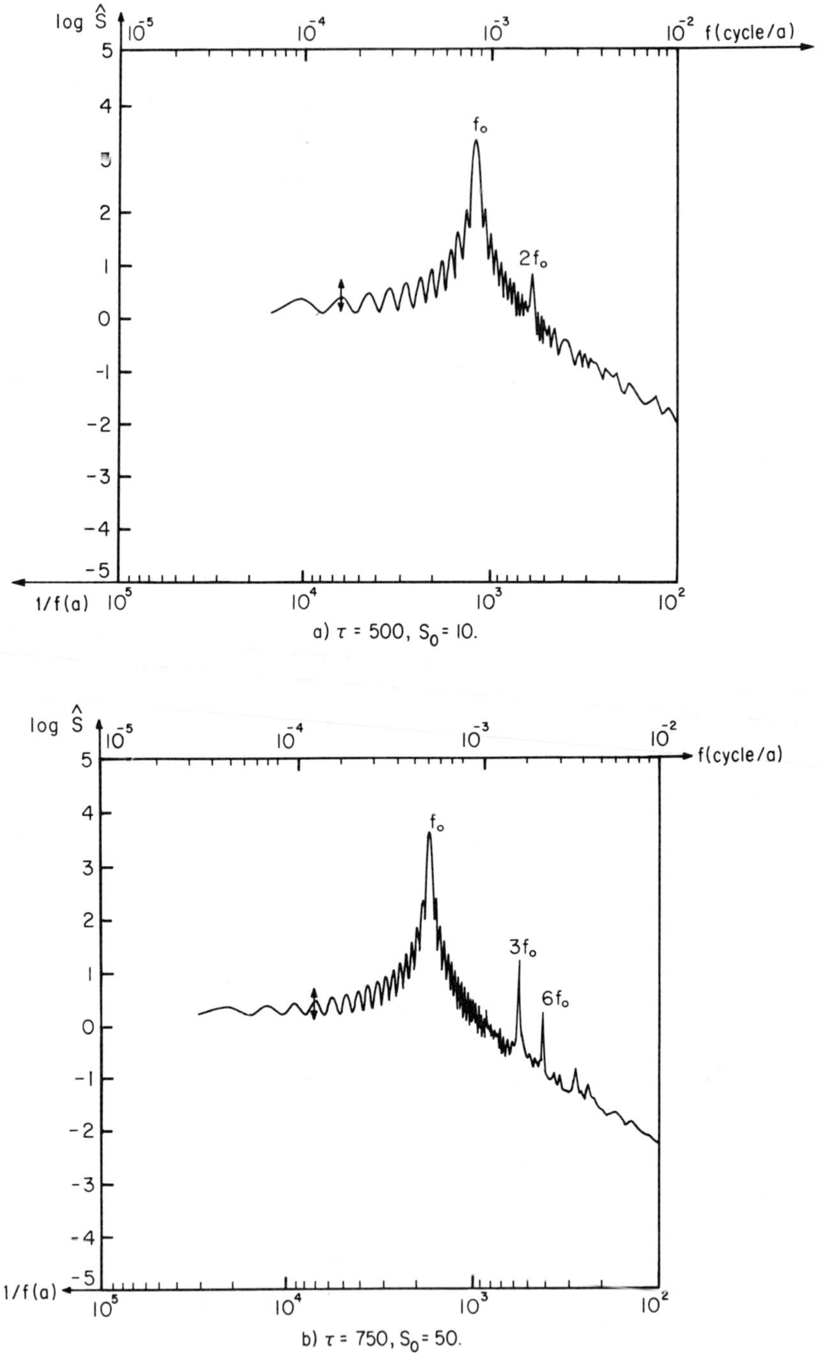

Fig. 10.16. Power spectra of the time series in Figure 10.15. Both axes are logarithmic, with the abscissa marked in frequencies $f(a^{-1})$ at the top and in periods 1/f (a) at the bottom (after Bhattacharya *et al.*, 1982).

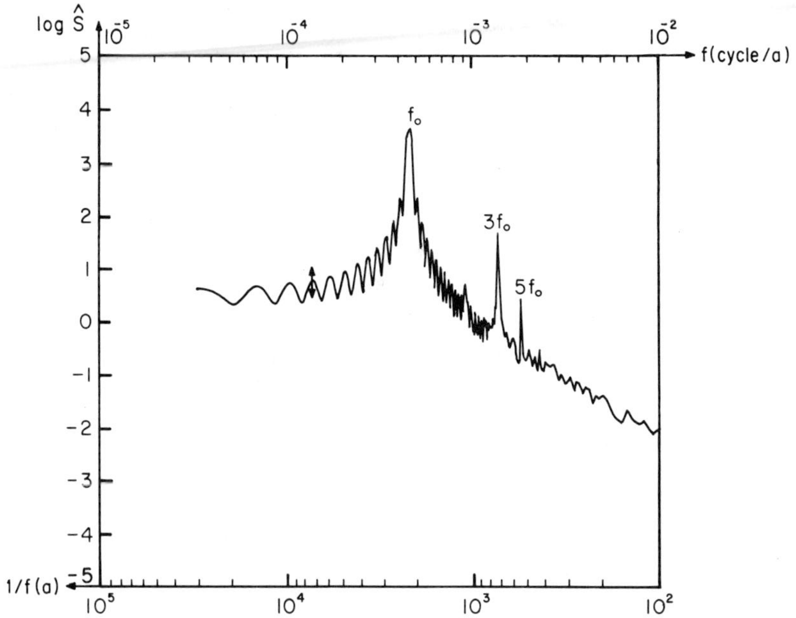

c) $\tau = 1000$, $S_0 = 100$.

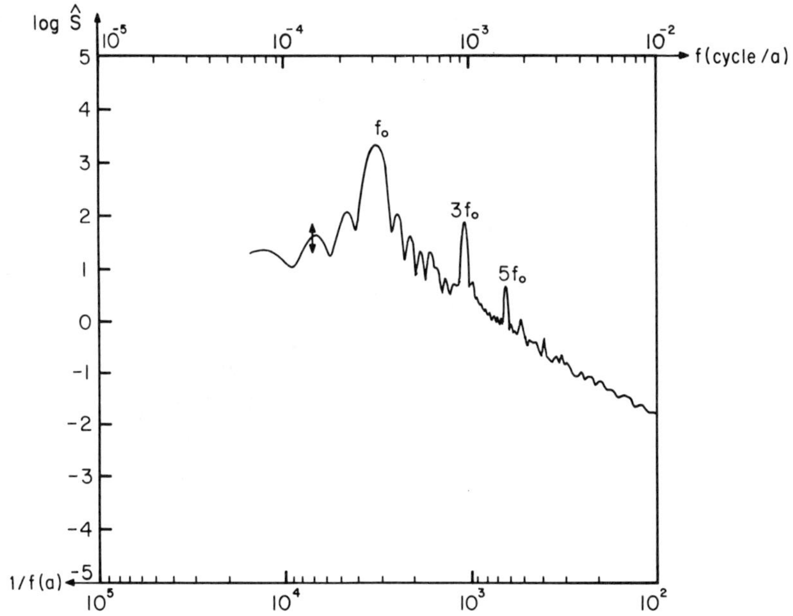

d) $\tau = 1500$, $S_0 = 100$.

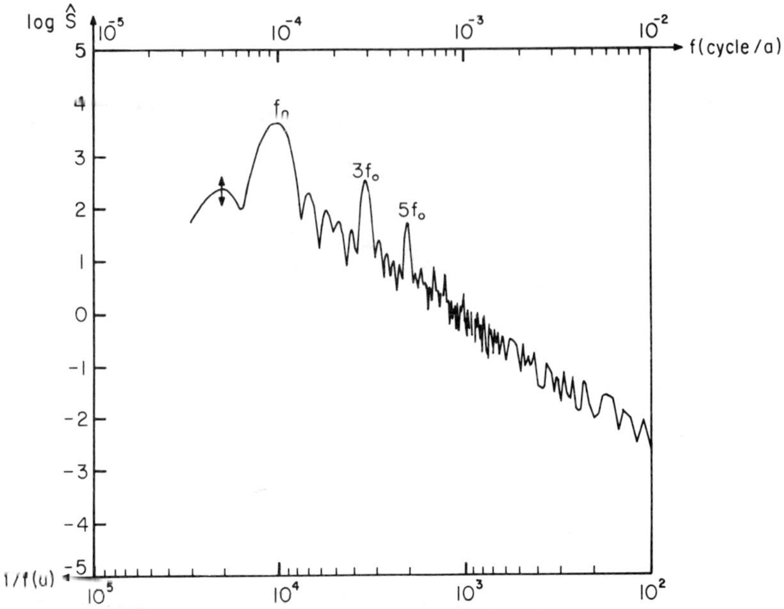

e) τ = 5000, S_0 = 500.

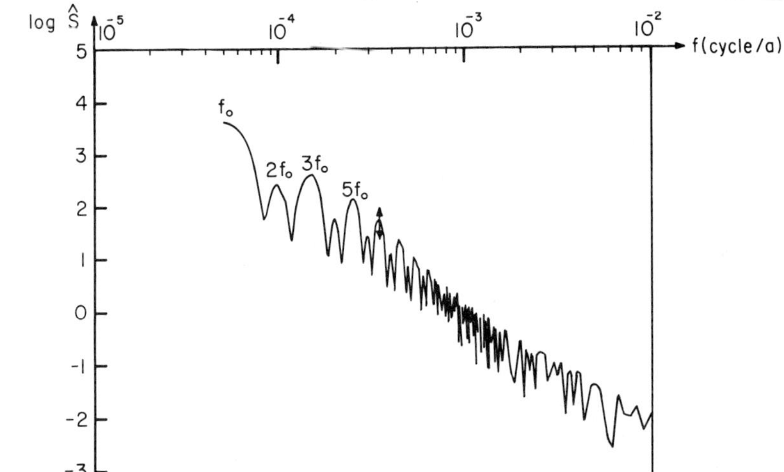

f) τ = 10000, S_0 = 500.

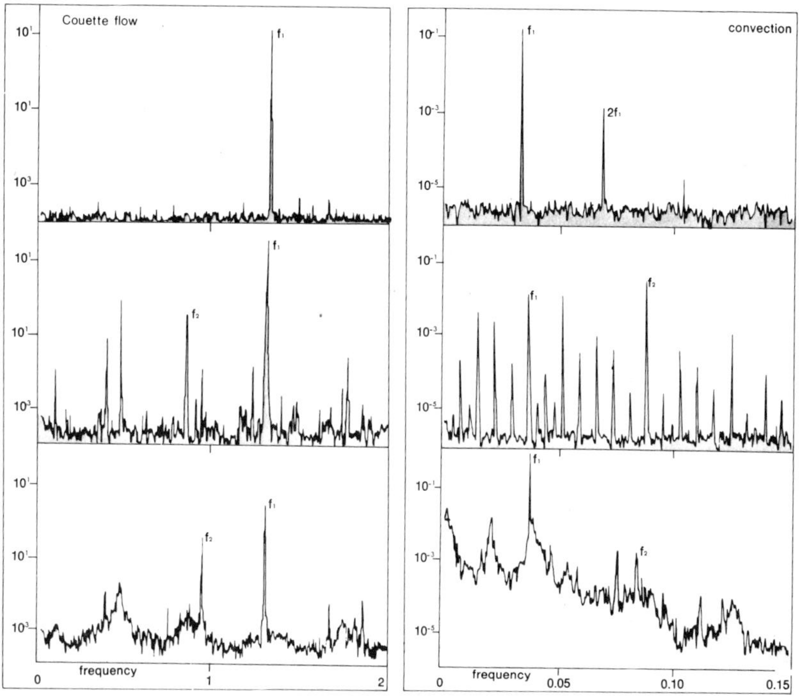

Fig. 10.17. Power spectra for measured time series in two fluid dynamical experiments in which transition from periodic to aperiodic behavior is apparent as a parameter varies (from Ruelle, 1980). On the left, Couette flow between two coaxial cylinders as the relative rotation speed increases. On the right, Rayleigh-Bénard convection between two horizontal plates as the heating from below increases (compare Section 5.4).

so that Fourier transforming Eq. (10.28) yields

$$\hat{u}(f;\omega) = \frac{1}{\lambda + if}\,\hat{\eta}(f;\omega).$$

The spectral density $\hat{S}(f)$ therefore, using an ensemble average rather than the ergodically equivalent time average of (10.52), is

$$\hat{S}(f) = E\,\hat{u}\,\hat{u}^* = \frac{1}{\lambda^2 + f^2}\,E\,\hat{\eta}\,\hat{\eta}^*.$$

Due to the white-noise property (10.29b) of $\eta(t;\omega)$ and using formula (10.32a) for its covariance, the spectral density of Brownian motion is

$$\ddot{S}(f) = \frac{\tau_*/2\pi}{\lambda^2 + f^2}, \tag{10.54}$$

which asymptotes to a straight line with slope (-2) in log-log coordinates.

The spectral density (10.54) decays monotonically with increasing frequency, which justifies the name "red noise" often given to the random process (10.28), by comparison to white noise, which has a constant spectral density τ_*/π. The relevance of Brownian motion concepts to the explanation of red-noise-like climatological spectra was suggested by Hasselmann (1976) and Mitchell (1976). Further references will be found in Section 10.5.

Figure 10.16f shows that the same asymptotic slope of continuous spectral density as given by (10.54) can be obtained in a *deterministic* climate model. Thus nonlinear interactions between a few degrees of freedom, acting on the time scale of interest itself, can yield the same type of irregular behavior as sustained perturbations produced by many more degrees of freedom active on much shorter time scales.

We shall not discuss the large-amplitude, irregular oscillations of this model between the "present climate" $T_1(x)$ and the "ice-age" climate $T_3(x)$. They are similar to those in Figure 10.12a, although caused by the deterministic delay, rather than by periodic forcing and stochastic perturbations.

The previous results are not realistic in the sense of capturing all the most important climatic mechanisms and simulating in detail climatic time series. They present, rather, tools for the inquiry into more complete models, as well as some qualitative features of climatic behavior which should be reproduced by such models.

10.5. Bibliographic Notes

We shall forgo the pleasure of citing exact references for the floods of Noah and Utnapishtim, or for the inferred connections between the two and seasonal floodings in prehistoric Mesopotamia. Numerous other ancient texts, Egyptian, Chinese and Indian, record climatic change or allude to it, as does the monumental work of the Greek historian Herodotus (especially its second volume).

Standard references for descriptive climatology, in particular the classification of regional climates, are Köppen (1923), Thornthwaite (1933) and Trewartha (1954). The transition between descriptive climatology, on the one hand, and climate dynamics, on the other, is provided by texts on physical climatology, which study quantitatively the energy and water balance; globally and regionally (Budyko, 1963; Lamb, 1972, 1977; Landsberg, 1967; Sellers, 1965). An up-to-date quantitative account of the present climate is provided by Levitus (1982) for the ocean and by Oort (1983) for the atmosphere. The words "dynamics" and "climate" probably appeared together in print for the first time in a volume edited by R. L. Pfeffer (1960).

Section 10.1. London and Sasamori (1971) compare their own, earlier calculations, based on conventional data (London, 1957), with those of Von der Haar and Suomi (1971; VHS in the caption of Figures 10.2 and 10.3), based on the newly incoming meteorological satellite data. Global radiative flux values slightly different from those in Figure 10.1 are given by the U.S. Committee for GARP (1975, p. 18). More recent satellite and recent data for both the solar constant and atmospheric fluxes are given by Jacobowitz *et al.* (1984), Stephens *et al.* (1981) and Wilson *et al.* (1980). A quick overview of earth radiation budget literature is given by Wiscombe (1983).

Using such recent data and the corresponding radiative flux computations, the uncertainty reflected in Figures 10.2 and 10.3 can be reduced, but not eliminated. This observational difficulty can only augment the importance of qualitative, theoretical results in climate dynamics which are either independent of very precise quantitative considerations or help reduce the quantitative uncertainty (Ghil, 1981c; Saltzman, 1983).

Section 10.2. Earlier references to quantitative models of the energy-balance type are given by North (1981) and by Saltzman (1978). It is clear, however, that the simultaneous and independent publication of the articles by Budyko (1969) and Sellers (1969), two recognized leaders in physical climatology, came at the right time and produced an increasing stream of contributions along the same lines. In addition to the articles already cited in the text, important advances were made by Faegre (1972), Schneider and Gal-Chen (1973) and Drazin and Griffel (1977).

Our exposition of 0-D EBMs benefited from Crafoord and Källén (1978). Successive review articles of the exploding literature were written by Schneider and Dickinson (1974), who place EBMs in a hierarchy of climate models of increasing complexity, Saltzman (1978), who provides systematic derivations of zonally-averaged equations from the full dynamic and thermodynamic equations, Ghil (1981a), whose review was used extensively in this section, and North *et al.* (1981), whose account of these models, with many additional details, can be considered as definitive. In particular, EBMs with seasonally varying insolation, and with some zonal resolution (two space dimensions) are reviewed by North *et al.* (1981), Saltzman (1978), and Schneider and Dickinson (1974).

The absence of explicit fluid dynamics from EBMs is not to be taken lightly. It comes back to haunt us in the representation of eddy diffusion as a function of T and ∇T only. This representation, or "parameterization", in the parlance of atmospheric and oceanic modelers, cannot be done rigorously or with uniform accuracy. For attempts at the most reasonable parameterization see Green (1970), Haidvogel and Held (1980), and Stone (1972); further references can be found in Ghil (1981a,c).

A systematic study of climate sensitivity for equilibrium models can be undertaken by the use of fluctuation-dissipation relations (see Bell, 1985, for a review), or by the adjoint method of functional analysis and nuclear reactor engineering (see Cacuci and Hall, 1984, and Marchuk, 1975, for applications to climate models). The fact that sensitivity analysis depends on the nature of the model under study, whether stationary, periodic or aperiodic, is emphasized by Ghil (1984a), who also illustrates methods of sensitivity analysis for nonstationary models (see also Section 12.1).

Section 10.3. The calculus of variations for solutions of ODEs and PDEs is a classical topic and it suffices to cite Courant and Hilbert (1953, Chapter IV). Its modern developments led to the Morse theory of global analysis (Smale, 1980, pp. 117-127) and include in particular the mountain pass lemma (Nirenberg, 1981).

The theory of stochastic differential equations (SDEs), originating with Einstein (1905), advanced through the classical articles collected by Wax (1954), and is well presented at a comfortable level by Arnold (1974). The idea of representing "weather" phenomena on smaller spatial and shorter time scales as additive stochastic perturbations to the large-scale equations of slow climate evolution is due to Hasselmann (1976) and Mitchell (1976).

The formulation of Hasselmann was immediately applied to EBMs by Fraedrich (1978) and Lemke (1977). Multiplicative noise, in which coefficients rather than an additive term are random, was studied in EBMs by Dalfes *et al.* (1983), Nicolis and Nicolis (1979) and Robock (1978).

The question of exit in finite time is important in the theory of SDEs in general; references are Jona-Lasinio (1985) and Schuss (1980, Chapters 7 and 8). Gardiner (1983) and Matkowsky *et al.* (1984) sharpen and extend the result of Kramers.

There are two basic approaches to computing the exit time. One studies the time evolution of the probability distribution $p(x,t)dx$ for the solution of the SDE to be in the interval $(x,x + dx)$ at time t. This probability density is governed by the Fokker-Planck equation associated with the SDE, which is a parabolic PDE with as many spatial independent variables as the original (system of) SDE(s) had dependent variables. In general, this equation has to be solved numerically. This is the approach used by Hasselmann and by C. Nicolis and G. Nicolis in climate dynamics, and it becomes prohibitively difficult as the number of variables increases.

The other approach starts with the equation for the expected exit time $\overline{\theta}(x;x_0,x_1)$ from the interval $[x_0,x_1]$ for a solution starting at the point x in the interval. This equation is essentially the adjoint of the steady-state Fokker-Planck equation for $p(x,+\infty)$. It can be solved by singular perturbation methods for the spatial operator's lowest eigenvalue (Schuss, 1980, Section 7.6), by a method of rays, analogous to geometric optics, for the paths of equilibrium likelihood (Ludwig, 1975), and by a minimax principle, due to Ventzel and Freidlin, applied to these paths (see Jona-Lasinio, 1985, and references therein). This second approach is used by Benzi, Sutera and associates in climate dynamics.

The computational difficulties associated with both approaches correspond to the metascientific principle of conservation of arduousness: after all, stochastic perturbations were supposed to represent an infinite number of degrees of freedom, in the simplest possible form. Beside the technical difficulties in applications with many slow, deterministic variables, fundamental difficulties associated with the Gaussian approximation for "small" noise also arise (Kushner, 1984). Essentially, the latter difficulties have to do with the fact that approximations involving very long exit times conflict with those involving much shorter diffusion times, i.e., the diffusion time, for which (10.32c) holds is still quite short compared to the exit time given by (10.39). The uniform validity of asymptotics used in deriving rigorously either formula is therefore questionable in many applications.

Section 10.4. The albedo modification, to include cloud effects, and the delay parameter needed to account for slow changes in continental ice masses are due to Ghil and Bhattacharya (1979). References for the physical mechanisms involved in these cloud-albedo effects are given in Bhattacharya *et al.* (1982), and in Somerville and Remer (1984).

The variational principle for 1-D EBMs was introduced by Ghil (1976) and studied further by North *et al.* (1979). It plays a crucial role in the investigation of stochastic perturbations of 0-D EMBs by Benzi, C. Nicolis, G. Nicolis and Sutera, since exit problems are considerably easier for so-called gradient systems, in which a pseudo-potential can be constructed (Gardiner, 1983). The use of such a potential for stochastic perturbations of a 1-D parabolic PDE is illustrated by Jona-Lasinio (1985).

The ideas of stochastic resonance were developed independently by Benzi *et al.* (1982) and Nicolis (1982), using the slightly different tools outlined before. It is important to realize that both approaches rely on approximations which are not known to be valid when the pseudo-potential is a function of time. The heuristic discussion of Eq. (10.47) and the numerical results of Figures 10.12 and 10.13 are the only evidence available for this very intriguing mathematical phenomenon. A rigorous proof for the existence of a periodic probability distribution in such cases, as well as some experimental or observational evidence from a natural system would be quite gratifying.

Standard references for delay-differential, or functional differential equations (FDEs), are Driver (1977), Hale (1977; see also the excellent introduction of the earlier, 1971 edition) and MacDonald (1978). Recent results on periodic and aperiodic solutions appear in Farmer (1982) and in Mallet-Paret and Nussbaum (1984).

The spectral analysis of time series, whether produced by deterministic models, stochastic processes or by observations of natural systems, is a topic of great theoretical and practical importance. As we saw in Sections 5.5, 6.4 and 6.6, one can define for strange attractors with nice properties an invariant measure, with respect to which almost all trajectories are ergodic. Hence the difference between truly stochastic and deterministically chaotic time series is less clear-cut than one might think (Guckenheimer, 1982). For a concise introduction to the main ideas of spectral analysis we recommend Hannan (1960). The tricks of the trade can be learned from Bloomfield (1976), Jenkins and Watts (1968), and Pestiaux and Berger (1984).

CHAPTER 11

GLACIATION CYCLES: PHENOMENOLOGY AND SLOW PROCESSES

The previous chapter dealt with the climatic system's radiation balance, which led to the formulation of equilibrium models. Slow changes of these equilibria due to external forcing, internal fluctuations about an equilibrium, and transitions from one possible equilibrium to another one were also studied.

Climatic records exist on various time scales, from instrumental records on the time scale of months to hundreds of years, through historical documents and archaeologic evidence, to geological proxy records on the time scale of thousands to millions of years. These records indicate that climate varies on all time scales in an irregular fashion. It is difficult to imagine that a model's stable equilibrium, whether slowly shifting or randomly perturbed, can explain all of this variability.

A summary of climatic variability on all time scales appears in Figure 11.1, due to M. Mitchell. The most striking feature is the presence of sharp peaks superimposed on a continuous background. The relative power in the peaks is poorly known; it depends of course on the climatic variable whose power spectrum is plotted, which is left undefined in the figure. Furthermore, phenomena of small spatial extent will contribute mostly to the the high-frequency end of the spectrum, while large spatial scales play an increasing role towards the low-frequency end.

Many phenomena are believed to contribute to changes in climate. As discussed in Section 6.1, anomalies in atmospheric flow patterns affect climate on the time scale of months and seasons. On the time scale of tens of millions of years, plate tectonics and continental drift play an important role. Variations in the chemical composition of the atmosphere and oceans are essential on the time scale of billions of years, and significant on time scales as short as decades.

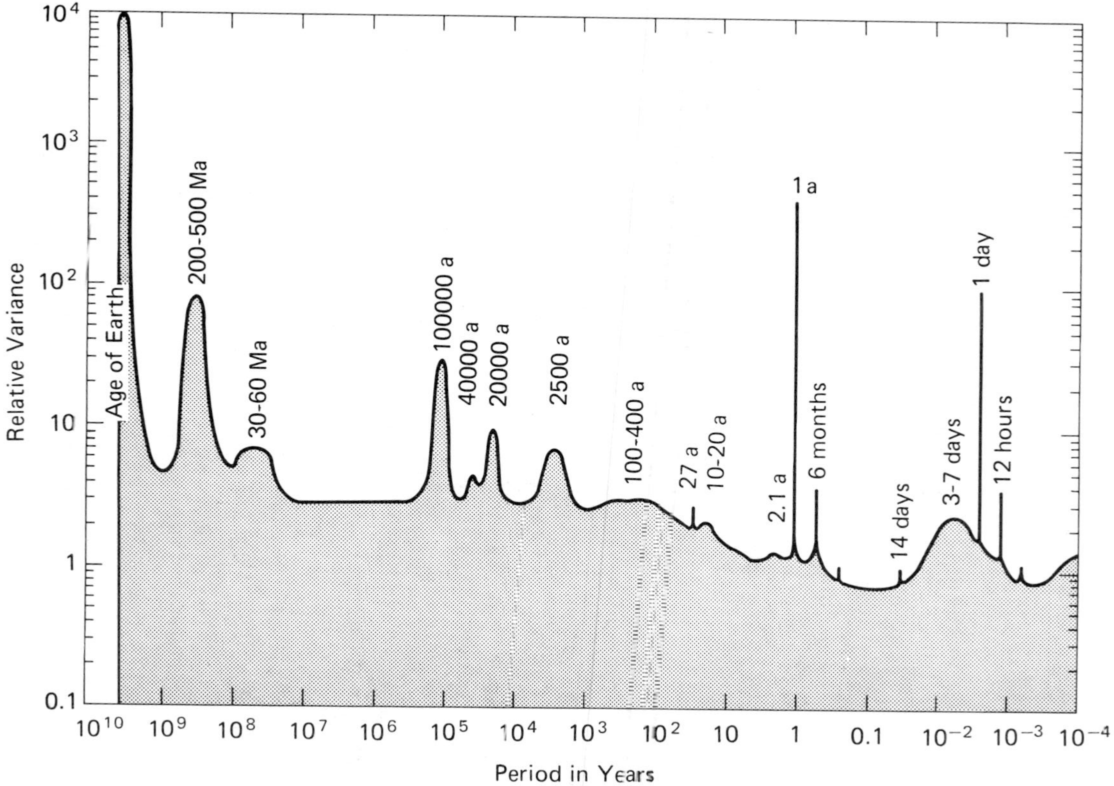

Fig. 11.1. Estimate of global climatic variability on all time scales (after Mitchell, 1976). The position of the peaks indicated is relatively well known, but not their relative height and width. Units of time are abbreviated as 1 year = 1a and 10^6 years = 1Ma.

The appropriate definition of the climatic system itself depends on the phenomena one is interested in, which determine the components of the system active on the corresponding time scale. No single model could encompass all temporal and spatial scales, include all the components, mechanisms and processes, and thus explain all the climatic phenomena at once.

Our goal in this chapter will be much more limited. We shall concentrate on the most striking phenomena to occur during the last two million years of the Earth's climatic history, the Quaternary period, namely on glaciation cycles. The time scale of these phenomena ranges from 10^3 to 10^5 years. We shall attempt to describe and model in the simplest way possible the components of the climatic system active on these time scales -- atmosphere, ocean, continental ice sheets, the Earth's upper strata -- and their nonlinear interactions.

In Section 11.1, we sketch the discovery of geological evidence for past glaciations, review geochemical methods for the study of deep-sea cores and describe the phenomenology of glaciation cycles as deduced from these cores. A near-periodicity of roughly 100,000 years dominates continuous records of isotopic proxy data for ice volume, with smaller spectral peaks near 40,000 and 20,000 years, as suggested in Figure 11.1. The records themselves are rather irregular and much of the spectral power resides in a continuous background.

In Section 11.2 we give a brief introduction to the dynamics of valley glaciers and large ice masses. The rheology of ice is reviewed and used in deriving the approximate geometry of ice sheets. The slow evolution due to small changes in mass balance of an ice sheet with constant profile is modeled next. A simplified, but temperature-dependent formulation of the hydrologic cycle and of its effect on the ice mass balance is given. We study multiple equilibria of the ice-sheet model thus formulated and their stability, pointing out similarities with the study of energy-balance models in Section 10.2. A hysteresis phenomenon occurs in the transition from one equilibrium solution to another as temperature changes.

In Section 11.3 we study the deformation of the Earth's upper strata under the changing load of ice sheets. The rheology of lithosphere and mantle is reviewed. Postglacial uplift data and their implications for this rheology are outlined. A simple model of creep flow in the mantle is used to derive an equation for maximum bedrock deflection under an ice sheet, and for the way this deflection affects the mass balance of the sheet.

In Section 11.4, we give references for climatic variability on various time scales and for its modeling. Geological information on the Quaternary and on earlier glaciations is cited, along with sources for stable isotope ratios as paleoclimatic proxy data. References for glaciology in general and for modeling of ice sheets and sea ice are discussed. Work on interactions between radiation, global temperature, the hydrologic cycle and ice mass balance is mentioned. The effect of basal sliding and of ice shelves on ice-sheet flow is noted. Knowledge of the elastic, short-term behavior of the solid Earth, and of its viscous, long-term behavior is reviewed briefly. Various models for mantle response to ice loading and for mantle convection are compared.

The equations derived and analyzed in this chapter for ice flow and bedrock response will lead, when combined with an equation for radiation balance from the previous chapter, to a system of differential equations which govern stable, self-sustained, periodic oscillations. Changes in the orbital parameters of the Earth on the Quaternary time scales provide small changes in insolation. These quasi-periodic changes in the system's forcing will produce forced oscillations of a quasi-periodic or aperiodic character, to be studied in Chapter 12. The power spectra of these oscillations show the above mentioned peaks with periodicities near 100,000 years, 40,000 years and 20,000 years, as well as the continuous background apparent in Figure 11.1.

11.1. Quaternary Glaciations, Paleoclimatic Evidence

History of the Problem. Changes in the extent of Alpine glaciers have been known to Swiss mountaineers for many generations. Huge erratic boulders lying in the low, currently unglaciated valleys, and parallel striations of permanently exposed, flat rock surfaces were easily associated with like boulders carried on the surface of glaciers flowing in the higher valleys, and with the aspect of present-day glacier beds temporarily exposed by unusual summer melting or deep cracks. The boulders and striations spoke clearly to the unbiased eye of times when the glaciers extended further down into the valleys, and filled them to the crest of their dividing ranges.

Until the early nineteenth century, however, the nascent science of geology had another explanation for these and related observations: the Biblical flood. It was only with difficulty that a number of courageous scientists, led by Louis Agassiz, faced the facts and gradually convinced their colleagues of the past extent of glaciations and their alternation in time with warmer periods, such as the one in which we live now.

Eventually, the geology of the Quaternary period became intimately linked with the study of the changing extent of continental ice sheets. In the Northern Hemisphere, the Laurentide ice sheet at various times covered most of Canada and New England, and much of the Middle West, Great Plains and Rockies. The Fenno-Scandian ice sheet extended into Eastern and Central Europe, without quite linking up with the smaller Alpine ice cap, mentioned before.

In the Southern Hemisphere, the Antarctic Ice Sheet was well established and reached its present volume before the Quaternary. The Antarctic continent being surrounded by oceans, this ice sheet maintained a nearly constant extent and volume throughout the Quaternary (Mercer, 1983). Smaller ice sheets, however, developed over parts of Australia and New Zealand, and extended out from the Andes. In between its extreme advances, the ice in both hemispheres retreated to areas comparable to those it occupies today: Antarctica, Greenland, the Canadian Archipelago, and small mountain glaciers.

On the whole, the Quaternary in its entirety has to be considered an ice age when compared with the mean temperature of much longer, completely ice-free periods of the Earth's history, such as the Mesozoic era, the age of Dinosaurs, which lasted for 160 million years. Other ice ages occurred in the Earth's past, 2.3 billion, 1 billion, 700, 450 and 270 million years ago. All the episodes for which the presence of ice is recorded and temporally resolved in the geological record seem to show higher climatic variability than the entirely ice-free episodes. The total duration of the ice-free episodes in the geological past seems to be much longer than that of episodes where ice of variable extent was present.

Plate tectonics, continental drift and orographic changes certainly play a role in creating geographic situations in which land ice can appear and develop. We shall not concern ourselves, however, with the time scales of tens of millions of years on which these phenomena are important. Ice will be assumed to be present in the system and the variations in its extent will be studied, along with the variations in temperature with which it interacts.

Geochemical proxy data. Variations in continental ice extent leave clear traces in the geological record. Besides the previously mentioned erratic boulders and striations, classical stratigraphic methods record glacial till, the coarse, unstratified debris left behind when glaciers melt, alternating with fine, stratified, interglacial deposits. The uplift of the Earth's crust as ice sheets melt is recorded in shifting shore lines along the Baltic Sea and the East Coast of North America. The pollen of temperate-climate plants alternates in statigraphic sequences with that deposited during cold climates.

But the most important, detailed evidence of glaciations started to accumulate with the advent of geochemical, isotopic methods in the 1950s. Long piston cores raised from the bottom of the sea by oceanographic research vessels contain fossil shells of micro-organisms living in the ocean, among which foraminifera are particularly wide-spread and well-studied. The calcium carbonate in the shells of bottom-dwelling, or benthic, foraminifera contains oxygen whose *isotopic* abundance *ratio* $R = {}^{18}O/{}^{16}O$ reflects to a large extent the same ratio in the water from which they precipitated. The isotopic abundance ratio of a sample, R_{sample}, is commonly represented by

$$\delta^{18}O = ((R_{sample}/R_{std}) - 1) \times 1000, \qquad (11.1)$$

where R_{std} is a reference standard. The standard for water is called Standard Mean Ocean Water (SMOW); a slightly different standard (Pee Dee Belemnivella, or PDB) is used for the carbonate. The normal abundance of the heavy ${}^{18}O$ isotope in water is only 0.2% approximately and it varies little, hence the use of the factor 1000 in the formula (11.1) for $\delta^{18}O$. Currently available mass spectrometers determine $\delta^{18}O$ in calcium carbonate with a nominal precision of ± 0.05 promil.

As part of the hydrologic cycle, water molecules which evaporate from the surface of the oceans contain preferentially the light isotope ${}^{16}O$. If these molecules precipitate into high-latitude ice sheets and are fixed there, the ${}^{16}O$ cannot return to the ocean as it does during warm episodes, with runoff and river water. Hence the ocean becomes impoverished in the light, abundant isotope ${}^{16}O$ when large ice sheets are present, so that $\delta^{18}O$ of ocean water, and of benthic foraminiferal shells deposited in it, is higher during glacial episodes.

The mixing time of the world ocean is of the order of 1000 years, and bottom waters never have an entirely uniform $\delta^{18}O$. Moreover, the isotopic ratio of the shells differs from one species to another, is influenced by the temperature of the ambient water, as well as by its $\delta^{18}O$, and is not quite in thermochemical equilibrium with the water. Still, the $\delta^{18}O$ of the micro-fossils in deep-sea cores appears to be positively correlated with the ice volume and hence constitutes a relatively (reliable) proxy indicator of global ice volume on the time scale of thousands of years.

The nature of glaciation cycles. What is the evidence from deep-sea cores as to climatic change during the Quaternary? Figure 11.2 shows the $\delta^{18}O$ record of a deep-sea core from the western equatorial part of the Pacific Ocean.

The scale on top is simply the depth in the core. To translate this into a time scale, absolute dates for points along the core are necessary. Other isotopic methods, in particular the study of the potassium-argon ratio in lava flows, helped establish in the 1960s *polarity reversals* of the Earth's magnetic field. The current polarity is called normal, the opposite one is reversed. Reversals involve changes in the amplitude, as well as in the direction of the field and take about ten to twenty thousand years (see Section 7.1, especially Figure 7.1 there; also Clement and Kent, 1984, for further references). The first reversal can be dated to 730 thousand years ago, to within 20 thousand years.

Two additional, short episodes of normal polarity have occurred during the Quaternary. They are called the Jaramillo and Olduvai normal events. The beginning of the Olduvai event, 1.8 million years ago, is now agreed upon as the beginning of the Quaternary period. Our unit of time henceforth will be 1ka = 1000 years, so that the Quaternary is 1800ka long.

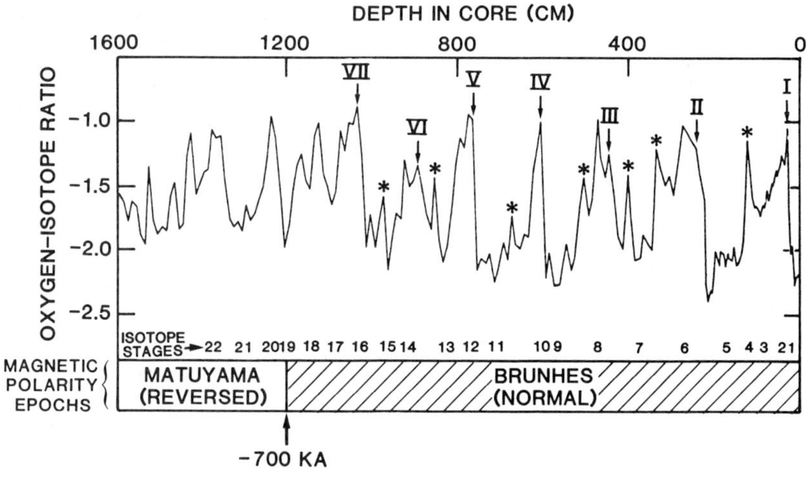

Fig. 11.2. Oxygen isotope record of deep-sea core V28-238 (after Imbrie and Imbrie, 1979).

Aside from the Brunhes/Matuyama polarity reversal a little over 700ka ago, no absolute dates were availabe to Shackleton and Opdyke (1973) for the analysis of the core in Figure 11.2. So a uniform sedimentation rate is assumed in the time scale at the bottom of the figure. The isotopic stages marked along this scale represent episodes of algebraically higher $\delta^{18}O$ ratios - even numbers, or lower ratios - odd numbers.

The record in Figure 11.2 shows an irregular evolution of $\delta^{18}O$, and hence ice volume, over its entire length. Prominent are sharp drops in ice volume approximately every 100ka, called *terminations* by Broecker and van Donk (1970) in their study of similar records. The growth of ice volume in between terminations seems more gradual in certain records, particularly in continental stratigraphic sequences (Kukla, 1969), leading to the idea of a roughly sawtooth-like shape of *glaciation cycles*.

The "terminations" are not equally spaced, nor do the segments of record between terminations look alike: irregular variability on time scales shorter than 100ka is evident in Figure 11.2. A few sharp spikes as high as the terminations, and many smaller ones, occur repeatedly, but aperiodically. Long plateaus with very little high-frequency variability appear in the $\delta^{18}O$ curve during some of the major cycles, but not others. These plateaus are mostly near the mean value of the record $\delta^{18}O \simeq -1.5$ promil.

Figure 11.3 shows the power spectrum, due to Hays, Imbrie and Shackleton (1976), of a combined $\delta^{18}O$ record taken from two cores in the southern part of the Indian Ocean, one of which goes back 450ka. The most prominent peak is at 106ka, with smaller peaks at 43ka, 24ka and 19ka. These peaks are superimposed on a red-noise-like continuous background as discussed in Section 10.4.

Other deep-sea cores and other proxy indicators, both isotopic and micropaleontological, all seem to show irregular variations like Figure 11.2, with a major near-periodicity close to 100ka, terminations, spikes and plateaus. Their power spectra show peaks near 100ka, 40ka and 20ka, as well as occasional additional peaks. These are superimposed on a continuous background which decreases from low to high frequencies.

<u>Modeling</u>. The amplitude of glaciation cycles, as inferred from the proxy data, is of a few degrees in temperature and of a factor of two or three in global ice extent. We saw in Sections 10.4 and 10.5 that such variations of temperature and ice volume cannot be accounted for by the

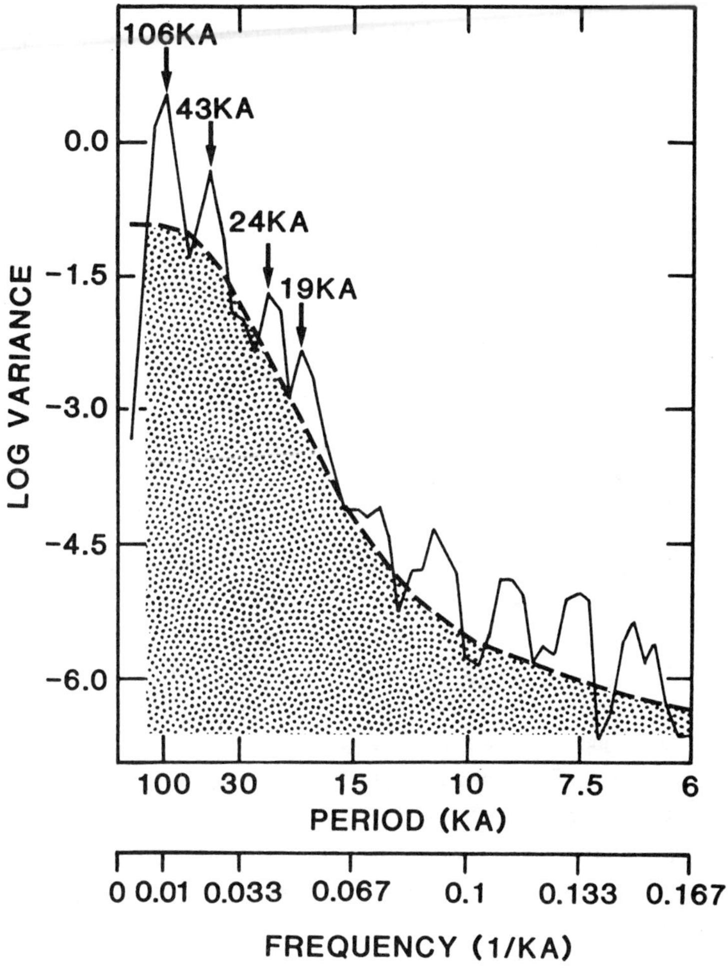

Fig. 11.3. Power spectrum of a patched $\delta^{18}O$ record based on deep-sea cores RC11-120 and E49-18 (after Imbrie and Imbrie, 1979).

quasi-equilibrium response of energy-balance models to small changes in the external radiation budget.

The ice-albedo feedback central to our previous modeling is but the simplest way that the presence of ice in the system acts on its temperature evolution. The fixed delay in the ice's response to temperature decrease used in Section 10.4 is clearly an oversimplification.

In the next two sections we shall deal therefore in more detail with the mass balance and visco-plastic flow of ice sheets, and with the response of the Earth's lithosphere and mantle to the ice load. The interaction between mass balance and radiation balance will lead in Chapter 12 to self-sustained oscillations in the model system, with a period near 10ka and amplitudes of the order of magnitude described before. The "present climate" equilibrium solution of the previous chapter is shown to transfer its stability to such periodic solutions.

11.2. Ice-sheet Dynamics

The study of snow and ice masses, their mass balance, phase transitions, thermal structure and flow, is an important, well-developed geophysical discipline. In this section we cover only a few salient features of large ice masses which are crucial to their paleoclimatological behavior. The presentation follows to a large extent Paterson (1980; 1981, Chapters 3 and 9) and Weertman (1964, 1976).

Cryodynamics. The flow of Alpine valley glaciers is a fact of experience for anybody who has walked on or near them. Surface velocities have typical values of tens to hundreds of meters per year, and exhibit spatial patterns which can be stationary in time or change from one season and year to another. As a result of changes in mass balance and flow pattern, the tongue of temperate-latitude valley glaciers can move up or down by several meters, and sometimes tens of meters, per year. Much larger velocities along the surface and at the tip can be attained by surging valley glaciers over rather short episodes separated by relatively quiescent intervals of considerably longer, but irregular duration.

Being remote and difficult of access, less is known about the oceanically confined ice sheets existing today, and even less about the freely expanding continental ice sheets of the past. It is reasonable to assume that some features of ice mass behavior are independent of size to a certain extent, and do not change in geological time.

Valley glaciers had been thoroughly observed by the beginning of the century, but the rheology of ice flow took longer to develop: glacier models in the first half of the century were based on a viscous flow assumption. These models could not explain, as we shall see, certain observed features of the flow. In order to examine ice flow rheology as it is understood today, we start by reviewing briefly basic stress-strain relations in solids and liquids.

Constitutive relations. The mechanical description of a continuous medium is given in terms of the stress and stain tensors at every point of the continuum. A *stress* τ is a force per unit area and a *strain* e

is a deformation per unit length. In Cartesian coordinates (x_1, x_2, x_3), the component τ_{ij} of the stress tensor is the stress in the direction x_i acting on an elementary surface with normal x_j. If $i = j$ the stress is normal, and is also called a tension or negative pressure. If $i \neq j$ then τ_{ij} is a shear stress. The notation for the strain tensor (e_{ij}) is similar. An important property of both the stress and strain tensors is that they are symmetric, $\tau_{ij} = \tau_{ji}$ and $e_{ij} = e_{ji}$.

Figure 11.4a shows schematically the *constitutive relation* between stress and strain for a metal. At low stress, the metal is elastic, i.e., it obeys Hooke's empirical law

$$\tau = Ee, \tag{11.2}$$

where E is called Young's modulus of elasticity. Eq. (11.2) holds for all components of the stress and strain tensors, $\tau_{ij} = Ee_{ij}$, but we drop the subscripts for the sake of conciseness and legibility.

The linear relation (11.2) changes as τ approaches the value τ_0, which depends on the metal and is called the *yield stress*. For $\tau = \tau_0$ the metal can undergo arbitrarily large deformations without any increase in the stress: the metal flows very slowly, or creeps, in a perfectly *plastic* manner.

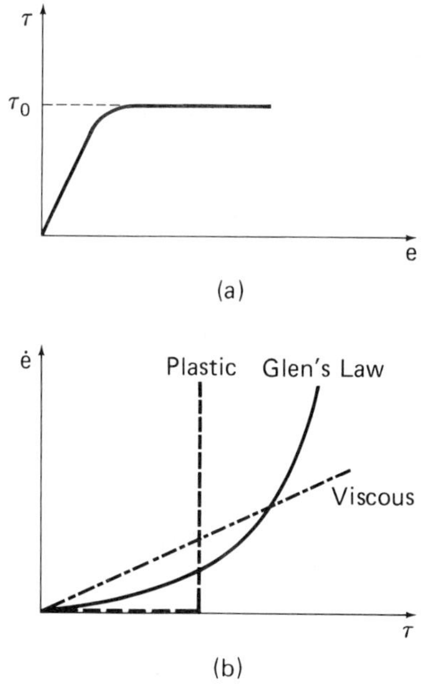

Fig. 11.4. Constitutive relations for a metal (a) and for ice (b) (after Paterson, 1980).

Nothing is said in Figure 11.4a about the rate at which the large displacements for $\tau = \tau_0$ occur. In many problems concerning engineering structures (Prager, 1959), one seeks only a limiting or stationary solution and the actual flow is not important. Most valley glaciers, however, have a temperature very close to the melting point and their flow occurs in the creep regime near or at the yield stress.

Figure 11.4b shows different constitutive relations or "flow laws" used for creeping ice, with strain rate $\dot{e}$ plotted versus stress τ. A perfectly viscous fluid obeys Newton's relation

$$\tau = \nu\dot{e}, \tag{11.3}$$

where ν is viscosity (dash-dotted line). Notice that the relation (11.3) holds only for the shearing components of stress and strain rate, $i \neq j$. It also holds, with one proviso, for the so-called deviatoric normal components, $\pi_{ii} = \tau_{ii} + p$, where the thermodynamic pressure p is minus the mean normal stress, $p = -\Sigma_{i=1}^{3}\tau_{ii}$. The proviso that (11.3) hold for π_{ii} is that the divergence of the velocity field v in the fluid vanish, $\nabla\cdot v \equiv \frac{1}{2}\Sigma_{i=1}^{3}\dot{e}_{ii} = 0$ (e.g., Goldstein, 1960, pp. 39 and 88); this condition is certainly met to a very good approximation in ice.

Otherwise Eq. (11.3) appears very similar to (11.2), with $\dot{e}$ replacing e. But the physical interpretation is quite different: a perfectly elastic solid will only undergo a finite deformation for finite stress, and will not be deformed any further at constant stress. A viscous liquid, on the other hand, cannot sustain a given, nonzero shear stress or deviatoric normal stress, without deformation; it is the rate of this deformation which is proportional to the applied stress.

We mentioned already that early glacier models used (11.3) as their flow law. This could not explain the large differences between summer and winter in observed surface velocities for certain valley glaciers. Important progress was made by Orowan (1949) and Nye (1951) in replacing Newtonian viscosity by perfect plasticity (heavy dashed line in Figure 11.4b) as a flow law for ice.

Ice, being a polycrystalline solid, has many features in common with metals. The realization of this similarity around 1950 played a key role in the development of ice rheology. Glen (1955) found in laboratory experiments that the strain rate $\dot{e}$ of ice had a power-law dependence on stress τ,

$$\dot{e} = A\tau^n. \tag{11.4}$$

The form of this dependence is reasonably well established and can be explained in broad outline by a study of deformations in ice crystals.

But different experiments found values of n from 1.5 to 4.2, $n = 3$ being most commonly used. While n is constant for a given sample with temperature T, A depends exponentially on T. Measured strain rates for a given stress and temperature can differ by an order of magnitude due to differences in n between samples, and even larger differences in measured strain rate appear as a result of changes in sample temperature.

Glen's law, shown in Figure 11.4b (heavy solid line) uses $n = 3$ and $A = 5 \times 10^{-16} s^{-1} (kPa)^{-3} = 0.28 \times 10^{-15}\ yr^{-1}\ Pa^{-3}$, where $1\ bar = 10^5 Pa = 100\ kPa$. The Newtonian viscous flow law (dash-dotted line in the figure) corresponds to $n = 1$ and uses $\nu = A^{-1} = 8 \times 10^{13} kg\ m^{-1} s^{-1}$

The fact that n is larger than one for ice flow explains the important effect that even small changes in stress can have on surface velocities. Thus the weight added by snowfall in winter to the Aletsch glacier in Switzerland can produce a twofold increase in surface velocity relative to the summer.

As n increases, the flow law (11.4) tends to plastic behavior. Writing $A = A_1 \tau_0^{-n}$, it is clear that for A_1 = const.

$$\dot{e} \to 0 \quad \text{for} \quad \tau < \tau_0, \tag{11.5a}$$

$$\dot{e} \to +\infty \quad \text{for} \quad \tau > \tau_0, \tag{11.5b}$$

as $n \to +\infty$. Thus $\dot{e}$, as well as e itself, are indeterminate at $\tau = \tau_0$, the yield stress. As seen from Figure 11.4b (heavy dashed line), the yield stress of ice is near $\tau_0 = 100 kPa$.

Given the uncertainties in n and A, perfect plasticity is a convenient approximation for many purposes, albeit a crude one. We shall use it in the sequel to derive the approximate geometry of ice sheets.

Ice-sheet geometry. The ice sheets existing today in Greenland and the Antarctica, as well as smaller ice caps in Iceland and the Canadian archipelago, have height profiles which are close to parabolas. This is shown in Figure 11.5.

Consider now the equilibrium of an ice sheet on flat bedrock, of infinite extent in the y direction, so that $\partial/\partial y \equiv 0$, and symmetric about $x = 0$, as shown in Figure 11.6. Its surface profile is given by $z = h(x)$. We consider the equilibrium of the right half of the ice sheet.

On the (y,z)-plane of symmetry of the ice sheet, only hydrostatic pressure is acting, proportional to $(H-z)$. Along the base, it is reasonable to assume that the ice is at yield stress, τ_0. The effect of atmospheric pressure on the surface is negligible, and no other forces are

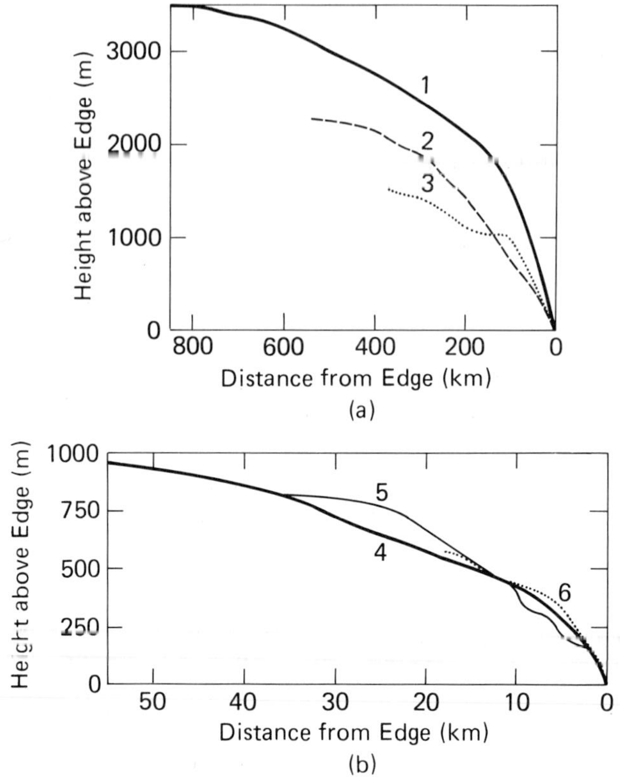

Fig. 11.5. Surface profiles of ice sheets and ice caps (from Paterson, 1980). a) East Antarctica (1) and Greenland (West (2), East (3)); (b) Vatnajökull, Iceland (4), Devon Island (5) and Baffin Island (6).

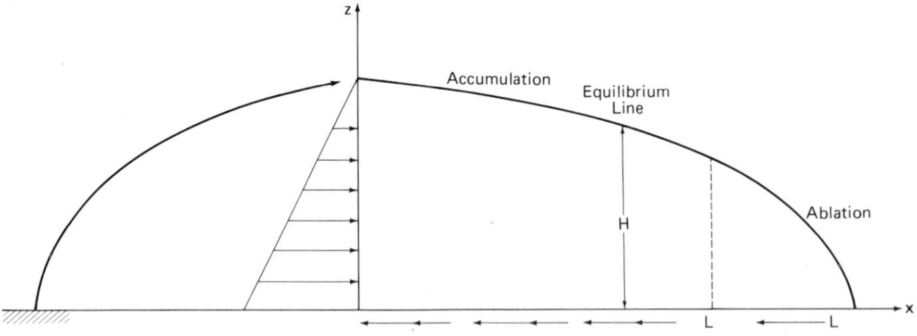

Fig. 11.6. Cross-section through an idealized ice sheet (after Orowan, 1949).

acting, so that the horizontal balance of forces on half the ice sheet is

$$\frac{1}{2}\rho_i g H^2 = \tau_0 L. \tag{11.6}$$

Here ρ_i is the density of the ice, g the acceleration of gravity, $H = h(0)$ the maximum height of the ice sheet and L its half-width.

Eq. (11.6) suggests a parabolic dependence of H on L and using reasonable numbers for ρ_i, τ_0 and L produces the maximum height of the Greenland or Antarctic ice sheet to a good approximation. This has encouraged many arguments (Nye, 1960) to apply the same balance of forces to an arbitrary section of the ice sheet between $x = x_0$ and $x = x_0 + dx$. Such arguments yield a formula

$$\tau(x, z = 0) = \rho_i gh \; dh/dx \tag{11.7a}$$

for the stress τ at the base of the sheet. Assuming

$$\tau(x, z = 0) \equiv \tau_0 \tag{11.7b}$$

gives the parabolic profile

$$(h/H)^2 = 1 - x/L. \tag{11.8}$$

Figure 11.7 shows a direct comparison between data for the Antarctic Ice Sheet (dots) and the parabola (11.8) (dashed line). The fit is not bad, considering the fact that both assumptions (11.7a,b) are far from being correct. It is well known that basal shear stresses are far from constant, and that the state of stress inside a solid which satisfies the limiting constitutive relations of Eqs. (11.5a,b) is very complicated (Duvaut and Lions, 1976, Ch. 5).

Using Glen's law (11.4), rather than perfect plasticity, and

$$2\dot{e}_{xz} = \partial u/\partial z,$$

where u is the flow velocity in the x-direction, yields after substitution into (11.7a)

$$\partial u/\partial z = \rho_i g(h - z) \; dh/dx$$

at a height $0 < z < H$. Further assumptions and simplifications then give the profile

$$(h/H)^{2+2/n} + (x/L)^{1+1/n} = 1. \tag{11.9}$$

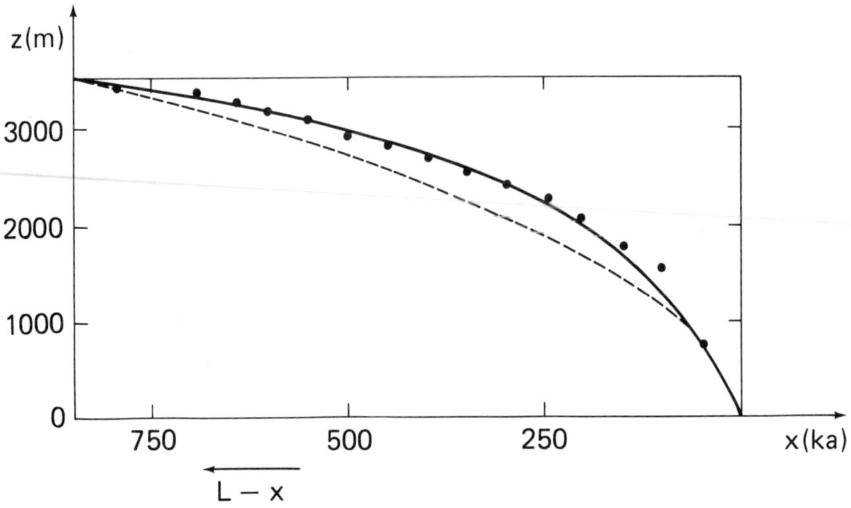

Fig. 11.7. Profile of Antarctic ice sheet from station Mirny to station Vostok (dots) and analytic approximations (after Paterson, 1980).

For $n \to +\infty$, Eq. (11.9) reduces to (11.8).

The profile (11.9), for $n = 3$, is shown as the solid line in Fig. 11.7. It produces a somewhat better fit to the given set of data. Comparison with the other data in Figure 11.5, however, makes it obvious that the nonuniformities in stress distribution in any given ice sheet will preclude a simple formula like (11.8) or (11.9) from producing a complete and accurate description of the profile.

Our main concern in the modeling of Quaternary ice sheets will be the relation between volume V and areal extent S of the ice. Given that $H^2 \sim L$ from (11.6), $S \sim L^2$ and $V \sim L^2H$, one would expect a relation of the form $\log(V/V_0) \cong (5/4)\log(S/S_0)$ for ice sheets of nearly parabolic profile. Paterson (1972) finds indeed a linear fit

$$\log_{10}V = 1.23(\log_{10}S - 1), \quad 3 \le \log_{10}S \le 7.2,$$

for present-day ice sheets, caps, and domes, with a standard deviation of the actual data from this regression line of only 0.1, or 12% of volume. Given our lack of detailed knowledge about Quaternary ice sheets, we retain therefore (11.8) as a suitable approximation for their profile.

A Quaternary ice-sheet model. It was seen in the previous section that total ice volume in the Southern Hemisphere varied but little during the Quaternary. We consider in the sequel, following Weertman (1976) and Källén *et al.* (1979), the evolution of a single ice sheet in the Northern Hemisphere. Its meridional cross section is shown in Figure 11.8.

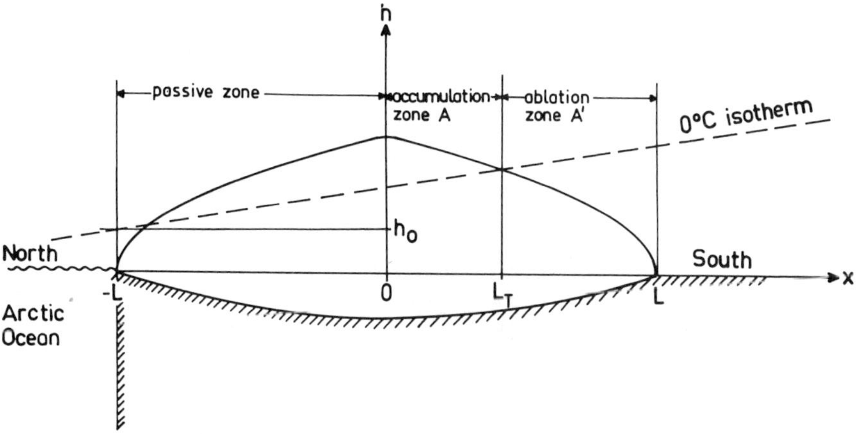

Fig. 11.8. Meridional cross section of a Quaternary ice sheet (from Källén *et al.*, 1979).

The width of the ice sheet is assumed to be sufficient, as in Eqs. (11.6) and (11.8), to ignore variations in the longitudinal direction. The northern edge of the ice sheet is attached to the coastline of the Arctic Ocean, past which it cannot move due to flow of the ice mass excess into ice shelves and calving into icebergs. These are the major mechanisms of ice mass loss observed today along the coasts of Greenland and the Antarctica.

The parabolic profile (11.8) suggests North-South symmetry of the ice sheet, as in Figure 11.6. The vertical h-axis in Figure 11.8 is thus an axis of symmetry, the total meridional extent of the ice sheet being 2L.

The total albedo of the Earth's land masses, α_{land}, will be taken as a linear function of L,

$$\alpha_{land} = \alpha_0 + \alpha_1 L, \tag{11.10a}$$

where α_0 is the albedo of ice-free land, for which $L = 0$, and

$$\alpha_{land,max} = \alpha_0 + \alpha_1 L_{max} \tag{11.10b}$$

for the maximum possible extent of the ice sheet. Since $2L_{max}$ for Quaternary ice sheets was still small when compared to the radius of the Earth, the curvature of the Earth is neglected.

The main purpose of the present discussion is to formulate an equation for the change in L, and hence in land albedo, as a function of temperature. Land, however, occupies only a fraction $\gamma \simeq 0.3$ of the total surface of the earth. The albedo of the ocean, according to the arguments of Section 10.2, is dominated by the presence or absence of sea ice. Continental ice sheets change their extent slowly, over hundreds and thousands of years. Sea ice, however, has huge seasonal changes of areal extent, by a factor of two in the Northern Hemisphere, and a factor of 6-8 in the Southern Hemisphere (Barry, 1980; Hibler, 1980). Hence, it can be safely assumed that the mean annual change in sea-ice extent responds instantaneously to that of global temperature T.

For the albedo of the ocean, α_{oc}, we shall use therefore the same ramp function of temperature as in Table 10.1. This is shown as the dash-dotted line in Figure 11.9. $T_{\alpha,u}$ corresponds to an ocean with minimal ice extent, $T_{\alpha,\ell}$ to an ocean with maximum ice extent.

Returning to the ice sheet in Figure 11.8, it is assumed that bed-rock deflection under the load of the ice is instantaneous, and given by the density ratio of the ice ρ_i to that of the Earth's upper strata, ρ_u, $\rho_i/\rho_u = 1/3$. Thus (2.3) of the total thickness of the ice sheet is above

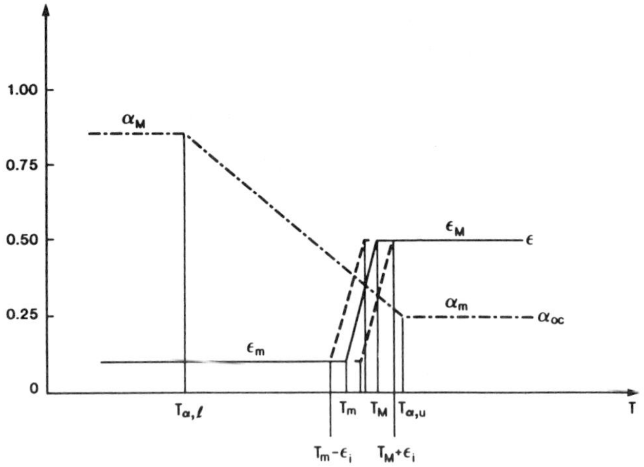

Fig. 11.9. The oceanic albedo $\alpha_{oc}(T)$ (dash-dotted line) and the accumulation-to-ablation ratio $\varepsilon(T)$ (solid line) as a function of global, mean, annual temperature T (from LeTreut and Ghil, 1983). The two parallel dashed lines are explained in Section 12.3.

mean sea level (MSL), taken as the x-axis, and (1/3) below. Combining this with Eqs. (11.6, 11.8) yields for the height of the ice sheet above MSL, denoted here by $\tilde{h}(x)$,

$$\tilde{h}(x) = \lambda^{1/2}(L - |x|)^{1/2}, \tag{11.11a}$$

where $\lambda = (4/3)\tau_0/\rho_i g$ is a parameter with the dimension of length. For reasonable values of τ_0, according to (11.7b) and the subsequent discussion, $\lambda = O(10m)$. It is convenient to normalize $\tilde{h}(x)$ by $\lambda^{1/2}$, so that

$$h(x) = \lambda^{-1/2}\,\tilde{h}(x). \tag{11.11b}$$

The order of magnitude notation, $\lambda = O(10\ m)$, should not be confused with the mathematical notation $f(x) = O(x^{\alpha})$ used in Parts I and III. Here and subsequently it means that λ is closer to 10m than to 1m or to 100m, while $\lambda \cong 10m$ would be used to indicate that λ does not differ from 10m by more than 10 or 20 percent.

The snowline. The northern half of the ice sheet in Figure 11.8 is assumed to be in passive mass balance with the southern half, due to ice calving at its oceanic edge. The southern half of the sheet is divided into an *accumulation zone*, where snow falls and solidifies into ice, and an *ablation zone*, where the ice melts and runs off as water. The two zones are separated by the mean annual *snowline*. The snowline is assumed to be the intersection of the ice sheet's surface with the 0°C isotherm.

The mean annual 0°C isotherm is a good approximation to the line of permanent snow on today's mountain ranges, and to the firnline on today's valley glaciers. It rises linearly with decreasing latitude for various North-South oriented ranges and has a slope of roughly 1km/1000km. As global climate changes, the local 0°C isotherm will also move accordingly.

The 0°C isotherm in Figure 11.8 has the equation

$$h_s(x;T) = h_0(T) + s(x + L). \tag{11.12a}$$

Here T is the global, mean annual temperature, h_0 is the height of the isotherm at the coast line of the Arctic Ocean, $x = -L$, and s is the slope of the isotherm. The 0°C isotherm is assumed to move parallel to itself as T changes, with $s = \text{const.}$ and

$$h_0(T) = \beta(T - T_{00}), \tag{11.12b}$$

where β and T_{00} are constants.

T_{00} is the globally averaged temperature at which the 0°C isotherm would intersect the Arctic coastline. We take $T_{00} = T_{\alpha,u} = 283K$, so that the minimal sea-ice extent of Figure 11.9 corresponds to the Arctic Ocean being ice covered. To determine β, we further assume that the intersection of the 0°C isotherm with MSL is located at the southern edge of the sea ice when the maximum extent of the sea ice is reached, $\alpha_{oc} = \alpha_M$, which occurs when the global temperature T equals $T_{\alpha,\ell}$. This position of the snowline is also taken to correspond to the maximum extent of the ice sheet, $L = L_{max}$. Hence

$$h_s(L_{max}; T_{\alpha,\ell}) = 0,$$

yielding

$$\beta = \frac{2s(\alpha_{land,max} - \alpha_0)}{\alpha_1(T_{00} - T_{\alpha,\ell})}. \tag{11.12c}$$

We take for simplicity $\alpha_{land,max} = \alpha_M$.

Mass balance of the ice sheet. The meridional extent of the accumulation zone is L_T, that of the ablation zone is $L - L_T$. L_T is determined as the value of $x > 0$ for which $h(x)$ from (11.11) equals $h_s(x;T)$ from (11.12):

$$L_T = s^{-2}\{(2s^2L + sh_0 + 1/4)^{1/2} - (s^2L + sh_0 + 1/2)\}, \tag{11.13}$$

where $h_0 = h_0(T)$ is given by (11.12b,c).

Given L_T, the mass balance of the ice sheet is governed by

$$\dot{V} = aL_T - a'(L - L_T), \tag{11.14}$$

where $\dot{V} = dV/dt$ is the rate of change of the ice volume V per unit width, a is the accumulation rate per unit meridional extent and unit width, and a' the corresponding ablation rate. The ratio $\varepsilon = a/a'$ is clearly as important to the mass balance (11.14) as $\alpha = \alpha(T)$ was to the radiation balance (10.2,10.3). A value often used in studies of present-day valley glaciers is $\varepsilon = 1/3$.

The accumulation-to-ablation ratio ε is closely connected to the activity of the seasonal hydrological cycle: heavy snowfall in high and middle latitudes corresponds to large evaporation from the low- and mid-latitude ocean. Likewise, strong melting corresponds to large runoff. Detailed modeling of the hydrological cycle in a simple, zonally or globally averaged ice-sheet model is as difficult as that of cloud cover in an energy-balance model. We assume therefore, consistent with the level of complexity of the models to be discussed here, that $\varepsilon = \varepsilon(T)$.

The assumed dependence of ε on T is shown as the solid line in Figure 11.9: $\varepsilon = \varepsilon_m < 1/3$ for $T < T_m$, $\varepsilon(T)$ increases linearly between T_m and T_M, and $\varepsilon = \varepsilon_M > 1/3$ for $T > T_M$. The increase of ε with T reflects the simple observation that it is the warmer winters which have the heavier snowfalls, due to increased evaporation from the ocean. The assumption that this effect exceeds in annual mean that of the increased melting during summer was first formulated in the context of ice age theory by Simpson in the 1930s (see Section 11.4). The ramp function $\varepsilon(T)$ will be called the *precipitation-temperature feedback*, by analogy with the ice-albedo feedback of the previous chapter. Supportive evidence for the increase of net annual accumulation with temperature, from ice-core geochemistry and from modeling studies using a detailed hydrological cycle, is reviewed by LeTreut and Ghil (1983).

Multiple equilibria of the ice sheet. At equilibrium, Eq. (11.14) reduces to

$$L = [1 + \varepsilon(T)]L_T. \tag{11.15a}$$

Together with (11.13), this yields an equation for L as a function of T at equilibrium,

$$\frac{L}{L^*} = \frac{1}{2}\left\{1 - \frac{h_0}{2h^*} \pm \left(1 - \frac{h_0}{h^*}\right)^{1/2}\right\}, \tag{11.15b}$$

where

$$L^*(T) = \frac{\varepsilon(T)[1+\varepsilon(T)]}{s^2[2+\varepsilon(T)]^2}, \quad h^*(T) = \frac{\varepsilon(T)}{4s[2+\varepsilon(T)]}. \tag{11.15c,d}$$

The plot of (11.15b) is shown in Figure 11.10 for $\varepsilon = \varepsilon_M = 0.5$ (solid curve) and $\varepsilon = \varepsilon_m = 0.1$ (dashed curve). The values of L^* and h^* shown in the figure correspond to $\varepsilon = \varepsilon_M$ in Eqs. (11.15c,d).

No equilibrium ice sheet is possible for $T > T_{00} + h^*/\beta$, i.e., $L = 0$ is the only solution of (11.15) in this case. For $T < T_{00}$ there is also a single equilibrium size L, with $L > L^*$. In between, $T_{00} < T < T_{00} + h^*/\beta$, there are three possible solutions. The situation is thus reminiscent of Figure 10.6 in the previous chapter, with a saddle-node bifurcation occurring at $T = T_{00} + h^*/\beta$ and a cusp bifurcation at $T = T_{00}$.

To study the stability of these solutions, we return to the evolution equation (11.14) for the volume V of the ice sheet of unit width. V can be expressed as a function of L by using (11.11) and remembering that the total thickness of the ice sheet is $(3/2)h(x)$, which yields $V = 2\lambda^{1/2}L^{3/2}$ and hence

$$\dot{L} = \frac{a'}{3(\lambda L)^{1/2}}[(1 + \varepsilon)L_T - L]. \tag{11.16}$$

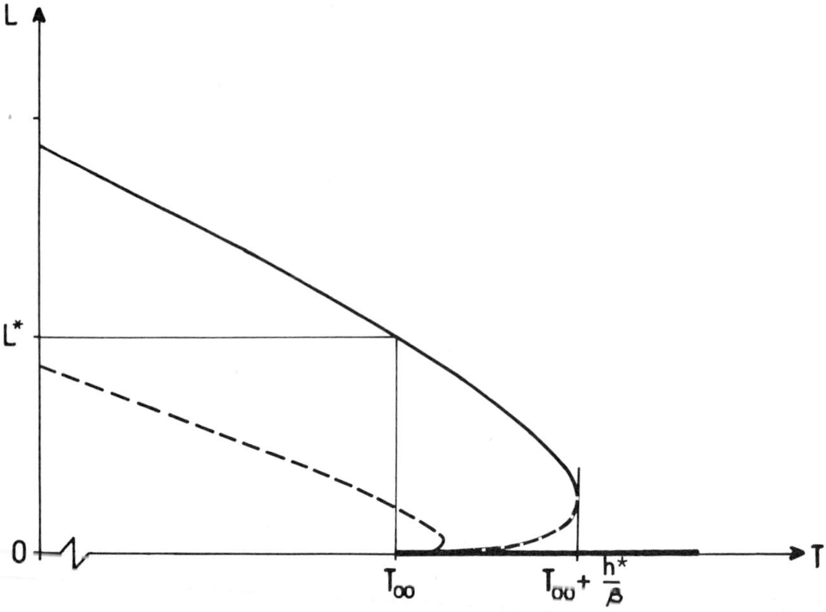

Fig. 11.10. Equilibrium ice-sheet size L as a function of temperature T (from Källén *et al.*, 1979).

Linearizing (11.16) as in (10.4) shows that the upper and lower branches in Figure 11.10 are linearly stable, and the middle one unstable (short dashes in the figure for $\varepsilon = \varepsilon_M$ curve).

Nonlinear stability and stochastic perturbations of (11.16) can be studied as in Section 10.3, by introducing a variational principle. The stability of the branches, nonlinear as well as linear, corresponds to physical intuition, with the stable branches having decreasing ice volume as temperature increases. The middle branch's instability again conforms to the mountain pass lemma for the quasi-potential of Eq. (11.16), as in Section 10.4.

It is interesting that the ice-sheet model (ISM) governed by Eq. (11.16), like the EBM (10.12), exhibits a *hysteresis* phenomenon: lowering temperature slowly from some high value at which ice is totally absent down to $T = T_{00}$, or just below, will not produce any ice at all. At that point ice extent will grow rapidly, and monotonically, in accordance with the nonlinear stability analysis sketched in the preceding paragraph, up to the value L^* indicated in Figure 11.10. Conversely, if temperature is now increased slowly, ice extent will decrease, slowly at first, following the solid curve in the figure. Then, when T reaches $T_{00} + h^*/\beta$, it will decrease rapidly to zero.

This increase and decrease along different paths in phase-parameter space is called hysteresis. We shall encounter such a phenomenon again for oscillatory, rather than stationary, solutions in the next chapter (Section 12.2), where it will be connected with the change in character of glaciation cycles between Early and Late Pleistocene (Section 12.7).

Evolution equation of the ice sheet. We wish to reformulate Eq. (11.16) for a nondimensional ice extent by normalizing L and L_T. From the discussion of Eq. (11.15) and Figure 11.10, it is clear that the appropriate scaling length is $L^*(T)$. For s = const., L^* is a monotone increasing function of $c > 0$. Hence we scale by

$$L_M^* = \frac{\varepsilon_M(1+\varepsilon_M)}{s^2(2+\varepsilon_M)^2}, \qquad (11.17a)$$

so that

$$\ell = L/L_M^*, \quad \ell_T = L_T/L_M^*. \qquad (11.17b,c)$$

The evolution equation then becomes

$$c_L\dot{\ell}\ \ \ell^{-1/2}\{[1 + \varepsilon(T)]\ell_T - \ell\}. \qquad (11.18a)$$

The e-folding relaxation time of ℓ to the equilibrium value (11.15a) is roughly $2c_L$, with

$$c_L = (3/a')(\lambda L_M^*)^{1/2}. \qquad (11.18b)$$

Using order of magnitude estimates for the quantities in (11.18b) of $a' = 0$ (3m/a), $\lambda = 0$ (10m) and $L_M^* = 0$ (10^7m) yields $c_L = 0$ (10^4a). Taking into account basal sliding, due to the presence of water at the ice-bedrock interface, considerably reduces this value of c_L (see Section 11.4 for references), and we shall use subsequently $c_L = 0(10^3\text{a})$.

In fact, it is likely that the flow properties of expanding ice sheets differ from those of shrinking ice sheets, and that different values of c_L should be used when $\dot{\ell} > 0$ or $\dot{\ell} < 0$. Given the uncertainties about accumulation and ablation rates during the Quaternary, shear stress at the bed, as well as flow law and geometry of the ice sheets, we shall not distinguish between the two possible types of ice-sheet behavior. Different, but constant, values of c_L will be used, and the effect of such differences in c_L on model behavior will be studied.

The model to be investigated in Section 12.1 couples the radiation balance of Eq. (10.2), using the land albedo of (11.10) and ocean albedo of Figure 11.9, with the ice mass balance of Eq. (11.18). Before studying this model, we shall complete our description of the physical mechanisms active on the Quaternary time scale by discussing isostatic adjustment of the bedrock to the ice load.

11.3. Geodynamics

In its most general meaning, geodynamics designates the study of motions in the Earth's interior. Thus in particular the fluid dynamics entering the terrestial dynamo problem in Part III could be included under this heading. Here we shall restrict our attention to the motions of the Earth's upper strata relevant to glaciation cycles.

Rheology of the Earth's upper strata. The stratification of the solid Earth was already discussed in Section 7.1: the *core* extends from the center to a radius of about 3500 km, and is made up of the inner, solid core, from 0 to 1400 km, and the outer, liquid core, from 1400km to 3500km. Chapters 7, 8 and 9 dealt with the flow in the outer core and its connection with the Earth's magnetic field.

The structure of the Earth's upper strata, between the core and the surface, is more complicated, and various descriptions, based on different criteria, exist. *Mineralogically*, one distinguishes between *crust* and *mantle*, with crustal rocks having smaller density. The boundary between crust and mantle is the Mohorovičic discontinuity, or "Moho." The typical thickness of oceanic crust is about 6km, while continental crust is about 35km thick. Within the mantle there also exist phase transitions at depths of about 400km and 650km.

Two other descriptions are based on mechanical properties of the crust and of different layers in the mantle. In order to understand better the issues involved in these descriptions, we shall examine the rheology of crust and mantle first. Our main interest is in the isostatic adjustment of the Earth's upper strata to varying ice loads. The presentation follows Lliboutry (1973), Peltier (1982) and Turcotte (1979).

In Figure 11.4 we have considered the constitutive relations for an elastic solid, a perfectly viscous liquid and the visco-plastic creep law of ice. The crust and mantle behave on short time scales, of minutes to hours, like an elastic solid, and therefore propagate rapidly fluctuating seismic waves, both pressure and shear waves.

On Quaternary time scales, there is strong evidence that at least part of the mantle behaves like a viscous fluid. Shifts in relative sea levels around the shores of Fennoscandia and North America are documented. These shifts seem to accompany over thousands of years the melting of the last major ice sheets, and they continue today. The continuation of displacement after the load has been removed is an indication, as we shall see, of visco-elastic behavior.

Many composite constitutive relations exist which model materials with elastic, instantaneous response to changes in stress, as well as viscous "memory" effects. The simplest relation of this type governs the Maxwell "body",

$$\frac{1}{E}\dot{\tau} + \frac{1}{\nu}\tau = \dot{e}. \qquad (11.19)$$

When $(1/\nu) \to 0$, the Maxwell body becomes a Hooke solid, cf. (11.2); when $(1/E) \to 0$ it becomes a Newton liquid, cf. (11.3).

The Maxwell body can be represented by a linear spring with constant E, coupled in series with a viscous dashpot having a damping constant ν. This two-element device will have a displacement $e(t)$ of its two end points relative to each other, governed by (11.19), when subject to an axial "force" $\tau(t)$.

Given such a simple laboratory realization of (11.19), one can think of two standard experiments. One is to impose a constant force $\tau = \tau_1$ at time $t = 0$ to the apparatus at rest:

$$e(-0) = 0, \qquad \tau = \tau_1\eta(t), \qquad (11.20a,b)$$

where $\eta(t)$ is the unit step function, or Heaviside function,

$$\eta(t) = 0, \qquad t < 0, \qquad (11.20c)$$

$$\eta(t) = 1, \qquad t \geq 0; \qquad (11.20d)$$

the force τ_1 is thus applied at $t = 0$ and maintained indefinitely.

For experiment (11.20), Eq. (11.19) becomes

$$\dot{e} = (\tau_1/E)\delta(t) + (1/\nu)\eta(t), \qquad (11.21a)$$

where $\delta(t) = \dot{\eta}(t)$ is the generalized function called Dirac's δ-function, $\delta(t) = \lim_{\theta\to 0} \delta_\theta(t)$, with

$$\delta_\theta(t) = 0, \qquad 0 < \theta < |t| \qquad (11.21b)$$

$$\delta_\theta(t) = \frac{1}{2\theta}, \qquad 0 < |t| < \theta. \qquad (11.21c)$$

Integrating (11.21) yields

$$e(t) = \frac{\tau_1}{E} + \frac{\tau_1}{\nu}t, \qquad t > 0: \qquad (11.22)$$

the displacement e takes its elastic value τ_1/E at once, and continues growing in a linear, viscous fashion. Experiment (11.20) is therefore called a *creep experiment*, cf. Figure 11.4a. The linearly growing viscous part of the displacement becomes comparable to the instantaneous, elastic part after a time $O(\nu E^{-1})$.

The second experiment is to impose a displacement $e = e_1$ at $t = 0$ and maintain it indefinitely. One immediately obtains

$$\tau(t) = Ee_1\exp\{-(E/\nu)t\}, \qquad t > 0, \tag{11.23}$$

with $\tau(0) = Ee_1$. For a constant strain, the stress relaxes to zero with a characteristic time ν/E, the *Maxwell time*. The second standard experiment is thus called a relaxation experiment, and it determines exactly the characteristic time separating, cf. (11.22), predominantly elastic from predominantly viscous responses.

The Maxwell time for the mantle is of the order of 200 years. A loading history for postglacial uplift would be somewhat different from, and more complicated than the standard experiments above. Figure 11.11 shows uplift histories from the Laurentide ice sheet, the largest one, the Fennoscandian sheet, two times smaller in linear dimensions, and the

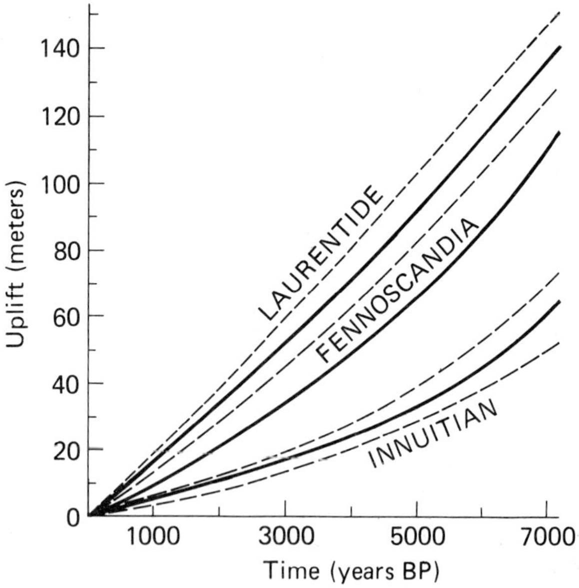

Fig. 11.11. Postglacial uplift in the center of the Laurentide, Fennoscandian and Innuitian uplift zones. Approximate limits of uncertainty, where known, are shown by dashed lines (after Walcott, 1973).

Innuitian sheet, which was centered in the northern part of the Canadian Archipelago, contiguous to the Laurentide sheet and four times smaller in linear dimensions.

Lithosphere and mantle. To connect our discussion of the Maxwell body, Eqs. (11.19 - 11.23), with Figure 11.11, it is easiest to consider first a geophysical realization of the laboratory device discussed before. This realization is illustrated schematically in Figure 11.12.

The *lithosphere*, as we shall see, comprises the crust and the uppermost stratum of the mineralogically defined mantle. The load of the ice sheet produces a displacement of the lithosphere and mantle under it. The portion of the lithosphere and mantle immediately below and nearby the ice sheet transmit forces to the far-field upper strata. The lithosphere is shown in Figure 11.12 as a spring, transmitting elastic stresses, while the mantle is shown as a dashpot, transmitting viscous stresses.

In fact, both lithosphere and mantle have complicated rheologies, which are temperature-dependent and nonlinear. On short time scales, the elastic behavior of the Earth's upper strata is relatively well known, due to the wealth of data available from the regular seismic network, as well as from many special measurements. Inferences about the long-time behavior

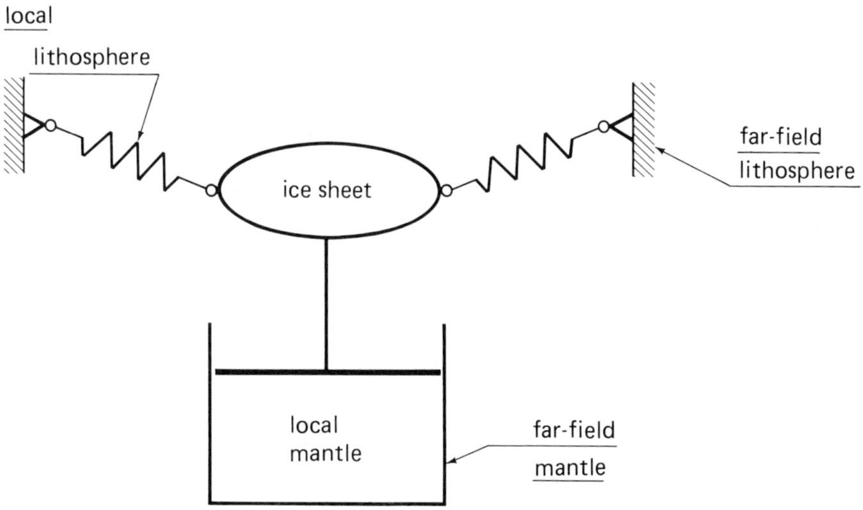

Fig. 11.12. A schematic representation of the mechanical interaction between ice load, lithosphere and mantle.

are based on very incomplete and disparate data, like those shown in Figure 11.11.

We are ready now to sketch the two mechanical descriptions of the Earth's upper strata mentioned at the beginning of this section. The purely *elastic* description, based on observed seismic-wave travel paths and travel times, distinguishes between lithosphere, asthenosphere and mesosphere. The oceanic lithosphere is approximately 100km thick, increasing in thickness from the mid-oceanic ridges where it is generated, to the oceanic trenches usually located near continental plate margins, where it is subducted. The continental lithosphere is approximately 200km thick, with large, irregular variations due to orogenic and sedimentary processes, and to erosion. The thickness of the lithosphere given above is based on a solid-liquid boundary at an isotherm of 1200° ± 100° C.

The *asthenosphere* is a layer about 200km thick under the lithosphere in which seismic wave velocities are lower and shear waves are damped much more rapidly than in the lithosphere. The rest of the mantle, down to 2900km, is the *mesosphere*, in which seismic waves propagate again better than in the asthenosphere. The boundary between asthenosphere and mesosphere is probably the phase discontinuity at 400km.

The second mechanical description deals with rheological properties of the upper strata on geological time scales. The mantle behaves mostly as a viscous fluid on geological time scales longer than its Maxwell time of 200 years. Convective flow in the mantle on time scales of 10^8 years is believed to be connected with the relative displacement of lithospheric plates, causing *continental drift*. On the shorter, Quaternary time scales, the position of continents is nearly constant, however, and we are only interested in the mantle's viscosity from the point of view of isostatic adjustment.

The middle part of the lithosphere, under the Moho and above the 600°C ± 100°C isotherm, bears elastic stresses on geological time scales, as well as on the short time scales of seismic wave propagation. The crust itself, above the Moho, is considerably weaker on geologic time than the middle, elastic lithosphere. The lower part of the lithosphere, between the 600°C and the 1200°C isotherms, exhibits stress relaxation as in Eq. (11.23). Plastic creep also plays a role in lithospheric regions of stress concentration, near orogenic belts or subduction zones.

Within the mantle, a discontinuity in viscosity is currently believed to occur near the phase discontinuity at 650km and it is customary to distinguish between the *upper mantle*, above this depth, and the *lower mantle*, down to the core. Isostatic uplift data have been used since the 1930s to attempt inferring the viscosity of the upper and lower mantle.

The simplest mantle models used in these attempts have been of either the thin-channel or the half-space type. In the one, a large viscosity jump was assumed across a boundary like the one at 650km, and the lower mantle was taken simply as rigid. In the second, a uniform or slowly varying mantle viscosity was assumed down to the core, which was taken at infinity (see Section 11.4).

The prevailing opinion today is that the upper mantle has a mean viscosity of 10^{22}P, in cgs units, with $1P = 1g\ cm^{-1}s^{-1}$. For comparison, the viscosity of air is $O(10^{-4}P)$ and that of water is $O(10^{-2}P)$. The lower mantle probably has a mean viscosity of the order of 10^{23}P (Hager *et al*., 1985). This viscosity jump is considerably smaller than the orders of magnitude previously assumed in thin-channel models, but still quite appreciable; it is comparable to that between a layer of lubricating oil at high temperatures and one at low temperatures.

There are many indications that isostatic uplift does not follow a simple exponential law, as suggested by Eq. (11.23). Two or more relaxation times are apparent on curves like those of Figure 11.11, with uplift occurring faster at first and slower at a later time. The shortest relaxation times are O(1ka), the longest are O(10ka), within the accuracy of the data (e.g., Lliboutry, 1973, Fig. 10 and 14).

Multiple relaxation times can be explained in the simplest manner by two or more Maxwell-type two-element devices in parallel, each contributing its own relaxation time. The previously discussed stratification of a linearly viscous mantle provides essentially this type of explanation. As the number and accuracy of uplift data and related paleo-indicators of geodynamics increases, knowledge of the viscous structure of the earth on geological time scales will become more detailed and reliable, following in the footsteps of the short-time elastic structure. In particular, it will become possible to identify strong lateral inhomogeneities, now recognized in the elastic structure, and which must play a crucial role in the geodynamic response to the plate-size and sub-plate-size ice loading.

A simple model of isostatic adjustment. In the derivation of the ice-sheet model of Eqs. (11.11 - 11.18) it was assumed that instantaneous isostatic adjustment, i.e., adjustment to a gravitationally stable configuration, occurs according to the density ratio ρ_i/ρ_u of ice to mantle. The accelerations in the simple mechanism of Figure 11.12 are in fact negligible (LeTreut and Ghil, 1983, Appendix A), justifying the term *isostasy*, but the velocities are not. In the remainder of this section, we shall incorporate some of the knowledge about lithosphere and mantle discussed above into the model to be studied in Section 12.4. Here and in the sequel we refer to the lithosphere according to its thermal definition, which is sufficient for both its short- and long-term elastic properties. The mantle refers therewith to that part of the mineralogically defined mantle which lies below the 1200°C isotherm. Our presentation follows Ghil and LeTreut (1981) and LeTreut and Ghil (1983).

The ice sheet of Figure 11.8 is now shown in Figure 11.13 suspended on the elastic lithosphere. Near the load, the lithosphere acts in flexure, like a bending beam or thick plate. At distances larger than $O(10^3 km)$, it acts more like a spring or membrane, cf. Figure 11.12, and this action is much weaker (Turcotte *et al.*, 1981). From the point of view of the simple, zero-dimensional model we shall study, there is no need to distinguish between the two types of elastic response of the lithosphere.

This elastic response has an effect on the relative position of the 0° C isotherm and the ice sheet, and hence on the net mass budget of the sheet. The effect can be written as

$$\Delta h = k(h + \zeta), \tag{11.24a}$$

where Δh is the vertical distance between the point of contact of cryosphere and lithosphere, shown by an open circle in Figure 11.13, and mean sea level (MSL). Eq. (11.12b) will be modified therewith to read

$$h_0 = \beta(T - T_{00}) + \Delta h. \tag{11.24b}$$

The elastic spring constant k is $O(10^{-2})$, i.e., a bedrock deflection of O(10 m) for a change in ice thickness O(1km). The total mass of the ice is expressed in (11.24a) as the sum of a contribution above the con-

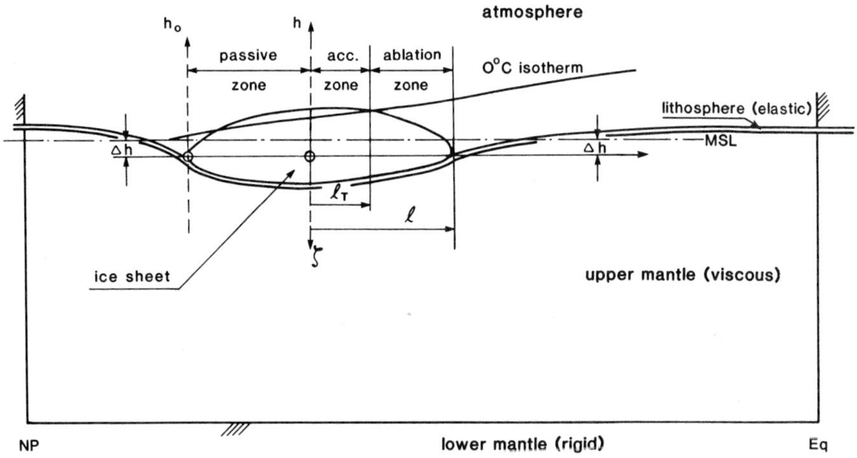

Fig. 11.13. Meridional cross section through the Earth's upper strata (after LeTreut and Ghil, 1983).

tact with the bedrock, h, and a contribution below this contact, ζ. The common origin of the h-axis, pointing up, and of the ζ-axis, pointing down, is marked by another open circle in the figure.

To study the effect of mantle viscosity on the distribution of ice mass between h and ζ, we make a number of simplifying assumptions. First, only the shorter adjustment times will be taken into account, for reasons to be discussed in Section 12.4. This is the same as assuming the lower mantle to be rigid. Next, the deflection of the bedrock will be assumed small compared to the depth H of the upper mantle, $\zeta \ll H$, since $\zeta = O(10^{-1}\text{km})$ and $H = O(10^2\text{km})$. Finally, the wave lengths κ^{-1} to be considered are large, $\kappa H < 1$, i.e., we discuss only large ice sheets, not small ice caps. Notice that the viscosity effects to be modeled below, and the preceding elasticity effects, can be superimposed in a linear rheological model, cf. (11.19).

At each point of the mantle, the hydrostatic pressure is

$$p_0 = P + \rho_u g\zeta,$$

where P is the sum of the pressure due to the atmosphere and to the lithosphere. P will be considered constant in space and time. The density of the upper mantle is ρ_u.

Let u be the velocity in the x-direction, and w in the ζ-direction. The momentum equations and the continuity equation, using the isostatic assumption, can be written as

$$\eta \, \Delta u = \frac{\partial(p - p_0)}{\partial x}, \tag{11.25a}$$

$$\eta \, \Delta w = \frac{\partial(p - p_0)}{\partial \zeta}, \tag{11.25b}$$

$$\frac{\partial u}{\partial x} + \frac{\partial w}{\partial \zeta} = 0, \tag{11.25c}$$

where η is a viscosity coefficient and $p - p_0$ the departure from hydrostatic pressure. Eqs. (11.25) are the steady-state Navier-Stokes equations for flow with negligible acceleration, also known as the Stokes equations.

Introducing the Fourier transformed variables F, G and Π by

$$u(x,\zeta) = \int_0^\infty F(\zeta,\kappa) \sin \kappa x \, d\kappa,$$

$$w(x,\zeta) = \int_0^\infty G(\zeta,\kappa) \cos \kappa x \, d\kappa,$$

$$p(x,\zeta) - p_0(\zeta) = \int_0^\infty \Pi(\zeta,\kappa) \cos \kappa x \, d\kappa,$$

leads to a new system:

$$F'' - \kappa^2 F = -(\kappa/\eta)\Pi, \tag{11.26a}$$

$$G'' - \kappa^2 G = \eta^{-1} \, \Pi', \tag{11.26b}$$

$$G' + \kappa F = 0, \tag{11.26c}$$

where $(\ )' = d(\ )/d\zeta$. The general solution of Eqs. (11.26) is

$$F = -(A + B + B \, \kappa\zeta)e^{\kappa\zeta} + (C - D + D \, \kappa\zeta)e^{-\kappa\zeta}, \tag{11.27a}$$

$$G = (A + B \, \kappa\zeta)e^{\kappa\zeta} + (C + D \, \kappa\zeta)e^{-\kappa\zeta}, \tag{11.27b}$$

$$\Pi = 2\eta\kappa(B \, e^{\kappa\zeta} + D \, e^{-\kappa\zeta}). \tag{11.27c}$$

The constants of integration A, B, C, D depend on the horizontal wave number κ and are determined by the boundary data.

The boundary conditions for (11.26) are a no-flow condition at the contact with the "rigid" lower mantle $\zeta = H$,

$$F(H) = G(H) = 0. \tag{11.28a}$$

and a no-slip and prescribed-stress condition at the contact with the cryosphere, $\zeta = \tilde{\zeta}$,

$$F(\tilde{\zeta}) = 0, \tag{11.28b}$$

$$\tau_{\zeta\zeta} = 2\eta \frac{\partial w}{\partial \zeta} - p. \tag{11.28c}$$

Here $\tau_{\zeta\zeta}$ is the vertical, normal stress on the mantle due to the cryosphere, which can depend on time t. From our assumptions it follows in particular that $\kappa\tilde{\zeta} << 1$ and hence (11.28b) can be simplified to read

$$F(0) = 0. \tag{11.28d}$$

We use Eqs. (11.27, 11.28) now to determine A, B, C and D. Writing out (11.28c) and substituting for p yields

$$2\eta \frac{\partial w}{\partial \zeta} - p_0 - \int_0^\infty \Pi(\zeta) \cos \kappa x \, d\kappa = -\rho_i g(\tilde{h}(x) + \tilde{\zeta}(x)) - P$$

or

$$\int_0^\infty (2\eta\kappa f + \Pi) \cos \kappa x \, d\kappa = \rho_i g \, \tilde{h}(x) + (\rho_i - \rho_u) \, g \, \tilde{\zeta}(x), \tag{11.29}$$

with ρ_i the density of the ice and $\tilde{h} + \tilde{\zeta}$ the height of the ice sheet above the bedrock (Figure 11.13). After some algebraic manipulation, we obtain

$$A + C = -\frac{\kappa^3 H^3}{2} (A - C).$$

Then, with the slow time dependence imposed by variations in ice load,

$$\frac{\partial \tilde{\zeta}}{\partial t} \simeq w(x,0) = \int_0^\infty (A(\kappa) + C(\kappa)) \cos \kappa x \, d\kappa$$

$$= -\frac{H^3}{2} \int_0^\infty (A(\kappa) - C(\kappa))\kappa^3 \cos \kappa x \, d\kappa$$

and from (11.29)

$$-\int 2\eta\kappa(A(\kappa) - C(\kappa)) \cos \kappa x \, d\kappa = \rho_i g \, \tilde{h}(x) + (\rho_i - \rho_u) g \, \tilde{\zeta}(x).$$

Introducing the inverse viscosity coefficient

$$D = \frac{\rho_u g \, H^3}{4\eta} \tag{11.30a}$$

and the density ratio

$$q = \rho_i/\rho_u, \tag{11.30b}$$

yields the equation

$$\frac{1}{D}\frac{\partial \tilde{\zeta}}{\partial t} = (1-q)\frac{\partial^2 \tilde{\zeta}}{\partial x^2} - q\frac{\partial^2 \tilde{h}}{\partial x^2} . \qquad (11.31)$$

We shall derive now from (11.31) a zero-dimensional equation for the maximum local deflection ζ_0 of the bedrock $\tilde{\zeta}(x)$. Such an equation can then be coupled with Eq. (11.16) or (11.18) for the ice extent and Eq. (11.24) for the snowline/bedrock-deflection feedback. For the purposes of this derivation, we make one more simplifying assumption.

We assume the ice sheet has a doubly-parabolic profile as in Figure 11.8 or 11.13, where h_0 is the maximum height of the ice sheet above the point of contact with the bedrock, $\zeta_0(t) = \tilde{\zeta}(0,t)$ is its maximum depth under the contact, and $2L$ is its latitudinal extent. Under these assumptions, (11.31) becomes

$$\dot{\zeta}_0 = -D(1-q)\frac{\zeta_0}{L^2} + Dq\frac{h_0}{L^2} , \qquad (11.32)$$

where $(\ )^{\cdot} = d(\ )/dt$.

For a zero-dimensional model, the exact form of the profile only enters the formula relating ice volume and ice extent. As long as the form of the profile does not vary too much in time, it will only change slightly the value of some numerical coefficients in the equations. We shall allow such coefficients to vary during our model analysis by amounts compatible with the uncertainty in their numerical value.

The equation for the cross-sectional area V of the doubly-parabolic ice sheet, in terms of h_0 and ζ_0, is

$$V = (4L/3)(h_0 + \zeta_0) . \qquad (11.33)$$

Previously, in deriving (11.6), it had been assumed that $\zeta_0 = h_0/2$, which yielded (11.18) in that case.

Differentiation of (11.33), using (11.32) and the parabolic-profile assumption, leads to two coupled equations for ℓ and $\zeta = \zeta_0(\lambda L_M^*)^{-1/2}$:

$$c_L\dot{\ell} = \{1 + (2/3)(L_M^*)^{-1/2}\zeta\,\ell^{-1/2}\}^{-1}\{(3/2)\ell^{-1/2}[(1+\varepsilon(T))\ell_T(T,\ell) - \ell]$$

$$(2/3)c_L\,\ell^{1/2}(A\zeta\ell^{-2} + B\ell^{-3/2}) , \qquad (11.34a)$$

$$\dot{\zeta} = A\zeta\ell^{-2} + B\ell^{-3/2} , \qquad (11.34b)$$

where

$$A = -\frac{D(1-q)}{4(L_M^*)^2} , \quad B = \frac{Dq}{4(L_M^*)^2} . \qquad (11.34c,d)$$

Eq. (11.34b) models the effect of ice extent ℓ on local bedrock deflection ζ, and (11.34a) shows how ζ modifies the evolution of ice extent ℓ with respect to the instantaneous adjustment assumed in (11.18). The coefficients A and B depend on the shape of the ice sheet and on the mantle properties D and q.

Eqs. (11.18) and (11.34) are the simplest expression of ice-sheet dynamics, and of coupled cryodynamics and geodynamics respectively. We shall study in Section 12.1 the free oscillations of a model coupling radiation balance (10.2) with (11.18) via the ice-albedo feedback. In Section 12.4 forced oscillations of a model including (10.2) and (11.34) will be studied.

11.4. Bibliographic Notes

For an excellent review of climatic variability on various time scales we refer to Crowley (1983). More detailed accounts can be found in Berger *et al.* (1984), from the special perspective of the astronomical theory (see Section 12.3), Berger (1981), Gribbin (1978), Lamb (1977, Part III) and U.S. Committee for GARP (1975, Chapter 4 and Appendix A). A handy instant reference for dates of and correlations between orogenic, i.e., mountain-building, events, stratigraphic classification and paleobiologic and climatologic events is the Geological Time Table of van Eysinga (1975).

For the modeling of climatic change on various time scales, some of the references above contain appropriate sections or chapters. To these we should add Gates (1979) and Joint Organizing Committee (1975). More references on general circulation models (GCMs), with high resolution in three spatial dimensions, are given in Section 4.6. Radiative-convective models (RCMs), with vertical resolution only (compare Sections 10.1 and 10.2), are reviewed by Ramanathan and Coakley (1978).

Section 11.1. The classical text on the Quaternary and its glaciations is Flint (1971). A beautiful account of the history of ice age research in popular form is given by Imbrie and Imbrie (1979). Frisch (1980) describes the continuity of human presence and of climatic change in the Alpine valleys where this history starts.

Early glaciations are described by Crowley (1983) and Gribbin (1978, Section 1.1). The development of stable isotope studies in paleoclimatology is outlined by Flint (1971; Chapter 27) and by Imbrie and Imbrie (1979; Chapter 11). Major contributions were made by Urey (1947), Epstein *et al.*

(1951) and Emiliani (1955). Technical reviews are provided by Dansgaard (1981), Duplessy (1978) and Shackleton (1981).

Section 11.2. A monumental text covering the whole field of glaciology is Lliboutry (1964/65). Paterson (1981) strikes an excellent balance between theory and observations, on the one hand, and between the competing theories, on the other. Colbeck (1980) contains contributions on ice sheets and ice shelves (by Paterson), on temperate valley glaciers (by C. F. Raymond), on sea ice (by Hibler), on seasonal snow (by D. H. Male), on avalanches (by R. I. Perla), and others. A recent single-author text is by Hutter (1983).

The modeling of large ice masses is reviewed by Budd and Radok (1971), and by Oerlemans and van der Veen (1984), who also emphasize the internal and surface thermodynamics of cold, polar ice sheets, in which temperature can drop well below freezing and modify flow properties. Spatial resolution in latitude is included in the models of Birchfield *et al.* (1981) and of Pollard (1983), which bear the same, relatively simple relation to the zero-dimensional model (11.34) as the one-dimensional model (10.12) had to (10.2). Stability of the similarity solution for ice-sheet flow subject to Glen's law in two and three dimensions is analyzed by Halfar (1983): in the limit of perfect plasticity the general similarity solution reduces to Orowan's parabolic profile, and the similarity solution is asymptotically stable with respect to arbitrary small perturbations which leave the volume of the ice invariant.

Evidence for large fluctuations in sea-ice extent on interannual and decadal time scales is given by Barry (1983) and by Zwally *et al.* (1983). Modeling of sea ice is discussed by Hibler (1980) and by Untersteiner (1983) for models with two- and three-dimensional spatial resolution, and by Saltzman *et al.* (1981) for a zero-dimensional model based on somewhat different ideas than the spatially-resolved models.

Simpson (1934, 1938) advanced the apparently paradoxical view that an increase in the solar radiation received by the Earth would bring about an ice age. His investigation of radiative processes in the atmosphere, and of their deviation from black-body radiation, was very advanced for that time; but many of the quantitative observations of radiation, as well as the paleoclimatological evidence available then, were much less accurate than they are now (see Sections 10.1 and 11.1).

The assumption used here, and confirmed by the deep ice-core studies of Andrews *et al.* (1974), Dansgaard *et al.* (1982), Lorius *et al.* (1979)

and others mentioned in LeTreut and Ghil (1983), is distinct from Simpson's thesis: we postulate that net snow accumulation goes down as system *temperature, not* solar *insolation*, goes down. As will be shown in Sections 12.1 and 12.4, global temperature, ice volume and insolation are not in phase or in opposition of phase, respectively, but have complicated phase relationships which change with time. The present assumption is therefore a refinement of Simpson's striking idea, made more compatible with the present quantity and quality of paleoclimatological data.

Paterson (1981, p. 158 ff.) discusses thoroughly many simplifying assumptions made in deriving ice sheet geometry and evolution. The most serious one is probably that of solid friction at the bedrock, with the shear stress equal to the yield stress. In fact, melt water due to very high pressure exists at the base of ice sheets, and most of the ice flow in Greenland and the Antarctica today occurs through very rapidly moving streams with liquid friction at the base. Although the controversies about basal sliding laws (Paterson, 1981, p. 113 ff.) have generated enough heat to melt a small ice cap, the net effect of lubrication, as well as that of unstably anchored ice shelves (Paterson, 1981, p. 171 ff.) is likely to give faster response times of ice sheets to global temperature changes than otherwise estimated.

Section 11.3. A general introduction to geodynamics from the point of view of continuum physics is given by Turcotte and Schubert (1982). The elementary facts of rheologic life can be found in Bland (1960), for instance.

The laterally homogeneous, i.e., spherically symmetric part of the Earth's short-time, elastic structure is relatively well known from seismology (Gilbert and Dziewonski, 1975). Recent increases in the density of digital seismographic stations permit now some resolution of lateral heterogeneities in the elastic properties of lithosphere and mantle. Large differences in seismic wave propagation with respect to geographic location, as well as depth, are apparent down to 670 km, and are not correlated in a simple way with ocean-continent contrasts or other known geologic features (Dziewonski and Anderson, 1984).

By comparison, data available to infer even the spherically symmetric part of the Earth's long-time, viscous structure are few and far between. The main data refer to relative sea levels and come from the post-glacial uplift of Fennoscandia, Greenland, Hudson Bay, the Arctic Archipelago, the East Coast of the United States and Lake Bonneville (Cathles, 1975). Additional data from residual, non-isostatically adjusted gravity anomalies and, more tentatively, from residual polar drift (Peltier, 1982), can be used to deduce the depth-dependent viscosity of the upper strata. Smaller ice masses produce more localized uplifts, and hence help probe the upper mantle, while data from larger ice masses are useful in probing down into the motions, and hence viscosity, of the lower mantle.

Shoreline data from the Baltic were known first, and it was in order to explain these data that both half-space models (Haskell, 1935; Vening-Meinesz, 1937) and thin-channel models (Van Bemmelen and Berlage, 1935; Crittenden, 1963) were originally developed. These models predict a dependence of relaxation time θ on horizontal wave number of the deformation κ with $\theta \sim \kappa$ and $\theta \sim \kappa^{-2}$, respectively. As data corresponding to different κ became availabe, McConnell (1968) realized that the viscosity of the mantle does not have to be constant throughout or jump to infinity below a certain depth, but that two or more layers of different, finite viscosity may be used to explain the data. The dependence of θ on κ is now believed to first decrease with κ, as in channel models, then bottom out and increase with κ, as in constant-viscosity models, with a range of θ between 1ka and 10ka over the available range of roughly 100km $\leq 1/\kappa <$ 10,000 km (Peltier, 1982, Figure 13).

Solid-solid phase discontinuities exist at a depth of 420 km and of 670 km, and elastic properties change at these two depths, especially the latter. But this does not necessarily imply a similar jump in viscosity at these depths. In fact, thin-channel models were originally restricted to a depth of 100 km, based on the earlier knowledge of a thin layer of mantle, which is seismically "softer" or "weaker"; hence the term "asthenosphere", used also by some authors for the entire upper mantle. According to present knowledge, the elastically defined asthenosphere does not reach deeper than about 400km, while the viscously defined upper mantle reaches to a depth of about 650km.

The existence and value of a viscosity jump between upper and lower mantle is important for modeling isostatic rebound. It is even more important in modeling the mantle convection which supports plate drift at the surface. Among many other unsolved questions concerning flow in the mantle, that of the vertical extent of convection cells (i.e., does convection penetrate the lower/upper mantle discontinuity or not, "one cell" or "two cells", see for instance Dziewonski and Anderson, 1984, Figure 1) is particularly intriguing.

CHAPTER 12

CLIMATIC OSCILLATORS

In Section 11.1 we reviewed some of the geological evidence for glaciation cycles during the Quaternary period. Large changes in global ice volume and changes of a few degrees in global mean temperature have occurred repeatedly over the last two million years. It is these changes we would like to investigate in the present chapter with the help of very simple models.

These simple models do not represent the definitive formulation of a theory for climatic variability on the time scales of interest. They are used merely to illustrate some ideas we believe to be basic for an understanding of this variability, an understanding which is still in early stages of development. Comments on other models and related ideas will be made in Section 12.7.

In Section 12.1, we formulate and analyze a coupled model of two ordinary differential equations, for global temperature and global ice volume. The equations govern radiation balance (Section 10.2) and ice-sheet flow (Section 11.2), respectively. This model exhibits self-sustained oscillations with an amplitude comparable to that indicated by the records (Section 11.1) and a period of $O(10\text{ ka})$, where 1ka = 1000 years. Phase relations between temperature and ice volume, and their role in the oscillation's physical mechanism, are investigated.

Exchange of stability between equilibria (Section 10.2) and limit cycles (Section 12.1) in models with an arbitrary number of dependent variables and spatial dimensions is studied in Section 12.2. The distinction is made between a stable limit cycle which grows slowly in amplitude from zero as a parameter is changed (direct or supercritical Hopf bifurcation) and sudden jumps from zero to finite amplitude (reverse or subcritical Hopf bifurcation). Structural stability and the special role of limiting, structurally unstable, homoclinic and heteroclinic orbits is discussed.

In Section 12.3 we introduce the geometry and kinematics of orbital changes in the Earth's motion around the Sun, from the perspective of the small insolation changes they generate. Eccentricity, obliquity and precession of the Earth's orbit are defined. A few fundamental concepts of celestial dynamics and the associated mathematical methods are reviewed. We report currently accepted results on the periodicities of insolation changes during the Quaternary period: 19ka and 23ka for precession, 41ka for obliquity, and 100 ka and 400 ka for eccentricity. Their action on the climatic system's radiation balance and hydrologic cycle is modeled.

In Section 12.4 we take up the effects of this action upon the climatic oscillator of Section 12.1, augmented by a third equation, governing bedrock response to ice load (Section 11.3). The free oscillations of this model are found to differ but little from those of the previous one. Eccentricity forcing is shown to produce a very small or very large response according to whether the system operates in an equilibrium or in an oscillatory mode.

In Section 12.5 we study the internal mechanisms by which forcing at one or more frequencies can be transferred through the system to additional frequencies, as well as to the climatic spectra's continuous background. Entrainment results in the system's free frequency becoming locked onto an integer or rational multiple of a forcing frequency. Loss of entrainment leads to aperiodic changes in system response. Combination tones are linear combinations with integer coefficients of the forcing frequencies; prominent among them in paleoclimatic spectra appear to be those with periods near 15 ka, 13 ka and 10ka, and their harmonics. As a result of hydrologic and insolation forcing at the orbital frequencies, the model considered here produces spectral lines at many of the observed frequencies, superimposed on a continuous background associated with aperiodic, irregular terminations of glaciated episodes.

In Section 12.6, we consider the predictability and reproducibility of climatic time series. The presence of multiple spectral lines and of the continuous background has important consequences for predictability, or the lack thereof. It is argued that irretrievable loss of predictive skill over a time interval O(100 ka) is intrinsic in the aperiodic nature of Quaternary climate changes.

Section 12.7 discusses extensively parallel and future lines of research in theoretical climate dynamics. General references for nonlinear oscillations and Hopf bifurcation are given, along with related results and reviews in the climatological literature. Formal conceptual models using the mathematical framework of Boolean delay equations are described as exploratory tools in formulating more complex quantitative climate models. The potential role of relaxation oscillations in terminations and in their irregular occurrence is mentioned.

The history of celestial mechanics in its theoretical and practical aspects is outlined. References for the theory of conservative and of dissipative perturbations to integrable Hamiltonian systems are given, and an assessment of the current accuracy and deficiencies in orbital calculations is offerred.

Forced nonlinear oscillations are reviewed and the importance of bistability and hysteresis effects is discussed. A scenario of successive transitions from the Tertiary to the Quaternary and from Early to Late Pleistocene is proposed, incorporating and extending the results of Chapters 10 and 12. References for combination tones, entrainment, detrainment and nonlinear resonance are given, recalling connections with music, hearing, celestial mechanics, electronics and fluid dynamics. Stochastic perturbations of periodic solutions and synchronization in the presence of noise are also discussed.

Different perspectives on predictability are outlined and the connection with spectral properties of a time series is emphasized. We close with a discussion of coupled oscillators distributed in space, and possible explanations of phase differences in temperature and other mean, climatic properties between various regions of the globe.

12.1. Free Oscillations of the Climatic System

We start by formulating a very simple model coupling the energy balance of the atmosphere-ocean system with the mass balance of continental ice sheets. The presentation in this section follows Källén *et al.* (1979) and Ghil and Tavantzis (1983).

The governing equations. In Chapter 10 it was shown that energy-balance models explain certain features of the Earth's radiation budget and temperature distribution. For the purposes of the present discussion, we shall write the global, annually averaged radiation balance of the Earth in a form slightly different from Eqs. (10.2, 10.3), as

$$c_T\dot{T} = Q\{1 - [\gamma(\alpha_0 + \alpha_1 L_M^* \ell) + (1-\gamma)\alpha_{oc}(T)]\} - \kappa(T - T_\kappa). \tag{12.1}$$

Here C_T is the heat capacity of the atmosphere-ice-ocean system, which is dominated by that of the ocean. The thermal overturning time of the ocean is $O(1\text{ ka})$; for the sake of definiteness we shall take $c_T^* = c_T/\kappa = 1$ ka, and time t will be measured in units of 1 ka.

T is the global, mean annual temperature, and Q the incoming solar radiation. This radiation is reflected by the model's surface features, which are shown in Figure 12.1.

There are two ways in which (12.1) differs from (10.2). First, in distinguishing between land albedo, which depends explicitly on ice-sheet extent L, and ocean albedo, which depends on temperature T as in (10.2). A fraction γ of the surface is land, and its albedo increases proportionally to the meridional extent L of the model's ice sheet, cf. (11.10). This dimensional extent is scaled by L_M^*, $L = \ell L_M^*$, cf. (11.17), so that ℓ is of order 1. The oceanic albedo $\alpha_{oc}(T)$ is given by Figure 11.9, and is the same as $\alpha = \alpha(\overline{T})$ used in (10.2) for the total albedo.

Second, the outgoing radiation is taken for simplicity as a linear function of temperature, $-\kappa(T - T_\kappa)$. We saw at the end of Section 10.2 that such a linearization does not modify qualitatively the results of energy-balance models. The same is true for the coupled models to be discussed here.

To complete the description of the model, we need an equation for ice-extent evolution. Such an equation, (11.18), was derived in Section 11.2. We repeat it for convenience here:

$$c_L\dot{\ell} = \ell^{-1/2}\{[1 + \varepsilon(T)]\ell_T(T,\ell) - \ell\}. \tag{12.1b}$$

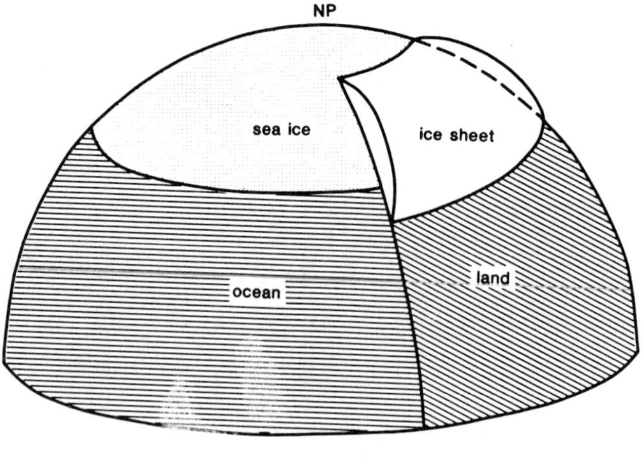

Fig. 12.1. Surface features of the model's radiation balance (after Källén *et al.*, 1979).

The time c_L it takes ice extent to relax to its equilibrium value at constant T, Eq. (11.15), is also $0(1\text{ ka})$. The precipitation-temperature feedback $\varepsilon(T)$ is shown in Figure 11.9. The position of the snowline $\ell_T = \ell_T(T,\ell)$ is given by Eqs. (11.13, 11.17) and shown in Figure 11.8.

For simplicity, we analyze fist in detail the behavior of the model governed by Eqs. (12.1a,b), using instantaneous isostatic adjustment. The equation for bedrock deflection derived in Section 11.3, Eq. (11.34b), will be coupled with this model in Section 12.4.

The model's phase plane. Eqs. (12.1a,b) form a system of two coupled, nonlinear ordinary differential equations (ODEs), which we rewrite in non-dimensional form as

$$\dot{\theta} = P(\theta,\ell), \tag{12.2a}$$
$$\dot{\ell} = \mu R(\theta,\ell); \tag{12.2b}$$

here $\theta = T/T_s$, with T_s a characteristic value close to the present climate, time has been scaled by c_T^*, and $\mu = c_T^*/c_L$. The functions P and R do not depend explicitly on time, so that the system is *autonomous*.

For an autonomous system of two ODEs, the solutions are easy to visualize as *trajectories* in the (θ,ℓ)-phase plane. At each point $Q_0 = (\theta_0,\ell_0)$ in the plane, the vector $(P,\mu R)$ is tangent to the solution $Q(t) = (\theta(t),\ell(t))$ passing through that point. Since $P(\theta,\ell)$ and $R(\theta,\ell)$ are piecewise continuously differentiable for $\theta > 0$ and $\ell > 0$, the solution through an arbitrary Q_0 in the first quadrant is unique, and trajectories cannot intersect, or be tangent to each other, in finite time (compare Section 5.4). Furthermore, trajectories are invariant under shifts in time: a solution passing through Q_0 at time $t = t_0$, and a solution passing there at $t = t_1 > t_0$ describe the same trajectory.

One can think of the vector field $(P,\mu R)$ as the velocity field of a two-dimensional flow. In the autonomous case, this is a steady-state flow, and the trajectories are the *streamlines* of this flow.

Points where $P = R = 0$ play a special role in the phase plane. They represent stationary solutions of (12.2). These points are shown in Figure 12.2; they are Q_1, Q_2 and Q_s.

The lines in the phase plane along which the vector field $(P,\mu R)$ has a given, constant slope are called *isoclines*. The isocline $P(\theta,\ell) = 0$, along with the flow is vertical, and the isocline $R(\theta,\ell) = 0$, along with it is horizontal, are shown as solid lines in the figure. These isoclines are piecewise continuous due to the piecewise linear character of $\alpha_{oc}(T)$ and $\varepsilon(T)$, respectively. The values $T_{\alpha,\ell}$, T_m and $T_{\alpha,u} = T_M$ of Figure 11.9 are shown as dashed lines in Figure 12.2.

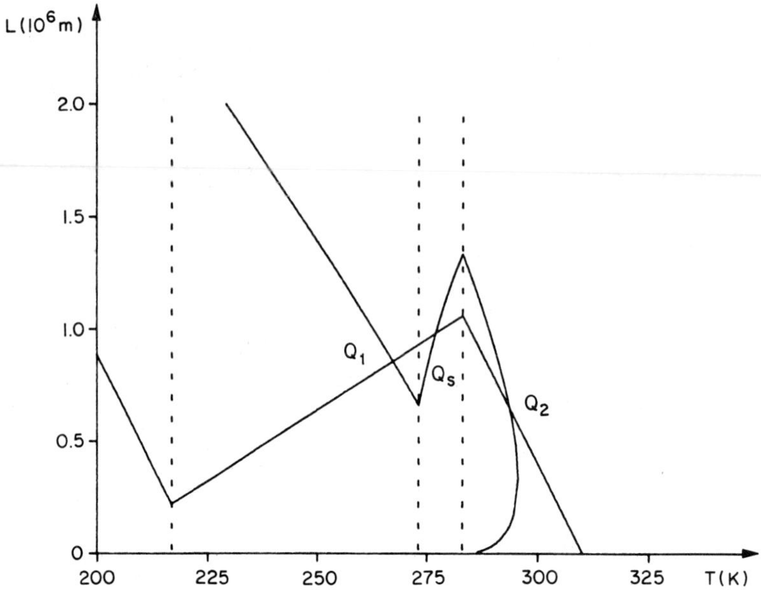

Fig. 12.2. The (T,L)-phase plane of the climate model (12.1) (from Ghil and Tavantzis, 1983).

The stationary points Q_1, Q_2 and Q_s are the intersections of the isoclines $P = 0$ and $R = 0$. They are also called *critical points* of the vector field $(P,\mu R)$, since they organize the entire flow around them, as we shall see.

Near an isolated critical point Q_c, like those in Figure 12.2, one can linearize the vector field, and consider the linear system

$$\begin{pmatrix} \dot{x} \\ \dot{y} \end{pmatrix} = \begin{pmatrix} P_x(0,0) & P_y(0,0) \\ R_x(0,0) & R_y(0,0) \end{pmatrix} \begin{pmatrix} x \\ y \end{pmatrix}, \tag{12.3}$$

where we took for simplicity $\mu = 1$, and $x = \theta - \theta_c$, $y = \ell - \ell_c$, $P_x = \partial P/\partial x = \partial P/\partial\theta$, and so on. The solutions of this system are of the form

$$x = C_1 e^{\lambda t}, \quad y = C_2 e^{\lambda t}. \tag{12.4}$$

Substitution of (12.4) yields nontrivial solutions provided λ is an eigenvalue of the matrix in (12.3),

$$(P_x - \lambda)(R_y - \lambda) - P_y R_x = 0. \tag{12.5}$$

The nature of the flow of the linear system (12.3) is given by the character of the roots $\lambda = \lambda_r + i\lambda_i$ of (12.5). The classification of critical points with unequal eigenvalues $\lambda_+ \neq \lambda_-$, is illustrated in

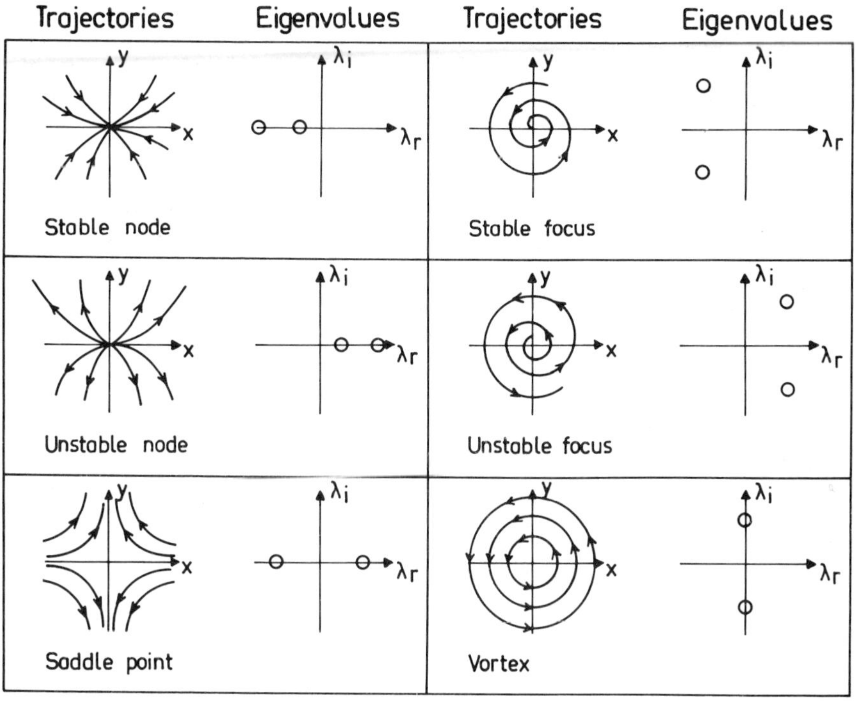

Fig. 12.3. Classification of phase-plane flows for a linear, autonomous sytem of two ODEs (from Källén *et al.*, 1979).

Figure 12.3. Solid lines in the (x,y)-phase plane show sample trajectories, and the arrows indicate the direction of the flow. The open circles in the (λ_r,λ_i)-diagrams give the corresponding eigenvalues.

The importance of classifying the flows for linear systems is that they also represent the possible flows near *isolated* critical points of nonlinear systems. It suffices to consider continuous deformations of the x and y axes in Figure 12.3 into smooth curves. As the entire flow is deformed by the nonlinearity, the topological structure of the flow is preserved.

More precisely, isolated critical points with the same number of negative λ_r are preserved under differentiable deformations. A linear node might become a nonlinear focus, and vice versa, since for either critical point both $\lambda_r^{(+)} < 0$ and $\lambda_r^{(-)} < 0$. But neither can become a saddle point, for which $\lambda_- < 0 < \lambda_+$.

Thus we are justified in classifying the critical points of the original, nonlinear system (12.2) in the terms defined for linear systems. The isolated critical points of a nonlinear autonomous system are *nodes*, *saddles* or *foci* according to whether the linearization of the system at the given point has eigenvalues as illustrated in Figure 12.3. In particular, we shall see that Q_1 and Q_2 are saddles, while Q_s is a focus.

Notice that the *position* of the critical points in the phase plane does not depend on the value of μ, the relaxation time ratio. Point Q_s is the one closest to the present climate, and we therefore took its coordinate T_s as the characteristic value for T in (12.2).

Critical points of an arbitrary system of ODEs may fail to be isolated if the lines P = 0 and R = 0 coincide over an interval of finite length. Sometimes an infinity of critical points can accumulate, tending to a limit point. Such examples, however, are rather artificial and structurally unstable (compare the discussion of Figure 10.6 in Section 10.2). The situation for system (12.2), with a finite number of isolated critical points, is that of general interest.

In Figure 12.3, it is easy to see that the x axis separates the trajectories which flow towards the saddle point into two regions. As they approach this point, the trajectories turn away and flow to infinity, separated again by the y axis. Thus a trajectory always stays in the same quadrant.

The curves which correspond to the x and y axes for a nonlinear saddle are called therefore *separatrices*. They are themselves trajectories, along which the flow will take an infinite time to get into or out of the critical point, cf. (12.4). The inflowing separatrix is the saddle's *stable manifold*, the outflowing one the *unstable manifold*, as seen already in Sections 5.4 and 6.4.

The (θ,ℓ)-phase plane of system (12.2) is shown in Figure 12.4. The separatrices of Q_1 and Q_2 are shown as solid lines, with arrows distinguishing the stable from the unstable manifold of either saddle. The sense of rotation around the focus Q_s is also shown. For comparison with Figure 12.2, the vertical and horizontal isoclines P = 0 and R = 0 are indicated by dashed lines.

<u>Structural stability and closed trajectories</u>. Consider the set S in Figure 12.4 bounded above by the stable manifold of Q_1 and below by the stable manifold of Q_2. For any point $Q_0 = (\theta_0,\ell_0)$ lying in S, the trajectory passing through Q_0 will stay forever in S. The set S plays here the same role as the ball $\mathcal{B}$ in Section 5.4: trajectories which enter it cannot exit. In fact, all trajectories in S tend to approach Q_s.

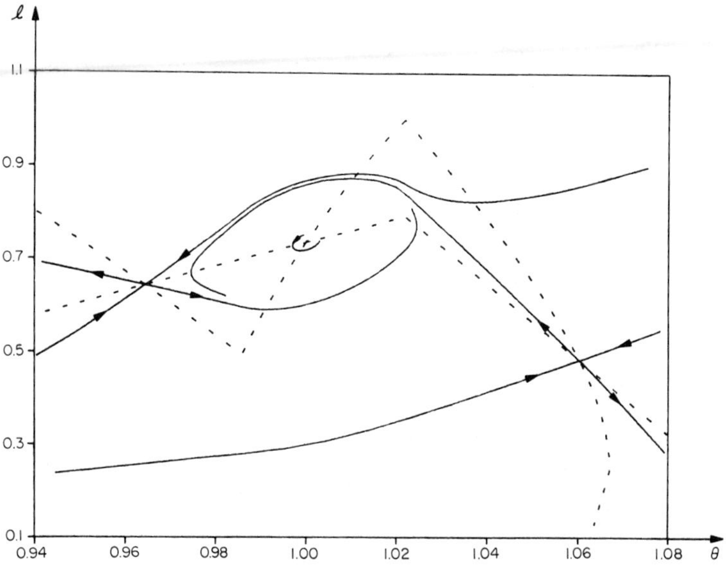

Fig. 12.4. Stable and unstable manifolds of the model's saddles (from Ghil and Tavantzis, 1983; $\mu = 1.76$).

To see what happens as trajectories approach Q_s, we have to return to the linearized system (12.3), where we reintroduce $\mu \neq 1$, so that $\dot{y}$ is replaced by $(1/\mu)\dot{y}$:

$$\dot{x} = a_j x + b_j y, \tag{12.6a}$$

$$(1/\mu)\dot{y} = c_j x + d_j y. \tag{12.6b}$$

Here $a_j = P_\theta(\theta_j, \ell_j)$, with $j = 1, 2$ or s, and similar expressions for b_j, c_j and d_j. The eigenvalues λ of the Jacobian matrix $\partial(P,R)/\partial(\theta,\ell)$ at Q_1, Q_2 or Q_s are given by

$$\lambda_\pm^{(j)} = \frac{1}{2}\{a_j + \mu d_j \pm [(a_j + \mu d_j)^2 - 4\mu J_j]^{1/2}\}, \tag{12.7a}$$

where J_j is the determinant of the Jacobian matrix,

$$J_j = a_j d_j - b_j c_j. \tag{12.7b}$$

For Q_1 the Jacobian determinant J_1 is negative. Since $\mu > 0$, both eigenvalues $\lambda_+^{(1)}$ and $\lambda_-^{(1)}$ are real, and of opposite sign, $\lambda_-^{(1)} < 0 < \lambda_+^{(1)}$. The same is true for Q_2. Hence both Q_1 and Q_2 are saddle points, independently of the value of μ.

At Q_s, the situation is slightly more complicated. The quantities a_s and J_s are positive, and d_s is negative. Eq. (12.7) shows that, as μ increases from zero, the eigenvalues λ_+ and λ_- start out real, positive and distinct. At $\mu \simeq 0.902$, they coalesce and split into a complex conjugate pair in the right half-plane.

At the value $\mu = \mu_0$,

$$\mu_0 = -a_s/d_s \simeq 1.758,$$

the eigenvalues cross the imaginary axis from right to left. Thus Q_s is a focus near $\mu = \mu_0$: for $\mu > \mu_0$ it is stable, and loses its stability for $\mu < \mu_0$. As μ continues to increase, λ_+ and λ_-, now in the left half-plane, coalesce on the negative axis, at $\mu \simeq 34.66$, after which they are real, negative and distinct.

Before examining in more detail the nonlinear phase flow at and near $\mu = \mu_0$, it is interesting to consider the limit $\mu \to +\infty$. This limit corresponds dimensionally to $c_L \to 0$ for $c_T \neq 0$. In this case the land albedo, as well as the ocean albedo, become instantaneous functions of temperature, as in Section 10.2 and Figure 11.10. In the neighborhood of $T = T_s$, or $\theta = 1$, Eq. (12.2) reduces to a linear, diagnostic equation

$$y = -(c_s/d_s)x, \tag{12.8a}$$

and substitution into (12.6a) then yields

$$\dot{x} = (a_s - b_s c_s/d_s)x. \tag{12.8b}$$

Comparison with Eq. (10.4) and Figure 10.4 suggests that Eq. (12.8b) should have a negative coefficient, so that T_s be stable in the limiting case at hand. A little calculation (Ghil, 1984a) shows that indeed $a_s - b_s c_s/d_s$ is negative for the numerical values used in Eqs. (12.1). Thus the coupled energy-balance/ice-sheet model under study here is a consistent extension of the pure energy-balance models discussed in Chapter 10. Furthermore, the "sensitivity" $Q\partial T_s/\partial Q$ of this model's equilibrium climate to changes in insolation is 85K, or 0.85K for one percent change in Q, in accordance with the values suggested at the end of Section 10.2.

What happens to the flow in the neighborhood of Q_s for μ near μ_0? For $\mu > \mu_0$, Q_s is a stable focus, so that all the trajectories approaching from farther away in S will just join smoothly with the trajectories spiraling into Q_s according to the local picture given by (12.4) and (12.7).

For $\mu < \mu_0$, Q_s is an unstable focus, so that locally trajectories spiral away from it, but in the same counterclockwise sense of rotation as before (cf. Figure 12.4). The flow at some distance from Q_s is still spiraling in, rotating also counterclockwise. The continuity of the flow implies that the rotating outflow and the rotating inflow have to join smoothly. This can only happen along a closed trajectory, or *limit cycle*, which attracts all trajectories from the interior as well as from the exterior region bounded by it.

Properties of the periodic solution. This closed trajectory represents a periodic solution of (12.2) (compare Sections 5.3, 6.3 and 6.4). Its size, shape and period depend on the value of μ. Near $\mu = \mu_0$, the limit cycle is very small and its period $\tau(\mu)$ is close to τ_0,

$$\tau_0 = 2\pi/\text{Im}\ \lambda_+^{(s)}(\mu_0) \simeq 5.2.$$

The dependence of the limit cycle's size $\Theta = \theta_{max} - \theta_{min}$, and of its period τ on the relaxation time ratio μ is shown in Figure 12.5. Figure 12.5 shows $\Theta(\mu)$, with the axis $\Theta = 0$ corresponding to the equilibrium solution $\theta \equiv \theta_s$, $\ell \equiv \ell_s$. The labels *s* and *u* stand for stable and unstable branches, respectively.

For μ near μ_0 and Θ small, the periodic branch is unstable, $\mu_0 < \mu < \mu_1 \simeq 1.767$. At $\mu = \mu_1$ the branch turns and becomes stable. The stable limit cycle exists for $\mu_\infty < \mu < \mu_1$, $\mu_\infty \simeq 1.579$. It is the stable portion of the periodic branch which is of physical interest, and we shall consider its properties below.

Figure 12.5 gives $\tau = \tau(\mu)$. The period increases along the unstable portion of the branch to $\tau \simeq 6$ ka, in dimensional units. Along the stable portion it increases first slowly, reaching $\tau \simeq 7.5$ ka near $\mu = 1.60$. Afterwards it has a vertical asymptote, with $\tau \uparrow +\infty$ as $\mu \downarrow \mu_\infty$. This behavior near $\mu = \mu_\infty$ will be explained at the end of the section.

Figure 12.6 shows the evolution of the periodic solution for a point $\ell(\Theta,\mu)$ on the stable branch in Figure 12.5. The solid line is $T(t;\mu)$ and the dashed line is $\ell(t;\mu)$.

The amplitude of the oscillations is $O(10\ \text{K})$ in global temperature T and $O(1)$ in ice extent ℓ. The oscillations occur in the absence of any changes in model parameters, in particular for constant Q. How do the oscillations then sustain themselves? The answer is essentially that, because of feedbacks between albedo, temperature and precipitation, the climatic system as modeled by Eqs. (12.1), has the capacity to retain a fraction which varies in time from the total amount of insolation Q available.

Figure 12.6 shows a lag of about one quarter of the period $\tau = O(10\ \text{ka})$ between temperature and ice extent, with T leading ℓ. This quarter-phase lag is closely connected with the mechanism of the oscillation, as already suggested by Eq. (10.48).

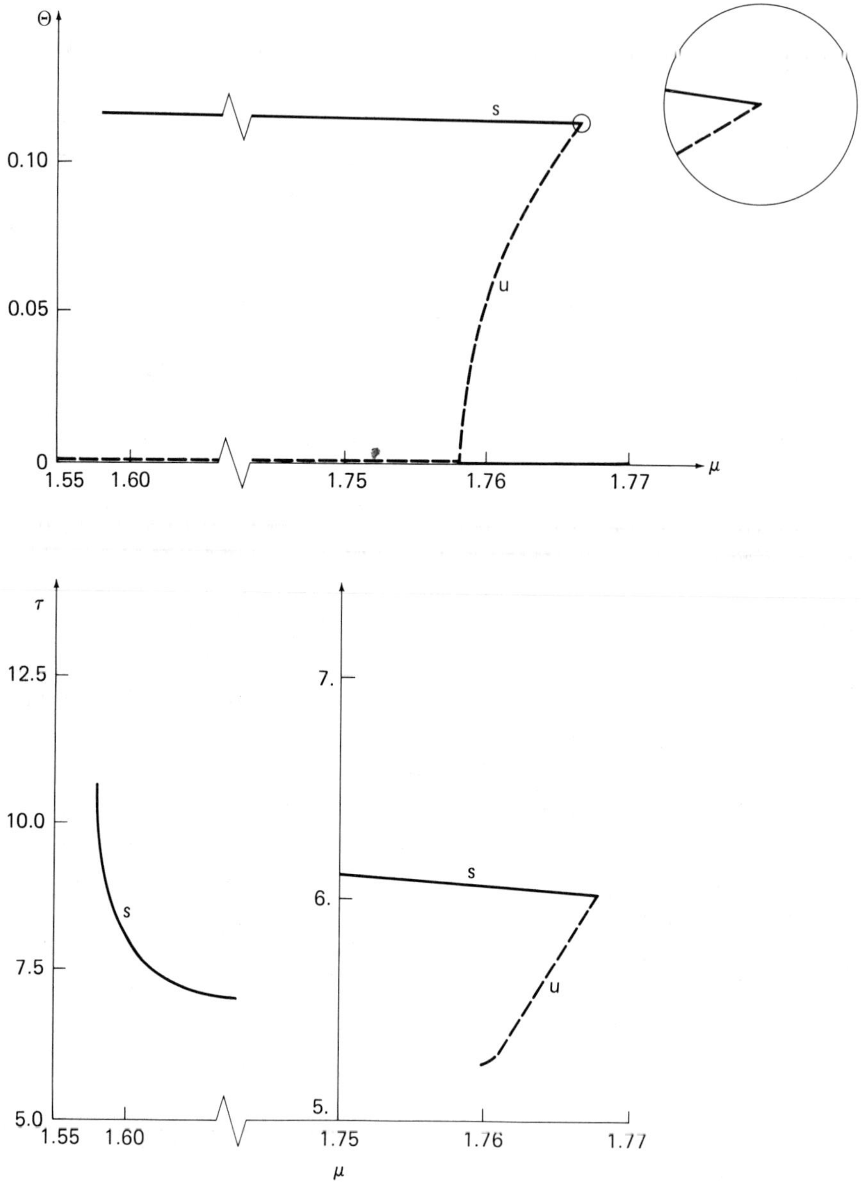

Fig. 12.5. Hopf bifurcation curve. (a) Θ -- size of the periodic orbit; (b) τ -- period (after Ghil and Tavantzis, 1983).

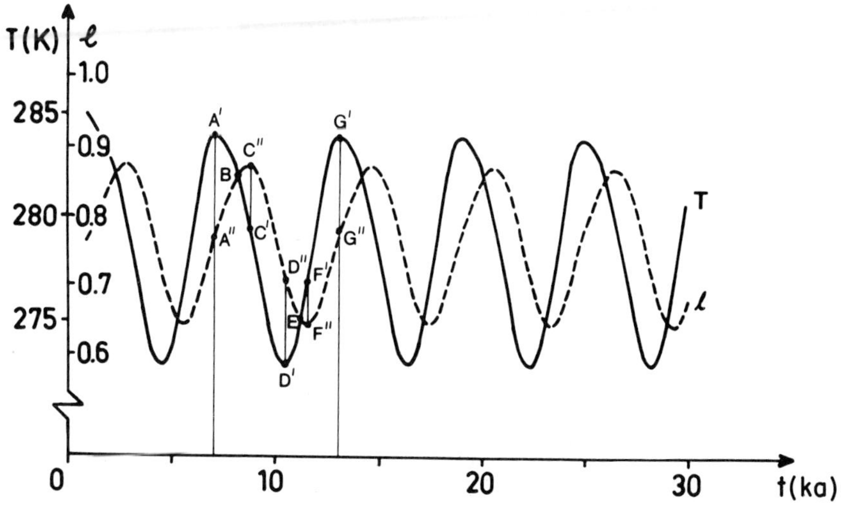

Fig. 12.6. Self-sustained oscillation of a coupled energy-balance/ ice-sheet model (from Ghil, 1981b).

When T is near its maximum (point A' in Figure 12.6), the hydrologic cycle is most active and, according to our assumptions, snow accumulates making ℓ grow (point A"). As a result of the growth in ℓ, albedo increases and temperature starts to decrease, gradually cutting back evaporation and precipitation (B' = B). The ice sheet is still expanding meridionally, as the snow mass accumulated along its surface flows though the interior of the sheet to its periphery (B" = B). Temperature keeps sinking (C'), while the ice reaches it maximum extent (C"), further lowering temperature (D').

Due to small snow accumulation, continued ablation causes the ice sheet to start shrinking (D"). This makes temperature rise again (E' = E), leading to a renewal of snow accumulation. The increased accumulation competes with the continued shrinking, due to the delay in the plastic flow of the ice (E" = E). As the ice sheet starts to expand again (F"), the temperature continues to rise due to the thermal delay in the system (F'). This in turn accelerates the hydrologic activity and the continued expansion of the ice sheets, completing the cycle (G',G").

The mechanism described above can be written in simplified shorthand as

$$\dot{T} \cong -\alpha, \quad \alpha \cong m, \tag{12.9a,b}$$

$$\dot{m} \cong p, \quad p \cong T. \tag{12.9c,d}$$

Here approximation signs stand for rough proportionality, i.e., a change in the quantity on the left-hand side occurs mostly due to the quantity on the right-hand side, and in the direction given by the sign. No particular form of the dependence is indicated, as would be by a linear proportionality. The new quantities are ice mass m and precipitation p, while T and α are temperature and albedo, as before. Eqs. (12.9b,d) are the short-hand representation of the ice-albedo and precipitation-temperature effects, or cross-feedbacks, respectively.

Eliminating α and p yields

$$\dot{T} \cong -m, \tag{12.10a}$$

$$\dot{m} \cong T. \tag{12.10b}$$

Eqs. (12.10) represent the standard form of a linear oscillator, cf. also (10.50). In fact, the oscillation of Figure 12.6 is nonlinear, but at the most basic level it can be understood in terms of the two opposing feedbacks of (12.10).

The self-sustained oscillation above should not be taken literally as a model of climatic evolution on Pleistocene time scales. Slight changes in insolation have occurred on this time scale and will be discussed in Section 12.3. Their interaction with the climatic system, however, seems to be conditioned by the potential for free oscillations highlighted by the model presented here. In the next section we study therefore a fairly general phenomenon which leads to such oscillations in nonlinear systems of autonomous ODEs.

12.2. Hopf Bifurcation

Center manifold and exchanges of stability. We start by considering a system of n smooth autonomous ODEs depending differentiably on a parameter μ. The situation to be investigated is that in which, at an isolated critical point $\underset{\sim}{x} = \underset{\sim}{x}_0$, and for $\mu = 0$, two eigenvalues of the linearization about $\underset{\sim}{x}_0$ are pure imaginary, $\lambda_1 = i\omega = -\lambda_2$, and all other eigenvalues have negative real parts, Re $\lambda_j < 0$, $j = 3,4,\ldots,n$.

Under these circumstances, for $|\mu|$ small, the system's phase flow will tend to an invariant, two-dimensional manifold, called the *center manifold*. This manifold is tangent at $\underset{\sim}{x}_0$ to the plane spanned by the two eigenvectors of the linearized system which are associated with λ_1 and λ_2. All the action is carried by this manifold, to which every trajectory near $\underset{\sim}{x}_0$ is asymptotic. Precise and even more general statements of the center manifold theorem can be found in Carr (1981), and references therein, along with the proofs.

This theorem justifies the importance of studying in detail the situation for $n = 2$. It is convenient to work with the complex variable $z = x + iy$, following Arnold (1983). The canonical equation for the situation of interest then becomes

$$\dot{z} = (\mu + i\omega)z + c(z\,\bar{z})z, \tag{12.11}$$

with $\bar{z} = x - iy$. Here c and ω are real, nonzero constants, and we take $\omega > 0$ for the sake of definiteness, while μ is a real parameter. If (12.11) were written as a system of two real equations, for x and y, the eigenvalues of its linear part would be exactly $\mu \pm i\omega$.

To investigate what happens as μ crosses zero, we let $z = \rho^{1/2}e^{i\phi}$. It is easy to verify that, for $\mu = 0$, $\dot{\phi} = \omega$, so the flow around the critical point $z = 0$ is counterclockwise, as long as c is sufficiently small. The radial variable $\rho = z\bar{z} \geq 0$ satisfies

$$\dot{\rho} = 2\rho(\mu + c\rho). \tag{12.12}$$

The critical points of (12.12) are

$$\rho = 0,$$

for all values of μ and c, and

$$\rho = -\mu/c,$$

which will be positive only if μ and c are of opposite signs. We consider first the case $c < 0$. Notice in fact that by a suitable scaling of time one can take $\omega = 1$, and scaling z one can also take $c = -1$ or $c = +1$, according to the situation under consideration.

The situation for $c < 0$ is summarized in Figure 12.7. The vector field of (12.12) for $\mu < 0$ is shown in Figure 12.7a. The only critical point is $\rho = 0$, and it is stable. Figure 12.7b shows the vector field of (12.12) for $\mu > 0$. The origin is now unstable, and the flow is into $\rho = -\mu/c$.

Figure 12.7c shows how the picture changes as μ crosses zero. To the left, $z = 0$ is a stable focus. To the right, the periodic solution $z = \rho^{1/2}e^{i\omega t}$ is stable. The amplitude $|z|$ of the periodic solution is a parabolic function of the parameter μ, $|z| = (-\mu/c)^{1/2}$. Thus the stability of the stationary solution is smoothly transmitted to the periodic solution.

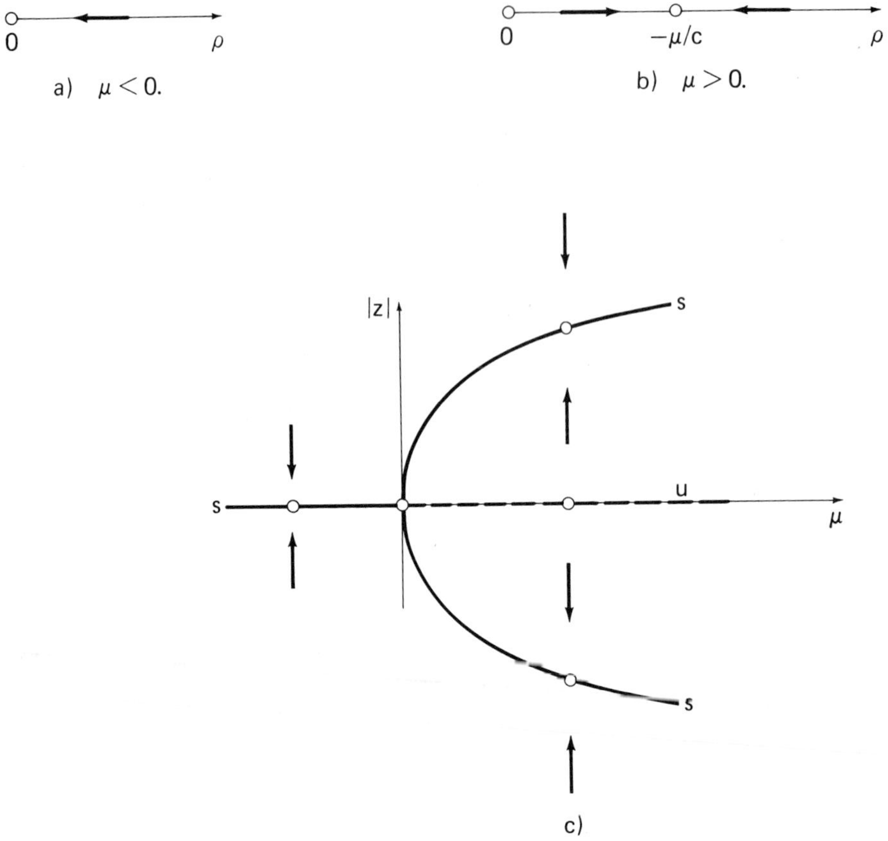

Fig. 12.7. Supercritical Hopf bifurcation, $c < 0$ (after Arnold, 1983).

The case $c > 0$ is summarized in Figure 12.8. Now two critical points of (12.12) exist for $\mu < 0$, and it is $\rho = 0$ which is stable (Figure 12.8a). For $\mu > 0$ only $\rho = 0$ is present, and it is unstable (Figure 12.8b).

Figure 12.8c shows that the stationary solution $z = 0$ still loses its stability at $\mu = 0$, but the periodic solution which grows out of the point $(z = 0, \mu = 0)$ is itself unstable. The fact that the periodic branch grows in the direction of the stable stationary branch of solutions is the cause of its instability. Changing μ into $(-\mu)$ just produces mirror images of Figures 12.7c and 12.8c, respectively.

The exchange of stability between stationary and periodic solutions was already known to Poincaré in the 1880s and had been analyzed in detail for systems of two autonomous ODEs by Andronov and co-workers in the 1930s. It was Hopf who extended its study in the 1940s to systems of partial differential equations and saw its importance in hydrodynamic

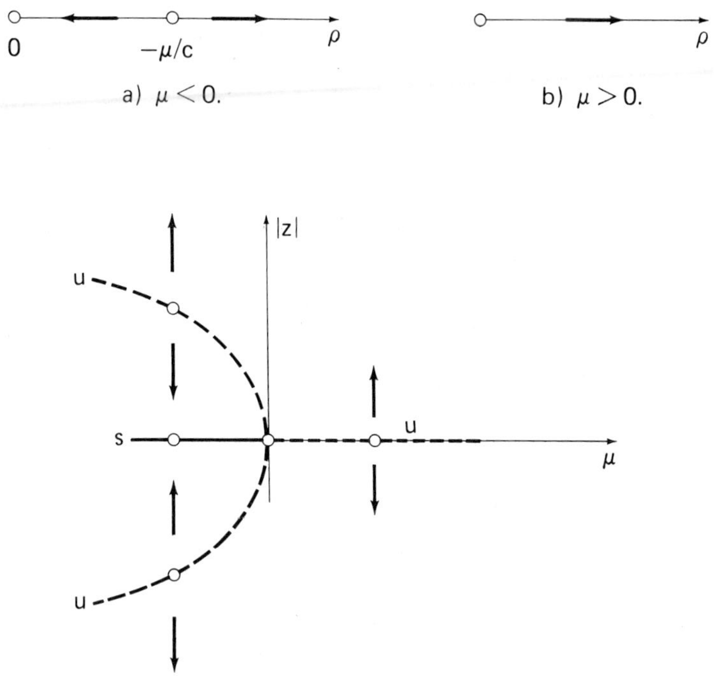

c)

Fig. 12.8. Subcritical Hopf bifurcation, c > 0 (after Arnold, 1983).

stability problems. The development of a periodic solution branch from a stationary solution branch, as a parameter crosses a critical value, is therefore commonly referred to as *Hopf bifurcation*.

When the periodic branch is stable and coexists with the unstable portion of the stationary branch, the bifurcation is called *supercritical*, as in Figure 12.7. In the opposite case, Figure 12.8, it is called *subcritical*. The latter is the situation for the climatic oscillator we studied in the previous section, cf. Figure 12.5. How does the periodic branch regain stability in this case?

We saw in Figure 12.5 that the periodic branch turns, and becomes stable after the turn. The general situation is described by the following equation:

$$\dot{z} = (\mu + i\omega)z + cz^2\bar{z} + dz^3\bar{z}^2. \tag{12.13}$$

Here μ, ω and c are as before, and d is another constant.

Now $\rho = z\bar{z}$ satisfies an ODE with cubic right-hand side,

$$\dot{\rho} = 2\rho(\mu + c\rho + d\rho^2). \tag{12.14}$$

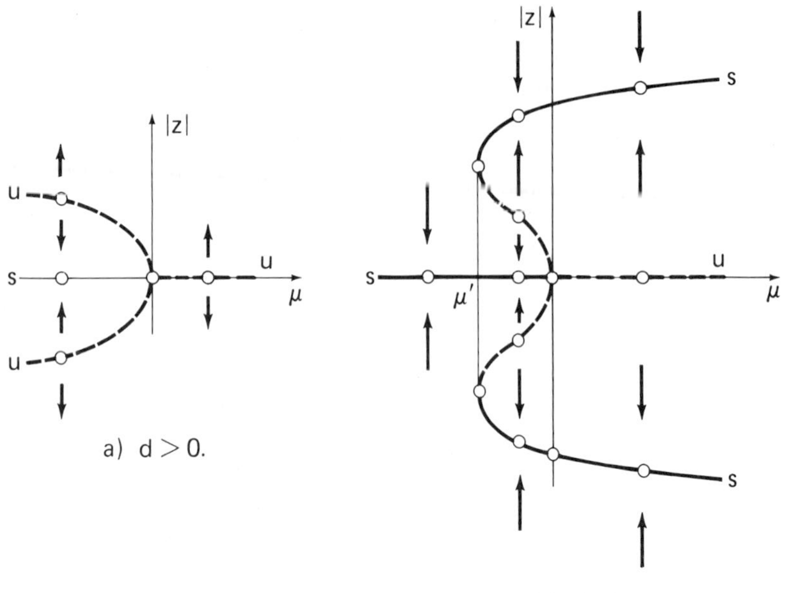

Fig. 12.9. Subcritical Hopf bifurcation in the presence of higher nonlinearity (after Arnold, 1983).

This ODE has one, two or three real, nonnegative critical points, according to the sign and value of μ, c and d. Notice that scaling z allows us to fix either c or d, but not both. One can then fix $c = 1$ for the case of interest, and think of d as a second parameter, in addition to μ.

Figure 12.9 illustrates what happens as d changes sign. Figure 12.9a, for $d > 0$, is the same as Figure 12.8c, i.e., an unstable periodic solution coexists with the stable stationary solution for $\mu < 0$, and no stable solution exists for $\mu > 0$. For $d < 0$, however, Figure 12.9b shows the unstable periodic branch turning to the right and gaining stability.

In a physical problem governed by (12.13), or in a numerical computation of its solutions, one would expect jumps between the stationary solution and the periodic one, for the range of parameters where both are stable. More precisely, if $d < 0$ and the system is at $z = 0$, it will jump to the periodic solution as μ increases beyond zero. If then μ is decreased beyond the value μ' say, where the periodic branch ceases to exist, the system will jump back to $z = 0$.

This hysteresis phenomenon in transitions between a stable stationary and a stable periodic solution is analogous to the situation discussed for bistable stationary solutions in connection with Figures 6.5, 10.6 and 11.10. For mechanical oscillators, one speaks of hard excitation when the Hopf bifurcation is subcritical ($d < 0$, Figure 12.9b), as opposed to soft excitation when it is supercritical (Figure 12.7c). In the presence of stochastic perturbations, jumps between the coexisting stable branches in Figure 12.9b are also possible for $\mu' < \mu < 0$, as in Sections 10.3 and 10.4.

Local and global Hopf bifurcation in applications. The behavior of the phase flow near an isolated critical point is called *local*. For any fixed parameter value away from bifurcation, we saw that it is described by the linearization about that point. A bifurcation in phase-parameter space can also be discovered by purely linear analysis.

To find out in a practical problem with complicated nonlinearities, like the one illustrated here, whether a critical point undergoes Hopf bifurcation, it suffices to check two linear conditions. The first one is that for $\mu = \mu_0$ the two eigenvalues are purely imaginary,

$$\lambda_r = 0, \quad \lambda_i \neq 0; \tag{12.15a,b}$$

the second is that they cross the imaginary axis with nonzero "speed", i.e.

$$(d\lambda_r/d\mu)_{\mu=\mu_0} \neq 0. \tag{12.16}$$

If $n \geq 3$, one has also to check that the other eigenvalues have negative real parts at $\mu = \mu_0$.

The super- or subcriticality of the periodic branch, and hence its stability, depend however on the model's nonlinearities, cf. Eqs. (12.11-12.14). They refer to *global* properties of the flow, away from the critical point.

The explicit computation of smooth transformations of variables which would exhibit the leading-order nonlinearities of a system like (12.1) in the so-called *normal forms* (12.11) or (12.13) is in general quite laborious. Alternative analytic approaches occupy many pages of Marsden and McCracken (1976), and are not any easier to carry out in practice. We refer therefore to the appendix of Ghil and Tavantzis (1983) for a rather straightforward qualitative approach. In this approach, based on the Morse lemma of global analysis (see also Section 10.5 for further references), one only needs to compute numerically, with controlled accuracy, one turn of a trajectory around the critical point, and to repeat this calculation for a few given points in phase-parameter space.

The picture of Figure 12.9b and its verification for system (12.2) explain the flow in a large neighborhood of Q_s. Still, Figure 12.4 suggests that the interaction between the system's stable limit cycle and the separatrices of Q_1 and Q_2 might be quite interesting for the flow still farther away from Q_s. We show schematically how these interactions change with μ in Figure 12.10.

For large μ, Q_s is a stable focus, and the separatrices of Q_1 and Q_2 spiral into it (Figure 12.10g). Next comes a range of μ where a stable and an unstable limit cycle coexist with the stable Q_s (Figure 12.10f). For smaller μ only the stable limit cycle is left, with the separatrices spiraling onto it. From here on, Q_s is an unstable focus (Figure 12.10e).

A critical value of $\mu \simeq 1.621$ is reached where the unstable manifold of Q_2 runs into Q_1 (Figure 12.10d). Such a heteroclinic *saddle-to-saddle separatrix* is structurally unstable and hence cannot be computed numerically. It is very important however, since for smaller μ the trajectories outside the stable manifold of Q_1 will no longer approach Q_s, but will rather escape towards much lower values of θ. Still, the stable limit cycle continues to exist (Figure 12.10c).

The next transition in the phase flow is associated with the limit cycle merging into a *homoclinic orbit* which reconnects the unstable manifold of Q_1 to the stable one (Figure 12.10b). The time it takes to get out of Q_1 and back into it is infinite, as noticed in connection with Eq. (12.4).

Such a homoclinic orbit was already apparent in Figures 5.6 and 6.12. It is also structurally unstable, but explains the lengthening of periods as $\mu \downarrow \mu_\infty \simeq 1.579$ in Figure 12.5b: as the stable limit cycle gets closer and closer to the homoclinic orbit, the solution spends more and more time near Q_1. The longer residence near a critical point results not only in the lengthening of the total period of the oscillation, but also in its shape becoming that of a sawtooth or square wave, like in Figures 10.15e,f, rather than staying sinusoidal, as in Figure 12.6.

Sustained oscillations in which abrupt changes alternate with more gradual ones are called *relaxation oscillations*. They were first analyzed by B. Van der Pol in a triode circuit and are now known to occur in a wide variety of biological, chemical, mechanical and physical systems. We shall return to their potential significance for climate dynamics in

(a) $\mu < 1.5790 \cdots$

(b) $\mu = 1.5790 \cdots$

(c) $1.5790 \cdots < \mu < 1.6207 \cdots$

(d) $\mu = 1.6207 \cdots$

(e) $1.6207 \cdots < \mu < 1.7583 \cdots$

(f) $1.7583 \cdots < \mu < 1.76735 \cdots$

(g) $1.76735 \cdots < \mu$

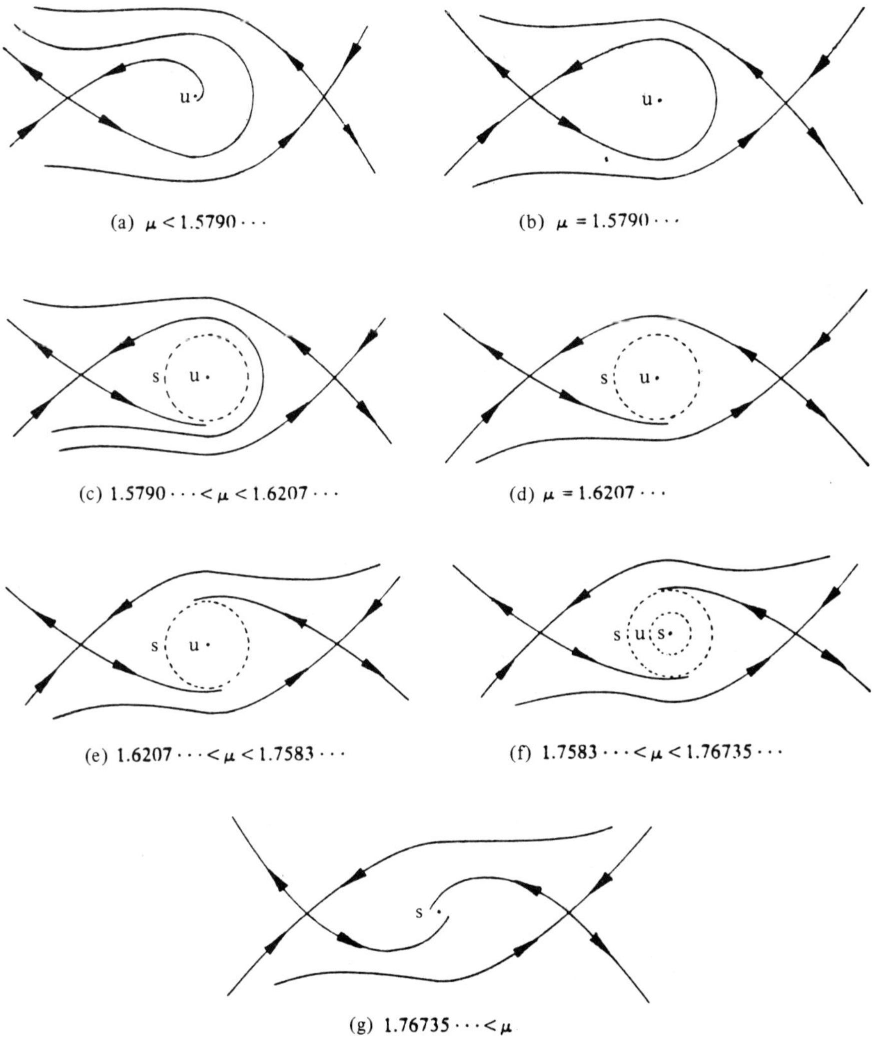

Fig. 12.10. Phase-plane changes with μ for the oscillatory climate model (12.2) (from Ghil and Tavantzis, 1983).

Section 12.7. To complete the discussion of global behavior in system (12.2), one notices in Figure 12.10a that all solutions starting near Q_s tend to lower θ, and the model loses its physical validity.

After this analysis of free oscillations in a climate model, we turn in the next section to the effect of small changes in forcing. The presence of free oscillations, as we shall see, greatly enriches the effect of the quasi-periodic forcing on climatic response.

12.3. Orbital Changes and Their Climatic Effect

Since the mid-nineteenth century and the beginnings of ice age theory (cf. Section 11.1), many hypotheses have been advanced about their causes. The global scale of the phenomenon has prompted a search for causes on the scale of the planet and even of the solar system.

The first half of the nineteenth century was dominated scientifically by the triumph and popularization of celestial mechanics in the work of Lagrange, Laplace and LeVerrier. It is not surprising therefore that speculation about causes for ice ages turned to changes in insolation due to the changing distance between Earth and Sun. Indeed, the most important changes in the climatic system's external conditions, on the time scale of 10-1000 ka, are probably those related to the Earth's *orbital elements*, which we define and discuss in the following paragraphs.

Orbital changes. In the absence of any other celestial bodies, the Earth would describe an elliptic orbit with the Sun as one of the foci: the shape and position in space of this ellipse, called the ecliptic, would be fixed. Furthermore, if the Earth were a perfectly spherical, homogeneous body, its axis of rotation would have a fixed direction in an inertial frame of reference. In particular, it would make a constant angle with the normal to the plane of the ecliptic, which in turn is fixed in space in the presence of Sun and Earth only.

The gravitational attraction of the other planets in the solar system perturbs the Earth's elliptic motion around the Sun. The masses of the planets are small compared to that of the Sun and hence these perturbations are small. It is practical, therefore, to describe the motion of the Earth as an ellipse whose orbital elements vary slowly in time.

The choice of elements is somewhat arbitrary, and depends on the problem under consideration. When dealing with the particular problem of insolation changes, it is customary to choose *eccentricity* e in order to describe changes in the *shape* of the ellipse. In defining the ellipse's *position* in space, one uses the fact that the direction of the axis of rotation with respect to the fixed stars is mostly affected by the gravitational pull of Sun and Moon on the Earth's equatorial bulge. Hence, the Earth's axis of rotation describes a nearly circular cone in space, under the influence of the Sun and Moon, with small deviations from circularity due to interference by the other planets. As a result, the angle of tilt of the axis of rotation on the ecliptic, or *obliquity* ε, reflects the effect of the planets on the position of the ecliptic in space, almost independently of the motion of the axis itself. The obliquity is chosen therefore as one of two angular elements necessary in order to describe the slow change in the position of the ecliptic. The second angle is the inclination i on a plane of reference.

In order to describe the slowly varying position of the Earth's elliptic orbit within the plane of the ecliptic, one uses the *precession* angle $\tilde{\omega}$ of the perihelion -- the point on the ellipse closest to the Sun -- with respect to a point of intersection of the ecliptic with the same plane of reference. In fact, the total precession is due to a combination of the luni-solar effect above on the Earth's mass distribution, on the one hand, and of the planetary effect on its total mass, on the other.

The position and tangential velocity of the Earth itself on its elliptic orbit, within a year, need not be considered here, since they are averaged out in the mechanical theories valid on the time scale under investigation. The other four orbital elements, such as the semi-major axis of the ecliptic, necessary in order to describe the position and velocity of the center of the Earth in space and that of its axis of rotation, vary on even slower time scales than eccentricity, inclination, obliquity and precession, and can therefore be considered as fixed on the Quaternary time scale, in a first approximation.

The elements described above are not actually used as such in orbital computations. To give a feeling for the massive size of the literature in celestial mechanics over the last two centuries, and for the role played in it by transformations of coordinates alone, the reader is urged to leaf through the encyclopedic work of Hagihara (1970-1975). An elementary presentation of the geometric and kinematic aspects of the subject is given by Danby (1962).

As a result of the interactions between the nine major planets and the Sun, the orbital elements of all planets, in particular those of the Earth, change in a *quasi-periodic* manner. A quasi-periodic function $F(t)$ can be represented by

$$F(t) = F_n(\omega_1 t, \omega_2 t, \ldots, \omega_n t), \tag{12.17}$$

with F_n being τ_j-periodic of period $\tau_j = 2\pi/\omega_j$ in each one of its $n \geq 2$ arguments; the n periods are by definition rationally independent, i.e., there is no longer period which is an integer multiple of all of them. Even for n as large as two or three, a quasi-periodic function can actually look rather irregular.

In fact, the behavior of quasi-periodic solutions to nonlinear differential equations is very difficult to compute accurately. Even the *three-body problem*, in which the motion of the Earth around the Sun is perturbed by a single planet, Jupiter, cannot be solved either in closed form or with prescribed accuracy. A large body of modern mathematics has been created in the effort to solve this problem, and to settle the stability question for the planetary system, a question which already preoccupied Newton.

Various methods have been devised to compute approximate solutions to the planetary N-body problem. Most of these methods are based on the smallness of the masses of the planets compared to that of the Sun, on the smallness of the planetary eccentricities, and on the smallness of the mutual angles of inclination of the planets' ecliptics. They use perturbations of the orbits about the hypothetical quasi-periodic phase-space flow of the planetary system represented by the independent revolutions of each planet around the Sun, in the absence of mutual attraction between the planets. Such a flow is called *integrable*, since the equations of motion can be integrated explicitly to yield the Keplerian periodic revolution of each planet separately.

Perturbation methods in celestial mechanics fall into two broad categories (Goldstein, 1980, Section 11.6). Time-dependent methods consider in fact perturbations of the initial-value problem for the system of ODEs governing planetary motion. The most widespread one is *variation of constants*, where the "constants" are the orbital elements of the Earth, say, which would be constant in the absence of the other planets. The slow variation of these elements is the result of the small gravitational attraction of the other planets. The effect of the Earth on the other planets is neglected altogether in this method.

Variation of constants is entirely appropriate for the computation of the short-term orbital changes called *ephemerides* and used in classical navigation. Its validity, however, is limited in time in the same way in which a truncated power series expansion of purely periodic motions fails to capture the periodicity, no matter how high the truncation.

Time-independent perturbation methods attempt to follow the deformation of the entire phase-space torus on which the unperturbed, quasi-periodic flow lies, as the small parameters increase from zero. This is similar to the study of the purely periodic solution branch as a function of the parameter μ in the previous section; its one-dimensional "torus" was simply the limit cycle.

The perturbation study of higher-dimensional tori is much more difficult than that of Hopf bifurcation, due to the appearance of *resonances* between the frequencies, as the small parameters change. Classical methods for this study, such as *perturbation of coordinates*, arbitrarily neglect certain terms in the perturbation expansion. Even so, they lead to asymptotic series for the elements which were shown already by Poincaré in the 1880s to be nonconvergent. This lack of convergence arises because near resonances between independent frequencies lead to small divisors in the coefficients of the series expansions.

It is only in the 1950s and 1960s that Kolmogorov, Arnold and Moser were able to show that the problem of resonances can be avoided under certain technical conditions. This provided an abstract proof that the motion of the planetary system is actually quasi-periodic with nonzero probability. So far, the methods of KAM theory, due to their considerable complexity, have not produced realistic computations of planetary orbits.

We shall only quote therefore the results of existing calculations based on perturbation of coordinates. For the duration of the Quaternary, the *mean* periodicities of the Earth's orbital elements obtained by these calculations are probably rather accurate, although the periodicities themselves are slowly changing due to the neglected terms. As a result of these and other inaccuracies, the phases of the elements are likely to be in error by as much as 180° beyond 1000 ka ago (Applegate *et al.*, 1986).

It should be noted that these estimates are based on intercomparisons of different methods. No observational verification exists until now of long-term astronomical computations. In fact, accurately computing the effect of orbital changes on the climate would be important for celestial mechanics. If paleoclimatological proxy data (Section 11.1) could be combined with a few absolute dates to determine a geological time scale with known error bars, this would allow the first observational verification of long-term orbital changes. Such a verification would turn celestial mechanics, which has become a branch of pure mathematics, back into a branch of physics, permitting a serious evaluation of non-gravitational effects, such as the solar wind or tidal dissipation, on the mechanics of the planetary system.

Insolation changes and their climatic effect. Tables 12.1, 12.2 and 12.3 (after Berger, 1978) give the first ten terms each in an asymptotic series expansion of trigonometric type for the obliquity, the precessional parameter $e \sin \tilde{\omega}$, and the eccentricity e, respectively. The terms are ordered by decreasing amplitude (second column), with the period of the term in the last column.

Table 12.1. Series expansion of obliquity variations (after Berger, 1978).

Term	Amplitude (")	Period (years)
1	-2462.22	41000
2	- 857.32	39730
3	- 629.32	53615
4	- 414.28	40521
5	- 311.76	28910
6	308.94	41843
7	- 162.55	29678
8	- 116.11	40190
9	101.12	42354
10	- 67.69	30365

Table 12.2. Series expansion of the precessional parameter $e \sin \tilde{\omega}$ (after Berger, 1978).

Term	Amplitude	Period (years)
1	0.0186080	23716
2	0.0162752	22428
3	-0.0130066	18976
4	0.0098883	19155
5	-0.0033670	19261
6	0.0033308	23293
7	-0.0023540	18873
8	0.0014002	16907
9	0.0010070	28818
10	0.0008570	19445

Table 12.3. Series expansion of the eccentricity e (after Berger, 1978).

Term	Amplitude	Period (years)
1	0.01102940	412885
2	-0.00873296	94945
3	-0.00749255	123297
4	0.00672394	99590
5	0.00581229	131248
6	-0.00470066	2305441
7	-0.00254464	102535
8	0.00231485	1306618
9	-0.00221955	136412
10	0.00201868	603630

The series used in variation of constants are so-called *d'Alembert series* $\Sigma_j C_j \cos \mathcal{D}_j$ (Brouwer and Clemence, 1961, p. 291; Hagihara, 1972, Vol. II, part 1, p. 129). Any argument $\mathcal{D}_j$ is a linear combination with integer coefficients $j_1, j_2, \ldots, j_n$ of "fast-periodic" elements, such as angular precession $\tilde{\omega}$ or annual position λ on its ecliptic of each of the planets involved, $j = (j_1, j_2, \ldots, j_n)$. The coefficients C_j are monomials in the perturbing masses, the eccentricities and the mutual inclinations (tan I) of the ecliptics of the planets involved, with the masses (m') expressed in solar units. The sum of the powers of the constant masses is the *order* of the term, the sum of the powers of the very slowly varying eccentricities and inclinations its *degree*.

Similar series are used in perturbation of coordinates. The amplitudes of the terms in Tables 12.1-12.3 and the periods are thus not constant in time, but they vary slowly on Quaternary time scales.

In Table 12.2 the largest term has a period of 41 ka, and the next term is three times smaller. Hence the Earth's obliquity has a clear 41 ka periodicity over the Quaternary.

Angular precession $\tilde{\omega}$ (not shown) has a similar dominant period, due to the luni-solar effect, of 22 ka. In insolation calculations, however, it is the precessional parameter $e \sin \tilde{\omega}$ which plays the important role. Table 12.2 shows the first four terms to be dominant, with the fifth term three times smaller than the fourth. The first two and the next two terms can be grouped into a periodicity of 23 ka and one of 19 ka, respectively. The "splitting" of the 22 ka angular period into these two precessional periodicities is the result of multiplication by the eccentricity, and of the so-called *addition theorem*, which allows the product of two trigonometric functions to be replaced by a sum of two trigonometric functions whose arguments are the sum and difference of the two original arguments.

Table 12.3 shows that the d'Alembert series of the eccentricity has very slowly decreasing amplitudes. The first term has a period of 413 ka, too long to be detected with statistical confidence in Quaternary records. The next four terms all have periods between 95 ka and 131 ka, and similar amplitudes.

How do orbital changes affect the climatic system? The first thing to notice is that their direct effect is very small. Among the three insolation-affecting parameters above, eccentricity is the only one which actually modifies the globally and annually averaged amount of insolation Q received at the top of the atmosphere. This change, however, is proportional to $(\Delta e)^2/2$, Δe being the change in eccentricity, which over the Quaternary has not exceeded 0.04. Thus ΔQ is only about one part in a thousand over the same period. As noticed in Sections 10.2, 10.3, 10.5 and 11.1, the quasi-equilibrium response of a mean-annual, zonally or globally averaged energy-balance model (EBM) to such an insolation change would not exceed a fraction of one degree.

Obliquity and precession have no net, globally and annually averaged effect at all. They only change the contrast between seasons. For obliquity this change of contrast is the same in both hemispheres, for precession the contrast is largest in one hemisphere when it is smallest in the other.

The precessional effect $e \sin \tilde{\omega}$ is the same at all latitudes within one hemisphere, while the obliquity effect increases with latitude. As a result, insolation changes due to obliquity may dominate at higher latitudes, while precessional changes may dominate at lower latitudes. Changes of insolation due to either effect averaged over a half year and a hemisphere are larger than the direct effect of Δe, but still quite small.

The quest for a mechanism which might translate minute yearly and small seasonal insolation changes into large changes in global temperature and ice volume has been focused primarily on the hydrological cycle and the annually averaged net accumulation rate of snow. Large snow accumulation in winter or less ablation in summer would both increase the ice sheets' mass over one year, while less snowy winters and warmer or less cloudy summers would both act in the opposite direction.

Considerable controversy has existed over just which season and which latitudes, or zones, would be most decisive in the net annual ice balance. Adhémar and Croll in the nineteenth century thought that high-latitude winter was crucial, while Köppen early in this century believed it was mid-latitude summer (Imbrie and Imbrie, 1979, Ch. 8). More recent arguments even encompassed low latitudes, where precessional effects dominate (ibid., Ch. 12). The complex phase relationship between orbital parameters and climatic response was emphasized by Held (1982).

In the absence of a convincing answer to the quandary of critical latitude and season, we shall let the accumulation-to-ablation rate $\varepsilon(T)$ vary in time, according to the orbital periodicities. Within the degree of simplicity and abstraction of the model to be studied here, this seems to be a reasonable assumption.

More precisely, the extreme values ε_m and ε_M in Fig. 11.9 are kept fixed. The only thing allowed to vary is the mean annual temperature

T_m at which ε starts to increase with T, and the temperature T_M at which it reaches its maximum value. Thus, instead of being constant in time, as in Section 11.2, T_m and T_M vary quasi-periodically,

$$T_m(t) = \overset{\circ}{T}_m\left\{1 + \sum_1^5 \varepsilon_j \sin\left(\frac{2\pi}{\tau_j} t + \phi_j\right)\right\}, \tag{12.18a}$$

$$T_M(t) = \overset{\circ}{T}_M\left\{1 + \sum_1^5 \varepsilon_j \sin \frac{2\pi}{\tau_j} t + \phi_j\right)\right\}, \tag{12.18b}$$

with periods τ_j and phases ϕ_j. The phases ϕ_j will at first be taken equal to each other, $\phi_j = 0$, $1 \leq j \leq 5$.

The range of the changes in ε given by (12.18) is indicated by the two dashed lines in Figure 11.9. The ramp portion of $\varepsilon(T)$ can move parallel to itself in time. Hence ε becomes a function of time t both via the temperature variable $T(t)$ and via the prescribed effective hydrologic temperatures $T_m(t)$ and $T_M(t)$,

$$\varepsilon = \varepsilon(T(t);\ T_m(t),\ T_M(t)); \tag{12.18c}$$

it reduces in the autonomous case of Section 12.1 to

$$\varepsilon(T) = \varepsilon(T(t);\ \overset{\circ}{T}_m,\ \overset{\circ}{T}_M).$$

We take, in order of increasing periodicity, the following approximate values for the orbital periods:

$$\tau_1 = 19\text{ka},\ \tau_2 = 23\text{ka},\ \tau_3 = 41\text{ka},\ \tau_4 = 100\text{ka},\ \tau_5 = 400\text{ka}. \tag{12.18d}$$

The first two periodicities, τ_1 and τ_2, correspond to the precessional effect, $e \sin \tilde{\omega}$, with the precessional angular frequency 1/22 ka modulated by an eccentricity frequency near 1/100 ka. The third period, τ_3, is associated with the obliquity, while τ_4 and τ_5 are approximate, simplified periodicities related to eccentricity changes. We shall also allow the insolation Q to vary according to

$$Q = Q_0\left\{1 + \delta_4 \sin\left(\frac{2\pi}{\tau_4}\right) + \delta_5 \sin\left(\frac{2\pi}{\tau_5} + \psi_5\right)\right\}. \tag{12.18e}$$

In the model of the following three sections, changes in $\varepsilon(T)$ and Q with time represent the net effect of orbital variations, integrated over latitude and season. These changes do not appear in the model equations as additional terms on the right-hand side, but rather as variations in the coefficients. In this sense, orbital forcing is different from periodic or quasi-periodic forcing in many mechanical and electronic oscillators. It could be called multiplicative, or *parametric*, rather than additive *forcing* (compare also Section 6.3).

12.4. Forced Oscillations: Nonlinear Resonance

Model description. The model to be used here differs from that of Section 12.1 by the inclusion of Eq. (11.34b) for bedrock adjustment. The dynamic coupling between ice-mass growth and isostatic sinking, Eqs. (11.24, 12.1b), modifies the mass balance of the ice sheet by displacing the snowline along the upper surface of the ice. Indeed, the position of the 0°C isotherm (Figure 11.13) relative to mean sea level (MSL) depends on the elastic response of the lithosphere to the ice load, cf. Eqs. (11.24a,b). Likewise, the position of the ice surface relative to MSL depends on the local bedrock deflection ζ, governed by the mantle's viscous response to the ice load, cf. Eqs. (11.34a-d). Thus, both the mantle's viscosity and the lithosphere's elasticity participate in determining the snowline and therewith the ice sheet's mass balance.

These two effects, elastic and viscous, can be combined into an ice-load/net-accumulation cross-feedback,

$$\dot{p} \cong -m, \tag{12.19a}$$

where we use the short-hand notation of Eqs. (12.9, 12.10). Here p is the net accumulation of snow on the ice sheet, which is reduced as the total mass m increases and the sheet sinks into the bedrock. As a result of this sinking, the snowline moves poleward, reducing the accumulation area A and hence the total accumulation, while both ablation area and rate are increased.

Coupling the load-accumulation effect of Eq. (12.19a) with the precipitation-temperature effect of Eq. (12.9c), which we repeat here as

$$\dot{m} \cong p, \tag{12.19b}$$

can also lead to an oscillatory mechanism, independent of the one studied in Section 12.1. Whether a self-sustained oscillation actually occurs in Eqs. (11.34) keeping T fixed depends on the parameter values used, in particular on the viscous relaxation time of the mantle.

In discussing Figures 11.11 and 11.12 of Section 11.3, it was pointed out that postglacial uplift data and various geodynamical models used to explain these data yield multiple relaxation times. In the present section we shall use a single relaxation time O(1 ka), cf. Eqs. (11.34c,d), for which the oscillator denoted by (12.19) is damped. A careful analysis of this oscillator, given in its complete, nonlinear form by Eqs. (11.34) with T fixed, can be carried out along the lines of Section 12.3; it will probably give parameter ranges for which the oscillation is self-sustained (see, for instance, preliminary results in this direction by Birchfield and Grumbine, 1985, and by Peltier and Hyde, 1984). The coupling of oscillators

(12.9) and (12.19) when both are self-sustained would be an interesting theoretical undertaking, which might explain further features of Quaternary paleoclimatic proxy data.

To summarize, the model studied in this section is governed by three coupled ODEs, which we repeat here for convenience:

$$\dot{T} = Q(t)\{1 - \gamma(\alpha_0 + \alpha_1 L_M^* \ell) - (1-\gamma)\alpha_{oc}(T)\} - \kappa(T - T_\kappa), \tag{12.20a}$$

$$\dot{\ell} = \{1 + (2/3)(L_M^*)^{-1/2}\, \zeta\ell^{-1/2}\}^{-1}\{(3/2)\mu\, \ell^{-1/2}[1 + \varepsilon(T,t)]\ell_T(T,\ell,\zeta) - \ell]$$
$$- (2/3)\, \ell^{-1/2}(A\zeta\ell^{-2} + B\ell^{-3/2})\}, \tag{12.20b}$$

$$\dot{\zeta} = A\zeta\ell^{-2} + B\ell^{-3/2}. \tag{12.20c}$$

Recall that time t is measured in units of $c_T^* = 1$ ka and $\mu = c_T^*/c_L$. All symbols have been explained in connection with Eqs. (11.34, 12.1) and Figure 11.9. The insolation $Q(t)$ and the accumulation-to-ablation ratio $\varepsilon(T,t) = \varepsilon(T(t);\, T_m(t),\, T_M(t))$ will be allowed to vary according to Eqs. (12.18).

For comparison with Eqs. (12.2), we write (12.20) in more compact notation:

$$\dot{T} = P(T,\ell,t;\underset{\sim}{\delta}), \tag{12.21a}$$

$$\dot{\ell} = R(T,\ell,\zeta,t;\mu,k,s,\underset{\sim}{\varepsilon}'), \tag{12.21b}$$

$$\dot{\zeta} = S(\ell,\zeta;\, D,q). \tag{12.21c}$$

Notice that all three variables, global temperature T, ice extent ℓ and bedrock deflection ζ are coupled in the ice-sheet evolution equation (12.21b), which represents the tie between the two oscillation mechanisms (12.10) and (12.19).

The parameters $\underset{\sim}{\delta} = (\delta_4,\delta_5)$, and $\underset{\sim}{\varepsilon}' = (\varepsilon_1,\varepsilon_2,\varepsilon_3,\varepsilon_4,\varepsilon_5)$ represent the intensity of the insolation and of the hydrologic forcing, respectively. We know from the orbital calculations summarized in Section 12.3 that $\delta_4 + \delta_5 \leq 0.001$ and that $\underset{\sim}{\varepsilon}'$ is also small.

The lithospheric elasticity parameter $k = O(10^{-2})$ and slope of the 0°C isotherm $s = \sigma \times 10^{-3} m^{-1/2}$ were kept constant in Sections 12.1 and 12.2 at $k = 0$ and $\sigma = 0.3$. In Ghil and LeTreut (1981) it was shown that for values of the density ratio $q = \rho_i/\rho_u$ near the usually accepted $q = 0.33$, and values of the nondimensional viscosity coefficient L_M^{*2}/Dc_T corresponding to the upper mantle and the shorter isostatic relaxation times, system (12.21) with $k = 0$ and $\sigma = 0.3$ exhibited self-sustained oscillations very similar to those discussed in detail in the previous sections. Such an oscillation is illustrated in Figure 12.11.

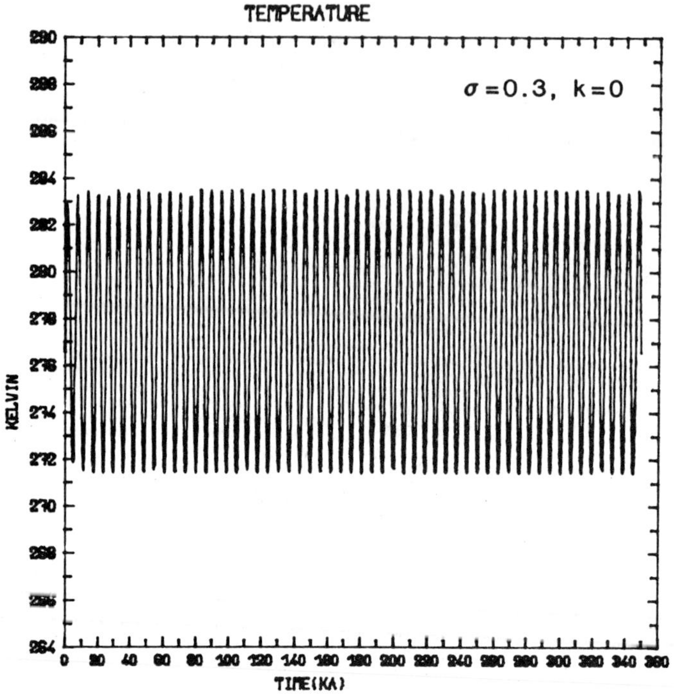

Fig. 12.11. Self-sustained oscillation of the climate model governed by Eqs. (12.21) (from LTG).

The parameter values used in Figure 12.11 and kept constant in the sequel are:

$$
\begin{aligned}
&\lambda = 14 \text{ m}, \qquad \gamma = 0.3, \\
&\alpha_0 = \alpha_m = 0.25, \quad \alpha_1 = 4.1 \times 10^{-7} \text{m}^{-1}, \quad \alpha_M = 0.85, \\
&T_{\alpha,\ell} = 217 \text{ K}, \quad \mathring{T}_m = 273 \text{ K}, \quad T_{\alpha,u} = \mathring{T}_M = 283 \text{ K}, \\
&\varepsilon_m = 0.1, \quad \varepsilon_M = 0.5, \quad \kappa = 1.74 \text{ Wm}^{-2}\text{K}^{-1}, \\
&T_\kappa = 154 \text{ K}, \quad T_{00} = 283 \text{ K}, \quad 1/\mu = 0.7, \\
&D = 40 \text{ km}^2/\text{yr}, \quad q = 0.33.
\end{aligned}
\tag{12.22}
$$

The values of k and σ, and the nonzero values of the forcing parameters δ_4, δ_5, $\varepsilon_1,\ldots,\varepsilon_5$ will be indicated in each subsequent figure.

The remainder of this section follows LeTreut and Ghil (1983; LTG hereafter) and will study the forced oscillations of the model governed by Eqs. (12.21, 12.22), as model parameters k and σ, and the forcing, change. Again, this study is undertaken as an illustration of the mathematical methods which we believe are useful in this type of problem.

Some of the results, however, are probably of direct physical relevance to an explanation of the paleoclimatological spectra introduced in Section 11.1.

Free oscillations. A nonzero elasticity parameter, $k \neq 0$, tends to damp the model's oscillations seen in Figure 12.1. For $\sigma = 0.3$ and $k = 0.03$, the self-sustained oscillations decay to a stable steady state $Q_s' = (T_s, \ell_s, \zeta_s) \simeq (281\text{ K}, 0.79, 0.44)$, which is the counterpart in the present model of the focus Q_s in Figure 12.4.

The first two eigenvalues of the system's linearization about Q_s' are complex conjugate, as in (12.7), for μ near 1.5. The third eigenvalue is real and negative, so that Q_s' is attractive in the ζ direction, for all parameter values considered here, and the model's center manifold is almost tangent to the (T,ℓ)-plane near Q_s'; i.e., the normal to the center manifold at Q_s' forms a very small angle with the ζ-axis (see Ghil and LeTreut, 1981, Figure 3). The sign of the real part of the complex conjugate pair becomes negative as k increases from zero, turning the generalized focus Q_s' into a stable one.

The effect of the slope parameter σ is similar to that of k. When σ decreases from $\sigma = 0.3$, say $\sigma = 0.28$ and $k = 0$, model oscillations increase out of the physically realistic range, cf. Figure 12.10. An increase in σ stabilizes the model. Hence larger k and smaller σ act in opposite directions.

At fixed μ, the period τ_0 of the self-sustained oscillation depends on k and σ, $\tau_0 = \tau_0(k,\sigma)$. The value corresponding to Figure 12.11 is $\tau_0 \simeq 6.25$ ka, or $f_0 = 1/\tau_0 \simeq 0.16\text{ ka}^{-1}$. For $k = 0.03$ and $\sigma = 0.28$, $f_0 \simeq 0.18\text{ka}^{-1}$. Recall that for decreasing values of μ, arbitrarily long periods obtain. As we shall see in the sequel, the behavior of the forced oscillations does not depend very much on the exact value of τ_0. We restrict ourselves here to values of $f_0 = O(10^{-1}\text{ka}^{-1})$, and will comment on much lower values in Section 12.7.

Notice that the time interval displayed in Figure 12.11 is much longer than in Figure 12.6. In all subsequent figures model solutions will be shown for 700 ka, corresponding to the Late Pleistocene, which is better known from paleoclimatological proxy records. The comparisons made with such records (Section 11.1) will be of a qualitative, rather than quantitative nature. The reason for this approach, and for the comparisons emphasizing the spectral domain over the time domain, will become apparent in the sequel, and will be summed up in Section 12.6.

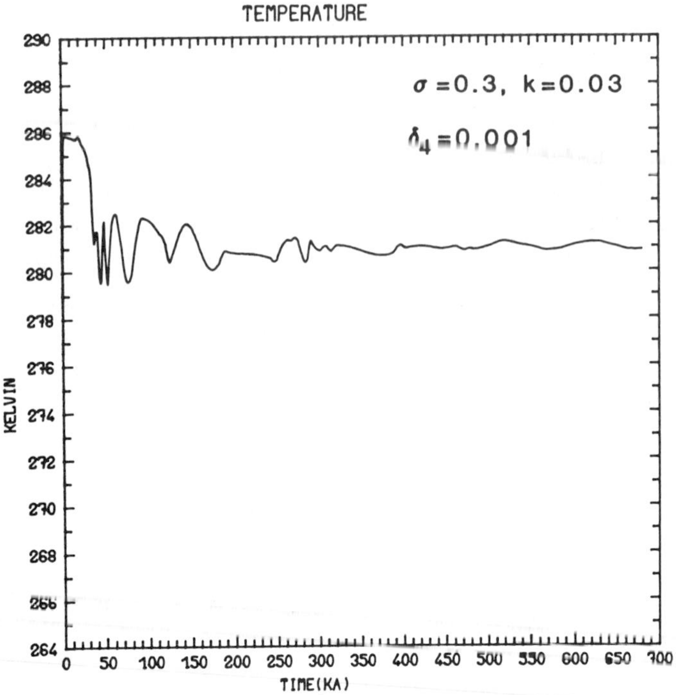

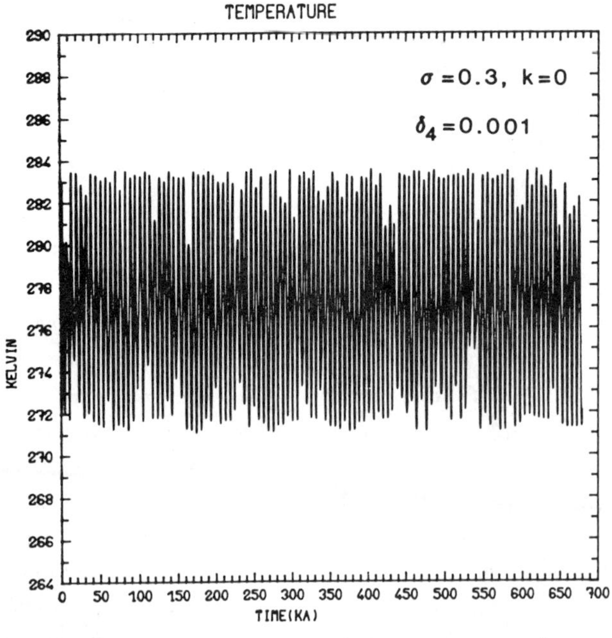

Fig. 12.12. Model response to eccentricity forcing. a) Quasi-equilibrium response; b) resonant response (from LTG).

Nonlinear resonance. To discuss the response of our climate model to external forcing, it is convenient to introduce the frequencies $f_j = 1/\tau_j$, $j = 1,2,3,4,5$. Notice that $f_0 > f_1 > f_2 > \ldots > f_5$.

Model response to insolation forcing is the easiest to understand and hence is studied first. We know from Section 12.3 that the largest value of $\Delta Q/Q$ over the Quaternary is one promil (one tenth of one percent), and that the correct response of an energy-balance model (EBM), governed by Eq. (10.2) or (10.12), to such a change in insolation is of approximately 0.1K, while actual changes in mean temperature over this period were of a few degrees Kelvin.

For the present model, response to insolation forcing with $\delta_4 = 0.001$ 0.001 in Eq. (12.18e) depends dramatically on the presence of self-sustained oscillations (Figure 12.12b) or their absence (Figure 12.12a). For $(\sigma,k) = (0.3, 0.03)$, the model's quasi-equilibrium response has the small amplitude which could be inferred from its equilibrium sensitivity of approximately 1 K for $\Delta Q/Q_0 = 1$ percent, similar to that of EBMs (see Section 12.1, following Eq. (12.8)). When $(\sigma,k) = (0.3, 0)$, as in Figure 12.11, the model responds resonantly, with the amplitude of a few degrees of its self-sustained oscillation at f_0, modulated by that of the forcing at f_4.

The same difference between quasi-equilibrium response and resonant response obtains for values of the insolation forcing both larger and smaller than $\delta_4 = 0.001$. Model response depends continuously on parameters, except at bifurcation points, where the nature of the response changes. In the sequel, model behavior will be illustrated at specific values of the internal and forcing parameters, with the understanding that it is representative of an open neighborhood of the given point in parameter space. This structural stability of model behavior is very important, given the lack of certainty in the actual values of relevant parameters, especially in the amplitudes of hydrologic forcing.

It is well known that a nonlinear oscillator like (12.21) can respond resonantly when the forcing frequency f_4 is rather different from its own frequency f_0 (see Section 12.7 for references). To understand better how this nonlinear resonance occurs for such large "detuning" as $f_0 - f_4$ or even $f_0 - f_5$, we shall consider the resonance mechanism in further detail.

12.5. Entrainment, Combination Tones, and Aperiodic Behavior

Obliquity forcing: entrainment and combination tones. As a first step in the study of our model's resonance mechanism, we verify that changes in the hydrologic cycle, Eqs. (12.18a,b), also lead to a resonant response, even at constant Q. Figure 12.13 illustrates a series of model responses to obliquity forcing, $\varepsilon_3 \neq 0$.

For all the solutions given in Figure 12.13, $(\sigma,k) = (0.3,0)$, as in Figure 12.11. Figure 12.13a shows the temperature evolution $T(t)$ for $\varepsilon_3 = 0.0025$. The modulation of the free oscillations with amplitude $O(10\ K)$ is similar to that in Figure 12.12b. Figure 12.13b gives the corresponding spectral density $\hat{T}(f)$, in logarithmic units. The nonlinear eigenfrequency f_0 dominates the spectrum, followed by the forcing frequency f_3. Additional frequencies visible in the figure are a harmonic of the forcing, $2f_3$, and the *combination tones* $f_0 + f_3$, $f_0 - f_3$ and $f_0 - 2f_3$. Harmonics and combination tones are also a characteristic of nonlinear system response to additive or multiplicative periodic forcing.

In Figure 12.13c, $\varepsilon_3 = 0.0035$ and both amplitude and character of the temperature oscillations appear rather similar to Figure 12.13a. A close inspection of the spectrum $\hat{T}(f)$ in Figure 12.13d shows, however, a remarkable difference: the peak at $f_0 \simeq 0.16\ \text{ka}^{-1}$ has been replaced by a peak at $f_0 \simeq 0.17\ \text{ka}^{-1}$, which corresponds to $7f_3$. Smaller peaks at $2f_3$ and $6f_3$ are also visible.

Between the two values of ε_3 in Figure 12.13b and in Figure 12.13d, *entrainment* has taken place -- the frequency of the free oscillation has locked onto the frequency of the forcing, or more precisely, onto a harmonic of the forcing. This phenomenon, also called *frequency locking*, is common in nonlinear oscillations (cf. Section 12.7). Its simplest form was observed first by Huygens who noticed that two clocks, which had slightly different pendulum periods when far apart, would synchronize when brought close together. Mathematically, entrainment means that over a whole interval of parameter values, the response frequency is equal to a constant, rational multiple of the forcing frequency, rather than changing as the parameter changes.

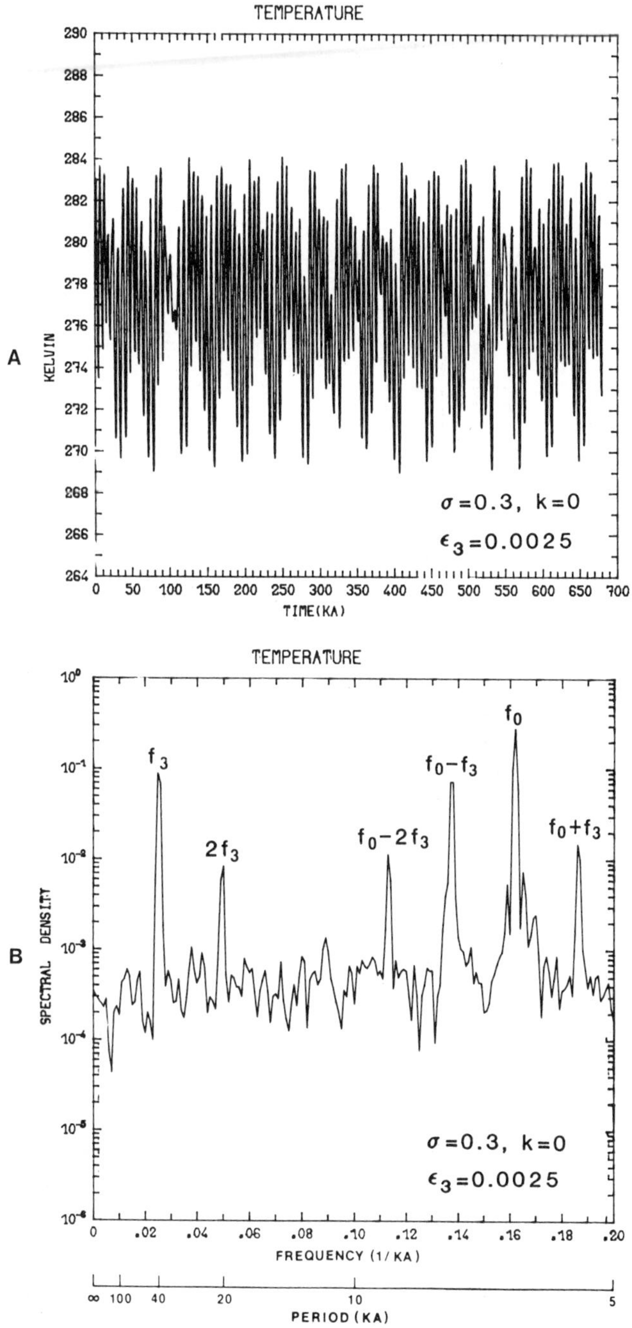

Fig. 12.13. Model response to obliquity forcing. a, c, e) Temperature evolution; b, d, f) power spectrum of temperature (from LTG).

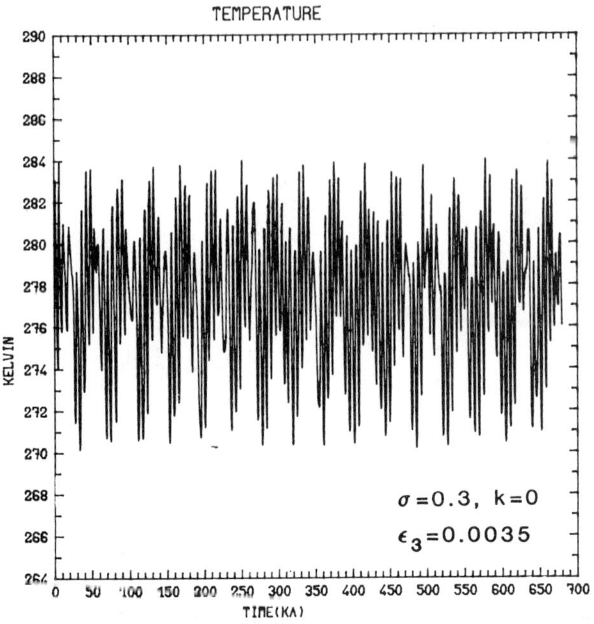
TEMPERATURE
KELVIN
TIME(KA)
σ=0.3, k=0
ε3=0.0035

C

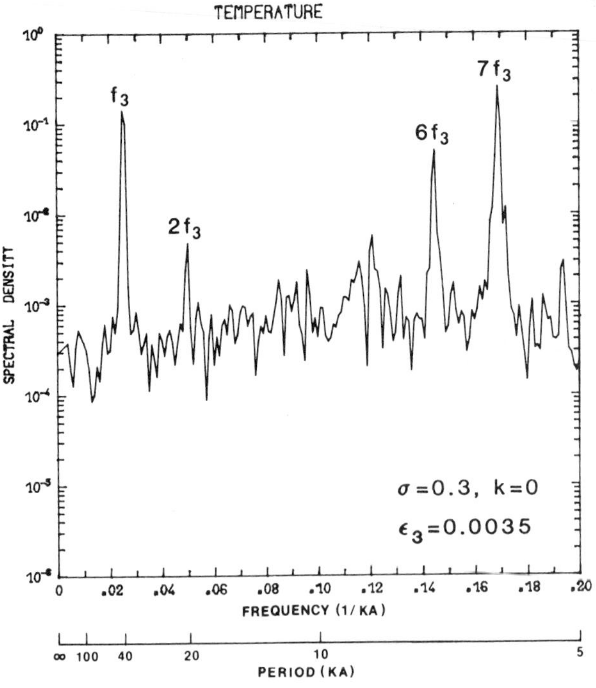
TEMPERATURE
SPECTRAL DENSITY
FREQUENCY (1/KA)
PERIOD (KA)
f3
2f3
6f3
7f3
σ=0.3, k=0
ε3=0.0035

D

E

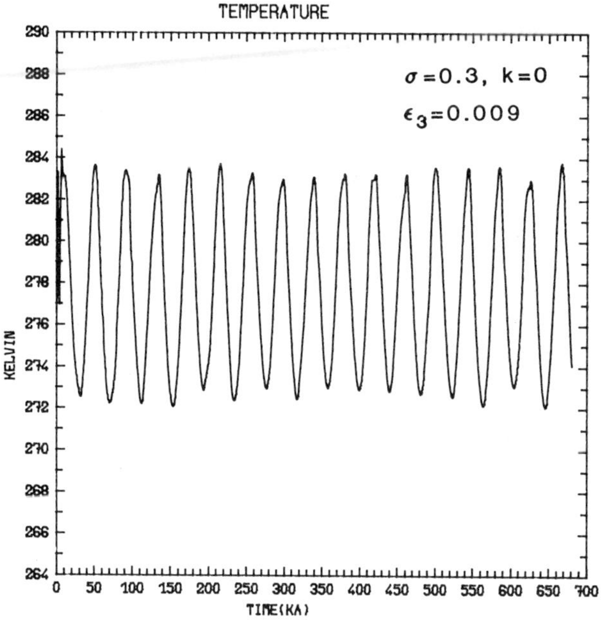
TEMPERATURE
σ=0.3, k=0
ε3=0.009
KELVIN
290
288
286
284
282
280
278
276
274
272
270
268
266
264
0 50 100 150 200 250 300 350 400 450 500 550 600 650 700
TIME(KA)

F

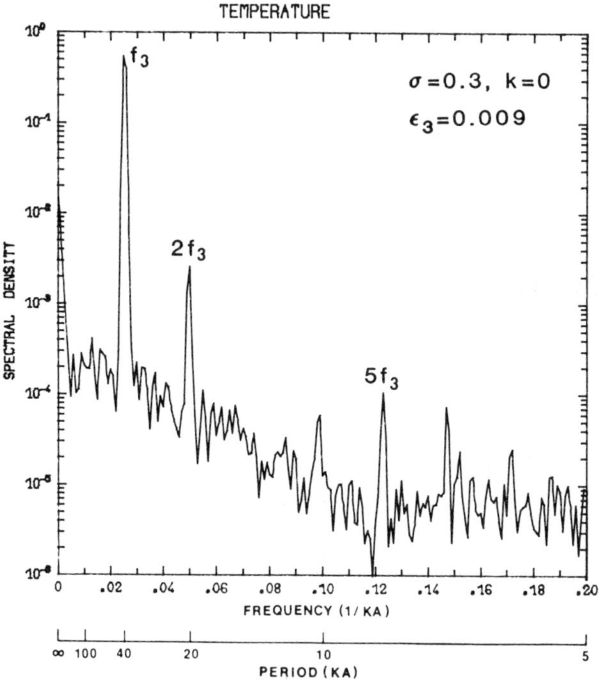
TEMPERATURE
σ=0.3, k=0
ε3=0.009
f3
2f3
5f3
SPECTRAL DENSITY
0 .02 .04 .06 .08 .10 .12 .14 .16 .18 .20
FREQUENCY (1/KA)
∞ 100 40 20 10 5
PERIOD (KA)

In the case observed by Huygens, as well as in the entrainment of most living organisms to the diurnal light cycle, *phase locking* occurs along with frequency locking. In many other oscillators, however, frequency locking does not necessarily imply synchronization of phases. The complexity of the climatic spectra discussed in Section 11.1, and the presence of the continuous background, seem to preclude simple phase locking between response and forcing for the system at hand. We shall return to this question in Section 12.6.

For $\varepsilon_3 = 0.009$, Figure 12.13 shows oscillations with an amplitude noticeably smaller than in Figures 12.13a and 12.13c, and with a dominant period of 41 ka. These oscillations, however, are not perfectly periodic: a slight irregularity is clearly visible. This is reminiscent of Figures 10.15e,f. It was shown in Section 10.4 that even slight aperiodicity of the temperature evolution is associated with a *continuous* component of its *spectral density*, which increases with decreasing frequency. Figure 12.13f shows, besides the now dominant peak at f_3, a continuous background which looks like that produced by random *red noise*, cf. Eq. (10.54). Upon this background one can still distinguish, as in Figures 10.16d-f, some harmonics of the fundamental, namely $2f_3$, $4f_3$, $5f_3$ and $6f_3$.

We have herewith another example of deterministic transition to aperiodic, or chaotic, behavior, which occurs for a value of ε_3 between 0.0035 and 0.009. The mechanism of transition here is loss of entrainment, which has also been observed experimentally in fluid systems (cf. Figure 10.17), as well as numerically and analytically in various model equations. Notice in Figure 12.13f that the peak $7f_3$ to which f_0 was entrained is about to disappear into the background noise.

Similar results, with entrainment producing a large number of sharp peaks in the spectrum, and its loss leading to a continuous, red-noise type background, were obtained for other forcing frequencies. Aperiodic model behavior is observed for forcing amplitudes as small as $\varepsilon_j = 0.001$ or $\delta_4 = 0.001$. In fact, aside from a red-noise type continuous background, decreasing with frequency to the right of the forcing peak which dominates the spectrum, one obtains also "band-limited chaos": a high, almost flat plateau in spectral density near f_0, left behing by a "cock's comb" of combination tones $f_0 \pm kf_5$, $1 \leq k \leq 3$, as it is first entrained by harmonics of f_5, then detrained as in Figure 12.13f.

Precessional forcing: more combination tones. We have seen the effects of single-frequency forcing in insolation (Figure 12.12), as well as in the hydrologic cycle (Figure 12.13), at the eccentricity frequency f_4 and the obliquity frequency f_3, respectively. It is natural to inquire next about the effects of multiple-frequency forcing, in particular at the two precessional frequencies, f_1 and f_2.

A series of numerical experiments with $(\sigma,k) = (0.3,0)$ and $\varepsilon_1 = \varepsilon_2 = 0.001$, 0.003 and 0.005 (LTG, Figures 8a-f, not shown here) give results very similar to those in Figure 12.13: at $\varepsilon_1 = \varepsilon_2 = 0.001$ the f_0-peak is most prominent, with smaller peaks at f_1, f_2 and $f_0 \pm (f_1-f_2)$. At $\varepsilon_1 = \varepsilon_2 = 0.003$, entrainment of f_0 to a fractional harmonic of the forcing, $(15/4)f_2$, obtains. At $\varepsilon_1 = \varepsilon_2 = 0.005$, detrainment has occurred, the high-frequency peaks at $f \geq 0.1\ \text{ka}^{-1}$ have disappeared, and the peaks at f_1 and f_2 are dominant, superimposed on low-frequency, red-noise-type background.

We present in Figure 12.14 the results for $(\sigma,k) = (0.28,0.05)$. Nonzero lithospheric elasticity has the effect (LTG, Figure 7; $\sigma = 0.3$, $k = 0.03$, $\varepsilon_3 = 0.005$, not shown here) of leading to a slower growth and more rapid decay of ice sheets, giving a sawtooth appearance to the ice-extent and bedrock evolution curves, while the temperature evolution is still roughly sinusoidal.

The model's self-sustained oscillation has a frequency of $f_0 \simeq 0.155\text{ka}^{-1}$ for the parameter values in the figure. At $\varepsilon_1 = \varepsilon_2 = 0.005$, this oscillation has been entrained already. The ice sheet evolution $\ell(t)$ in Figure 12.14a shows an irregular behavior with intermittent, large-amplitude fluctuations every 100 ka, approximately. The behavior for $k = 0$, $\sigma = 0.3$, discussed before, is also modulated at roughly 100 ka, but is less intermittent.

Figure 12.14b gives the power spectrum of the ice volume, $\hat{V}(f)$ for the solution in Figure 12.14a, where V is proportional to $\ell(\zeta + \ell^{1/2})$, and the spectrum is normalized. The most prominent peak in the entire spectrum is at $f_1 - f_2$. With our choice of approximate values for f_1 and f_2, this peak has a period of 109 ka.

Recall from the discussion of insolation changes that the precessional effect $e \sin \tilde{\omega}$ had two apparent periods, of approximately 19 ka and 23 ka, due to the modulation of angular precession $\tilde{\omega}$ by the eccentricity e in the range 95 ka to 130 ka. The climatic system can thus resynthesize and focus, by its combination tones, a response to the

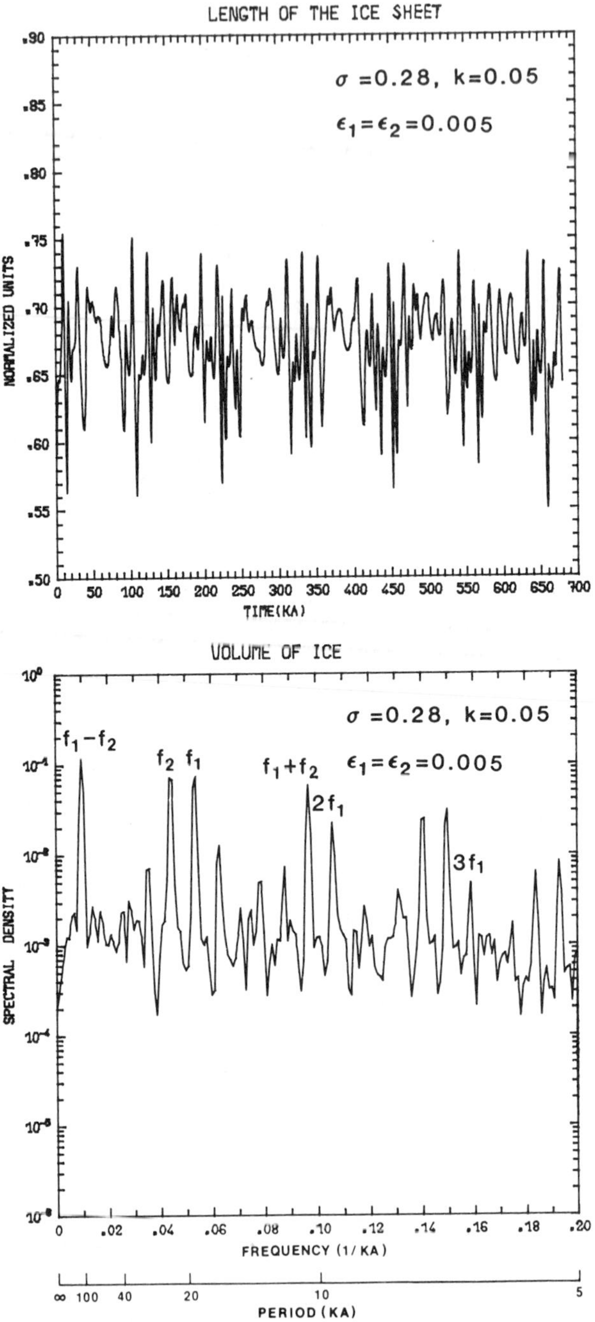

Fig. 12.14. Model response to precessional forcing. a) Ice-extent evolution; b) power spectrum of ice volume (from LTG).

original, diffuse frequency range of eccentricity changes, which are hidden in the precessional effect.

After the peaks at f_1 and f_2, the next largest peak in Figure 12.14b is at $f_1 + f_2$, which corresponds to a period of roughly 10 ka. This peak is flanked by two peaks at $2f_1$ and $2f_2$. Peaks between 9.5 ka and 11.5 ka appear in various marine and continental records with high sedimentation rates (Pestiaux and Duplessy, 1985, and references therein). The stratigraphic, geochemical and micropaleontological analysis of such records will lead in the near future to more refined Quaternary time scales and more reliable proxy records of glaciations. Careful harmonic analysis will then permit the high-resolution discrimination between peaks suggested by the theoretical spectra presented here.

It is also interesting from the point of view of the climatological interpretation of proxy data that the temperature spectrum $\hat{T}(f)$ for the solution in Figure 12.14 (LTG, Figure 9c, not shown here) has marked dif-differences from $\hat{V}(f)$. The largest peaks in $\hat{T}(f)$ are at f_1 and f_2. The next largest peak is at $f_1 + f_2$, with smaller peaks at a distance f_1-f_2 away on either side. Another such triplet of peaks is visible at higher frequencies, as in Figure 12.14b: $f_1 + 2f_2$, $2f_1 + f_2$ and $3f_1$. But the peak at $f_1 - f_2$ is not present in the spectrum of temperature, while it dominates the spectrum of ice-volume. If proxy records reflect combinations of ice-volume evolution and temperature evolution, the same will hold for their power spectra. Differences between the power spectrum of $\delta^{18}O$ and that of micropaleontological indicators are apparent in various deep-sea cores, for instance.

Sharp peaks, aperiodicity and terminations. We have examined so far the effect of insolation forcing at each eccentricity frequency, as well as that of mean-annual hydrologic dorcing at the obliquity frequency, and at the two precessional frequencies. The effects of insolation changes and of forced changes in the hydrologic cycle are essentially independent of each other and can be conceptually superimposed. Thus for instance the nonlinearly resonant peak near 100 ka in Figure 12.14b, caused by precessional forcing, and the resonant peak near 100 ka implied in Figure 12.12b, caused by eccentricity forcing, will mutually reinforce each other to yield the considerable spectral power near 100 ka apparent in paleoclimatological spectra. It is interesting, however, to see the combined effect of all three frequencies which affect the hydrologic cycle exclusively, f_1, f_2 and f_3.

For the sake of definiteness, we choose the values of ε_1, ε_2 and ε_3 which correspond approximately to July insolation at 65°N. That is, the ratio is $\varepsilon_1:\varepsilon_2:\varepsilon_3 = 3:4:2$, with the amplitude taken close to those previously considered, i.e., $\varepsilon_1 = 3 \times 10^{-3}$, etc. This is of course arbitrary, since the present model avoids the changes of forcing with latitude and season, by integrating their net effect over the globe and the year. The choice is only made for comparison purposes with models which might include more details of insolation distribution in space and time. Figure 12.15a shows the evolution of bedrock deflection. Terminations are clearly visible at roughly 100 ka, 200 ka, 330 ka, 520 ka and 640 ka. The slow growth of ice sheets does not, according to paleoclimatic records, occur monotonically, but rather in successive rises and smaller falls, cf. Figure 11.2. This feature of small oscillations superimposed on a large trend is well captured by the model solution in Figure 12.15a.

Occasionally, one or two spikes appear in the solution between a termination and the subsequent slow growth, for instance at about 220 ka in Figure 12.15a. Such a spike is apparent in the Pacific core V28-238, as well as in the combined record of cores RC11-120 and E49-18 from the Indian Ocean, during isotopic stage 7 (Figures 11.2 and 11.3). A similar spike is now believed to have occurred during the last deglaciation, where one distinguishes between termination IA and IB (Duplessy *et al.*, 1981). The exact position, amplitude and duration of this spike, associated with the Younger Dryas event of European continental stratigraphy, is still a matter of debate, but paleoclimatologists tend to concur more and more as to its existence.

The lack of exact periodicity in the terminations, as well as the irregularity of smaller peaks and plateaus within each major cycle, suggest again the presence of a complex spectrum. Figure 12.15b shows the Fourier spectrum of the total ice volume for the solution at hand. It is dominated by the peaks at f_1, f_2, f_1-f_2 and f_1+f_2. A number of combination tones lying between f_1-f_2 and f_2 are as prominent as the peak at f_3.

The striking feature of Figure 12.15b is the presence, as in Figure 12.13f, of the continuous background, decreasing with frequency (stippled in the figure). This background is produced by the deterministic nonlinear interactions in the model, rather than by any stochastic forcing.

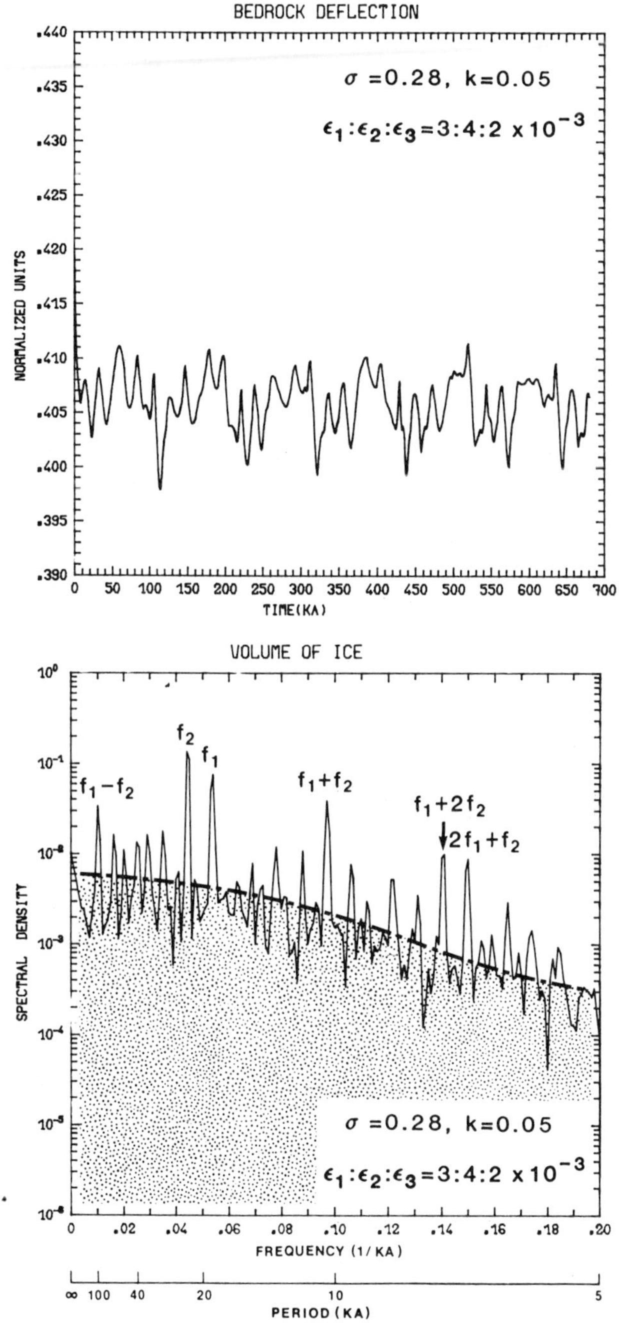

Fig. 12.15. Model response with irregular terminations and sharp peaks on a continuous background. a) Bedrock evolution; b) power spectrum of ice volume (from LTG).

The deterministic model analyzed here reproduces well the most important qualitative features of Quaternary climate evolution, as discussed in Section 11.1. Hence, it is likely that at least part of the continuous spectral density observable in paleoclimatic records is due to internal variability on the time scale of glaciation cycles, rather than to mere passive filtering of higher-frequency weather noise by the slow components of the climatic system. Changes in sedimentation rate of marine deposits, and post-depositional processes, such as bioturbation, also contribute to the power in the continuous spectrum of deep-sea proxy records (Pestiaux, 1984).

The temperature spectrum (not shown) has the same frequency-dependent continuous background. It also shows peaks at f_1, f_2 and f_1+f_2, but show no peak at f_1-f_2. On the other hand, two peaks at f_3 and at $f_3 - f_1 + f_2$ are prominent. The latter corresponds to a period of roughly 66 ka.

Model solutions for $T(t)$ and $V(t)$ were also translated into simulated isotopic records in ice cores and deep-sea cores by H. LeTreut and J. Jouzel (personal communication, 1983). A simple, robust model of fractionation processes taking place during the hydrologic cycle (Jouzel and Merlivat, 1984) is used for this purpose (compare Section 11.1). The power spectra of the simulated $\delta^{18}O$ records also show sharp peaks at $f_2 + f_3 \cong 1/15$ ka and $f_1 + f_3 \cong 1/13$ ka.

The relative power in each peak changes from one simulated record to another, but their frequencies are the same. Some of these and other combination tones are found, along with a few of their harmonics, in four carefully dated and analyzed rapid-sedimentation cores from the Indian Ocean (Pestiaux, 1984; Pestiaux and Duplessy, 1985). These results and related ones suggest in particular that the spectral peak at $\tau \cong 2.5$ ka (Figure 11.1) which appears for instance in ice cores from Greenland and the Antarctica (Benoist *et al.*, 1982; Dansgaard *et al.*, 1982), might be the fourth harmonic of the $(f_1 + f_2)$ combination tone emphasized in this section, with $\tau = 10.4$ ka/4 = 2.6 ka.

Nonlinear resonance, harmonics, combination tones and deterministic aperiodicity thus appear to explain many characteristics of paleoclimatic time series and of their Fourier spectra. An important feature of these results is that they are insensitive to the exact nature of the system's self-sustained oscillations, their causes, or their periods. These oscillations only act as an amplifier of the forcing, and lead to a transfer of the system's internal variability from the high frequencies, $O(10^{-1}ka^{-1})$, at which most physical mechanisms involved are active, to the low frequencies, $O(10^{-2}ka^{-1})$, where they are observed in paleoclimatic records. We turn now to the consequences of quasi-periodicity and aperiodicity for the predictability of climate on the time scales under discussion.

12.6. Periodicity and Predictability of Climate Evolution

The search for periodicity in geophysical time series is obviously motivated by the desire to understand, as well as to predict. Constant behavior is the most predictable; it is without surprises, but is seldom encountered in nature. The next most predictable type of behavior is purely periodic.

Phase errors and frequency errors. For periodic behavior it becomes important to ascertain the phase of the phenomenon. An error of phase determination at one time, however, will merely result in a similar error at any other time, past or future. Not so for an error in period: even the smallest error in period will result, after a sufficient number of periods, in a completely erroneous forecast, or hindcast, as the case may be.

For a quasi-periodic function with two or more rationally independent periods, things become rapidly more complicated. Errors in any period will result in complete loss of predictive skill after a small multiple of the shortest period involved. We restrict ourselves in the sequel to the type of error which is both less severe and more inevitable -- phase error.

Phase errors can arise in the forcing, as well as in the internal oscillation of the system. The initial values $P_0 = (T_0,\ell_0,\zeta_0)$ used for the model variables so far were equal to

$$P_1 = (T_1,\ell_1,\zeta_1) = (276 \text{ K}, 0.65, 0.42), \tag{12.23a}$$

i.e., $P_0 = P_1$. In order to study the effect of the internal oscillations' phase on the predictability of model variables, the points $P_0 = P_j$, $j = 2,3,4$, were also used as initial data, where

$$P_2 = (280\ \mathrm{K},\ 0.65,\ \zeta_1),$$
$$P_3 = (272\ \mathrm{K},\ 0.70,\ \zeta_1), \qquad (12.23b)$$
$$P_4 = (284\ \mathrm{K},\ 0.80,\ \zeta_1)$$

The points P_1, P_2, P_3, P_4 are approximately at a quarter phase from each other along the limit cycle of the model. The phase of the forcing is varied by taking $\phi_j = k\pi/2$, $k = 0,1,2,3$ in Eqs. (12.18a,b). So far $\phi_j = 0$ in all solutions discussed.

Figure 12.16 illustrates the differences between the solution in Figure 12.13, with $P_0 = P_1$ and $\phi_3 = 0$, and those obtained for $\phi_3 = \pi/2$ (Figure 12.16a) and for $P_0 = P_2$ (Figure 12.16b) respectively. The two frequencies present in the solution are f_0 and f_3. They are actually not rationally independent, but the corresponding common period, close to 250 ka, is sufficiently long for the very small background noise to prevent exact repetition of solution values after this common "period." Thus the resulting solution appears quasi-periodic in the time domain.

Comparison between Figures 12.16a and 12.13a shows a clear phase shift of the longer, modulating period τ_3, as expected. But the high frequency oscillations within each long period are not at all alike for two corresponding periods, say $t_0 \leq t \leq t_0 + \tau_3$ in Figure 12.13a and $t_0 - \tau_3/4 \leq t \leq t_0 + 3\tau_3/4$ in Figure 12.16a. For instance the "quiet episode" near 100 ka when $\phi_3 = 0$ has no "shifted" counterpart in Figure 12.16a, nor does the quiet episode near 300 ka when $\phi_3 = \pi/2$ have its nearby counterpart in Figure 12.13a.

This observation is confirmed by Figure 12.16b, in which $P_0 = P_2$, while $\phi_3 = 0$ again, as in Figure 12.13a. The initial difference between the temperature values does not disappear with time, given the identical forcing, but continues to produce differences between the two figures for the entire duration of the display. The "missing spikes" in Figure 12.16b just before 200 ka and just after 250 ka have no counterpart in Figure 12.13a, while the solution in Figure 12.13a is more quiescent near 700 ka than in Figure 12.16b.

Figure 12.17 shows the actual difference between the temperature evolution in Figure 12.16a $T(t;\phi_3 = 0)$ and that in Figure 12.13a, $T(t;\phi_3 = \pi/2)$. The peak-to-peak amplitude of the difference is twice that of either solution. This maximum is reached within 100 ka, i.e., rougly twice the forcing period. The difference exhibits the aspect of "beats", with the two frequencies present producing quiet intervals, like that between 260 ka and 400 ka, and agitated episodes, like that near 100 ka or 550 ka.

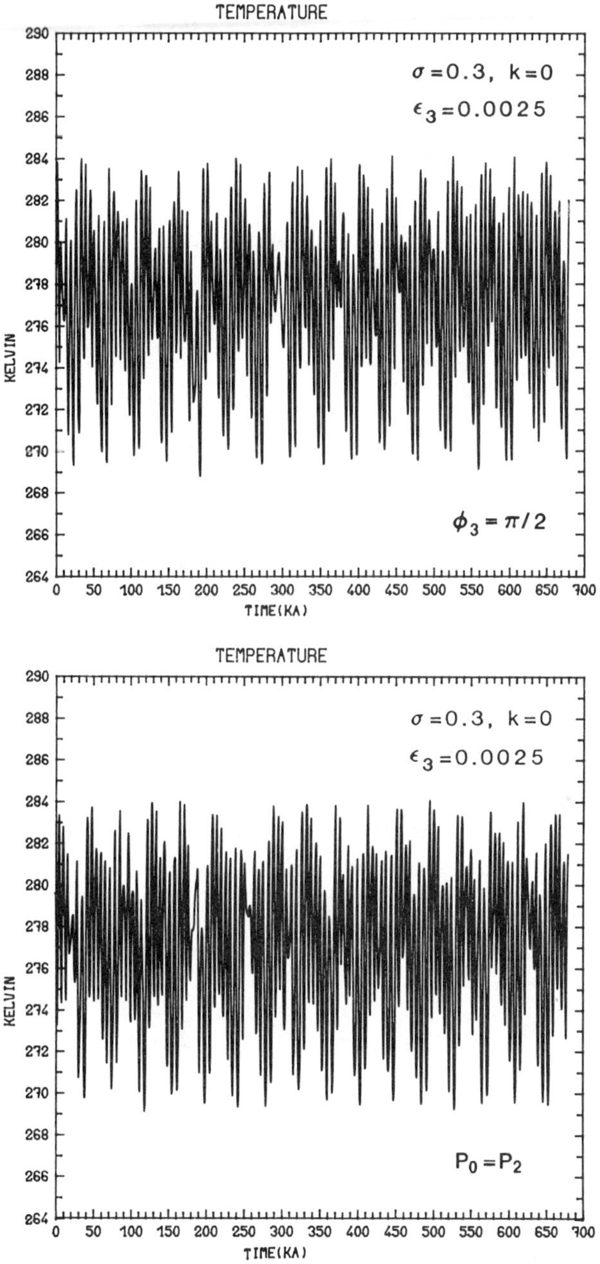

Fig. 12.16. Effect of phase errors on a quasi-periodic model solution. a) Error in the forcing; b) error in the initial data (from LTG).

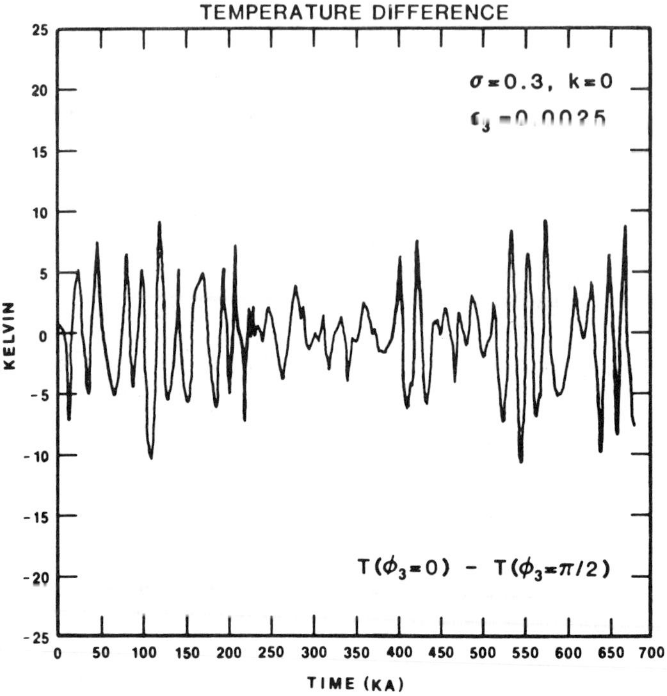

Fig. 12.17. Difference between two solutions distinguished by the phase of the forcing (after Ghil, 1985b).

Similar results obtain by taking the difference between the solution in Figure 12.16b and that in Figure 12.13a.

The corresponding temperature spectra for the three solutions above (not shown) are practically indistinguishable from each other. All the peaks which are marked in Figure 12.13b appear in the identical location and with like magnitude for the solutions in Figure 12.16a,b. Thus spectral information characterizes solution behavior, independently of phase information.

Measures of predictability. Various measures of predictive skill in the time domain ("pointwise predictability", see Sections 4.5 and 6.4) can be used to make the difficulties illustrated above more precise. The most natural one in the present context is the lagged correlation of solutions, R(s):

$$R(s) = \int_{\tau_0}^{\tau_0+\tau} T(t)\ T(t+s)dt \Big/ \int_{\tau_0}^{\tau_0+\tau} T^2(t)dt. \tag{12.24}$$

In principle, $R(s)$ depends on the length of the averaging interval, τ, and on its initial point τ_0. If τ is sufficiently long, and the solutions take on all possible values under the given conditions, both dependences are very weak.

For a constant solution, $R \equiv 1$. For a periodic or quasi-periodic solution, $R(s)$ is itself a periodic or quasi-periodic function. For random white noise, our measure of predictive skill (12.24) will be a delta-function concentrated at the origin, $R(s) = \delta(s)$. Hence $R(s)$ takes on the correct behavior in the two extreme cases, of perfect predictability and of complete lack of predictability. Its behavior in the intermediate case of quasi-periodicity is also suggestive.

Figure 12.18 shows the behavior of the lagged correlation function for a quasi-periodic and an aperiodic solution. We saw that for a spectral density distribution which is perfectly flat, i.e., white noise, $R(s) = \delta(s)$. For a spectral density decreasing with frequency, like red noise, $R(s)$ will be an exponential function. Similar results would obtain if the initial conditions were random and the averaging in (12.24) were with respect to these initial data, rather than with respect to time (cf. Eqs. (10.29b, 10.32b) here, and Appendix B of Bhattacharya *et al.*, 1982).

In all solutions of (12.21) considered so far, there is a small component of continuous spectral density, due to numerical noise in the computations. This component leads to the initial, rapid decay of the correlation function in Figure 12.18a with lag time. Such a decay of predictability with time will be present in any numerical integration, and the more complicated the model, the stronger this purely numerical decay will be. That is, increasing the number of physical components or spatial dimensions in the model will lead to greater loss of predictive skill due to numerical inaccuracies, rather than reduce this loss of skill, even if the model were perfect in every other respect.

The decay in Figure 12.18b is much larger. After the small, rapid decay caused by the numerical component of the spectral density, the exact model's true aperiodicity takes over in reducing predictive skill. The correlation time of this portion of the continuous spectral density is longer, and hence the larger portion of the predictability decay occurs more slowly, over a time $O(10^2 \text{ka})$. Finally, the pure-line part of the spectrum leaves a periodic correlation function of amplitude $O(0.1)$ at large lag times.

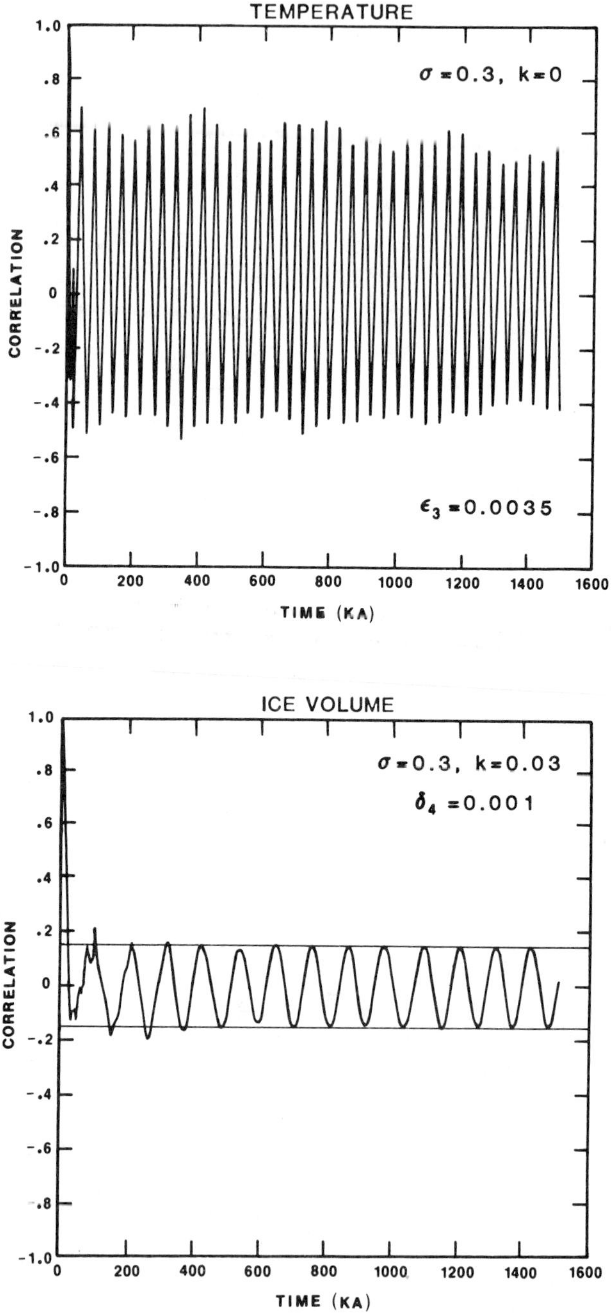

Fig. 12.18. Lagged correlations for a quasi-periodic solution (a), and an aperiodic solution (b) (after Ghil, 1985b).

These results concerning the damped periodicity of correlations are again model-independent. They can be obtained from the power spectrum of the data themselves by the Wiener-Khinchin formula (10.53). Hence, detailed simulation with arbitrarily small, prescribed errors of the Earth's paleoclimatic history is not possible, based on the limited amount and accuracy of proxy data available.

On the other hand, much can be learned about climatic mechanisms, and about orbital changes, by studying paleoclimatic records in the spectral domain. The distribution of power between the peaks and the continuous background, as well as the exact position of the peaks and their power relative to each other, give tell-tale indications about the climate's internal workings and the external changes which affect it.

12.7. Bibliographic Notes

Section 12.1. The classical theory of *nonlinear oscillations* is presented in books by Hayashi (1964), Minorsky (1962), and Stoker (1950). Connections with the modern point of view of dynamical systems and bifurcation theory appear in Andronov *et al.* (1966) and Howard (1979). The books of Arnold (1973, 1983) and of Guckenheimer and Holmes (1983) are written entirely from the latter point of view.

Most of the references above contain examples of nonlinear oscillations from biology, chemistry, mechanics, electromagnetism and other parts of the physical sciences. Reviews of oscillatory models of climatic change are given by Ghil (1981b, 1985b), Ghil and Saltzman (1984) and Saltzman (1983).

The climatic variables and nonlinear interactions among them introduced in Chapter 11 and studied here have been incorporated in a slightly different way, or in combination with additional variables and physical or chemical mechanisms, into other climatic oscillators by Birchfield and Grumbine (1985), Oerlemans (1982), Peltier and Hyde (1984), Saltzman *et al.* (1981, 1982) and Sergin (1979). Among the mechanisms used in this context are carbon dioxide (CO_2) uptake in the biosphere and ocean, alternating with an increased greenhouse effect due to higher CO_2 concentrations in the atmosphere (Saltzman and co-workers), ice-sheet thermodynamics interacting with ice-mass flow to produce increased basal sliding (Oerlemans; compare also Section 11.4), the effect of hypothetical massive, ocean-basin-sized ice shelves on continental ice sheets (Saltzman *et al.*; Section 11.4) and the interaction between deep-ocean circulation and surface temperature and ice extent (Ghil *et al.*, 1986). For the observational evidence in support of each one of these mechanisms we refer to the articles cited.

The complexity of the climatic system should be evident from the preceding chapters and from the paragraph above. This complexity is accompanied by an insufficient knowledge of numerous parameter values which are necessary to the formulation of quantitative models of climatic change; even the sign of certain feedbacks, positive or negative, is often in question, and when that is known, the order of magnitude of associated coefficients is uncertain (compare, for instance, Saltzman *et al.*, 1982, with their 1981 paper). Therefore, a qualitative, preliminary analysis of climatic interactions, as suggested by Eqs. (12.9, 12.10, 12.19) appears desirable.

A systematic framework for the development and analysis of *formal conceptual models* of climatic change has been elaborated recently (Dee and Ghil, 1984; Ghil and Mullhaupt, 1985). It consists in the mathematical theory of Boolean delay equations (BDEs), which originated in the kinetic logic of the geneticist R. Thomas (1973, 1979), applied first to climate dynamics by C. Nicolis (1982b). The exposition of BDE theory and of its climatic applications (Ghil, 1984b; Ghil *et al.*, 1986) is outside the scope of the present book.

Suffice it to say here that variables are only allowed a few discrete values, such as high or low temperature, high, medium or low ice volume, strong or weak thermohaline circulation in the ocean. The interactions between variables are expressed as logical functions (connectives), and the interaction between each pair of variables is assumed to take a certain length of time until it achieves its logical effect; these delays are distinct in general for each pair of variables. Given a set of variables, connectives and delays, model behavior can be stationary, periodic, quasi-periodic or aperiodic. BDE models are much easier to formulate and somewhat easier to analyze than models governed by differential equations. The only parameters present are the delays and sometimes knowing the order of magnitude of these suffices to determine qualitatively solution behavior. Hence the study of such formal conceptual models of climatic change might be a useful step in the formulation of more elaborate, quantitative models.

Section 12.2. The first book dedicated entirely to *Hopf bifurcation* and its applications to ordinary and to partial differential equations is Marsden and McCracken (1976). We followed essentially the presentation of Arnold (1983). Especially useful in complementing this presentation will be Arnold (1973), Carr (1981), Guckenheimer and Holmes (1983), Howard (1979) and Iooss and Joseph (1980).

The references above also contain information on homoclinic and heteroclinic orbits. Many of the general references to Section 12.1 describe *relaxation oscillations*, which have been studied most thoroughly in connection with the Van der Pol oscillator (e.g., Guckenheimer and Holmes, 1983, Section 2.1). A monograph entirely devoted to such fast-slow oscillations, and to their applications in physics and biology, is Grasman (1986).

In Section 12.5, discussion of Figure 12.14, it was shown that lithospheric elasticity can lead from a roughly sinusoidal shape of oscillations to a sawtooth-shaped evolution of ice volume and bedrock deflection. Relaxation oscillations also obtain in Section 12.2 for a large ratio of ice-flow relaxation time to thermal relaxation time (small $\mu = c_T^*/c_L$) and in Section 10.4 for large phase lags between ice volume and temperature (Figure 10.15). These model results suggest that relaxation phenomena play a role in the sharp terminations of glaciated periods (Figure 11.2; see also Imbrie and Imbrie, 1980, and Pollard, 1984). The fact that relaxation oscillations, being highly nonlinear, become more easily aperiodic than sinusoidal oscillations was already known to Van der Pol (1940). This fact is at least consistent with the irregular occurrence of terminations emphasized in Sections 11.1 and 12.5.

Section 12.3. Nicole Oresme was already aware in the fourteenth century of the kinematic consequences of *quasi-periodicity* for celestial motions (Grant, 1961). He realized that a periodic and a quasi-periodic motion cannot be distinguished from each other during a finite period of observation. Oresme also knew that the motion of a point on a torus will describe a simple closed loop if the two angular velocities are commensurable, while the point's orbit will never close, but cover the surface of the torus in a dense way if the two velocities are incommensurable (ibid.; compare Arnold, 1983, Section 11).

The first concerns of astronomer-mathematicians in the 18th and first half of the 19th century were the calculation of ephemerides for navigation and the stability of the solar system in the sense of boundedness of the Earth's motion around the Sun. Euler and Lagrange formulated the method of *variation of constants* (Brouwer and Clemence, 1961, Chapter 11) for three bodies within the context of what is now known as Lagrangian mechanics. Laplace (1787) extended the method to a five-body solar system, the Sun and the inner four planets. He also studied the stability of this system's periodic motion. His popular presentation of celestial mechanics (1796) had a great impact on the physical and philosophical thought of the nineteenth century.

LeVerrier (1855) applied the Lagrange-Laplace method to obtain short term ephemerides for obliquity and eccentricity, which led him to the discovery of Neptune. This discovery provided strong support to Laplace's view of celestial mechanics, and of the universe. Variation of constants is entirely appropriate for short-term computations, of the order of a few years or decades (Arnold, 1983, pp. 142-144).

Long-term computations had to be based on time-independent methods, such as *secular perturbations* (Brouwer and Clemence, 1961, Chapters 13, 15 and 16). In this approach, fast-periodic terms are simply dropped from the perturbing potential, rather than being averaged in a systematic way (Arnold, 1983, Chapter 4; Buys and Ghil, 1984; Goldstein, 1980; Guckenheimer and Holmes, Sections 4.1-4.4; and most of the general references given for Section 12.1). Secular perturbation techniques were used by Newcomb (1895), Miskovic (1931), Brouwer and van Woerkom (1950), Sharaf and Budnikova (1967), and Bretagnon (1974) for very long term computations. These orbital computations were used in turn by Adhémar (1842), Croll (1867), Milankovitch (1941), Vernekar (1972) and Berger (1978) to calculate insolation changes which might influence climate on the time scale of 10^3-10^7 years.

Comparisons between the different calculations are given by Berger (1984), Berger and Pestiaux (1984) and Bretagnon (1984). Frequencies stay constant over the entire length of the time interval considered, 3200 ka, due to the nature of the perturbation series used, admitting of no additional frequencies or secular terms which might be present in the system itself. But the frequencies themselves change slightly from calculation to calculation and the amplitudes of the corresponding spectral peaks change from one 800 ka-long subinterval to another, and from one calculation to another. Consequently, phase differences between the calculations compared increase with time as illustrated in Buys and Ghil (1984) for simple oscillators with known analytic solutions.

All of the methods applied so far to the long-term planetary problem suffer from a number of common shortcomings, both technical and fundamental. On the technical side, some planetary masses are still known with rather limited accuracy, the presence of the ninth planet, Pluto, is neglected, the more difficult luni-solar problem is coupled to the planetary one only kinematically, and not dynamically (with the exception of Bretagnon, 1984, where partial dynamical coupling is performed), the presence of thousands of minor planets, asteroids, moons, rings, comets and other small bodies is neglected (Brouwer and Clemence, 1961, p. 529), as are interstellar gas, tides, radiation pressure and relativistic effects (Vicente, 1983). Related to this, and on a more fundamental level, even the modern perspective of Kolmogorov-Arnold-Moser (KAM) theory only offers the existenceof neutrally stable solutions to a dissipationless, Hamiltonian perturbation problem.

Good presentations of *KAM theory* within the framework of Hamiltonian mechanics can be found in Arnold (1983, Chapter 4), and Moser (1973), along with the original references. Lichtenberg and Lieberman (1983) provide also many beautiful numerical illustrations of the theory. Since Poincaré (1892), the concern about stability of the solar system has shifted from ejection of a major planet onto a hyperbolic, comet-like orbit, to bounded, but aperiodic motion of all planets. KAM theory states that the major planets, considered as point masses, move quasi-periodically with nonzero probability. But the tori described in their phase space by each such motion do not separate regions of phase space where the motion is aperiodic, i.e., where it appears random or stochastic. Thus the system's representative point can move, or diffuse, all through phase space, unimpeded by the tori which survive the small perturbations of mutual attraction between planets. It is this so-called Arnold diffusion which becomes very intuitively plausible in the illustrative figures of Lichtenberg and Lieberman (1983).

The small *dissipative effects* on planets and satellites of tides, and of interplanetary dust drag, on the one hand, and of the solar-wind forcing, on the other, should be noticeable on the long time scales of interest here. Hence a theory of forced-dissipative perturbations, related to the model systems studied in Chapters 5 and 6, and in the rest of this chapter, would seem to be necessary, and to offer the bonus of asymptotically stable solutions, which are more easily computable than neutrally stable ones (Buys and Ghil, 1984). Such a mathematical theory, initiated by Holmes and Rand (1976), Murdock (1976), Stoker (1950, Section 4.10 and Appendix II) and others (see Guckenheimer and Holmes, 1983, Sections 4.5-4.8), can yield physically intuitive results about the asymptotic stability of low-order resonances (Wolansky, 1985), such as the 1:1 resonance between the rotation and revolution of the Moon, the near 3:2 ratio between the rotation and revolution periods of Venus, or the near 2:5 ratio between the revolutions of the giant planets Jupiter and Saturn (the "great inequality", where inequality in celestial mechanics means any term in a d'Alembert series).

Section 12.4. The N-body problem, discussed in Section 12.3 and its bibliographic notes, can be considered as the problem of N weakly coupled harmonic oscillators, each oscillator representing the Keplerian motion of a major planet around the Sun (Buys and Ghil, 1984). The weakest form of coupling is that in which the action of one oscillator upon another is represented as periodic forcing, ignoring completely the reaction of the other oscillator. Thus the study of *forced oscillators*, following upon that of free, autonomous oscillators, is an additional step towards the study of strongly interacting, fully coupled oscillators. All general references given for Section 12.1 study forced, as well as free oscillations, in nonlinear systems.

Stoker (1950, pp. 116-117) presents a systematic comparison of free and forced oscillations in linear and nonlinear systems, based on a similar table in Duffing's (1918) original book. One of the striking ways in which nonlinear resonance can differ from the linear case is *bistability and hysteresis* of the response (Stoker, 1950, Sections 4.1-4.5). We recall that the Hopf bifurcation in Eqs. (12.1, 12.2) was subcritical (Section 12.2), and that we alluded already in Section 11.2 to a possible connection between this fact and the transition between Early and Late Pleistocene.

The observational evidence points to an abrupt increase in climatic variability about 1000 ka ago (Shackleton and Opdyke, 1976; Pisias and Moore, 1981). The overall increase in variability is associated in particular with a sharp jump in the amplitude of the 100 ka peak (Pestiaux, 1984, Figure 4.9; Pestiaux and Berger, 1984, Figure 10). This peak may quite plausibly result from the resonance of the climatic system in its oscillatory mode (Figure 12.12).

Hence the following scenario (compare Emiliani and Geiss, 1957) suggests itself for the successive climatic transitions from Pliocene to Pleistocene and from Early to Late Pleistocene: As land masses moved towards more northerly positions, small ice caps formed on mountain chains and at high latitudes. These ice caps, due to their feedback on albedo, made climate more sensitive to insolation variations than it was in the total absence of ice. The response of the climatic system to such variations during the Early Pleistocene (2000 ka - 1000 ka ago) was still relatively weak, of a fraction of a degree centigrade in global temperature perhaps, in agreement with the quasi-equilibrium results of Section 10.2.

As ice caps passed, about 1000 ka ago, a certain critical size, the unforced system jumped from its stable equilibrium to its stable limit-cycle state (Figures 12.5 and 12.9), increasing dramatically the climate's total variability, to a few degrees centigrade in global temperature. Furthermore, resonant response became possible (see also Oerlemans, 1984, and Sergin, 1979), enhancing abruptly the amplitude of the peak at 100 ka, among others.

Section 12.5. *Combination tones* are a well-known feature of nonlinear oscillations. Perhaps the simplest treatment is given by Landau and Lifshitz (1960, Section 28), who also dispose in one section each of parametric resonance (ibid., Section 27) and nonlinear resonance to additive forcing (ibid., Section 29).

In music, *difference tones* were first discovered by the German organist Sorge (in 1745) and the Italian violinist Tartini (in 1754). Tartini also computed their pitch, but one octave too high. *Sum tones* were discovered by Helmholtz (1885, and references), who developed a rigorous theory.

The lowest difference tones are commonly stronger and thus more easily heard (ibid., p. 152 ff.) than sum tones, which explains the earlier discovery of the former. This situation recurs in ice age theory, where the $f_1 - f_2$ peak is much stronger than the $f_1 + f_2$ peak (Wigley, 1976), the higher peaks being obscured by post-depositional mixing and smoothing of the climatic signal (Pestiaux, 1984; Pestiaux and Berger, 1984).

In Section 12.4 we restricted ourselves to a short mantle relaxation time O(1 ka), which kept the ice-sheet/mantle oscillator (11.34, 12.19) passive. We also chose the nonlinear eigenfrequency of our EBM/ISM oscillator $f_0 = 1/\tau_0$ in the range near 0.1ka^{-1}, rather than $O(10^{-2}\text{ka}^{-1})$. By working with mantle relaxation times O(10 ka) (Birchfield and Grumbine, 1985; Peltier and Hyde, 1984), or with marine ice-sheet relaxation times O(10 ka) (Saltzman *et al.*, 1982), one can obtain nonlinear oscillator frequencies $\tau_0 = O(10^2\text{ka})$, for which the resonance near 100 ka exhibited here (Figure 12.12) is possibly even stronger. But the sum tones $f_1 + f_2 \cong 1/10\text{ka}^{-1}$, $f_1 + f_3 \cong 1/13\text{ka}^{-1}$, $f_2 + f_3 \cong 1/15\text{ka}^{-1}$, and their harmonics will be considerably weaker in such oscillators than in the ones exhibited here.

Given the complexity of the climatic system, it is quite plausible that self-sustained oscillations with periods both larger than 23ka and smaller than 19ka exist. The former will emphasize difference tones, while the latter bring out stronger sum tones of the precessional and obliquity frequencies. In any case, the existence and strength of sum tones predicted here seems to be borne out by careful spectral studies of paleoclimatic proxy records with high sedimentation rates (Pestiaux and Duplessy, 1985, and additional references there for high-frequency peaks near 10ka, 5ka and 2.5ka; D. Thompson, personal communication, 1984).

In celestial mechanics, Laplace observed first the appearance of combination tones (Minorsky, 1962, p. 462), in the presence of two perturbing frequencies of periodic motion (see Section 12.3). The striking analogy between this relatively sophisticated concept in music and in astronomy provides modern support to the Pythagorean idea of "music of the spheres" (Molchanov, 1969, and references therin). We remarked in this context in Section 12.3 that the eccentricity frequency f_4 splits the periodicity of the precessional angle $\tilde{\omega}$, namely 22ka, into the two periodicities of the precessional effect, $e \sin \tilde{\omega}$, approximately equal to 19 ka and 23 ka. In fact, there are four such periodicities, 19.0 ka, 19.2 ka, 22.4 ka and 23.7 ka (Table 12.2), since both f_4 and f_5 contribute to the splitting of the precessional effect. The eccentricity frequencies can be resynthesized, as we saw in Section 12.5, by the climatic system as combination tones of the precessional peaks.

It is interesting that Helmholtz attributed the nonlinearities evidenced by combination tones to the middle ear. Later neurophysiological studies, reviewed by Steele (1979), produced convincing evidence of additional nonlinear characteristics in the ear's response to external stimuli, but also indicated that the organic nature of the nonlinearity was much more subtle than originally postulated, and the matter is still not entirely settled. We hope that the study of the climatic system's oscillatory nonlinearities will lead more rapidly to a consensus than the study of auditory sensations.

Entrainment is discussed in all of the standard references on nonlinear oscillations mentioned before as well as in Hoppensteadt (1976, and references therein). It is a major theme of Winfree's (1980) book, where many beautiful examples and the corresponding references, experimental and theoretical, are given. Rigorous treatments of entrainment appear in Grasman (1986) and Kuramoto (1984).

In the engineering literature one speaks of the frequency of the free, self-sustained oscillation as autoperiodic, of the frequency of the forcing as heteroperiodic, and thus entrainment is also called heterodyning. Another term used, already mentioned in the main text, is frequency locking. This has to be distinguished from phase locking, or synchronization, which sometimes, but far from always, accompanies frequency locking. Still, many of these terms are used interchangeably in various texts and articles, but the exact meaning should be clear from the context. For the climatic oscillations under study, no phase locking seems possible, due to the aperiodic, continuous portion of the spectrum (Section 12.6) and to nonconstant phase differences between various climatic components and various parts of the globe (see below).

A particular case of successive entrainments is that in which two commensurable frequencies $f^{(1)}$ and $f^{(2)}$ are present in the system, say $f^{(1)} = p_1/q_1$ and $f^{(2)} = p_2/q_2$, with p_1, p_2, q_1, q_2 natural numbers and p_1/q_1, p_2/q_2 irreducible. In this case a *Farey sequence* of entrained common frequencies is $f^{(3)} = (p_1 + p_2)/(q_1 + q_2)$, or its reduced form, and so on, with

$$f^{(k)} = \frac{p_{k-1} + p_{k-2}}{q_{k-1} + q_{k-2}}$$

always in irreducible form. Then $f^{(k)}$ lies between $f^{(k-1)}$ and $f^{(k-2)}$, and the distances between successive elements of the sequence decrease, so that in general $f^{(k)}$ tends to an irrational number.

Van der Pol (1946) used Farey sequences to define musical scales. More recently such sequences and other cascades or "devil's staircases" of rational entrainments, leading to a quasi-periodic oscillation, have been found in fluid-dynamical, electronic and other experimental devices (Grasman, 1986, Sections 1.5, 3.3 and 3.4; Libchaber, 1985). As an interesting aside, Farey was a geologist and his number-theoretical statements were proved by Cauchy (see Hardy and Wright, 1954, pp. 36-37).

What happens after the limiting irrational frequency has been reached? Loss of entrainment, or *detrainment* as we called it in the text, represents yet another route to chaos (compare Section 6.6). This route, which is often encountered in experiment, is now actively being studied, analytically and numerically, in the idealized setting of families of maps of the circle to itself depending on two parameters (as opposed to the one-parameter routes reviewed in Section 6.6; see Aronson *et al.*, 1982; Feigenbaum *et al.*, 1982; Guckenheimer and Holmes, 1983, Chapter 7; Ostlund *et al.*, 1982). We remark in passing that system (12.21) possesses, for certain parameter values, the same unstable saddle-focus point already mentioned in Sections 5.3, 5.5, and 6.4 and 6.6 in connection with the Shilnikov homoclinic-orbit scenario of transition to turbulence (compare Figure 12.10 and the discussion of free oscillations in Section 12.4).

In Sections 10.3 and 10.4 we studied the effects of *stochastic perturbations* on equilibrium and quasi-equilibrium models. Effects of both multiplicative and additive noise on oscillatory models are of even greater interest. General references are Arnold (1974, Chapter 11), Jona-Lasinio (1985) and Schuss (1980, Section 7.3). The particular question of synchronization in the presence of noise, of great importance in electronic circuits, is studied extensively by Stratonovich (1967, vol. II, Part 2, especially Chapter 9; see also Kushner, 1984, and Ludwig, 1979).

In the context of climate modeling, this question has been studied numerically by Saltzman *et al.* (1981) and asymptotically near Hopf bifurcation points by Nicolis (1984a,b). The main result is that the probability distribution of the phase can be rendered sufficiently narrow for a simply-periodic, stochastically-perturbed oscillation driven by external, deterministic forcing with near-resonant frequency. Of course, such a result appears rather unlikely for oscillations whose deterministic spectrum contains already a few incommensurable lines and a continuous background.

Section 12.6. The basic *theory of prediction* for a given time series was developed by Wiener (1949; see especially Appendix C, by N. Levinson, for a very readable exposition), with major contributions by Khinchin and Kolmogorov. The Bochner-Khinchin-Wiener theorem, Eq. (10.53), plays a crucial role in this theory, as can be seen also from Hannan (1960).

The mathematical framework considered by Wiener is that of the time series $f(t)$ being given for $-\infty < t \leq 0$, with $f(t)$ noisy, i.e., $f(t) = g(t) + \varepsilon(t)$, where $g(t)$ is deterministic and $\varepsilon(t)$ is a (small) stochastic process. Prediction is finding $g(t)$ for $t > 0$. No connection is made with the dynamics which might generate $g(t)$ or $f(t)$.

Two modern approaches to the same problem are those of sequential estimation theory and of dynamical system theory. For the latter, numerous references have been given in Sections 5.5 and 6.6, and in the notes to Sections 12.1 through 12.3. Particularly relevant is the ergodic theory of dynamical systems (Katok, 1981; Moser, 1973, Chapter 3; Ruelle, 1981b, 1985).

Sequential estimation theory (Kalman, 1960; Kalman and Bucy, 1961; Jazwinski; 1970; Ghil *et al.*, 1981) only requires information about the finite past of the time series, $f(t)$: $t_0 < t \leq 0$, but uses the deterministic, randomly perturbed, dynamics which generated $f(t)$ (compare Eqs. (10.28, 10.46)). The point of view taken here adapts Wiener's classical results to these two modern approaches: the connection between spectra and correlations is fully exploited, while making allowance for the underlying dynamics.

The results presented in this chapter refer to a single forced oscillator, and comments were made about coupling of two such oscillators (cf. Eqs. (12.10, 12.19) and notes to Section 12.4). In fact, even many more global variables than three or four are quite insufficient to describe the changes of climate on Quaternary as well as on shorter time scales. One could conceive of regionally coupled oscillators, with a few variables for the tropical atmosphere and ocean (e.g., McWilliams and Gent, 1978; Cane and Zebiak, 1985) and a few for the high-latitude ice sheets, ocean, bedrock and marine ice in each hemisphere. But eventually it will be necessary to consider an infinity of coupled oscillators, i.e., to consider the coupling of nonlinear partial, rather than ordinary differential equations. One of the most striking phenomena in such systems is phase periodicity and aperiodicity.

Certain examples of *coupled reaction-diffusion equations* have been extensively studied, including the Fitzhugh-Nagumo model of impulse conduction in nerves, the oreganator model of the Belousov-Zhabotinsky (BZ) chemical reaction and the Kuramoto-Siwashinsky model of flame propagation (Howard, 1979; Winfree, 1980, Chapter 13). The easiest to visualize is the BZ reaction, in which the colors ("phases") change in time and space from red to blue and back, in a periodic or aperiodic fashion.

All these systems can be described by space-independent, coupled nonlinear terms, such as those suggested by Eqs. (12.10, 12.19) and embodied in Eqs. (12.1, 12.20), to which diffusive terms are added, such as those in Eq. (10.12) and in the one-dimensional ice-sheet models mentioned in Section 11.4. One may think of an oscillator governed by the system's space-independent part attached to each location, and interacting with the adjacent oscillator by a diffusive flux (see preliminary results of Bhattacharya *et al.*, 1982; Birchfield and Grumbine, 1985; Oerlemans, 1982; and Peltier and Hyde, 1984). Due to the analogous structure of the governing equations, it is quite likely that the same phenomena of phase chaos which appear in the previously cited biological and chemical examples would also appear in a spatially-dependent climatic oscillator, accompanied in this case by a quasi-periodic component. Phase differences in temperature and other meteorological variables between different regions of the globe are of course known to occur on all time scales, including the Quaternary (CLIMAP, 1984).

At this point, the reader is on his or her own. We hope they enjoyed some of the physical ideas put forth in this volume, and will find some of the mathematical tools useful. Beyond lies the elusive scientific truth, for them to discover.

REFERENCES

Ablowitz, M. J. and H. Segur, 1981. Solitons and the Inverse Scattering Transform. Society for Industrial and Applied Mathematics, Philadelphia, PA, 125 pp.

Adhémar, J. A., 1842. Revolutions de la Mer. Paris.

Alfvén, H., 1950. Origin of solar magnetic fields. Tellus, 2, 74-82.

Andrews, D. G., 1983. A finite-amplitude Eliassen-Palm theorem in isentropic coordinates. J. Atmos. Sci., 40, 1877-1883.

Andrews, D. G. and M. E. McIntyre, 1978. An exact theory of nonlinear waves on a Lagrangian mean flow. J. Fluid Mech., 89, 609-646.

Andrews, J. T., S. Funder, C. Hjort and J. Imbrie, 1974. Comparison of the glacial chronology of Eastern Baffin Island, East Greenland, and The Camp Century accumulation record. Geology, 2, 355-358.

Andronov, A. A., A. A. Vitt and S. E. Khaikin, 1966. Theory of Oscillators. Pergamon Press, Oxford/London, 815 pp.

Anufriyev, A. P. and V. M. Fishman, 1982. Magnetic field structure in the two-dimensional motion of a conducting fluid. Geomagn. Aeron., 22, 245-248.

Applegate, J. H., M. R. Douglas, Y. Gürsel, G. J. Sussman and J. Wisdom, 1986. The outer solar system for 200 million years. Astron. J., in press.

Arneodo, A., P. Coullet and C. Tresser, 1982. Oscillators with chaotic behavior: an illustration of a theorem by Shil'nikov. J. Stat. Phys., 27, 171-182.

Arnold, L., 1974. Stochastic Differential Equations: Theory and Applications. J. Wiley, New York, 228 pp.

Arnold, V. I., 1965. Sur la topologie des écoulements stationnaires des fluides. C. R. Acad. Sci. Paris, 261, 17-21.

Arnold, V. I., 1973. Ordinary Differential Equations. MIT Press, Cambridge, Mass./London, England, 280 pp.

Arnold, V. I., 1983. Geometrical Methods in the Theory of Ordinary Differential Equations. Springer-Verlag, New York/Heidelberg/Berlin, 334 pp.

Arnold, V. I., and E. I. Korkina, 1983. Magnetic field strength in a three-dimensional stationary flow of an incompressible fluid. Vest. Mosk. Yh-Ta. Ser. 1. Matem. Mehan., No. 3, 43-46.

Arnold, V. I., Ya. B. Zel'dovich, A. A. Ruzmaikin and D. D. Sokolov, 1981. A magnetic field in a stationary flow with stretching in Riemannian space. Sov. Phys. JETP, 54(6), 1083-1086.

Aronson, D. G., M. A. Chory, G. R. Hall and R. P. McGehee, 1982. Bifurcations from an invariant circle for two-parameter families of maps of the plane: a computer-assisted study. Commun. Math. Phys., 83, 303-354.

Austin, J. F., 1980. The blocking of middle latitude westerly winds by planetary waves. Quart. J. Roy. Meteorol. Soc., 106, 327-350.

Backus, G. E., 1958. A class of self-sustaining dissipative spherical dynamos. Ann. Phys., 4, 372-447.

Backus, G. E., 1975. Gross thermodynamics of heat engines in the deep interior of the Earth. Proc. Nat. Acad. Sci., USA, 72, 1555-1558.

Baer, F. and J. J. Tribbia, 1977. On complete filtering of gravity modes through nonlinear initialization. Mon. Wea. Rev., 105, 1536-1539.

Barcilon, A., and R. L. Pfeffer, 1979. Further calculations of eddy heat fluxes and temperature variances in baroclinic waves using weakly non-linear theory. Geophys. Astrophys. Fluid Dyn., 12, 45-60.

Barenblatt, G. I., G. Iooss and D. D. Joseph (eds.), 1983. Nonlinear Dynamics and Turbulence. Pitman, Boston/London/Melbourne, 356 pp.

Barry, R. G., 1980. Meteorology and climatology of the seasonal sea ice zone. Cold Regions Sci. Tech., 2, 133-150.

Barry, R. G., 1983. Arctic Ocean ice and climate: perspectives on a century of polar research. Ann. Assoc. Amer. Geogr., 73(4), 485-501.

Batchelor, G. K., 1967. An Introduction to Fluid Dynamics. Cambridge Univ. Press, Cambridge/London/New York/Sydney, 615 pp.

Bell, T. L., 1985. Climatic sensitivity and fluctuation-dissipation relations. In Ghil et al. (1985), 424-440.

Benettin, G., L. Galgani, A. Giorgilli and J.-M. Strelcyn, 1980. Lyapunov characteristic exponents for smooth dynamical systems and for Hamiltonian systems; a method for computing all of them. Part 2: Numerical application. Meccanica, 15, 21-30.

Bengtsson, L., 1981. Numerical prediction of atmospheric blocking - A case study. Tellus, 33, 19-42.

Bengtsson, L., M. Ghil and E. Källén (eds.), 1981. Dynamic Meteorology: Data Assimilation Methods. Springer-Verlag, New York/Heidelberg/Berlin, 330 pp.

Benoist, J. P., F. Glangeaud, N. Martin, J. L. Lacoume, C. Lorius and A. Oulahman, 1982. Study of climatic time series by time-frequency analysis. In Proc. ICASSP82, IEEE Press, New York, pp. 1902-1905.

Benzi, R., G. Parisi, A. Sutera and A. Vulpiani, 1982. Stochastic resonance in climatic change. Tellus, 34, 10-16.

Benzi, R., G. Parisi, A. Sutera and A. Vulpiani, 1983. A theory of stochastic resonance in climatic change. SIAM J. Appl. Math., 43, 565-578.

Berger, A. L., 1978. Long-term variations of daily insolation and Quaternary climatic changes. J. Atmos. Sci., 35, 2362-2367.

Berger, A. (ed.), 1981. Climatic Variations and Variability: Facts and Theories. D. Reidel, Dordrecht/Boston/London, 795 pp.

Berger, A., 1984. Accuracy and frequency stability of the Earth's orbital elements during the Quaternary. In Berger et al. (1984), pp. 3-39.

Berger, A. and P. Pestiaux, 1984. Accuracy and stability of the Quaternary terrestrial insolation. In Berget et al. (1984), pp. 83-111.

Berger, A., J. Imbrie, J. Hays, C. Kukla and B. Saltzman (eds.), 1984. Milantovitch and Climate: Understanding the Response to Astronomical Forcing, vols. I & II. D. Reidel, Dordrecht/Boston/Lancaster, 895 pp.

Bhattacharya, K., M. Ghil and I. L. Vulis, 1982. Internal variability of an energy-balance model with delayed albedo effects. J. Atmos. Sci., 39, 1747-1773.

Birchfield, G. E. and R. W. Grumbine, 1985. "Slow" physics of large continental ice sheets and underlying bedrock and its relation to the Pleistocene ice ages. J. Geophys. Res., 90, 11294-11302.

Birchfield, G. E., J. Weertman and A. T. Lunde, 1981. A paleoclimate model of northern hemisphere ice sheets. Quatern. Res., 15, 126-142.

Bjerknes, J., 1969. Atmospheric teleconnections from the equatorial Pacific. Mon. Wea. Rev., 97, 163-172.

Blackmon, M. (ed.), 1978. The General Circulation: Theory, Modeling and Observations. Notes from a Summer Colloquium. NCAR/CQ-6+1978-ASP, National Center for Atmospheric Research, Boulder, Co.

Bland, D. R., 1960. The Theory of Linear Viscoelasticity. Pergamon Press, Oxford/New York, 125 pp.

Bloomfield, P., 1976. Fourier Analysis of Time Series: An Introduction. J. Wiley, New York, 258 pp.

Boville, B., 1980. Amplitude vacillation on an f-plane. J. Atmos. Sci., 37, 1413-1423.

Braginsky, S. I., 1964a. Self-excitation of a magnetic field during the motion of a highly conducting fluid. Sov. Phys. JETP, 20, 726-735.

Braginsky, S. I., 1964b. Magnetohydrodynamics of the Earth's core. Geomagn. Aeron., 4, 698-712.

Braginsky, S. I., 1975. An almost axially-symmetric model of the hydrodynamic dynamo of the Earth. Geomagn. Aeron., 15, 149-156.

Braginsky, S. I., 1976. On the nearly axially-symmetric model of the hydrodynamic dynamo of the Earth. Phys. Earth Planet. Inter., 1, 191-199.

Bretagnon, P., 1974. Termes à longues périodes dans le système solaire. Astron. Astrophys., 30, 141-154.

Bretagnon, P., 1984. Accuracy of long term planetary theory. In Berger et. al. (1984), pp. 41-53.

Bretherton, F. P. and D. B. Haidvogel, 1976. Two-dimensional turbulence above topography. J. Fluid Mech., 78, 129-154.

Brillouin, L., 1953. Wave Propagation in Periodic Structures. Dover, New York, 255 pp.

Broecker, W. S. and J. van Donk, 1970. Insolation changes, ice volumes, and the O^{18} record in deep-sea cores. Rev. Geophys. Space Phys., 8, 169-197.

Brouwer, D. and A. J. J. van Woerkon, 1958. Secular variations of the orbital elements of the principal planets. Astron. Papers Amer. Ephemeris Naut. Almanac, 13, 81-107.

Brouwer, D. and M. Clemence, 1961. Methods of Celestial Mechanics. Academic Press, New York/London, 598 pp.

Budd, W. F. and U. Radok, 1971. Glaciers and other large ice masses. Rep. Prog. Phys., 34, 1-70.

Budyko, M. I. (ed.), 1963. Atlas Teplovogo Balansa. Meshduvedomstv. Geotiz. Komitet, Moscow.

Budyko, M. I., 1969. The effect of solar radiation variations on the climate of the Earth. Tellus, 21, 611-619.

Bullard, E. C., 1955. The stability of the homopolar dynamo. Proc. Camb. Phil. Soc., 51, 744-760.

Bullard, E. C. and H. Gellman, 1954. Homogeneous dynamos and terrestrial magnetism. Phil. Trans. Roy. Soc., A247, 213-278.

Burridge, D. M. and E. Källén (eds.), 1984. Problems and Prospects in Long and Medium Range Weather Forecasting. Springer-Verlag, Berlin, 274 pp.

Busse, F. H., 1975a. A necessary condition for the geodynamo. J. Geophys. Res., 80, 278-280.

Busse, F. H., 1975b. A model of the geodynamo. Geophys. J. Roy. Astron. Soc., 42, 437-459.

Busse, F. H., 1976. Generation of planetary magnetism by convection. Phys. Earth Planet. Inter., 12, 350-358.

Busse, F. H., 1978. Introduction to the theory of geomagnetism. In Roberts and Soward (1978), pp. 361-388.

Busse, F. H., 1981. Transition to turbulence in Rayleigh-Bénard convection. In Swinney and Gollub (1981), pp. 97-139.

Buys, M. and M. Ghil, 1984. Mathematical methods of celestial mechanics illustrated by simple models of planetary motion. In Berger et al. (1984), pp. 55-82.

Buzyna, G., R. L. Pfeffer and R. Kung, 1984. Transition to geostrophic turbulence in a rotating differentially heated annulus of fluid. J. Fluid Mech., 145, 377-403.

Buzzi, A., A. Trevisan and A. Speranza, 1984. Instabilities of a baroclinic flow related to topographic forcing. J. Atmos. Sci., 41, 637-650.

Cacuci, D. G. and M. C. G. Hall, 1984. Efficient estimation of feedback effects with application to climate models. J. Atmos. Sci., 41, 2063-2068.

Cane, M. A., 1983. Oceanographic events during El Niño. Science, 222, 1189-1195.

Cane, M. A. and S. E. Zebiak, 1985. A theory for El Niño and the Southern Oscillation. Science, 228, 1085-1087.

Carr, J., 1981. Applications of Centre Manifold Theory. Springer-Verlag, New York, 142 pp.

Cathles, L. M., 1975. The Viscosity of the Earth's Mantle. Princeton Univ. Press, Princeton, N. J., 388 pp.

Chamberlain, J. W., 1978. Theory of Planetary Atmospheres: An Introduction to Their Physics and Chemistry. Academic Press, New York/San Francisco/London, 330 pp.

Chandrasekhar, P., 1943. Stochastic problems in physics and astronomy. Rev. Mod. Phys., 15, 1-89; reprinted in Wax (1954), pp. 3-91.

Chandrasekhar, S., 1961. Hydrodynamic and Hydromagnetic Stability. Clarendon Press, Oxford, 654 pp.

Chang, J. (ed.), 1977. General Circulation Models of the Atmosphere. Methods Comput. Phys., 17, Academic Press, New York, 337 pp.

Charney, J. G., 1947. The dynamics of long waves in a baroclinic westerly current. J. Meteorol., 4, 135-163.

Charney, J. G., 1971. Geostrophic turbulence. J. Atmos. Sci., 28, 1087-1095.

Charney, J. G., 1973. Planetary fluid dynamics. In Morel (1973), pp. 97-351.

Charney, J. G. and J. G. DeVore, 1979. Multiple flow equilibria in the atmosphere and blocking. J. Atmos. Sci., 36, 1205-1216.

Charney, J. G. and P. G. Drazin, 1961. Propagation of planetary-scale disturbances from the lower into the upper atmosphere. J. Geophys. Res., 66, 83-109.

Charney, J. G. and D. M. Straus, 1980. Form-drag instability, multiple equilibria and propagating planetary waves in baroclinic, orographically forced, planetary wave systems. J. Atmos. Sci., 37, 1157-1176.

Charney, J. G. J. Shukla and K. C. Mo, 1981. Comparison of a barotropic blocking theory with observation. J. Atmos. Sci., 38, 762-779.

Childress, S., 1969. A class of solutions of the magnetodydrodynamic dynamo problem. In The Application of Modern Physics to the Earth and Planetary Interiors, S. K. Runcorn, ed., Wiley-Interscience, New York, pp. 629-648.

Childress, S., 1970. New solutions of the kinematic dynamo problem. J. Math. Phys., 11, 3063-3076.

Childress, S., 1977. Convective dynamos. In Problems of Stellar Convection, E. A. Spiegel and J. P. Zahn, eds., Lecture Notes in Physics 71, Springer-Verlag, New York, 195-224.

Childress, S., 1978. Notes of the Summer Study Program on Dynamo Models of Geomagnetism. Woods Hole Oceanographic Institute, Tech. Rep. WHOI-78-67, Woods Hole, Mass.

Childress, S., 1979. Alpha-effect in flux ropes and sheets. Phys. Earth Planet. Inter., 20, 172-180.

Childress, S., 1983a. Stationary induction by intermittent velocity fields. In Soward (1983), pp. 81-90.

Childress, S., 1983b. Macrodynamics of spherical dynamos. In Soward (1983), pp. 245-257.

Childress, S. and A. M. Soward, 1972. Convection-driven hydromagnetic dynamo, Phys. Rev. Lett., 29, 837-839.

Childress, S. and A. M. Soward, 1985. On the rapid generation of magnetic fields. In Chaos in Astrophysics, J. R. Buchler, G. M. Purdang and E. A. Spiegel (eds.), D. Reidel, Dordrecht/Boston,/London, pp. 223-244.

Clement, B. M. and D. V. Kent, 1984. A detailed record of the lower Jaramillo polarity transition from a Southern Hemisphere, deep-sea sediment core. J. Geophys. Res., 89B, 1049-1058.

CLIMAP Project Members, 1984. The last interglacial ocean. Quatern. Res., 21, 123-224.

Colbeck, S. C. (ed.), 1980. Dynamics of Snow and Ice Masses. Academic Press, New York, 468 pp.

Constantin, P. and C. Foias, 1985. Global Lyapunov exponents, Kaplan-Yorke formulas and the dimension of attractor for 2D Navier-Stokes equations. Comm. Pure Appl. Math., 38, 1-27.

Corby, G. A. (ed.), 1969. The Global Circulation of the Atmosphere. Royal Meteorological Society, London, 257 pp.

Coullet, P. H. and E. A. Spiegel, 1983. Amplitude equations for systems with competing instabilities. SIAM J. Appl. Math., 43, 776-821.

Courant, R. and D. Hilbert, 1953/1962. Methods of Mathematical Physics, vols. I (1953) & II (1962). Wiley-Interscience, New York/London/Sydney, 560 pp. + 830 pp.

Cowling, T. G., 1934. The magnetic field of sunspots. Mon Nat. Roy. Astron. Soc., 94, 39-48.

Cowling, T. G., 1957. Magnetodydrodynamics. Wiley-Interscience, New York, 115 pp.

Cox, A., 1969. Geomagnetic reversals. Science, 163, 237-245.

Crafoord, C. and E. Källén, 1978. A note on the condition for existence of more than one steady state solution in Budyko-Sellers type models. J. Atmos. Sci., 35, 1123-1125.

Craig, R. A., 1965. The Upper Atmosphere. Meteorology and Physics. Academic Press, New York/San Francisco/London.

Crittenden, M. D., Jr., 1963. effective viscosity of the earth derived from isostatic loading of Pleistocene Lake Bonneville. J. Geophys. Res., 68, 5517-5530.

Croll, J., 1867. On the excentricity of the earth's orbit, and its physical relations to the glacial epoch. Philos. Mag., 33, 119-131.

Crowley, T. J., 1983. The geologic record of climatic change. Rev. Geophys. Space Phys., 21, 828-877.

Cuong, P. G. and F. H. Busse, 1981. Generation of magnetic fields by convection in a rotating sphere I. Phys. Earth Planet. Inter., 24, 272-283.

Daley, R., 1981. Normal mode initialization. In Bengtsson et al. (1981), pp. 77-109.

Dalfes, H. N., S. H. Schneider and S. L. Thompson, 1983. Numerical experiments with a stochastic zonal climate model. J. Atmos. Sci., 40, 1648-1658.

Danby, J. M. A., 1962. Fundamentals of Celestial Mechanics. MacMillan, New York, 348 pp.

Dansgaard, W., 1981. Palaeo-climatic studies on ice cores. In Berger (1981), pp. 193-206.

Dansgaard, W., H. B. Clausen, N. Gundestrup, C. U. Hammer, S. F. Johnsen, P. M. Kristindottir and N. Reeh, 1982. A new Greenland deep ice core. Science, 218, 1273-1277.

Davey, M., 1981. A quasi-linear theory for rotating flow over topography. Part II: β-plane annulus. J. Fluid Mech., 99, 267-292.

Dee, D. and M. Ghil, 1984. Boolean difference equations, I: formulation and dynamic behavior. SIAM J. Appl. Math., 44, 111-126.

Dee, D. P., S. E. Cohn, A. Dalcher, and M. Ghil, 1985. An efficient algorithm for estimating noise covariances in distributed systems. IEEE Trans. Autom. Control, AC-30, 1057-1065.

Dole, R. M. and N. D. Gordon, 1983. Persistent anomalies of the extratropical Northern Hemisphere wintertime circulation: geographical distribution and regional persistence characteristics. Mon. Wea. Rev., 111, 1567-1586.

Drazin, P. G., 1978. Variations on a theme of Eady. In Roberts and Soward (1978), pp. 139-169.

Drazin, P. G. and D. H. Griffel, 1977. On the branching structure of diffusive climatological models. J. Atmos. Sci., 34, 1858-1867.

Driver, R. D., 1977. Ordinary and Delay Differential Equations. Springer-Verlag, New York, 501 pp.

Duffing, G., 1918. Erzwiengene Schwingungen bei veränderlicher Eigenfrequenz. F. Vieweg, Braunschweig.

Duplessy, J. -C., 1978. Isotope studies. In Gribbin (1978), pp. 46-67.

Dutton, J. A., 1976. The Ceaseless Wind, An Introduction to the Theory of Atmospheric Motion. McGraw-Hill, New York, 579 pp.

Duvaut, G. and J. L. Lions, 1976. Inequalities in Mechanics and Physics. Springer-Verlag, New York, 397 pp.

Dziewonski, A. M. and D. L. Anderson, 1984. Seismic tomography of the Earth's interior. American Scientist, Sept./Oct. issue, 483-494.

Eady, E. T., 1949. Long waves and cyclone waves. Tellus, 1, 33-52.

Eady, E. T., 1950. The cause of the general circulation of the atmosphere. Centennial Proc. Roy. Meteorol. Soc., London, pp. 156-172.

Eckmann, J. P., 1981. Roads to turbulence in dissipative dynamical systems. Rev. Mod. Phys., 53, 643-654.

Egger, J., 1978. Dynamics of blocking highs. J. Atmos. Sci., 35, 1788-1801.

Egger, J. and H. D. Schilling, 1983. On the theory of the long-term variability of the atmosphere. J. Atmos. Sci., 40, 1073-1085.

Einstein, A., 1905. On the movement of small particles suspended in a stationary liquid demanded by the molecular-kinetic theory of heat. Ann. d. Physik (Leipzig), 17, 549 ff., reprinted in Investigations on the Theory of the Brownian Movement, five articles by A. Einstein, R. Fürth (ed.) and A. D. Cowper (transl.), 1956, Dover Publ., New York, 122 pp.

Eliassen, A. and E. Palm, 1960. On the transfer of energy in stationary mountain waves. Geophys. Publ., 22, No. 3, 23 pp.

Ellsaesser, H. W., 1966. Expansion of hemispheric meteorological data in antisymmetric surface spherical harmonic (Laplace) series. J. Appl. Meterol., 5, 263-276.

Elsasser, W. M., 1946. Induction effects in terrestrial magnetism, I. Theory Phys. Rev., 69, 106-116.

Eltayeb, I. A., 1972. Hydromagnetic convection in a rapidly rotating fluid layer. Proc. Roy. Soc., A326, 229-254.

Eltayeb, I. A. and P. H. Roberts, 1970. On the hydromagnetics of rotating fluids. Astrophys. J., 162, 699-701.

Emiliani, C., 1955. Pleistocene temperatures. J. Geol., 63, 538-578.

Emiliani, C. and J. Geiss, 1957. On glaciations and their causes. Geol. Rundschau, 46(2), 576-601.

Epstein, S., R. Buchsbaum, H. Lowenstam and H. C. Urey, 1951. Carbonate-water isotopic temperature scale. Geol. Soc. Amer. Bul., 62, 417-425.

Errico, R., 1982. The strong effects of non-quasigeostrophic dynamic processes on atmospheric energy spectra. J. Atmos. Sci., 39, 961-968.

Faegre, A., 1972. An intransitive model of the earth-atmosphere-ocean system. J. Appl. Meteorol., 11, 4-6.

Farmer, J. D., 1982. Chaotic attractors of an infinite dimensional dynamical system. Physica, 4D, 366-393.

Farmer, D., J. Hart and P. Weidman, 1982. A phase space analysis of baroclinic flow. Phys. Lett., 91A, 22-24.

Farrell, B. F., 1983. Pulse asymptotics of three-dimensional baroclinic waves. J. Atmos. Sci., 40, 2202-2210.

Fautrelle, Y. and S. Childress, 1982. Convective dynamos with intermediate and strong fields. Geophys. Astrophys. Fluid Dyn., 22, 235-279.

Feigenbaum, M. J., 1978. Quantitative universality for a class of nonlinear transformations. J. Stat. Phys., 19, 25-52.

Feigenbaum, M. J., L. P. Kadanoff and S. J. Shenker, 1982. Quasiperiodicity in dissipative systems: a renormalization group analysis. Physica, 5D, 370-386.

Ferrel, W., 1856. An essay on the winds and the currents of the ocean. Nashville J. Medicine & Surgery, 11, 287-301. Reprinted in Prof. Papers Signal Serv., 12 (1882), 7-19.

Fjørtoft, R., 1950. Application of integral theorems in deriving criteria of stability for laminar flows and for the baroclinic circular vortex. Geophys. Publ., 17, No. 6, 52 pp.

Fjørtoft, R., 1953. On the changes in the spectral distribution of kinetic energy for two-dimensional nondivergent flow. Tellus, 5, 225-230.

Flint, R. F., 1971. Glacial and Quaternary Geology. J. Wiley, New York, 892 pp.

Fraedrich, K., 1978. Structural and stochastic analysis analysis of a zero-dimensional climate system. Quart. J. Roy. Meteorol. Soc., 104, 461-474.

Franceschini, V., C. Tebaldi and F. Zironi, 1984. Fixed-point limit behavior of N-mode truncated Navier-Stokes equations as N increases. J. Stat. Phys., 35, 387-397.

Frederiksen, J., 1983. Disturbances and eddy fluxes in Northern Hemisphere flows: instability of three-dimensional January and July flows. J. Atmos. Sci., 40, 836-855.

Frederiksen, J. S. and B. L. Sawford, 1981. Topographic waves in nonlinear and linear spherical barotropic models. J. Atmos. Sci., 38, 69-86.

Friedman, B., 1956. Principles and Techniques of Applied Mathematics. J. Wiley, New York, 315 pp.

Friedrichs, K. O., 1965. Perturbation of Spectra in Hilbert Space. Lecture Notes in Applied Mathematics, Vol. III, American Mathematical Society, Providcnce, R. I., 178 pp.

Frisch, M., 1980. Man in the Holocene. Harcourt Brace Jovanovich, New York, 120 pp.

Fultz, D., R. R. Long, G. V. Lwens, W. Bohan, R. Kaylor and J. Weil, 1959. Studies of Thermal Convection in a Rotating Cylinder with Some Pl Implications for Large-Scale Atmospheric Motions. Meteorol. Monogr., 4, American Meteorological Society, Boston, Mass., pp. 1-104.

Gall, R., R. Blakenslee and R. C. J. Somerville, 1979. Baroclinic instability and the selection of the zonal scale of the transient eddies of middle latitudes. J. Atmos. Sci., 36, 767-784.

Galloway, D. J., M. R. E. Proctor and N. O. Weiss, 1978. Magnetic flux ropes and convection. J. Fluid Mech., 87, 243-262.

Gardiner, C. W., 1983. The escape time in nonpotential systems. J. Stat. Phys., 30, 157-177.

Gaspard, P., R. Kapral and G. Nicolis, 1984. Bifurcation phenomena near homoclinic systems: a two-parameter analysis. J. Stat. Phys., 35, 697-727.

Gates, W. L. (ed.), 1979. Report of the JOC Study Conference on Climate Models: Performance, Intercomparison and Sensitivity Studies. GARP Publ. Series, No. 22, vols. I & II, World Meteorological Organization, Geneva, Switzerland, 1049 pp.

Gent, P. R. and J. C. McWilliams, 1982. Intermediate model soluttions to the Lorenz equations: strange attractors and other phenomena. J. Atmos. Sci., 39, 3-13.

Ghil, M. 1975. Steady-State Solutions of a Diffusive Energy-Balance Climate Model and Their Stability. Report IMM-410, Courant Institute of Mathematical Sciences, New York University, New York, 74 pp.

Ghil, M. 1976. Climate stability for a Sellers-type model. J. Atmos. Sci., 33, 3-20.

Ghil, M., 1978. Numerical methods in geophysical fluid dynamics. In Roberts and Soward (1978), pp. 499-521.

Ghil, M., 1981a. Energy-balance models: an introduction. In Berger (1981), pp. 461-480.

Ghil, M., 1981b. Internal climatic mechanisms participating in glaciation cycles. In Berger (1981), pp. 539-557.

Ghil, M., 1981c. Comments on "Seasonal simulation as a test for uncertainties in the parametrization of a Budyko-Sellers zonal climate model" (by D. G. Warren and S. H. Schneider). J. Atmos. Sci., 38, 666-669.

Ghil, M., 1984a. Climate sensitivity, energy balance models, and oscillatory climate models. J. Geophys. Res., 89D, 1280-1284.

Ghil, M., 1984b. Formal conceptual models of climatic change. Terra Cognita, 4, 336.

Ghil, M., 1985a. Future possibilities in objective analysis and data assimilation for atmospheric dynamics. In Proceedings of the First National Workshop on the Global Weather Experiment, U.S. Committee for GARP, National Academy Press, Washington, D.C., 794-802.

Ghil, M., 1985b. Theoretical climate dynamics: an introduction. In Ghil et al. (1985), pp. 347-402.

Ghil, M., 1986. Dynamics, statistics and predictability of planetary flow regimes. In Irreversible Phenomena and Dynamical Systems Analysis in the Geosciences, G. Nicolis (ed.), D. Reidel, Dordrecht/Boston/Lancaster, in press.

Ghil, M. and K. Bhattacharya, 1979. An energy-balance model of glaciation cycles. In Gates (1979), pp. 886-916.

Ghil, M. and H. LeTreut, 1981. A climate model with cryodynamics and geodynamics. J. Geophys. Res., 86, 5262-5270.

Ghil, M. and J. Tavantzis, 1983. Global Hopf bifurcation in a simple climate model. SIAM J. Appl. Math., 43, 1019-1041.

Ghil, M. and B. Saltzman, 1984. Oscillator models of climatic change. In Berget et al. (1984), pp. 859-866.

Ghil, M. and A. Mullhaupt, 1985. Boolean delay equations, II: periodic and aperiodic solutions. J. Stat. Phys., 41, 125-173.

Ghil, M., S. Cohn, J. Tavantzis, K. Buke and E. Isaacson, 1981. Applications of estimation theory to numerical weather prediction. In Bengtsson et al. (1981), pp. 139-224.

Ghil, M., R. Benzi and G. Parisi (eds.), 1985. Turbulence and Predictability in Geophysical Fluid Dynamics and Climate Dynamics. North-Holland, Amsterdam/Oxford/New York/Tokyo, 449 pp.

Ghil, M., A. Mullhaupt and P. Pestiaux, 1986. Deep water formation and Quaternary glaciations. Climate Dynamics, submitted.

Gilbert, F. and A. M. Dziewonski, 1975. An application of normal mode theory to the retrieval of structural parameters and source mechanisms from seismic spectra. Philos. Trans. Roy. Soc. (London), A278, 187-269.

Gill, A. E., 1982. Atmosphere-Ocean Dynamics, Academic Press, 662 pp.

Gilman, P., 1983. Dynamically consistent nonlinear dynamos driven by convection in a rotating spherical shell, II. Dynamos with cycles and strong feedbacks. Astrophys. J. Supp. Series, 53, 243-268.

Gilman, P. A. and J. Miller, 1981. Dynamically consistent nonlinear dynamos driven by convection in a rotating spherical shell. Astrophys. J. Supp. Series, 46, 211-238.

Glen, J. W., 1955. The creep of polycrystalline ice. Proc. Roy. Soc. (London), A228, 519-538.

Godske, C. L., T. Bergeron, J. Bjerknes and R. C. Bundgaard, 1957. Dynamic Meteorology and Weather Forecasting. Carnegie Institution, Washington, D. C., 800 pp.

Goldstein, H., 1980. Classical Mechanics, 2nd ed., Addison-Wesley, Reading, Mass., 672 pp.

Goldstein, S., 1960. Lectures on Fluid Mechanics. Wiley-Interscience, London/New York, 309 pp.

Gollub, J. P. and S. V. Benson, 1980. Many routes to turbulent convection. J. Fluid Mech., 100, 449-470.

Goody, R. M. and A. R. Robinson, 1966. A discussion of the deep circulation of the atmosphere of Venus. Astrophys. J., 146, 339-353.

Gottlieb, D. and S. A. Orszag, 1977. Numerical Analysis of Spectral Methods: Theory and Applications. Society for Industrial and Applied Mathematics, Philadelphia, Pa., 172 pp.

Grant, E., 1961. Nicole Oresme and the commensurability or incommensurability of the celestial motions. Arch. Hist. Exact Sci., 1 (1960-1962), 420-458.

Grasman, 1986. Mathematical Methods for Relaxation Oscillations in Physics and Biology. Springer-Verlag, New York, to appear.

Grebogi, C., E. Ott and J. A. Yorke, 1983. Crises, sudden changes in chaotic attractors, and transient chaos. Physica, 7D, 181-200.

Green, J. S. A., 1970. Transfer properties of the large-scale eddies and the general circulation of the atmosphere. Quart. J. Roy. Meteorol. Soc., 96, 157-185.

Greenspan, H., 1968. The Theory of Rotating Fluids. Cambridge, Univ. Press, Cambridge/London/New York/Sydney, 328 pp.

Gribbin, J. (eds.), 1978. Climatic Change. Cambridge Univ. Press, Cambridge/New York/Melbourne, 280 pp.

Gubbins, D., 1973. Numerical solutions of the kinematic dynamo problem, Phil. Trans. Roy. Soc., A274, 493-521.

Gubbins, D., 1977. Energetics of the Earth's core. J. Geophys., 43, 453-464.

Guckenheimer, J., 1982. Noise in chaotic systems. Nature, 298, 358-361.

Guckenheimer, J. and G. Buzyna, 1983. Dimension measurements for geostrophic turbulence. Phys. Rev. Lett., 51, 1438-1441.

Guckenheimer, J. and P. Holmes, 1983. Nonlinear Oscillations, Dynamical Systems, and Bifurcations of Vector Fields. Springer-Verlag, New York/Berlin/Heidelberg/Tokyo, 453 pp.

Gustafsson, N., 1981, A review of methods for objective analysis. In Bengtsson et al. (1981), pp. 17-76.

Hadley, G., 1735. Concerning the cause of the general trade-winds. Phil. Trans., 29, 58-62.

Hager, B. H., R. W. Clayton, M. A. Richards, R. P. Comer and A. M. Dziewonski, 1985. Lower mantle heterogeneity, dynamic topography and the geoid. Nature, 313, 541-545.

Hagihara, Y., 1970-1975. Celestial Mechanics, vols. I-IV. MIT Press, Cambridge, Mass., and Japan Society for the Promotion of Sciences, Tokyo.

Haidvogel, D. B. and I. M. Held, 1980. Homogeneous quasi-geostrophic turbulence driven by a uniform temperature gradient. J. Atmos. Sci., 37, 2644-2660.

Hale, J. K., 1977. Theory of Functional Differential Equations. Springer-Verlag, New York, 365 pp.

Halem, M., E. Kalnay, W. E. Baker and R. Atlas, 1982. An assessment of the FGGE satellite observing system during SOP-1. Bull. Amer. Math. Soc., 63, 407-426.

Halfar, P., 1983. On the dynamics of ice sheets 2. J. Geophys. Res., 88C, 6043-6051.

Haltiner, G. J. and R. T. Williams, 1980. Numerical Prediction and Dynamic Meteorology, 2nd ed., J. Wiley, New York, 471 pp.

Hannan, E. J., 1960. Time Series Analysis. Methuen, London/Barnes & Noble, New York, 152 pp.

Hardy, G. H. and E. M. Wright, 1954. An Introduction to the Theory of Numbers, 3rd ed., Clarendon Press, Oxford, 419 pp.

Hart, J. E., 1979. Finite amplitude baroclinic instability. Ann. Rev. Fluid Mech., 11, 147-172.

Hart, J. E., 1984. Laboratory experiments on the transition to baroclinic chaos. In Holloway and West (1984), pp. 369-375.

Hartman, P., 1973. Ordinary Differential Equations. J. Wiley/P. Hartman, Baltimore, 612 pp.

Haskell, N. A., 1935. The motion of a viscous fluid under a surface load, J. Physics, (N.Y.), 6, 265-269.

Hasselmann, K., 1961. On the non-linear energy transfer in a gravity-wave spectrum. J. Fluid Mech., 12, 481-500.

Hasselmann, K., 1976. Stochastic climate models, part I. Theory. Tellus, 28, 473-485.

Hayashi, C., 1964. Nonlinear Oscillations in Physical Systems. McGraw-Hill, New York, 392 pp.

Hays, J. D., J. Imbrie and N. J. Shackleton, 1976. Variations in the earth's orbit: pacemaker of the ice ages. Science, 194, 1121-1132.

Heirtzler, J. R., G. O. Dickson, E. M. Herron, W. C. Pitmann II and X. LePichon, 1968. Marine magnetic anomalies, geomagnetic field reversals, and motions of the ocean floor and continents. J. Geophys. Res., 73, 2119-2136.

Held, I. M., 1982. Climate models and the astronomical theory of the ice ages. Icarus, 50, 449-461.

Held, I. M., 1983. Stationary and quasi-stationary eddies in the extratropical troposphere: theory. In Hoskins and Pearce (1983), pp. 127-168.

Held, I. M. and M. J. Suarez, 1974. Simple albedo feedback models of the icecaps. Tellus, 36, 613-628.

Held, I. M., D. I. Linder and M. J. Suarez, 1981. Albedo feedbacks and the meridional structure of the effective heat diffusivity, and climate sensitivity: results from dynamic and diffusive models. J. Atmos. Sci., 38, 1911-1927.

Held, I. M., R. L. Panetta and R. T. Pierrehumbert, 1985. Stationary external Rossby waves in vertical shear. J. Atmos. Sci., 42, 865-883.

Helmholtz, H. L. F., 1885. On the Sensations of Tone as a Physiological Basis for the Theory of Music, 2nd English ed., Reprinted by Dover, New York (1954), 576 pp.

Hénon, M., 1966. La topologie des lignes de courant dans un cas particulier. C.R. Acad. Sci. Paris, 262, 312-314.

Henrici, P., 1963. Error Propagation for Difference Methods. J. Wiley, New York/London, 73 pp.

Hewitt, J. M., D. P. McKenzie and N. O. Weiss, 1975. Dissipative heating in convective flows. J. Fluid Mech., 68, 721-738.

Hibler, W. D., III, 1980. Sea ice growth, drift, and decay. In Colbeck (1980), pp. 141-209.

Hide, R., 1977. Experiments with rotating fluids. Quart. J. Roy. Meteorol. Soc., 103, 1-28; reprinted in Robert and Soward (1978), pp. 1-28.

Hide, R. and P. H. Roberts, 1961. The origin of the main geomagnetic field. in Physics and Chemistry of the Earth, Vol. 4, edited by Ahrens, L. H., K, Rankama, and S. K. Runcorn, Pergamon Press, London, pp. 25-98.

Hide, R. and P. J. Mason, 1975. Sloping convection in a rotating fluid. Adv. Physics, 24, 47-100.

Hide, R., N. T. Birch, L. V. Morrison, D. J. Shea and A. A. White, 1980. Atmospheric angular momentum fluctuations and changes in the length of day. Nature, 286, 114-117.

Hignett, P., 1983. Spectral characteristics of amplitude vacillation in a differentially heated rotating fluid annulus. Unpublished manuscript, 26 pp. + 10 figs.

Holloway, G., 1978. A spectral theory of nonlinear barotropic motion above irregular topography. J. Phys. Oceanogr., 8, 414-427.

Holloway, G. and B. J. West (eds.), 1984. Predictability of Fluid Motions. American Institute of Physics, New York, 612 pp.

Holmes, P. J. and D. A. Rand, 1976. The bifurcations of Duffing's equation: an application of catastrophe theory. J. Sound Vib., 44, 237-253.

Holton, J. R., 1975. The Dynamic Meteorology of the Stratosphere and Mesosphere. Meteorol. Monogr., 37, American Meteorological Society, Boston, Mass.

Holton, J. R., 1979. An Introduction to Dynamic Meteorology, 2nd ed. Academic Press, New York/San Francisco/London, 391 pp.

Hoppensteadt, F. C. (ed.), 1979a. Nonlinear Oscillations in Biology. American Mathematical Society, Providence, R. I., 253 pp.

Hoppensteadt, F. C., 1979v. Computer studies of nonlinear oscillations. In Hoppensteadt (1979a), pp. 131-139.

Hoskins, B. J., 1982. The mathematical theory of frontogenesis. Ann. Rev. Fluid Mech., 14, 131-151.

Hoskins, B. J., 1983. Dynamical processes in the atmosphere and the use of models. Quart. J. Roy. Meteorol. Soc., 109, 1-21.

Hoskins, B. J. and R. P. Pearce (eds.), 1983. Large-Scale Dynamic Processes in the Atmosphere. Academic Press, London/New York, 397 pp.

Howard, L. N., 1979. Nonlinear Oscillations. In Hoppensteadt (1979), pp. 1-67.

Hutter, K., 1983. Theoretical Glaciology, Material Science of Ice and the Mechanics of Glaciers and Ice Sheets. D. Reidel, Dordrecht/Boston/Tokyo, 500 pp.

Imbrie, J. and K. P. Imbrie, 1979. Ice Ages: Solving the Mystery. Enslow Publ., Short Hills, N. J., 224 pp.

Imbrie, J. and J. L. Imbrie, 1980. Modeling the climatic response to orbital variations, Science, 207, 943-953.

Iooss, G. and D. D. Joseph, 1980. Elementary Stability and Bifurcation Theory. Springer-Verlag, New York/Heidelberg/Berlin, 286 pp.

Isaacson, E. and H. B. Keller, 1966. Analysis of Numerical Methods. J. Wiley, New York/London/Sydney, 541 pp.

Itoh, H., 1985. The formation of quasi-stationary waves from a viewpoint of bifurcation theory. J. Atmos. Sci., 42, 917-932.

Jacobowitz, H., R. J. Tighe and the NIMBUS 7 ERB Experiment Team, 1984. The Earth radiation budget derived from the NIMBUS 7 ERB experiment. J. Geophys. Res., 89, 4997-5010.

Jazwinski, A. H., 1970. Stochastic Processes and Filtering Theory. Academic Press, New York, 376 pp.

Jenkins, G. M. and D. G. Watts, 1968. Spectral Analysis and its Applications. Holden-Day, San Francisco, 525 pp.

Joint Organizing Committee, 1975. The Physical Basis of Climate and Climate Modelling. GARP Publ. Ser., No. 16, World Meteorological Organization, Geneva, Switzerland, 265 pp.

Jona-Lasinio, G., 1985. Small stochastic perturbations of dynamical systems. In Ghil et al. (1985), pp. 29-41.

Jones, A. R., 1982. In Soward (1983), pp. 159-170.

Jones, C. A., N. O. Weiss and F. Cattaneo, 1985. Nonlinear dynamos: a complex generalization of the Lorenz equations. Physica, 14D, 161-176.

Jouzel, J. and L. Merlivat, 1984. Deuterium and oxygen-18 in precipitation: modeling of the isotopic effects at snow formation. J. Geophys. Res., 89D, 11749-11757.

Kadanoff, L. P., 1983. Roads to chaos. Physics Today, 1983(12), 46-53.

Källén, E., 1984. Bifurcation mechanisms and atmospheric blocking. In Burridge and Källén (1984), pp. 229-263.

Källén, E., C. Crafoord and M. Ghil, 1979. Free oscillations in a climate model with ice-sheet dynamics. J. Atmos. Sci., 36, 2292-2303.

Kalman, R. E., 1960. A new approach to linear filtering and prediction problems. Trans. ASME, 82D, 35-45.

Kalman, R. E., R. S. Bucy, 1961. New results in linear filtering and prediction theory. Trans. ASME, Ser. D: J. Basic Eng., 83, 95-108.

Kalnay-Rivas, E., 1973. Numerical models of the circulation of the atmosphere of Venus. J. Atmos. Sci., 30, 763-779.

Kalnay, E. and K. C. Mo, 1986. Mechanistic experiments to determine the origin of Southern Hemisphere stationary waves. Adv. Geophys., 29, pp. 415-442.

Kaplan, J. L. and J. A. Yorke, 1979. Preturbulence: a regime observed in a fluid flow model of Lorenz. Commun. Math. Phys., 67, 93-108.

Katok, A. B., 1981. Dynamical systems with a hyperbolic structure. In Three Papers on Dynamical Systems, J. Szücs (ed.), AMS Transl. (Ser 2), 116, American Mathematical Society, Providence, R. I., pp. 43-95.

Keller, J. B., 1969. Survey of the theory of turbulence. In Contemporary Physics, Vol. 1, International Atomic Energy Agency, Vienna.

Kennett, R. G., 1974. Convectively-driven dynamos. In Notes of the Summer Study Program in Geophysical Fluid Dynamics, Woods Hole Oceanographic Institution, Woods Hole, Mass., 94-117.

Köppen, W., 1923. Die Klimate der Erde. Berlin/Leipzig.

Kraichnan, R. H., 1976. Diffusion of weak magnetic fields by turbulence. J. Fluid Mech., 75, 657-676.

Krause, F., and P. H. Roberts, 1973. Some problems of mean-field electrodynamics. Astrophys. J., 181, 977-992.

Krishnamurti, T. N. and D. Subrahmanyan, 1982. The 30-50 day mode at 850 mb during MONEX. J. Atmos. Sci., 39, 2088-2095.

Kuhn, T. S., 1970. The Structure of Scientific Revolutions, 2nd ed. Univ. of Chicago Press, Chicago/London, 210 pp.

Kukla, J., 1969. The cause of the Holocene climate change. Geol. Mijnbouw, 48(3), 307-334.

Kuramoto, Y., 1984. Chemical Oscillations, Waves, and Turbulence. Springer-Verlag, Berlin/Heidelberg/New York/Tokyo, 156 pp.

Kushner, H. J., 1984. Robustness and approximation of escape times and large deviations estimates for systems with small noise effects. SIAM J. Appl. Math., 44, 160-182.

Lamb, H. H., 1972/1977. Climate: Present, Past and Future, vol. 1: Fundamentals and Climate Now (1972), 613 pp; vol. 2: Climatic History and the Future (1977), 835 pp. Methuen, London/Barnes & Noble, New York.

Landau, L. D. and E. M. Lifschitz, 1959. Fluid Mechanics, Addison-Wesley, Reading, Mass., 536 pp.

Landau, L. D. and E. M. Lifshitz, 1960. Mechanics. Pergamon Press, Oxford/London/New York/Paris, 165 pp.

Landsberg, H., 1967. Physical Climatology, 2nd ed. Gray, DuBois, Pa.

Lanford, O. E., 1982. The strange attractor theory of turbulence. Ann. Rev. Fluid Mech., 14, 347-364.

Laplace, P. S., 1787. Mémoire sur les inégalités séculaires des planetes et des satellites. Mém. Acad. Roy. Sci. Paris, pp. 1-50.

Laplace, P. S., 1796. Exposition du Système du Monde; 6th ed. (1835), Bachelier, Paris, 509 pp.; reprinted in Oeuvres Complètes de Laplace, vol. 6, Gauthier-Villars, Paris, 1894.

Larmor, J., 1919. How could a rotating body such as the sun become a magnet? Rep. Brit. Assoc. Adv. Sci., 1919, 159-160.

Legras, B. and M. Ghil, 1983. Ecoulements atmosphériques stationnaires, périodiques et apériodiques. J. Méc. Théor. Appl., no. spécial 1983 (2D turbulence, R. Moreau, ed.), 45-82.

Legras, B. and M. Ghil, 1984. Blocking and variations in atmospheric predictability. In Holloway and West (1984), pp. 87-105.

Legras, B. and M. Ghil, 1985. Persistent anomalies, blocking and variations in atmospheric predictability. J. Atmos. Sci., 42, in press.

Leith, C. E., 1978. Objective methods for weather prediction. Ann. Rev. Fluid Mech., 10, 107-128.

Leith, C. E., 1980. Nonlinear normal mode initialization and quasi-geostrophic theory. J. Atmos. Sci., 37, 958-968.

Lemke, P., 1977. Stochastic climate models, Part 3. Application to zonally averaged energy models. Tellus, 30, 97-103.

LeTreut, H., and M. Ghil, 1983. Orbital forcing, climatic interactions, and glaciation cycles. J. Geophys. Res., 88C, 5167-5190; 8623.

LeTreut, H., and M. Ghil, 1984. The predictability of glaciation cycles. Annals Glaciol., 5, 213-214.

LeVerrier, U. J. J., 1855. Recherches astronomiques. Ann. Obs. Impérial Paris, vol. II.

Levitus, S., 1982. Climatological Atlas of the World Ocean. NOAA Professional Paper 12, National Oceanic and Atmospheric Administration, Rockville, Md., 173 pp. + 17 microfiches.

Li, G. Q., R. L. Pfeffer and R. Kung, 1986. An experimental study of baroclinic flows with and without two-wave bottom topography. J. Atmos. Sci., submitted.

Libchaber, A., 1985. The onset of weak turbulence, an experimental introduction. In Ghil et al. (1985), pp. 17-28.

Lichtenberg, A. J. and M. A. Liebermann, 1983. Regular and Stochastic Motion. Springer-Verlag, New York/Heidelberg/Berlin, 499 pp.

Lindzen, R. S., B. Farrell and D. Jacqmin, 1982. Vacillations due to wave interference: applications to the atmosphere and to annulus experiments. J. Atmos. Sci., 39, 14-23.

Lliboutry, L., 1964/1965. Traité de Glaciologie; vol. I: Glace, Neige, Hydrologie Nivale (1964), vol. II: Glaciers, Variations du Climat, Sols Gelés (1965). Masson, Paris, 428 + 616 pp.

Lliboutry, L., 1973. Isostasie, propriétés rhéologiques du manteau supérieur. In Traité de Géophysique Interne, J. Coulomb and G. Jobert (eds.), vol. I (Sismologie et Pesanteur), Ch. 17, Masson, Paris, pp. 473-505.

London, J., 1957. A Study of the Atmospheric Heat Balance. Dept. of Meteorology and Oceanography, Final Report No. AF 19(122)-165, New York University, New York.

London, J. and T. Sasamori, 1971. Radiative energy budget of the atmosphere: In Man's Impact on the Climate, N. H. Matthews, W. W. Kellogg and G. D. Robinson (eds.), MIT Press, Cambridge, Mass., pp. 141-155.

Longuet-Higgins, M. S. and A. E. Gill, 1967. Resonant interactions between planetary waves. Proc. Roy. Soc. (London), A299, 120-140; 141-145.

Loper, D. E., 1978. The gravitationally powered dynamo. Geophys. J. Roy. Astron. Soc., 54, 389-404.

Loper, D. E. and P. H. Roberts, 1978. On the motion of an iron alloy containing a slurry, I. General theory. Geophys. Astrophys. Fluid Dyn., 9, 289-321.

Loper, D. E. and P. H. Roberts, 1980. On the motion of an iron alloy containing a slurry, II. A simple model, Geophys. Astrophys. Fluid Dyn., 16, 83-127.

Loper, D. E. and P. H. Roberts, 1981. A study of conditions at the inner-core boundary of the Earth. Phys. Earth Planet. Inter., 24, 302-307.

Lorenz, E. N., 1960a. Maximum simplification of the dynamic equations. Tellus, 12, 243-254.

Lorenz, E. N., 1960b. Energy and numerical weather prediction. Tellus, 12, 364-373.

Lorenz, E. N., 1962. Symplified dynamic equations applied to the rotating-basin experiments. J. Atmos. Sci., 19, 39-51.

Lorenz, E. N., 1963a. Deterministic nonperiodic flow. J. Atmos. Sci., 20, 130-141.

Lorenz, E. N., 1963b. The mechanics of vacillation. J. Atmos. Sci., 20, 448-464.

Lorenz, E. N., 1964. The problem of deducing the climate from the governing equations. Tellus, 16, 1-11.

Lorenz, E. N., 1967. The Nature and Theory of the General Circulation of the Atmosphere. World Meteorological Organization, Geneva, Switzerland, 161 pp.

Lorenz, E. N., 1980. Attractor sets and quasi-geostrophic equilibrium. J. Atmos. Sci., 37, 1685-1699.

Lorenz, E. N., 1983. A history of prevailing ideas about the general circulation of the atmosphere, Bull. Amer. Meteorol. Soc., 64, 730-734.

Lorenz, E. N., 1985. The growth of errors in prediction. In Ghil et al. (1985), pp. 243-265.

Lorius, C., L. Merlivat, J. Jouzel and M. Pourchet, 1979. A 30,000-yr isotope climatic record from Antarctic ice. Nature, 280, 644-648.

Lowes, F. J., 1974. Spatial power spectrum of the main geomagnetic field, and extrapolation to the core. Geophys. J. R. Astron. Soc., 36, 717-730.

Ludwig, D., 1975. Persistence of dynamical systems under random perturbations. SIAM Rev., 17, 605-640.

Ludwig, D., 1979. Stochastic modelling and nonlinear oscillations. In Hoppensteadt (1979a), pp. 127-129.

MacDonald, N., 1978. Time Lags in Biological Models. Springer-Verlag, New York, 112 pp.

Madden, R. A. and P. R. Julian, 1971. Detection of a 40-50 day oscillation in the zonal wind in the Tropical Pacific. J. Atmos. Sci., 28, 702-708.

Malkus, W. V. R., 1972. Reversing Bullard's dynamo. EOS, Trans. Amer. Geophys. Union, 53, 617.

Malkus, W. V. R. and M. R. E. Proctor, 1975. The macrodynamics of α-effect dynamos in rotating fluids. J. Fluid Mech., 67, 417-444.

Malguzzi, P. and P. Malanothe Rizzoli, 1984. Nonlinear stationary Rossby waves on nonuniform zonal winds and atmospheric blocking. Part I: the analytical theory. J. Atmos. Sci., 41, 2620-2628.

Mallet-Paret, J. and R. D. Nussbaum, 1984. Global continuation and complicated trajectories for periodic solutions of a differential-delay equation. Proc. 1983 AMS Symp. Nonlin. Functl. An., to appear.

Marchuk, G. I., 1975. Formulation of theory of perturbations for complicated models. Appl. Math. Optimization, 2, 1-33.

Marsden, J. E. and M. McCracken, 1976. The Hopf Bifurcation and its Applications, Springer-Verlag, New York, 408 pp.

Maschke, E. K. and B. Saramito, 1982a. On truncated series approximations in the theory of Rayleigh-Bénard convection. Phys. Lett., 88A, 154-156.

Maschke, E. K. and B. Saramito, 1982b. On the transition of turbulence in magneto-hydro dynamic models of confined plasmas. Phys. Scripta, T2/2, 410-417.

Matkowsky, B. J., Z. Schuss and C. Tier, 1984. Uniform expansion of the transition rate in Kramers' problem. J. Stat. Phys., 35, 443-456.

May, R., 1976. Simple mathematical models with very complicated dynamics. Nature, 261, 459-467.

McConnell, R. K., 1968. Viscosity of the mantle from relaxation time spectra of isostatic adjustment. J. Geophys. Res., 73, 7089-7105.

McIntyre, M. E., 1968. The axisymmetric convective regime for a rigidly bounded rotating annulus. J. Fluid Mech., 32, 625-655.

McIntyre, M., 1980. An introduction to the generalized Lagrangian-mean description of wave, mean-flow interaction. Pure Appl. Geophys., 118, 152-176.

McLaughlin, J. B. and P. C. Martin, 1975. Transition to turbulence in a statically stressed fluid system. Phys. Rev. A, 12, 186-203.

McWilliams, J. C., 1980. An application of equivalent modons to atmospheric blocking. Dyn. Atmos. Oceans, 5, 43-66.

McWilliams, J. C. and P. R. Gent, 1978. A coupled air and sea model for the tropical Pacific. J. Atmos. Sci., 35, 982-989, and 36 (1979), 181.

Mechoso, C. R., M. J. Suarez, K. Yamazaki, A. Kitoh and A. Arakawa, 1986. Numerical forecasts of anomalous atmospheric events during the winter of 1979. Adv. Geophys., 29, to appear.

Mercer, J. H., 1983. Cenozoic glaciation in the Southern Hemisphere. Ann. Rev. Earth Planet. Sci., 11, 99-132.

Merkine, L. O. and M. Shafranek, 1980. The spatial and temporal evolution of localized unstable baroclinic disturbances. Geophys. Astrophys. Fluid Dyn., 25, 157-190.

Milankovitch, M., 1941. Canon of Insolation and the Ice Age Problem (in German). Roy. Serb. Acad. Spec. Publ., 133, 1-633; Engl. transl.; Israel Program for Scientific Translations, Jerusalem, 1969, 482 pp. (available from Natl. Tech. Info. Svc., US Dept. of Commerce).

Minorsky, N., 1962. Nonlinear Oscillations. Van Nostrand, Princeton, N.J., 714 pp.

Miskovic, V. V., 1931. Secular variations in astronomical elements of the Earth's orbit (in Serbo-Croatian). Glas. Srp. Kral. Akad., 143 (70), 93-117.

Mitchell, J. L. and T. Maxworthy, 1985. Large-scale turbulence in the Jovain atmosphere. In Ghil et al. (1985), pp. 226-240.

Mitchell, J. M., Jr., 1976. An overview of climatic variability and its causal mechanisms. Quatern. Res., 6, 481-493.

Mitchell, K. E., and J. A. Dutton, 1981. Bifurcations from stationary to periodic solutions in a low-order model of forced dissipative barotropic flow. J. Atmos. Sci., 38, 690-716.

Miyakoda, K., C. T. Gordon, R. Caverly, W. F. Stern, J. Sirutis and W. Bourke, 1983. Simulation of a blocking event in January 1977. Mon. Wea. Rev., 111, 846-869.

Miyamoto, K., 1980. Plasma Physics for Nuclear Fusion. The MIT Press, Cambridge, Mass./London, 610 pp.

Moffatt, H. K., 1969. The degree of knottedness of tangled vortex lines. J. Fluid Mech., 35, 117-129.

Moffatt, H. K., 1976. Generation of magnetic fields by fluid motion. Adv. Appl. Mech., 16, 119-181.

Moffatt, H. K., 1978. Magnetic Field Generation in Electrically Conducting Fluids. Cambridge Univ. Press, Cambridge/London, 343 pp.

Moffatt, H. K., 1979. A self-consistent treatment of simple dynamo systems. Geophys. Astrophys. Fluid Dyn., 14, 146-166.

Morel, P. (ed.), 1973. Dynamic Meteorology. D. Reidel Publ. Co., Dordrecht-Holland/Boston, U.S.A., 622 pp.

Moser, J. K., 1973. Stable and Random Motions in Dynamical Systems, with Special Emphasis on Celestial Mechanics. Princeton Univ. Pres, Princeton, NJ, 198 pp.

Molchanov, A. M., 1969. The reality of resonance in the solar system. Icarus, 11, 104-110.

Murdock, J. A., 1976. Nearly-Hamiltonian systems in nonlinear mechanics: averaging and energy methods. Indiana Univ. Math. J., 25, 499-523.

Namias, J., 1950. The index cycle and its role in the general circulation. J. Meteorol., 7, 130-139.

Namias, J., 1968. Long range weather forecasting-history, current status and outlook. Bull. Amer. Meteorol. Soc., 49, 438-470.

Namias, J., 1982. Short Period Climatic Variations. Collected works of J. Namias, vols. I and II (1934-1974) and vol. III (1975-1982). Univ. of California, San Diego, 905 pp. + 393 pp.

Nayfeh, A. H., 1973. Perturbation Methods. J. Wiley, New York, 519 pp.

Newcomb, S., 1895. Elements of the Four Inner Planets and Fundamental Constants of Astronomy. Suppl. Amer. Ephemeris Naut. Almanac 1897. Govt. Printing Office, Washington, D.C.

Nicolis, C., 1982a. Stochastic aspects of climatic transitions - response to a periodic forcing. Tellus, 34, 1-9.

Nicolis, C., 1982b. A Boolean approach to climate dynamics. Quart. J. Roy. Meteorol. Soc., 108, 707-715.

Nicolis, C., 1984a. Self-oscillations, external forcings, and climate predictability. In Berger et al. (1984), pp. 637-652.

Nicolis, C., 1984b. Self-oscillations and predictability in climate dynamics - periodic forcing and phase locking. Tellus, 36A, 217-227.

Nicolis, C. and G. Nicolis, 1979. Environmental fluctuation effects on the global energy balance. Natue, 281, 132-134.

Nicolis, C. and G. Nicolis, 1981. Stochastic aspects of climatic transitions - additive fluctuations. Tellus, 33, 225-234.

Nirenberg, L., 1981. Variational and topological methods in nonlinear problems. Bull. Amer. Math. Soc. (New Series), 4, 267-302.

North, G. R., 1975a. Analytical solution to a simple climate model with diffusive heat transport. J. Atmos. Sci., 32, 1301-1307.

North, G. R., 1975b. Theory of energy-balance climate models. J. Atmos. Sci., 32, 2033-2043.

North, G. R., L. Howard, D. Pollard and B. Wielicki, 1979. Variational formulation of Budyko-Sellers climate models. J. Atmos. Sci., 36, 255-259.

North, G. R., R. F. Cahalan and J. A. Coakley, Jr., 1981. Energy balance climate models. Rev. Geophys. Space Phys., 19, 91-121.

Nye, J. F., 1951. The flow of glaciers and ice sheets as a problem in plasticity. Proc. Roy. Soc. (London), A207, 554-572.

Nye, J. F., 1969. The effect of longitudinal stress on the shear stress at the base of an ice sheet. J. Glaciol., 8, 207-213.

Oerlemans, J., 1982. Glacial cycles and ice sheet modelling. Climatic Change, 4, 353-374.

Oerlemans, J., 1984. On the origin of the ice ages. In Berger et al. (1984), pp. 607-611.

Oerlemans, J. and C. J. van der Veen, 1984. Ice Sheets and Climate. D. Reidel, Dordrecht/Boston/Lancaster, 217 pp.

Oopstegh, J. D. and H. M. van der Dool, 1980. Seasonal differences in the stationary response of a linearized primitive equation model: prospects for long-range weather forecasting? J. Atmos. Sci., 37, 2169-2185.

Oort, A. H., 1983. Global Atmospheric Circulation Statistics, 1958-1973. NOAA Professional Paper 14, National Oceanic and Atmospheric Administration, Rockville, Md., 180 pp. + 47 microfiches.

Orlanski, I., B. Ross, L. Polinsky and R. Shaginan, 1986. Advances in the theory of atmospheric fronts. Adv. Geophys., 28, to appear.

Orowan, E., 1949. Remarks at Joint Meeting of the British Glaciological Society, the British Rheologists' Club and the Institute of Metals. J. Glaciol., 1, 231-236; discussion, 236-240.

Ostlund, S., D. Rand, J. Sethna and E. Siggia, 1982. Universal properties of the transition from quasi-periodicity to chaos in dissipative systems. Institute for Theoretical Physics, Univ. of California, Santa Barbara, Preprint ITP-82-80, 108 pp. + 23 figs.

Palmén, E. and C. W. Newton, 1969. Atmospheric Circulation Systems, Their Structure and Physical Interpretation. Academic Press, New York/San Francisco/London, 603 pp.

Parker, E. N., 1955. Hydrodynamic dynamo models. Astrophys. J., 122, 293-314.

Parker, E. N., 1979. Cosmical Magnetic Fields. Clarendon Press, Oxford, 841 pp.

Paterson, W. S. B., 1972. Laurentide ice sheet: estimated volumes during Lake Wisconsin. Rev. Geophys. Space Phys., 10, 885-917.

Paterson, W. S. B., 1980. Ice sheets and ice shelves. In Colbeck (1980), pp. 1-78.

Paterson, W. S. B., 1981. The Physics of Glaciers, 2nd ed. Pergamon Press, Oxford, 380 pp.

Pedlosky, J., 1971. Geophysical fluid dynamics. In Reid (1971), vol. I, pp. 1-60.

Pedlosky, J., 1979. Geophysical Fluid Dynamics. Springer-Verlag, New York/Heidelberg/Berlin, 624 pp.

Pedlosky, J., 1981a. The nonlinear dynamics of baroclinic wave ensembles. J. Fluid Mech., 102, 169-209.

Pedlosky, J., 1981b. Resonant topographic waves in barotropic and baroclinic flows. J. Atmos. Sci., 38, 2626-2641.

Pedlosky, J. and C. Frenzen, 1980. Chaotic and periodic behavior of finite-amplitude baroclinic waves. J. Atmos. Sci., 37, 1177-1196.

Peltier, R., 1982. Dynamics of the ice age earth. Adv. Geophys., 24, 1-146.

Peltier, R. and W. Hyde, 1984. A model of the ice age cycle. In Berger et al. (1984), pp. 565-580.

Pestiaux, P., 1984. Approche Spectrale en Modélisation Paléoclimatique. Ph.D. Thesis, Univ. Catholique de Louvain, Louvain-la-Neuve, Belgium.

Pestiaux, P. and A. Berger, 1984. An optimal approach to the spectral characteristics of deep-sea climatic records. In Berger et al. (1984), pp. 417-445.

Pestiaux, P. and J.-C. Duplessy, 1985. Paleoclimatic variability at frequencies ranging from 10^{-4} cycle per year to 10^{-3} cpy: evidence for nonlinear behavior of the climate system. Quatern. Res., submitted.

Petterssen, S., 1956. Weather Analysis and Forecasting, 2nd ed., vol. I (Motion and Motion systems) and vol. II (Weather and Weather Systems). McGraw-Hill, New York/Toronto/London, 428 pp. + 266 pp.

Pfeffer, R. L. (ed.), 1960. Dynamics of Climate. Pergamon Press, New York/Oxford/London/Paris, 137 pp.

Pfeffer, R. L., G. Buzyna and R. Kung, 1980. Time-dependent modes of behavior of thermally-driven rotating fluids. J. Atmos. Sci., 37, 2129-2149.

Philander, S. G. H., T. Yamagata and R. C. Pacanowski, 1984. Unstable air-sea interactions in the tropics. J. Atmos. Sci., 41, 604-613.

Phillips, N. A., 1956. The general circulation of the atmosphere: a numerical experiment. Quart. J. Roy. Meteorol. Soc., 82, 123-164.

Phillips, N. A., 1973. Principles of large-scale numerical weather prediction. In Morel (1973), pp. 1-96.

Phillips, O. M., 1966. The Dynamics of the Upper Ocean. Cambridge Univ. Press, Cambridge/London/New York, 261 pp.

Pierrehumbert, R. T., 1984. Local and global baroclinic instability of zonally varying flow. J. Atmos. Sci., 41, 2141-2162.

Pisias, N. G. and T. C. Moore, Jr., 1981. The evolution of the Pleistocene climate; a time series approach. Earth Planet. Sci. Lett., 52, 450-458.

Platzman, G. W., 1960. The spectral form of the vorticity equation. J. Meterol., 17, 635-644.

Poincaré, H., 1892. Les Methodes Nouvelles de la Mécanique Céleste, 3 vols. Gauthier-Villars, Paris.

Pollack, J. B., C. B. Leovy, P. W. Greiman and Y. Mintz, 1981. A Martian general circulation experiment with large topography. J. Atmos. Sci., 38, 3-29.

Pollard, D., 1984. Some ice-age aspects of a calving ice-sheet model. In Berger et al. (1984), pp. 541-564.

Pomeau, Y. and P. Manneville, 1980. Intermittent transition to turbulence in dissipative dynamical systems. Commun. Math. Phys., 74, 189-197.

Proctor, M. R. E., 1977a. Numerical solutions of the nonlinear α-effect dynamo equations. J. Fluid Mech., 80, 769-784.

Proctor, M. R. E., 1977b. On Backus' necessary condition for dynamo action in a conducting sphere. Geophys. Astrophys. Fluid Dyn., 9, 89-93.

Proctor, M. R. E., 1979. Necessary conditions for the magnetohydrodynamic dynamo. Geophys. Astrophys. Fluid Dyn., 14, 127-145.

Ramanathan, V. and J. A. Coakley, 1978. Climate modeling through radiative convective models. Rev. Geophys. Space Phys., 16, 465-489.

Rand, D., 1982. Dynamics and symmetry. Predictions for modulated waves in rotating fluids. Arch. Rational Mech. An., 79, 1-37.

Rasmusson, E. M. and J. M. Wallace, 1983. Meteorological aspects of the El Niño/Southern Oscillation. Science, 222, 1195-1202.

Reid, W. H. (ed.), 1971. Mathematical Problems in the Geophysical Sciences, vol. I: Geophysical Fluid Dynamics, vol. II: Inverse Problems, Dynamo Theory and Tides. American Mathematical Society, Providence, R.I., 383 pp. + 370 pp.

Reinhold, B. B. and R. T. Pierrehumbert, 1982. Dynamics of weather regimes: quasi-stationary waves and blocking. Mon. Wea. Rev., 110, 1105-1145.

Rex, D. F., 1950a. Blocking action in the middle troposphere and its effect on regional climate. Part I: An aerological study of blocking action. Tellus, 2, 196-211.

Rex, D. F., 1950b. Part II: The climatology of blocking action. Tellus, 2, 275-301.

Rhines, P., 1981. Review of Pedlosky (1979). J. Fluid Mech., 110, 497-502.

Richtmyer, R. D. and K. W. Morton, 1967. Difference Methods for Initial-Value Problems, 2nd ed. Wiley-Interscience, New York, 405 pp.

Roads, J. O., 1982. Stable and unstable near-resonant states in multi-level, severely-truncated, quasi-geostrphic models. J. Atmos. Sci., 39, 203-224.

Robbins, K. A., 1977. A new approach to subcritical instability and turbulent transitions in a simple dynamo. Proc. Camb. Phil. Soc., 82, 309-325.

Roberts, G. O., 1970. Spatially periodic dynamos. Phil. Trans. Roy. Soc., A266, 535-558.

Roberts, G. O., 1972. Dynamo action of fluid motions with two-dimensional periodicity. Phil. Trans. Roy. Soc., A271, 411-454.

Roberts, P. H., 1967. An Introduction to Magnetohydrodynamics. American Elsevier, New York, 264 pp.

Roberts, P. H., 1971. Dynamo theory. In Reid (1971), Vol. 2, pp. 129-206.

Roberts, P. H., 1972. Kinematic dynamo models. Phil. Trans. Roy. Soc., A272, 663-98.

Roberts, P. H. and M. Stix, 1971. The turbulent dynamo: a translation of a series of papers by F. Krause, K.-H. Rädler and M. Steenback. Tech. Note 60, National Center for Atmospheric Research, Boulder, Colo.

Roberts, P. H. and A. M. Soward, 1972. Magnetohydrodynamics of the Earth's core. Ann. Rev. Fluid Mech., 4, 117-154.

Roberts, P. H. and M. Stix, 1972. α-effect dynamos, by the Bullard-Gellman formalism. Astron. Astrophys., 18, 453-466.

Roberts, P. H. and A. M. Soward (eds.), 1978. Rotating Fluids in Geophysics. Academic Press, London/New York/San Francisco, 551 pp.

Robinson, A. R., 1959. The symmetric state of a rotating fluid differentially heated in the horizontal. J. Fluid Mech., 6, 599-620.

Robinson, A. R. (ed.), 1983. Eddies in Marine Science. Springer-Verlag, Berlin/Heidelberg/New York/Tokyo, 609 pp.

Robock, A., 1978. Internally and externally caused climate change. J. Atmos. Sci., 35, 1112-1122.

Rochester, M. G., J. A. Jacobs, D. E. Smylie and K. F. Chong, 1975. Can precession power the geomagnetic dynamo? Geophys. J. Roy. Astron. Soc., 43, 661-678.

Rossby, C.-G.and collaborators, 1939. Relation between variations in the intensity of the zonal circulation of the atmosphere and the displacements of the semi-permanent centers of action. J. Marine Res., 2, 38-55.

Ruelle, D., 1980. Strange attractors. Math. Intelligencer, 3, 126-137.

Ruelle, D., 1981a. Differentiable dynamical systems and the problem of turbulence. Bull. Amer. Math. Soc. (New Series), 5, 29-42.

Ruelle, D., 1981b. Small random perturbations of dynamical systems and the definition of attractors. Commun. Math. Physics, 82, 137-151.

Ruelle, D., 1985. The onset of turbulence: a mathematical introduction. In Ghil et al. (1985), pp. 3-16.

Ruelle, D. and F. Takens, 1971. On the nature of turbulence. Commun. Math. Phys., 20, 167-192, and 23, 343-344.

Sadourny, R., 1985. Quasi-geostrophic turbulence: an introduction. In Ghil et al. (1985), pp. 133-158.

Saltzman, B., 1962. Finite amplitude free convection as an initial value problem - I. J. Atmos. Sci., 19, 329-341.

Saltzman, B., 1978. A survey of statistical-dynamical models of the terrestrial climate. Adv. Geophys., 20, 183-304.

Saltzman, B., 1983. Climatic systems analysis. Adv. Geophys., 25, 173-233.

Saltzman, B., A. Sutera and A. Evenson, 1981. Structural stochastic stability of a simple auto-oscillatory climatic feedback system. J. Atmos. Sci., 38, 494-503.

Saltzman, B., A Sutera and A. R. Hansen, 1982. A possible marine mechanism for internally generated long-period climate cycles. J. Atmos. Sci., 39, 2634-2637.

Savijärvi, H., 1984. Spectral properties of analyzed and forecast global 500mb fields. J. Atmos. Sci., 41, 1745-1754.

Schneider, S. H. and T. Gal-Chen, 1973. Numerical experiments in climate stability. J. Geophys. Res., 78, 6182-6194.

Schneider, S. H. and R. E. Dickinson, 1974. Climate modeling. Rev. Geophys. Space Phys., 12, 447-493.

Schuss, Z., 1980. Theory and Applications of Stochastic Differential Equations. J. Wiley, New York, 321 pp.

Sellers, W. D., 1965. Physical Climatology. Univ. of Chicago Press, Chicago/London, 272 pp.

Sellers, W. D., 1969. A climate model based on the energy balance of the earth-atmosphere system. J. Appl. Meteorol., 8, 392-400.

Sergin, V. Ya., 1979. Numerical modeling of the glacier-ocean-atmosphere global system. J. Geophys. Res., 84, 3197-3204.

Shackleton, N. J., 1981. Palaeoclimatology before our ice age. In Berger (1981), pp. 167-179.

Shackleton, N. J. and N. D. Opdyke, 1973. Oxygen isotope and paleomagnetic stratigraphy of equatorial Pacific core V28-238: oxygen isotope temperatures and ice volumes on a 10^5 and 10^6 year scale. Quatern. Res., 3, 39-55.

Shackleton, N. J. and N. D. Opdyke, 1976. Oxygen isotope and paleomagnetic stratigraphy of Pacific core V28-239 Late Pliocene to Latest Pleistocene. Geol. Soc. Amer. Memoir, 145, 449-463.

Sharaf, Sh. G. and N. A. Budnikova, 1967. Secular perturbations in the elements of the earth's orbit, which influence the climates of the geological past (in Russian). Byul. Inst. Teor. Astron., 11(4), 231-261; Engl. transl.: NASA Tech. Transl. F-12-467, 39 pp.

Shilnikov, L. P., 1965. A case of the existence of a denumerable set of periodic motions. Sov. Math. Dokl., 6, 163-166.

Shukla, J. and K. C. Mo, 1983. Seasonal and geographical variation of blocking. Mon. Wea. Rev., 111, 388-402.

Simmons, A. J. and B. J. Hoskins, 1979. The downstream and upstream development of unstable baroclinic waves. J. Atmos. Sci., 36, 1239-1254.

Simmons, A. J., J. M. Wallace and G. W. Branstator, 1983. Barotropic wave propagation and instability, and atmospheric teleconnection patterns. J. Atmos. Sci., 40, 1363-1392.

Simpson, G. C., 1934. World climate during the Quaternary period. Quart. J. Roy. Meteorol. Soc., 60, 425-471; 471-478.

Simpson, G., 1938. Ice ages. Nature, 141, 591-598.

Smale, S., 1980. The Mathematics of Time, Essays on Dynamical Systems, Economic Processes, and Related Topics. Springer-Verlag, New York/Heidelberg/Berlin, 151 pp.

Somerville, R. C. J. and L. A. Remer, 1984. Cloud optimal thickness feedbacks in the CO_2 climate problem. J. Geophys. Res., 89D, 9668-9672.

Sommerfeld, A., 1952. Electrodynamics. Academic Press, New York, 371 pp.

Soward, A. M., 1971a. Nearly symmetric kinematic and hydromagnetic dynamos. J. Math. Phys., 12, 1900-1906.

Soward, A. M., 1971b. Nearly symmetric advection. J. Math. Phys., 12, 2052-2062.

Soward, A. M., 1972. A kinematic dynamo theory of large magnetic Reynolds number dynamos. Phil. Trans. Roy. Soc., A272, 431-462.

Soward, A. M., 1974. A convection-driven dynamo, I. The weak-field case. Phil. Trans. Roy. Soc., A275, 611-651.

Soward, A. M., 1980. Bounds on turbulent convective dynamos. Geophys. Astrophys. Fluid Dyn., 15, 317-341.

Soward, A. M. (ed.), 1983. Stellar and Planetary Magnetism, Vol. 2, Fluid Mechanics of Astrophysics and Geophysics, Gordon and Breach, London, 376 pp.

Soward, A. M. and C. A. Jones, 1983. α^2-dynamos and Taylor's constraint. Geophys. Astrophys. Fluid Dyn., 27, 87-122.

Sparrow, C., 1982. The Lorenz Equations: Bifurcations, Chaos, and Strange Attractors. Springer-Verlag, New York/Heidelberg/Berlin, 269 pp.

Steele, C., 1979. Studies of the ear. In Hoppensteadt (1979a), pp. 69-91.

Steenback, M. F., F. Krause, and K.-H. Rädler, 1966. A calculation of the mean electromotive force in an electrically conducting fluid in turbulent motion, under the influence of Coriolis forces. Z. Naturforsch., 21A, 369-376.

Stephens, G., G. Campbell and T. Vonder Haar, 1981. Earth radiation budgets. J. Geophys. Res., 86, 9739-9760.

Stix, M., 1981. Theory of the solar cycle. Solar Phys., 74, 79-101.

Stix, M., 1973. Special αω-dynamos by a variational method. Astron. Astrophys., 24, 275-281.

Stoker, J. J., 1950. Nonlinear Vibrations in Mechanical and Electrical Systems. Wiley-Interscience, New York, 273 pp.

Stoker, J. J., 1957. Water Waves: The Mathematical Theory with Applications. Wiley-Interscience, New York, 567 pp.

Stone, P. H., 1972. A simplified radiative-dynamical model for the static stability of rotating atmospheres. J. Atmos. Sci., 29, 405-418.

Stone, P. H., 1975. The dynamics of the atmosphere of venus. J. Atmos. Sci., 32, 1005-1016.

Strang, G. W. and G. J. Fix, 1973. An Analysis of the Finite Element Method. Prentice-Hall, Englewood Cliffs, N. J., 306 pp.

Stratonovich, R. L., 1967. Topics in the Theory of Random Noise, 2 vols. Gordon and Breach, New York/London/Paris, 292 + 329 pp.

Sutera, A., 1981. On stochastic perturbation and long-term climate behaviour. Quart. J. Roy. Meterol. Soc., 107, 137-152.

Swinney, H. L. and J. P. Gollub (eds.), 1981. Hydrodynamic Instabilities and the Transition to Turbulence. Springer-Verlag, Berlin/Heidelberg/New York, 292 pp.

Taylor, F. W., et al., 1980. Structure and meteorology of the middle atmosphere of Venus: infrared remote sensing from the Pioneer orbiter. J. Geophys. Res., 85A, 7963-8006.

Taylor, J. B., 1963. The magnetohydrodynamics of a rotating fluid and the Earth's dynamo problem. Proc. Roy. Soc., A274, 274-283.

Thomas, R., 1973. Boolean formalization of genetic control circuits. J. Theoret. Biol., 42, 563-585.

Thomas, R. (ed.), 1979. Kinetic Logic: A Boolean Approach to the Analysis of Complex Regulatory Systems. Springer-Verlag, Berlin/Heidelberg/New York, 507 pp.

Thornthwaite, C. W., 1933. The climates of the Earth. Geogr. Rev. (New York), 23, 433-440.

Trenberth, K. E. and K. C. Mo, 1985. Blocking in the Southern Hemisphere. Mon. Wea. Rev., 113, 3-21.

Trewartha, G. T., 1954. An Introduction to Climate. McGraw-Hill New York.

Tung, K. K., and R. S. Lindzen, 1979. A theory of stationary long waves. I: A simple theory of blocking. Mon. Wea. Rev., 107, 714-734.

Turcotte, D. L., 1979. Flexure. Adv. Geophys., 21, 51-86.

Turcotte, D. L. and G. Schubert, 1982. Geodynamics: Application of Continuum Physics to Geological Problems. J. Wiley, New York, 450 pp.

Turcotte, D. L., R. J. Willemann, W. F. Haxby and J. Norberry, 1981. Role of membrane stresses in the support of planetary topography. J. Geophys. Res., 86, 3951-3959.

Untersteiner, N. (ed.), 1986. The Geophysics of Sea Ice. Plenum, New York, to appear.

Urey, H. C., 1947. The thermodynamic properties of isotopic substances. J. Chem. Soc. (London), 562-581.

U.S. Committee for GARP, 1975. Understanding Climatic Change, A Program for Action. National Academy of Sciences, Washington, D.C., 239 pp.

Ushiki, S., 1982. Central difference scheme and chaos. Physica, 4D, 407-424.

Vainstein, S. I. and Ya. B. Zeldovich, 1972. Origin of magnetic fields in astrophysics. Usp. Fiz. (SSSR), 106, 431-457 (Engl. trans. Sov. Phys. Usp., 15, 159-172).

Van Bemmelen, R. W. and H. P. Berlage, 1935. Versuch einer mathematischen Behandlung tektonischer Bewegungen unter besonderer Berücksichfigung der Undulationstheorie. Gerlands. Beitr. Geophys., 43, 19-55.

Van der Pol, B., 1940. Biological rhythms considered as relaxation oscillations. Acta. Mech. Scand. Suppl., 108, 76-87.

Van der Pol, B., 1946. Music and elementary theory of numbers. Music Rev., 7, 1-25.

Van Eysinga, F. W. B., 1975. Geological Time Table, 3rd ed., Elsevier, Amsterdam, 1 p.

Vautard, R., and B. Legras, 1986. Invariant manifolds, quasi-geostrophy and initialization. J. Atmos. Sci., 43, 565-584.

Vening-Meinesz, F. A., 1937. The determination of the earth's plasticity from the postglacial uplift of Scandinavia: isostatic adjustment. Proc. K. Ned. Akad. Wet. (Amsterdam), 40, 654-662.

Vernekar, A., 1972. Long-period global variations of incoming solar radiation. Meteorol. Monogr., 12(34), 21 pp. + tables.

Vicente, R. O., 1983. Instabilities in planetary systems. In Long-Time Prediction in Dynamics, C. W. Horton, Jr., L. E. Reichl and V. G. Szebehely (eds.), J. Wiley, New York, pp. 235-244.

Vickroy, J. G., and J. A. Dutton, 1979. Bifurcation and catastrophe in a simple forced, dissipative quasi-geostrophic flow. J. Atmos. Sci., 36, 45-52.

Vonder Haar, T. H. and V. E. Suomi, 1971. Measurements of the earth's radiation budget from satellites during a five-year period. J. Atmos. Sci., 28, 305-314.

Walcott, R. I., 1973. Structure of the Earth from glacio-isostatic rebound. Ann. Rev. Earth Planet. Sci., 1, 15-37.

Wallace, J. M. and D. S. Gutzler, 1981. Teleconnections in the geopotential height field during the Northern-Hemisphere winter. Mon. Wea. Rev., 109, 784-812.

Warn, T. and B. Brasnett, 1983. The amplification and capture of atmospheric solitions by topography: a theory of the onset of regional blocking. J. Atmos. Sci., 40, 28-38.

Wax, N., 1954. Selected Papers on Noise and Stochastic Processes. Dover Publ., New York, 337 pp.

Weertman, J., 1964. Rate of growth or shrinkage of nonequilibrium ice sheets. J. Glaciol., 5, 145-158.

Weertman, J., 1976. Milankovitch solar radiation variations and ice age ice sheet sizes. Nature, 261, 17-20.

Weiss, N. O., 1968. The expulsion of magnetic flux by eddies. Proc. Roy. Soc. (London), A293, 310-328.

Weiss, N. O., 1971. The dynamo problem. Quart. J. Roy. Astron. Soc., 12, 432-446.

Weiss, N. O., F. Cattaneo and C. A. Jones, 1984. Periodic and aperiodic dynamo waves. Geophys. Astrophys. Fluid Dyn., 30, 305-341.

Wetherald, R. T. and S. Manabe, 1975. The effect of changing the solar constant on the climate of a general circulation model. J. Atmos. Sci., 32, 2044-2059.

Whitham, G. B., 1974. Linear and Nonlinear Waves. Wiley-Interscience, New York, 636 pp.

Whysall, K. and R. Hide, 1984. A Seven-Week Fluctuation of the General Circulation of the Earth's Atmosphere: a Brief Survey. U.K. Meteorological Office, Tech. Memo. 0218402, 38 pp. + 16 figs.

Wiener, N., 1949. Extrapolation, Interpolation and Smoothing of Stationary Time Series, with Engineering Applications. M.I.T. Press, Cambridge, Mass., 163 pp.

Wigley, T. M. L., 1976. Spectral analysis and the astronomical theory of climatic change. Nature, 264, 629-631.

Wiin-Nielsen, A., 1979. Steady states and stability properties of a low-order barotropic system with forcing and dissipation. Tellus, 31, 375-386.

Willson, R. C., C. H. Duncan and J. Geist, 1980. Direct measurement of solar luminosity variation. Science, 207, 177-179.

Wiscombe, W., 1983. Atmospheric radiation: 1975-1983; in U.S. National Report to IUGG 1979-1982, Contributions in Meteorology, D. E. James and R. C. Taylor (eds.). Rev. Geophys. Space Phys., 21, 997-1021.

Wolansky, G., 1985. Dissipative Perturbations of Completely Integrable Systems, with Applications to Celestial Mechanics and to Geophysical Fluid Dynamics. Ph.D. Thesis, New York University, 212 pp.

Yoden, S., 1985. Bifurcation properties of a quasi-geostrophic, barotropic, low-order model with topography. J. Meteorol. Soc. Japan, 63, 535-546.

Zebiak, S. E., 1985. Tropical Atmosphere-Ocean Interaction and the El Niño/Southern Oscillation Phenomenon. Ph.D. Thesis, Massachusetts Institute of Technology, 261 pp.

Zeldovich, Ya. B., A. A. Ruzmaikin and D. D. Sokoloff, 1983. Magnetic Fields in Astrophysics. Gordon and Breach, New York, 365 pp.

Zwally, H. J., C. L. Parkinson and J. C. Comiso, 1983. Variability of Antarctic sea ice and changes in carbon dioxide. Science, 220, 1005-1012.

INDEX